Physical Chemistry: Theories, Models and Applications

Physical Chemistry: Theories, Models and Applications

Editor: Jamie Langdon

NY RESEARCH PRESS
New York

Published by NY Research Press
118-35 Queens Blvd., Suite 400,
Forest Hills, NY 11375, USA
www.nyresearchpress.com

Physical Chemistry: Theories, Models and Applications
Edited by Jamie Langdon

International Standard Book Number: 978-1-63238-577-2 (Hardback)

Cataloging-in-Publication Data

Physical chemistry : theories, models and applications / edited by Jamie Langdon.
p. cm.
Includes bibliographical references and index.
ISBN 978-1-63238-577-2
1. Chemistry, Physical and theoretical. 2. Chemistry. I. Langdon, Jamie.
QD453.3 .P49 2018
541--dc23

Printed in China.

Contents

Preface

Physical chemistry is a branch of chemistry that studies atoms, subatoms and particulate phenomena in order to understand concepts related to energy, force, motion, etc. It studies a number of theories to develop a better understanding of the effects of intermolecular forces, surface chemistry, colligative properties and electrochemical cells. This book elucidates new techniques and their applications in a multidisciplinary approach. It strives to provide a fair idea about this discipline and to help develop a better understanding of the latest advances within this field. Those in search of information to further their knowledge will be greatly assisted by this book.

After months of intensive research and writing, this book is the end result of all who devoted their time and efforts in the initiation and progress of this book. It will surely be a source of reference in enhancing the required knowledge of the new developments in the area. During the course of developing this book, certain measures such as accuracy, authenticity and research focused analytical studies were given preference in order to produce a comprehensive book in the area of study.

This book would not have been possible without the efforts of the authors and the publisher. I extend my sincere thanks to them. Secondly, I express my gratitude to my family and well-wishers. And most importantly, I thank my students for constantly expressing their willingness and curiosity in enhancing their knowledge in the field, which encourages me to take up further research projects for the advancement of the area.

Editor

1

Sulphur-impregnated flow cathode to enable high-energy-density lithium flow batteries

Hongning Chen[1], Qingli Zou[1], Zhuojian Liang[1,2], Hao Liu[3], Quan Li[3] & Yi-Chun Lu[1,2]

Redox flow batteries are promising technologies for large-scale electricity storage, but have been suffering from low energy density and low volumetric capacity. Here we report a flow cathode that exploits highly concentrated sulphur-impregnated carbon composite, to achieve a catholyte volumetric capacity 294 Ah l^{-1} with long cycle life (>100 cycles), high columbic efficiency (>90%, 100 cycles) and high energy efficiency (>80%, 100 cycles). The demonstrated catholyte volumetric capacity is five times higher than the all-vanadium flow batteries (60 Ah l^{-1}) and 3-6 times higher than the demonstrated lithium-polysulphide approaches (50-117 Ah l^{-1}). Pseudo-*in situ* impedance and microscopy characterizations reveal superior electrochemical and morphological reversibility of the sulphur redox reactions. Our approach of exploiting sulphur-impregnated carbon composite in the flow cathode creates effective interfaces between the insulating sulphur and conductive carbon-percolating network and offers a promising direction to develop high-energy-density flow batteries.

[1] Department of Mechanical and Automation Engineering, The Chinese University of Hong Kong, Shatin, N.T. 999077, Hong Kong SAR, China. [2] Shun Hing Institute of Advanced Engineering, The Chinese University of Hong Kong, Shatin, N.T. 999077, Hong Kong SAR, China. [3] Department of Physics, The Chinese University of Hong Kong, Shatin, N.T. 999077, Hong Kong SAR, China. Correspondence and requests for materials should be addressed to Y.-C.L. (email: yichunlu@mae.cuhk.edu.hk).

Energy storage system is a critical enabling factor for deploying unstable and intermittent renewable power sources, such as solar and wind power sources[1–6]. Redox flow batteries (RFBs) are promising technologies for large-scale electricity storage, owing to its design flexibility in decoupling power and energy capacity[7–10]. However, the RFBs have been suffering from low energy density and low volumetric capacity, which significantly decreases its competitiveness for both stationary and transportation applications[11–14].

Increasing the energy density and volumetric capacity of RFBs has been one of the major research efforts for RFBs and has motivated several new design concepts to date. Lu and Goodenough[15,16] have proposed a hybrid flow battery using an alkaline metal (for example, lithium metal) and aqueous redox-active species (for example, ferricyanide) as cathodes, separated by a glass ceramic membrane. This unique concept creates high cell voltage between the aqueous cathode and the most electronegative alkaline metal. However, the hybrid configuration using lithium metal limits the scalability of energy and power of the lithium-negative electrode[17], which can be achieved in an all-vanadium flow battery. The authors demonstrated a lithium-ferricyanide flow battery that operates at ~3.40 V with a coulombic efficiency >97%. Inspired by this concept, several groups have proposed to replace the ferricyanide redox couple with aqueous $FeCl_3/FeCl_2$ (ref. 18) or I^-/I_3^- (ref. 19) or aqueous polysulphide[20] to increase the concentration of the catholyte while keeping the solid lithium-metal anode for large cell voltage. One major drawback of the lithium/aqueous configuration is the need for a crack-free glass ceramic membrane, which is expensive (LICGC USD \$90 cm^{-2} (http://www.oharacorp.com/) versus Celgard separators <USD \$0.1 cm^{-2} (http://www.celgard.com/)) and highly resistive (low ionic conductivity $1.0 \times 10^{-4}\,S\,cm^{-1}$ versus $9 \times 10^{-3}\,S\,cm^{-1}$ for organic electrolytes[21]).

To increase the volumetric capacity of lithium flow batteries (Fig. 1a) and eliminate the need for a glass ceramic membrane, two major approaches were reported, as illustrated in Fig. 1b,c. First, Yang *et al.*[12] proposed the concept of using nonaqueous lithium polysulphide in the catholyte[12,22] (with a carbon fibre electrode current collector) and have demonstrated a catholyte volumetric capacity of 50 $Ah\,l^{-1}$ (108 $Wh\,l^{-1}$, average cell voltage 2.2 V against lithium metal)[12] using 5.0 M polysulphide catholyte. Inspired by this work, Fan *et al.*[23] replaced the carbon fibre electrode current collector with a carbon-percolating conducting network as a flow current collector (Fig. 1b), demonstrating Li-polysulphide flow batteries with a catholyte volumetric capacity of 117 $Ah\,l^{-1}$ (234 $Wh\,l^{-1}$, average cell voltage 2.0 V against lithium metal)[23]. In this approach, the solid sulphur phase was purposely avoided due to the extremely low solubility and insulating nature of the solid sulphur, which sacrifices the theoretical volumetric capacity. Second, Duduta *et al.*[13] proposed the concept of semi-solid Li-ion flow batteries utilizing solid intercalation materials in a carbon (Ketjen black (KB))-percolating conducting network (Fig. 1c). In this approach, the solid intercalation materials are mechanically mixed with the conducting additives and organic electrolytes to form a suspension, which can be circulated in the catholyte (for example, $LiCoO_2$) and anolyte (for example, $Li_4Ti_5O_{12}$) separated by a polymeric separator. Recently, researchers have employed a similar approach to develop a semi-solid $LiFePO_4$ flow cathode (19 $Ah\,l^{-1}$: 12.6 vol% $LiFePO_4$, achieved capacity 160 $mAh\,g^{-1}$)[24], and aqueous semi-solid $LiFePO_4$ flow cathode (16 $Ah\,l^{-1}$: 10.0 vol% $LiFePO_4$, achieved capacity 150 $mAh\,g^{-1}$)[25]. While the mechanically mixed intercalation semi-solid approach effectively increases the volumetric capacity of the flow batteries, the poor contacts between the large/insulating intercalation particles and the carbon network in the flow cathode become critical challenges[13].

In this work, we demonstrate a new flow-cathode concept that offers higher catholyte volumetric capacity compared with the above approaches, and alleviates contact issues between insulating active materials and the conductive carbon network. Our design employs sulphur-impregnated carbon (S/C) composite as a flow cathode (Fig. 1d) to achieve high-energy lithium-flow batteries with catholyte volumetric capacity ranging between 294 and 192 $Ah\,l^{-1}$ at various current densities with long cycle life (>100 cycles), high columbic efficiency (90–94.5% at 100th cycle) and energy efficiency (82–83.7% at 100th cycle). The demonstrated catholyte volumetric capacity is >5 times higher than the state-of-the-art vanadium flow battery (60 $Ah\,l^{-1}$)[3,26] and 3–6 times higher than the demonstrated Li-polysulphide approach (50–117 $Ah\,l^{-1}$)[12,23]. Physical origins responsible for the enhanced volumetric capacity and superior electrochemical reversibility will be discussed.

Results

Design considerations of the S/C flow cathode. The design of sulphur-impregnated carbon composite flow cathode offers three unique advantages compared with the existing

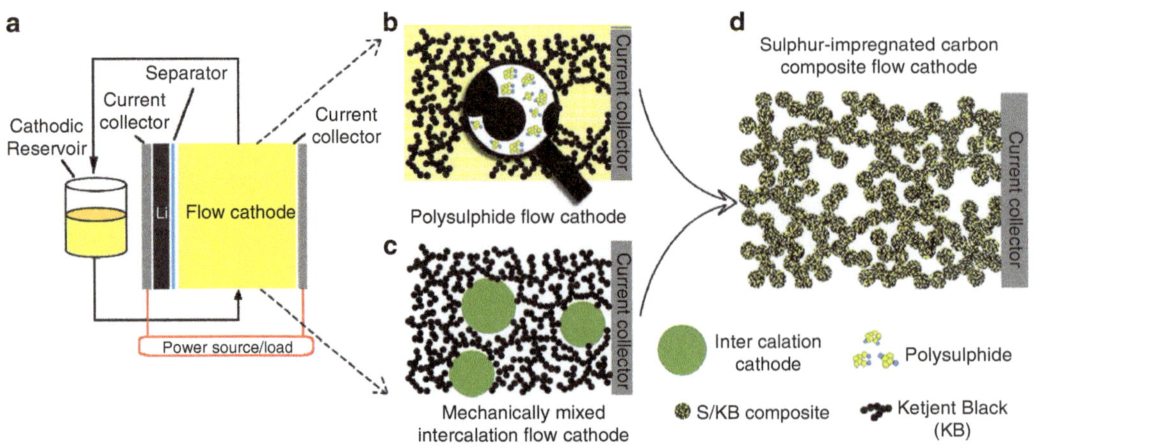

Figure 1 | Concept of a sulphur-impregnated carbon composite flow cathode. (**a**) Schematic representation of a nonaqueous lithium flow battery using (**b**) a polysulphide flow cathode[12,22,23]; (**c**) mechanically mixed intercalation flow cathode[13,24]; and (**d**) a new flow-cathode structure employing sulphur-impregnated Ketjen black composite suspended in the nonaqueous electrolyte described in this study.

Table 1 | Theoretical volumetric capacity of flow cathodes.

Flow cathode type	Theoretical volumetric capacity (Ah l^{-1})
Soluble polysulphide flow cathode	
5.0 mol S/L: Li_2S_8/Li_2S_4	67
2.5 mol S/L: Li_2S_8/Li_2S	117
Mechanically mixed intercalation flow cathode	
26 vol% $LiCoO_2$–0.8 vol% KB	178
12.6 vol% $LiFePO_4$–4.4 vol% KB	17
Mechanically mixed sulphur-carbon 'flow' cathode	
20 vol% S–26 vol% KB	Solid agglomerate—not flowable
Sulphur-impregnated carbon composite flow cathode	
20 vol% S–26 vol% KB (12.9 M [S])	670

KB, Ketjen black.
A comparison between a number of reported[13,23,24,27] high-energy-density nonaqueous flow cathodes and this work. All the volumetric percentages were calculated using tap densities of the materials for comparison ($LiCoO_2$: 2.4 g cm^{-3}, $LiFePO_4$: 0.78 g cm^{-3}, sulphur: 2.07 g cm^{-3}, KB: 0.12 g cm^{-3})[24,39–41].

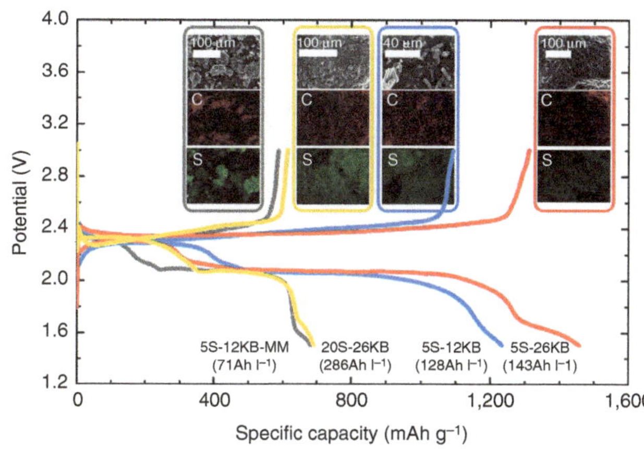

Figure 2 | Galvanostatic voltage profiles and SEM/EDX images. First galvanostatic discharge/charge profile of Li-suspension cells using sulphur-carbon catholytes of 5S-12KB-MM, 5S-12KB, 5S-26KB and 20S-26KB at C/9 (1C≡1,675 mA g^{-1}). The electrochemical tests were replicated for more than three times. Insets: SEM/EDX images of pristine catholytes of 5S-12KB-MM, 5S-12KB, 5S-26KB and 20S-26KB.

approaches[12,13,23,24,27]. First, it exploits solid elemental sulphur (S_8) as the active material instead of polysulphide (Li_2S_8/Li_2S_4) in the flow cathodes[12,23]. This increases the volumetric capacity of the flow cathode by 22.5% compared with polysulphide phase at a given S concentration. Previous studies[12,23] purposely avoided using solid sulphur due to its extremely low solubility and insulating nature. Here, we purposely go beyond the solubility limit of sulphur and exploit the solid phase to achieve high concentration of sulphur. Second, to address the concerns of insulating solid sulphur in the flow cathode, we impregnate sulphur with KB carbon rather than the conventional mechanical mixing approach reported for intercalation compounds[13,24,27]. We show that exploiting the sulphur-impregnated catholyte uniformly integrates the active materials (solid sulphur) and the conductive carbon-percolating network (KB), which enhances the utilization of insulating sulphur. The sulphur-impregnated carbon composite flow-cathode structure is inspired by well-established strategies for traditional solid sulphur electrodes[28–38]. Third, an unexpected advantage of using S/C composite is that the viscosity of the S/C composite catholyte is significantly lower than the mechanically mixed S+C catholyte (under the same S and carbon concentration, see Results). The reduced suspension viscosity significantly increases the maximum allowable concentration of sulphur in the catholyte, which boosts the theoretical volumetric capacity of the flow cathode. Table 1 compares the theoretical catholyte volumetric capacity of our approach with several reported nonaqueous flow cathodes (that is, the polysulphide flow cathode[12,23] and mechanically mixed intercalation cathode[13,24]). All the volumetric percentages were calculated using the tap density for comparison ($LiCoO_2$: 2.4 g cm^{-3}, $LiFePO_4$: 0.78 g cm^{-3}, sulphur: 2.07 g cm^{-3} and KB: 0.12 g cm^{-3}) (refs 24,39–41). The S/C composite catholyte concept offers higher volumetric capacity compared with the soluble polysulphide approach and the mechanically mixed intercalation approaches, owing to the increased active-material concentration and number of electrons transferred.

We examine the influences of (1) the mixing method of sulphur (mechanical mixing versus sulphur impregnation), (2) the concentration of carbon and (3) the concentration of sulphur on the electrochemical behaviour/performance of the S/C catholyte in Li-suspension cells (or 'zero-gap' cell, Supplementary Fig. 1). Figure 2 shows the first discharge/charge profiles and the scanning electron microscope/energy-dispersive X-ray spectroscopy (SEM/EDX) images of four sulphur–carbon catholytes including a mechanically mixed sulphur–carbon suspension with 5.0 vol% sulphur–12.0 vol% KB (that is, 5S–12KB-MM, 3.2 M [S]) and three sulphur-impregnated S/C composite suspensions (that is, 5.0 vol% sulphur–12.0 vol% KB (5S–12KB, 3.2 M [S]), 5.0 vol% sulphur–26.0 vol% KB (5S–26KB, 3.2 M [S]) and 20.0 vol% sulphur–26.0 vol% KB (20S–26KB, 12.9 M [S])).

High-energy catholyte enabled by sulphur impregnation. Impregnation of sulphur with the conductive carbon matrix improves the electrical conductivity of the catholyte and enhances the interfacial contacts between the insulating solid sulphur and the carbon matrix, compared with a mechanically mixed suspension. The gravimetric discharge capacity of the mechanically mixed catholyte (Fig. 2, 5S–12KB-MM, 700 mAh g^{-1}) is lower than that of the S/C composite catholyte (Fig. 2, 5S–12KB, 1,235 mAh g^{-1}) by ~50%. This suggests that the utilization of sulphur in the catholyte is enhanced by the uniform intermixing of S and C using sulphur impregnation. The SEM/EDX image of the 5S–12KB catholyte shows that the S and C atoms are evenly distributed and overlapped across the composite suspension. On the other hand, the S and C atoms are separated in the mechanically mixed suspension (5S–12KB-MM). We employed the electrochemical impedance spectroscopy (EIS) technique to probe the influence of the sulphur-mixing method on the ohmic (R_{ohm}), interfacial (R_{int}) and charge-transfer resistances (R_{ct}) of the S/C catholyte cells (Fig. 3). Figure 3a shows the EIS responses of 5S–12KB-MM and 5S–12KB after the 1st, 5th and 10th discharge/charge cycles with fitting results shown in Fig. 3c–e based on an equivalent circuit shown in Fig. 3b. Applying a well-recognized EIS physical model for Li-S/Li-ion batteries[23,42,43], the high-frequency intercept has been attributed to the cell ohmic resistance, the middle-frequency semi-circle has been attributed to the interfacial resistance/capacitance of the catholyte and the low-frequency semi-circle has been attributed to the charge-transfer resistance/pseudocapacitance of the catholyte[23,42,43]. First, the S/C composite (5S–12KB) shows a threefold decrease in the ohmic resistance compared with the mechanically mixed sample (Fig. 3c), suggesting that the electrical conductivity of the catholyte network is improved with the well-intermixed S/C composite compared with the microscopically separated S+C

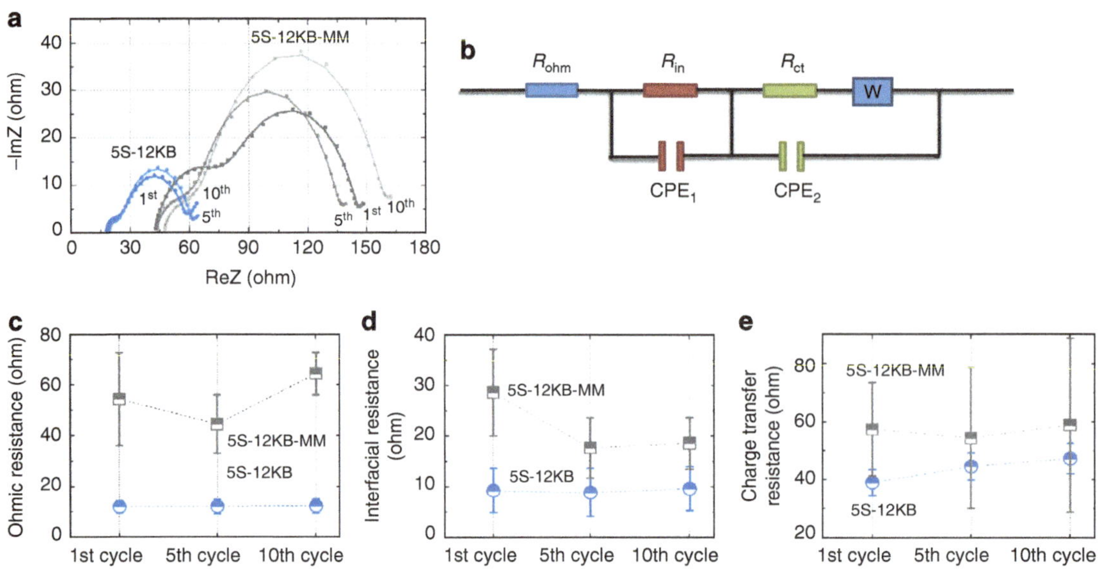

Figure 3 | EIS measurements of Li-suspension cells. (**a**) EIS data (dot symbols) and the fitting results (solid line) of 5S-12KB-MM and 5S-12KB Li-suspension cells after 1st, 5th and 10th discharge/charge cycles at C/9 (1C = 1,675 mA g^{-1}). (**b**) The equivalent circuit used to fit the EIS data shown in **a**. EIS-fitting results on (**c**) ohmic resistance, (**d**) interfacial resistance and (**e**) charge-transfer resistance for 5S-12KB-MM and 5S-12KB Li-suspension cells after 1st, 5th and 10th discharge/charge cycles. The EIS tests were replicated for three times.

mixtures. Second, the composite (5S–12KB) shows a 1.5–2-fold decrease in the interfacial resistance compared with the mechanically mixed sample (Fig. 3d). This can be attributed to the uniform intermixing of sulphur and carbon in the suspension, where small and accessible grains of sulphur are readily surrounded by the conductive KB. No significant difference can be concluded for the charge-transfer resistance (Fig. 3e) due to the large scattering of the mechanically mixed sample, which could be related to the inherent non-homogeneity of the mechanically mixed sample. Nevertheless, we show that impregnating insulating active materials with a conductive carbon matrix enhances the electrical conductivity and interfacial contacts of the catholyte, leading to the improved electrochemical performance of the catholyte. The subsequent cycles of the 5S–12KB and 5S–12KB-MM catholyte cells agree well with the observation in the first cycle discussed here (Supplementary Fig. 2).

The viscosity of the flow composite cathode (5S–12KB) was significantly lower than that of the mechanically mixed sample (5S–12KB-MM), as shown in Fig. 4. The reduced suspension viscosity may be related to the reduced porosity of the KB in the sulphur-impregnated KB slurry, compared with the untreated KB, due to the incorporation of sulphur particles. This effect is clearly demonstrated using suspensions with much higher sulphur content. We compare the rheological properties of 20.0 vol% sulphur–26.0 vol% KB composite (20S–26KB) suspension with a mechanically mixed suspension having 20.0 vol% sulphur–26.0 vol% KB (20S–26KB-MM). Interestingly, the sample obtained by mechanical mixing became a solid agglomerate (Supplementary Fig. 3a), whereas the S/C composite formed a smooth suspension (Supplementary Fig. 3b). In other words, the conventional mechanically mixed KB slurry with high concentration, larger than 13–26 vol%, is unable to form a colloidal solution and cannot flow. In contrast, the sulphur-impregnated KB slurry (20S–26KB) can form a uniform and flowable suspension at the same high KB and sulphur content (Supplementary Fig. 3). The significantly reduced suspension viscosity associated with using S/C composite permits much higher allowable concentration of sulphur, which increases the theoretical capacity of the S/C catholyte.

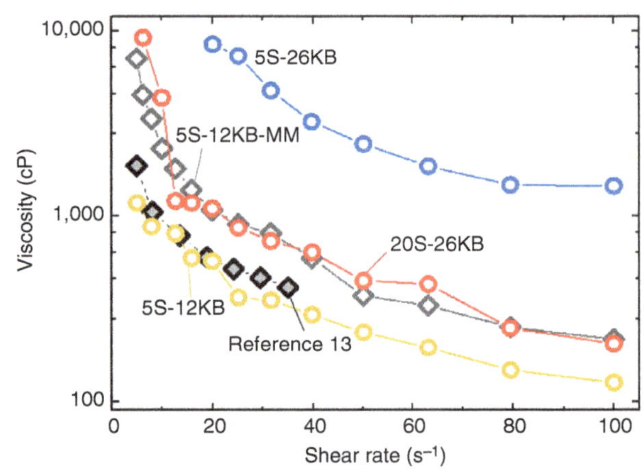

Figure 4 | Viscosity measurements of S/C catholytes. Viscosity versus shear rate of sulphur-carbon suspensions examined in this study in comparison with 22.4 vol% $LiCoO_2$ and 0.6 vol% KB in alkyl carbonate electrolyte reported by Duduta *et al.*[13] Note that the 20S-26KB-MM sample formed a solid agglomerate (Supplementary Fig. 3a) and cannot be used for the viscosity test.

Influence of the concentration of carbon. With the sulphur-impregnation approach, we increased the carbon content from 12 to 26 vol% while keeping the same sulphur content at 5 vol% (3.2 M [S]) to form the 5S–26KB catholyte. The first gravimetric discharge capacity of the 5S–26KB Li-suspension cell was higher than that of the 5S–12KB by ~20% (Fig. 2). There is no significant change in the EIS response for the ohmic resistance and interfacial resistance between 5S–12KB and 5S–26KB Li-suspension cells (Fig. 5a,b). This suggests that further increase in the carbon content from 12 vol% to 26% leads to no improvement in the ohmic and interfacial resistance. Surprisingly, the charge-transfer resistance of the 5S–26KB Li-suspension cell is higher than that of the 5S–12KB sample (Fig. 5c). In addition, the viscosity of the 5S–26KB suspension is significantly larger than that of the 5S–12KB suspension by 5–10 times (Fig. 4). We hypothesize that the increased charge-transfer resistance associated with high KB content results from the significantly increased suspension viscosity (~5–10-fold increase, Fig. 4). Further studies are on-going to understand the fundamental correlations between rheological and electrochemical properties of S/C composite catholytes.

Influence of the concentration of sulphur. The concentration of sulphur determines the theoretical volumetric capacity of the catholyte, as discussed in Table 1. We aim to maximize the volumetric concentration of sulphur without significantly increasing the viscosity of the catholyte. Here, we increased the concentration of sulphur from 5.0 vol% S (5S–26KB, 3.2 M [S]) to 20.0 vol% S (20S–26KB, 12.9 M [S]). Note that the concentration of KB (26 vol%) was chosen based on the minimum KB content inherently existing in the S/C composite (93.1 wt% sulphur and 6.9 wt% KB by thermogravimetric analysis, Supplementary Fig. 4) for a suspension having 20 vol% sulphur. The gravimetric discharge capacity of the 20S–26KB Li-suspension cell (Fig. 2, ~700 mAh g^{-1}) was lower than that of the 5S–26KB sample (Fig. 2 ~1,500 mAh g^{-1}) by >50%, which can be attributed to the lower rate of utilization of sulphur associated with higher concentration of sulphur. However, the 20S–26KB catholyte delivered a twofold increase in volumetric capacity (286 Ah l^{-1}) compared with the 5S–26 KB catholyte (128 Ah l^{-1}). This demonstrates the advantage of employing catholyte with high sulphur concentration in achieving high volumetric capacity. The 20S–26KB suspension forms uniform and flowable catholyte (Supplementary Fig. 3b) with comparable viscosity to that of the semi-solid flow catholyte reported by Duduta *et al.*[13] (<1 vol% KB with mechanically mixed 22.4 vol% $LiCoO_2$, Fig. 4). These observations support that the sulphur-impregnated slurry (20S–26KB) could work as a flowing catholyte. However, we note that the slurry property and flow ability will be gradually dictated more by the gravitational forces if KB and sulphur content are continuously increased. This is an important trade-off design consideration for the catholyte concept proposed in this study. Together, these observations highlight the unique advantage of using S/C composite catholyte, as it permits high concentration of active materials while it maintains proper viscosity in the suspension. Further optimization in increasing the concentration of sulphur and lowering the viscosity of the suspension may lead to further increase in the volumetric capacity and the electrochemical performance.

Electrochemical and morphological reversibility. To demonstrate the electrochemical reversibility of the S/C composite catholyte, we performed a cycling test of the 20S–26KB catholyte and examined the changes in the microstructure of the suspension via SEM at various discharge/charge stages. Figure 6a shows the first galvanostatic discharge/charge profile of the 20S–26KB Li-suspension cells at rates ranging from 2.5 to 6 mA cm^{-2}. Volumetric capacity ranging from 294 to 234 Ah l^{-1} was achieved based on the catholyte. To our knowledge, this system demonstrates one of the highest volumetric capacity of redox flow catholytes to date[12,23,24]. The superior volumetric capacity is resulted from the concentration of sulphur (20.0 vol% S or 12.9 M [S]) leading to high volumetric capacity 294–234 Ah l^{-1}. Figure 6b shows the cycling retention of the 20S–26KB catholyte at 4 and 6 mA cm^{-2} in volumetric capacity (left axis), coulombic efficiency (right axis) and energy efficiency (right axis). We note that the rates used in this work (2–6 mA cm^{-2}) are comparable to other nonaqueous semi-solid flow-battery chemistries (2–10 mA cm^{-2}, Fig. 9)[12,23,24], but significantly lower than the state-of-the-art aqueous flow-battery chemistries (for example, all-vanadium system 50–500 mA cm^{-2}, Fig. 9)[3,5,26]. Substantial decrease in volumetric capacity/efficiency only occurred after the first cycle, which could be attributed to the inefficiency of the lithium anode and/or the loss of polysulphide[44]. Since the 2nd cycle (Fig. 6b), the cells achieved volumetric capacity retention of 79% (4 mA cm^{-2}) and 85% (6 mA cm^{-2}) after 100 cycles. In addition, high coulombic efficiency (90.3–94.5%) and energy efficiency (82–83.7%) after 100 cycles were achieved. The high volumetric capacity and electrochemical reversibility of the S/C catholyte suggest that the employment of the solid sulphur composite in catholyte is a promising direction for high-energy-density Li flow batteries.

We exploited SEM to examine the changes in the microstructure at different stages of discharge/charge of the 20S–26KB Li-suspension cells, as shown in Fig. 6c–g. The SEM image of the pristine sample (before discharge, Fig. 6c) shows submicron pores with uniform features. The S and C atoms are uniformly intermixed, as evidenced from the EDX mapping shown in Fig. 2 (20S–26KB). Upon discharge to the end of the first discharge plateau (Fig. 6d), large clusters (~10–15 μm) appeared with film-like deposits, which partially close the pores of the composite. The deposits may be related to precipitated polysulphides. This confirms that reduction of solid sulphur occurs directly on the carbon network in the suspension. Upon full discharge (Fig. 6e), the discharge product forms a continuous film that covers the entire composite and closes the pores. The film is attributed to Li–S solid discharge products such as Li_2S (refs 28,30,44,45). An important open question arises as follows. Will the insulating discharge product/deposit/film be removed upon charging of the S/C suspension? (see further discussion in Fig. 7). Interestingly, upon charging to the first 50% of the capacity, small and scattered holes/pores/features reappeared (Fig. 6f), and finally, the porous features of the S/C composite recovered upon full recharge of the suspension (Fig. 6g). Highly reversible morphological evolution of the catholyte highlights that the Li–S electrochemical reactions occur reversibly in the S/C composite network, having solid

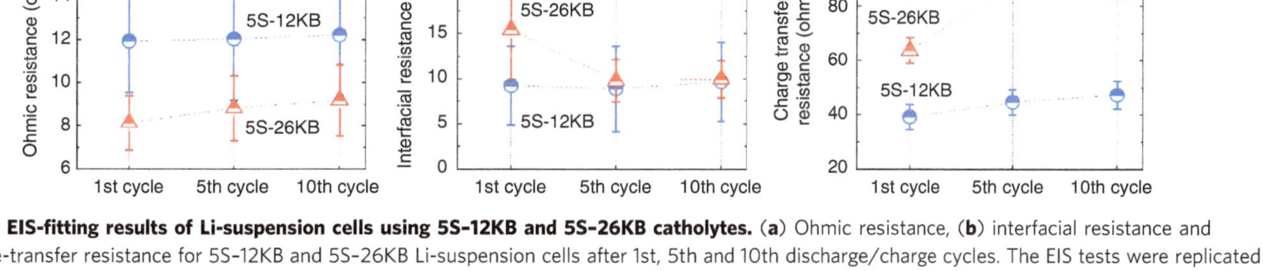

Figure 5 | EIS-fitting results of Li-suspension cells using 5S-12KB and 5S-26KB catholytes. (**a**) Ohmic resistance, (**b**) interfacial resistance and (**c**) charge-transfer resistance for 5S-12KB and 5S-26KB Li-suspension cells after 1st, 5th and 10th discharge/charge cycles. The EIS tests were replicated for three times.

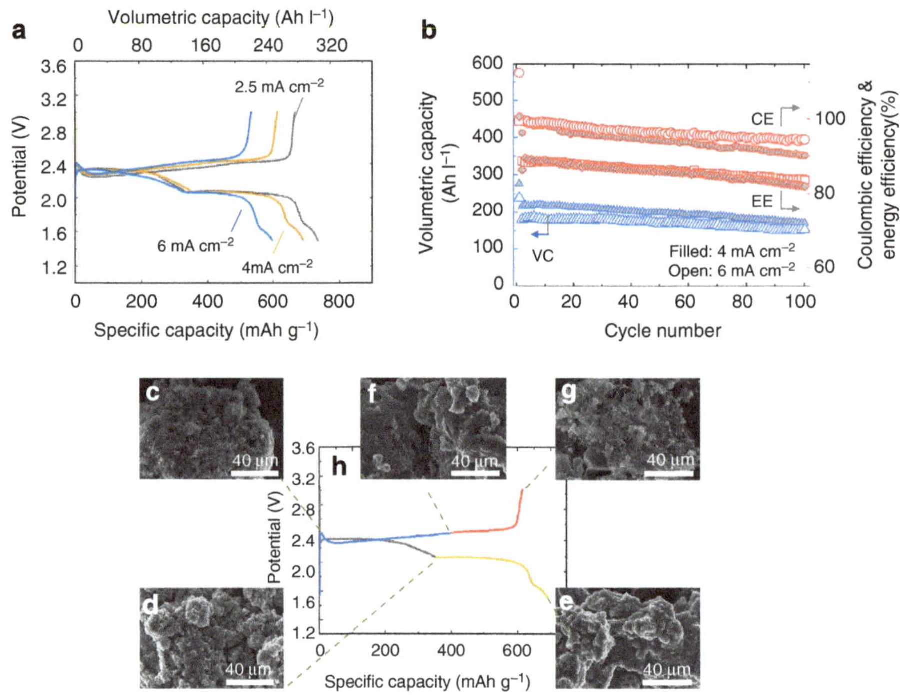

Figure 6 | Electrochemical performance and SEM images of 20S-26KB (12.9 M [S]) catholyte. (**a**) First galvanostatic discharge/charge profiles of Li-suspension cells using 20S-26KB S/C composite catholyte at 2.5, 4.0 (≡C/9) and 6.0 mA cm^{-2}. (**b**) Cycling retention in the volumetric capacity (VC, based on catholyte), energy efficiency (EE) and coulombic efficiency (CE) of the 20S-26KB catholyte cells cycling at 4.0 (≡C/9) and 6.0 mA cm^{-2}. (**c-g**) SEM images of the 20S-26KB catholyte at specified discharge/charge steps at 4 mA cm^{-2}. (**h**) Representative discharge/charge profile of the 20S-26KB catholyte at 4 mA cm^{-2} for illustration purpose. The electrochemical tests were replicated for more than three times.

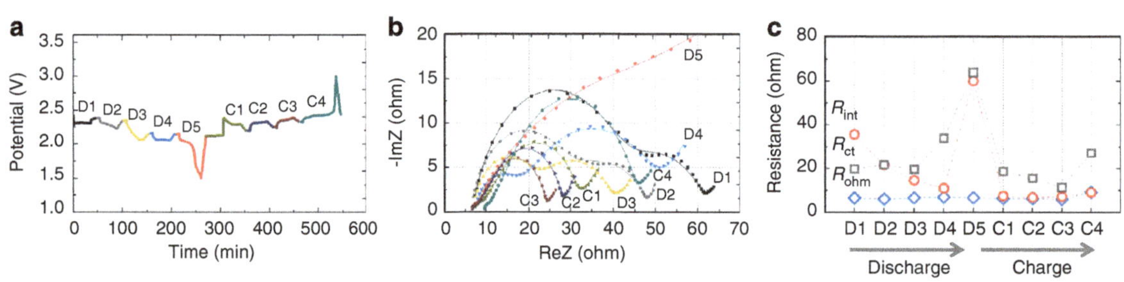

Figure 7 | Pseudo-*in situ* EIS measurement. (**a**) Discharge and charge steps of the Li-suspension cells using 20S-26KB S/C composite catholyte. (**b**) EIS measurements during discharge (D1-D5, at open-circuit voltage (OCV)) and charge (C1-D4, at OCV). Each EIS measurement was measured after resting at OCV for 20 min. (**c**) The EIS-fitting results for the three resistance components (R_{ohm}, R_{int} and R_{ct}) during discharge/charge based on the equivalent circuit shown in Fig. 3b. The pseudo-*in situ* EIS tests were replicated for more than three times.

sulphur as the starting material and solid Li_2S as the end products.

To examine the reversibility of various impedance processes, we employed pseudo-*in situ* EIS measurement as the reaction proceeds. Figure 7a shows the discharge and charge steps of the 20S–26KB Li-suspension cells inserted with five EIS measurements (Fig. 7b, D1–D5, at open-circuit voltage (OCV)) during discharge and four EIS measurements (Fig. 7b, C1–C4, at OCV) during charging, as well as the fitting results of the three resistance components (R_{ohm}, R_{int} and R_{ct})[23,42,43] during discharge/charge (Fig. 7c). The ohmic resistance of the catholyte is the smallest among the three processes and is insensitive to the discharge/charge steps (Fig. 7c). The interfacial resistance first decreased upon discharge to polysulphide phases (D1–D4) and markedly increased upon full discharge (D5). This suggests that the interfacial resistance decreases by transforming the insulating solid sulphur to soluble polysulphides, which improves the interfacial contacts between sulphur-containing species and carbon[22], but significantly increases due to the formation of insulating film-like Li_2S (refs 28,30,44,45). Reversibly, the interfacial resistance decreased after charging the catholyte to the polysulphide phase (C1–C3) and remained small at the end of the charging. Finally, no significant increase in the charge-transfer resistance was noticed at the early stage of discharge (D1–D3, polysulphides) until the later stage of discharge (D4–D5), which is attributed to the formation of insulating Li_2S solids, which blocks the charge transfer[42,46]. Reversibly, the charge-transfer resistance decreases upon charging of the catholyte, which is attributed to the formation of polysulphide species from decomposing the insulating Li_2S solid. Overall, the SEM and pseudo-*in situ* EIS analysis reveal superior morphological reversibility and electrochemical reversibility of the S/C catholyte reported in this study.

Flow cells under static and intermittent-flow modes. We characterize the electrochemical performance of the S/C catholyte

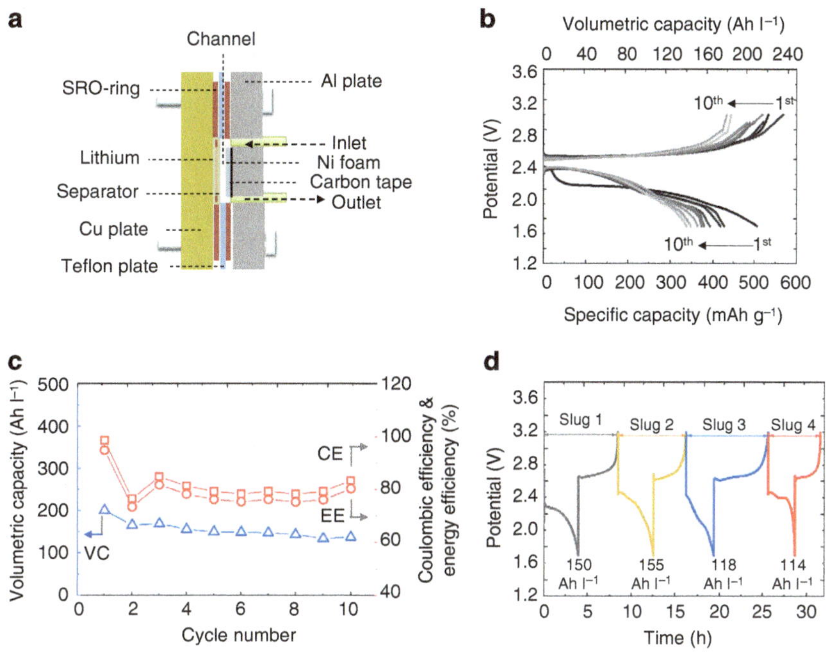

Figure 8 | Electrochemical performance in the 1-mm-channel-gap Li-flow cell. (**a**) Schematic illustration of the Li-flow cell configuration with a 1.0-mm-channel gap. (**b**) Galvanostatic discharge/charge cycle profiles (10 cycles) of the 20S-26KB catholyte in the Li-flow cell at a static mode at 2.5 mA cm^{-2}. (**c**) Cycling retention in the volumetric capacity (VC, based on catholyte), energy efficiency (EE) and coulombic efficiency (CE) of the 20S-26KB catholyte in the Li-flow cell at a static mode at 2.5 mA cm^{-2}. (**d**) Intermittent-flow tests of the 20S-26KB catholyte in the Li-flow cell at 4 mA cm^{-2}. The flow channel dimensions are 1 × 2 × 10 mm. The electrochemical tests were replicated for more than three times.

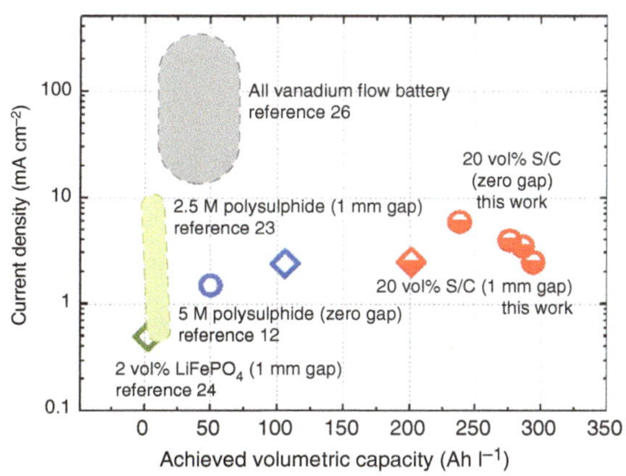

Figure 9 | Achieved catholyte volumetric capacity versus current density. Achieved catholyte volumetric capacities in this work (half-filled symbols) are compared with various reported work discussed earlier[12,23,24]. Open symbols were calculated from the work of Fan *et al.*[23], Yang *et al.*[12] and Hamlet *et al.*[24] The dotted green area denotes the range of the catholyte volumetric capacity (2.7-19 Ah l^{-1}) achieved and current density used in the work of Hamlet *et al.*[24] The dotted grey area denotes the state-of-the-art performance of all-vanadium flow batteries[26].

with a Li-flow cell configuration that adapts a 1.0-mm channel gap (Fig. 8a). Figure 8b shows the discharge/charge cycling profiles of the 20S–26KB catholyte in the Li-flow cell under static condition. Catholyte volumetric capacity of 221 Ah l^{-1} was achieved with coulombic efficiency and energy efficiency >80% (Fig. 8c). We further performed an intermittent-flow test of the S/C catholyte, as the intermittent-flow mode was suggested to be a more effective operation mode for the semi-solid flow battery[13]. Figure 8d shows four iterations of charge/discharge profiles of the 20S–26KB catholyte in an intermittent-flow mode at 4 mA cm^{-2}, demonstrating catholyte volumetric capacity between 110–155 Ah l^{-1}. The power output of the semi-solid sulphur flow cell is comparable to other nonaqueous semi-solid flow cells[13,24], but significantly lower than aqueous flow cells[3,5,26]. Further improvements on the single-cell stack design, electrical conductivity and interfacial contacts in the flow cathode are expected to improve the power output of the semi-solid sulphur flow system.

Discussion

We demonstrate a new flow cathode that employs a highly concentrated sulphur-impregnated carbon composite, and show high catholyte volumetric capacity (294–192 Ah l^{-1}) with long cycle life (>100 cycles), high columbic efficiency (90–94.5% at 100th cycle) and energy efficiency (82–83.7% at 100th cycle). Pseudo-*in situ* EIS and SEM techniques revealed superior electrochemical and morphological reversibility of the S/C catholyte. Figure 9 shows the achieved catholyte volumetric capacity reported in this work (both in ‘zero-gap’ and ‘1-mm-gap’ cells) in comparison with reported Li-flow-cathode chemistries at various current densities[12,23,24]. The highly concentrated (20 vol% S, 12.9 M [S]) sulphur-impregnated carbon composite catholyte introduced in this work demonstrates substantial improvement in volumetric capacity compared with other aqueous[3,5,26] and nonaqueous[12,23,24] flow chemistries (Fig. 9). In addition, the solid-sulphur flow cathode exhibits similar power capability compared with other nonaqueous semi-solid flow chemistries[13,24], but significantly lower power output compared with the aqueous all-vanadium flow batteries[3,5,26]. The hybrid configuration used in our study limits the scalability of the negative electrode[17], which can be improved by replacing the lithium metal with semi-solid negative flow catholytes such as the semi-solid silicon anolyte[27]. Our concept of impregnating insulating solid active materials with a conductive carbon network offers a promising direction to further increase the energy density and cycling reversibility of RFBs and semi-solid flow batteries.

Methods

Materials. All chemicals were used as received. Ketjen Black EC-600JD (KB) was received from AzkoNobel. Sulphur, lithium nitrate ($LiNO_3$), lithium perchlorate ($LiClO_4$), 1,3-dioxolane (DOL) and 1,2-dimethoxyethane(DME) were received from Sigma-Aldrich. Carbon paper (HCP010N) was received from Shanghai Hesen Electric Co. Ltd. Lithium foil was received from Shenzhen Meisen Electromechanical Co. Ltd.).

Preparation of the S/C composite. The S/C composite was prepared according to the well-established method for traditional solid sulphur electrodes[28–38]. Sulphur and KB were mixed with a mass ratio 22:1 in DME, followed by sonication using the SLPt Cell Disruptor (Branson, USA) for 30 min. The sulphur and KB mixture was first air dried at 25 °C for 2 h to remove most of the solvent followed by heat treatment at 155 °C for 6 h for sulphur impregnation. The chemical composition of the S/C composite was determined to be 93.1 wt.% sulphur and 6.9 wt.% KB by thermogravimetric analysis, Supplementary Fig. 4.

Preparation of the S/C suspension. All the suspensions reported herein were made with an electrolyte of 0.2 M $LiClO_4$ and 0.1 M $LiNO_3$ in DOL and DME (volume ratio 1:1). The active materials (sulphur or S/C composite) and KB were mixed in a 10-ml glass container with the electrolyte, followed by sonication using the SLPt Cell Disruptor for 10 min before testing.

Assembly of the Li-suspension cell or zero-gap cell. Supplementary Fig. 1 shows the structure of the Li-suspension cell or 'zero-gap' cell. One piece of lithium foil (20 mm diameter) was placed onto the bottom stainless steel cell body. Two Celgard 2325 separators (25 mm diameter) were placed on the lithium foil. Electrolyte (100 µl) (0.2 M $LiClO_4$ and 0.1 M $LiNO_3$ in DOL and DME (1:1 v/v)) was added on the separators. One piece of carbon paper (18 mm diameter) was placed on the separator act as a buffer layer to decrease the shuttle effect and enhance the conductivity, as suggested by the work of Manthiram *et al.*[47,48] Ten µl suspension was added on the surface of the carbon paper. A piece of nickel foam (18 mm diameter) was placed on the suspension as a current collector followed by a stainless steel spring and a polytetrafluoroethylene O-ring. The effective geometric surface area of the suspension was smaller than the carbon interlayer and current collector and was measured to be 0.2 cm^2. Two cell bodies (bottom and top) were separated by a polytetrafluoroethylene spacer to avoid a short circuit. The cell assembly process was conducted in an Ar-filled glove box (Etelux, <0.1 p.p.m. of H_2O and <0.1 p.p.m. O_2).

Assembly of the Li-flow cell or 1-mm-gap cell. Figure 8a shows the structure of the Li-flow cell or '1-mm-gap' cell. One piece of lithium foil (2 × 10 mm) was attached to a copper (Cu) cell body, which acted as a current collector for the negative electrode. Two Celgard 2325 separators (4 × 14 mm) were placed on the surface of the lithium foil followed by adding 50 µl of electrolyte. A teflon channel spacer (1 mm thickness) was placed between the copper cell body and a piece of pressed nickel foam was attached to an outer aluminium (Al) cell body as a current collector fixed by six bolts. Twenty µl catholytes were injected into the channel (1 × 2 × 10 mm) from the Al plate inlet by a SPLab02 syringe pump (Baoding Shenchen Precision Pump Co. Ltd., China). The Li-flow cell assembly process was conducted in an Ar-filled glove box (Etelux, <0.1 p.p.m. of H_2O and <0.1 p.p.m. O_2).

Electrochemical characterizations. All the electrochemical characterizations were performed using a VMP3 electrochemical testing unit (Bio-Logic). Galvanostatic charge/discharge tests were performed between 1.5 and 3.0 V versus Li^+/Li. Current density was calculated based on the effective geometric surface area of the catholyte in the 'zero-gap' cell (0.2 cm^2) and '1-mm-gap' cell (0.2 cm^2). The specific capacity was calculated from the mass of active materials of sulphur. The volumetric capacity was calculated from the total volume of the suspension. The energy density was calculated by integrating the discharge capacity and voltage normalized by the total volume of the suspension.

EIS measurements used sinusoidal voltage oscillations of 20 mV amplitude at the OCV of the cells. The oscillation frequencies were swept from 100 kHz to 100 mHz with three repetitions for every test. An equivalent circuit shown in Fig. 3b was used to fit all the EIS data by a nonlinear calculation method[23,42,43].

Material characterizations. SEM characterization was performed on Quanta 400 FEI. The SEM samples for different discharge/charge steps (Fig. 6c–g) were collected from cells disassembled at specified discharge/charge steps (Fig. 6c–g) in the glove box. The residual solvent was evaporated before SEM observing. The viscometric behaviours of various suspensions were measured using Physica MCR 301, Anton Paar. The viscosity was recorded as a function of the shear rate (increasing) and obtained from the RHEOPLUS software.

References

1. Yang, Z. G. *et al.* Electrochemical energy storage for green grid. *Chem. Rev.* **111,** 3577–3613 (2011).
2. Dunn, B., Kamath, H. & Tarascon, J. M. Electrical energy storage for the grid: a battery of choices. *Science* **334,** 928–935 (2011).
3. Skyllas-Kazacos, M., Chakrabarti, M. H., Hajimolana, S. A., Mjalli, F. S. & Saleem, M. Progress in flow battery research and development. *J. Electrochem. Soc.* **158,** R55–R79 (2011).
4. Barnhart, C. J., Dale, M., Brandt, A. R. & Benson, S. M. The energetic implications of curtailing versus storing solar- and wind-generated electricity. *Energy Environ. Sci.* **6,** 2804–2810 (2013).
5. Liu, J. *et al.* Materials science and materials chemistry for large scale electrochemical energy storage: from transportation to electrical grid. *Adv. Funct. Mater.* **23,** 929–946 (2013).
6. Wang, K. *et al.* Lithium-antimony-lead liquid metal battery for grid-level energy storage. *Nature* **514,** 348–350 (2014).
7. Huang, Q. & Wang, Q. Next-generation, high-energy-density redox flow batteries. *ChemPlusChem* doi:10.1002/cplu.201402099 (2014).
8. Wang, W. *et al.* Recent progress in redox flow battery research and development. *Adv. Funct. Mater.* **23,** 970–986 (2013).
9. Song, Z. & Zhou, H. Towards sustainable and versatile energy storage devices: an overview of organic electrode materials. *Energy Environ. Sci.* **6,** 2280–2301 (2013).
10. Manthiram, A., Fu, Y. & Su, Y.-S. In charge of the world: electrochemical energy storage. *J. Phys. Chem. Lett.* **4,** 1295–1297 (2013).
11. Menictas, C. & Skyllas-Kazacos, M. Performance of vanadium-oxygen redox fuel cell. *J. Appl. Electrochem.* **41,** 1223–1232 (2011).
12. Yang, Y., Zheng, G. & Cui, Y. A membrane-free lithium/polysulfide semi-liquid battery for large-scale energy storage. *Energy Environ. Sci.* **6,** 1552–1558 (2013).
13. Duduta, M. *et al.* Semi-solid lithium rechargeable flow battery. *Adv. Energy Mater.* **1,** 511–516 (2011).
14. Evers, S., Yim, T. & Nazar, L. F. Understanding the nature of absorption/adsorption in nanoporous polysulfide sorbents for the Li–S Battery. *J. Phys. Chem. C* **116,** 19653–19658 (2012).
15. Lu, Y. & Goodenough, J. B. Rechargeable alkali-ion cathode-flow battery. *J. Mater. Chem.* **21,** 10113–10117 (2011).
16. Lu, Y., Goodenough, J. B. & Kim, Y. Aqueous cathode for next-generation alkali-ion batteries. *J. Am. Chem. Soc.* **133,** 5756–5759 (2011).
17. Leung, P. *et al.* Progress in redox flow batteries, remaining challenges and their applications in energy storage. *RSC Adv.* **2,** 10125–10156 (2012).
18. Wang, Y., Wang, Y. & Zhou, H. A Li-liquid cathode battery based on a hybrid electrolyte. *ChemSusChem* **4,** 1087–1090 (2011).
19. Zhao, Y., Wang, L. & Byon, H. R. High-performance rechargeable lithium-iodine batteries using triiodide/iodide redox couples in an aqueous cathode. *Nat. Commun.* **4,** 1–7 (2013).
20. Li, N. *et al.* An aqueous dissolved polysulfide cathode for lithium-sulfur batteries. *Energy Environ. Sci.* 3307–3312 (2014).
21. Xu, K. Nonaqueous liquid electrolytes for lithium-based rechargeable batteries. *Chem. Rev.* **104,** 4303–4417 (2004).
22. Rauh, R. D., Abraham, K. M., Pearson, G. F., Surprenant, J. K. & Brummer, S. B. Lithium-dissolved sulfur battery with an organic electrolyte. *J. Electrochem. Soc.* **126,** 523–527 (1979).
23. Fan, F. Y. *et al.* Polysulfide flow batteries enabled by percolating nanoscale conductor networks. *Nano Lett.* **14,** 2210–2218 (2014).
24. Hamelet, S. *et al.* Non-aqueous Li-based redox flow batteries. *J. Electrochem. Soc.* **159,** A1360–A1367 (2012).
25. Li, Z. *et al.* Aqueous semi-solid flow cell: demonstration and analysis. *Phys. Chem. Chem. Phys.* **15,** 15833–15839 (2013).
26. Li, L. *et al.* A stable vanadium redox-flow battery with high energy density for large-scale energy storage. *Adv. Energy Mater.* **1,** 394–400 (2011).
27. Hamelet, S., Larcher, D., Dupont, L. & Tarascon, J.-M. Silicon-based nonaqueous anolyte for Li redox-flow batteries. *J. Electrochem. Soc.* **160,** A516–A520 (2013).
28. Yang, Y., Zheng, G. Y. & Cui, Y. Nanostructured sulfur cathodes. *Chem. Soc. Rev.* **42,** 3018–3032 (2013).
29. Ji, X. L., Lee, K. T. & Nazar, L. F. A highly ordered nanostructured carbon-sulphur cathode for lithium-sulphur batteries. *Nat. Mater.* **8,** 500–506 (2009).
30. Kim, J. *et al.* An advanced lithium-sulfur battery. *Adv. Funct. Mater.* **23,** 1076–1080 (2013).
31. Scrosati, B., Hassoun, J. & Sun, Y. K. Lithium-ion batteries. a look into the future. *Energy Environ. Sci.* **4,** 3287–3295 (2011).
32. Wang, D.-W. *et al.* Carbon-sulfur composites for Li-S batteries: status and prospects. *J. Mater. Chem. A* **1,** 9382–9394 (2013).
33. Zheng, G. Y., Yang, Y., Cha, J. J., Hong, S. S. & Cui, Y. Hollow carbon nanofiber-encapsulated sulfur cathodes for high specific capacity rechargeable lithium batteries. *Nano Lett.* **11,** 4462–4467 (2011).
34. Zheng, J. *et al.* Ionic liquid-enhanced solid state electrolyte interface (SEI) for lithium-sulfur batteries. *J. Mater. Chem. A* **1,** 8464–8470 (2013).

35. Manthiram, A., Fu, Y. & Su, Y.-S. Challenges and prospects of lithium-sulfur batteries. *Acc. Chem. Res.* **46,** 1125–1134 (2012).
36. Bauer, I., Thieme, S., Brückner, J., Althues, H. & Kaskel, S. Reduced polysulfide shuttle in lithium-sulfur batteries using Nafion-based separators. *J. Power Sources* **251,** 417–422 (2014).
37. Jin, Z., Xie, K. & Hong, X. Electrochemical performance of lithium/sulfur batteries using perfluorinated ionomer electrolyte with lithium sulfonyl dicyanomethide functional groups as functional separator. *RSC Adv.* **3,** 8889–8898 (2013).
38. Li, X. L. *et al.* Optimization of mesoporous carbon structures for lithium-sulfur battery applications. *J. Mater. Chem.* **21,** 16603–16610 (2011).
39. Ying, J., Jiang, C. & Wan, C. Preparation and characterization of high-density spherical LiCoO2 cathode material for lithium ion batteries. *J. Power Sources* **129,** 264–269 (2004).
40. Evers, S. & Nazar, L. F. New approaches for high energy density lithium-sulfur battery cathodes. *Acc. Chem. Res.* **46,** 1135–1143 (2013).
41. King, J. A. *et al.* Electrical conductivity and rheology of carbon-filled liquid crystal polymer composites. *J. Appl. Polym. Sci.* **101,** 2680–2688 (2006).
42. Deng, Z. F. *et al.* Electrochemical impedance spectroscopy study of a lithium/sulfur battery: modeling and analysis of capacity fading. *J. Electrochem. Soc.* **160,** A553–A558 (2013).
43. Zhang, S. S., Xu, K. & Jow, T. R. Electrochemical impedance study on the low temperature of Li-ion batteries. *Electrochim. Acta* **49,** 1057–1061 (2004).
44. Aurbach, D. *et al.* On the surface chemical aspects of very high energy density, rechargeable Li-sulfur batteries. *J. Electrochem. Soc.* **156,** A694–A702 (2009).
45. Ji, X. L. & Nazar, L. F. Advances in Li-S batteries. *J. Mater. Chem.* **20,** 9821–9826 (2010).
46. Barchasz, C., Lepretre, J.-C., Alloin, F. & Patoux, S. New insights into the limiting parameters of the Li/S rechargeable cell. *J. Power Sources* **199,** 322–330 (2012).
47. Su, Y. S. & Manthiram, A. A new approach to improve cycle performance of rechargeable lithium-sulfur batteries by inserting a free-standing MWCNT interlayer. *Chem. Commun.* **48,** 8817–8819 (2012).
48. Su, Y. S. & Manthiram, A. Lithium-sulphur batteries with a microporous carbon paper as a bifunctional interlayer. *Nat. Commun.* **3,** 1166–1172 (2012).

Acknowledgements

The work described in this paper was substantially supported by a grant from the Research Grants Council of the Hong Kong Special Administrative Region, China, under Theme-based Research Scheme through Project No. T23-407/13-N, and partially supported by project RNEp1-13 of the Shun Hing Institute of Advanced Engineering, The Chinese University of Hong Kong.

Author contributions

Y.-C.L. conceived the project. Y.-C.L. and H.C. designed the experiments. H.C. performed experiments and analysed the data. Q.Z. conducted scanning electron microscopy. Z.L. designed and fabricated the 'zero-gap' Li-suspension cell. H.L. assisted with electrochemical measurements. Q.L. provided scientific advice. Y.-C.L. and H.C. wrote the manuscript. All authors edited the manuscript.

Additional information

Competing financial interests: The authors declare no competing financial interests.

2

Unassisted photoelectrochemical water splitting exceeding 7% solar-to-hydrogen conversion efficiency using photon recycling

Xinjian Shi[1,2,*], Hokyeong Jeong[3,*], Seung Jae Oh[3], Ming Ma[4], Kan Zhang[1], Jeong Kwon[4], In Taek Choi[5], Il Yong Choi[3], Hwan Kyu Kim[5], Jong Kyu Kim[3] & Jong Hyeok Park[1]

Various tandem cell configurations have been reported for highly efficient and spontaneous hydrogen production from photoelectrochemical solar water splitting. However, there is a contradiction between two main requirements of a front photoelectrode in a tandem cell configuration, namely, high transparency and high photocurrent density. Here we demonstrate a simple yet highly effective method to overcome this contradiction by incorporating a hybrid conductive distributed Bragg reflector on the back side of the transparent conducting substrate for the front photoelectrochemical electrode, which functions as both an optical filter and a conductive counter-electrode of the rear dye-sensitized solar cell. The hybrid conductive distributed Bragg reflectors were designed to be transparent to the long-wavelength part of the incident solar spectrum ($\lambda > 500$ nm) for the rear solar cell, while reflecting the short-wavelength photons ($\lambda < 500$ nm) which can then be absorbed by the front photoelectrochemical electrode for enhanced photocurrent generation.

[1] Department of Chemical and Biomolecular Engineering, Yonsei University, 50 Yonsei-ro, Seodaemun-gu, Seoul 120-749, Republic of Korea. [2] Department of Mechanical Engineering, Stanford University, Stanford, California 94305, USA. [3] Department of Materials Science and Engineering, Pohang University of Science and Technology, 77 Cheongam-ro, Nam-gu, Pohang 790-784, Korea. [4] Department of Chemical Engineering, School of Chemical Engineering and SKKU Advanced Institute of Nano Technology (SAINT), Sungkyunkwan University, Chunchun-dong, Suwon 440-746, Korea. [5] Global GET-Future Lab, Department of Advanced Materials Chemistry, Korea University, 2511 Sejong-ro, Sejong 339-700, Korea. * These authors contributed equally to this work. Correspondence and requests for materials should be addressed to J.K.K. (email: kimjk@postech.ac.kr) or to J.H.P. (email: lutts@yonsei.ac.kr).

Conventional photoelectrochemical (PEC) tandem devices for unassisted solar water splitting systems consist of a front photoelectrode for water splitting and a rear photovoltaic cell[1,2]. Since the design of a tandem cell configuration for water splitting was first proposed[1–3], a variety of studies have been reported[4,5], and various photoanode[4,6]/photovoltaic[5,7–13] combinations have been tested to optimize the performance and efficiency. For example, Shi *et al.* reported a wireless monolithic photoanode/dye-sensitised solar cell (DSSC) tandem device[8] that could realize unassisted solar water splitting with a high solar-to-hydrogen (STH) efficiency (5.7%). Nevertheless, the efficiency must be further improved to greater than 10% to reach a viable level for commercialization. To achieve an STH efficiency above 10%, the photoelectrode of the front cell in tandem devices needs to be absorptive to effectively harvest photons from the sun while also being transparent enough to feed the rear cell for unassisted operation.

Typically, the short-wavelength component of solar radiation is mainly absorbed by the PEC photoelectrode, and the remaining long-wavelength component can be harvested by the rear solar cell (for example, DSSC). Increasing the transparency of the PEC photoelectrode for the rear cell results in a reduced photocurrent generation from the front PEC cell. This results in an unsatisfactory compromise with regard to the transparency to find an acceptable balance between the light absorption by the front and rear photoelectrodes via the use of a semi-transparent PEC photoelectrode, resulting in an unavoidable limited utilization of the solar spectrum by the rear cell[6–9]. Incorporating an optical filter called a distributed Bragg reflector (DBR)—consisting of multiple layers of alternating materials with high- and low-refractive indices that can reflect the short-wavelength component of solar radiation for photon recycling by the PEC photoelectrode while transmitting the long-wavelength component for the rear solar cell—into the tandem cell would be very beneficial for an effective utilization of sunlight. However, the conventional dielectric DBR framework has a non-conductive nature, which limits its applicability for monolithic wireless tandem systems in which it should act as a conductive substrate for counter-electrode of the rear solar cell as well. A number of studies on the combination of two thin-film materials that are conductive and transparent, yet have high refractive-index contrast have been performed[14–16], however, their application in PEC tandem cell with desired optical and electrical properties is still immature and needed to be further studied.

Here, we demonstrate unassisted PEC solar water splitting tandem devices with photon-recycling hybrid conductive DBRs (cDBRs). The hybrid cDBR structures consist of alternating dense and porous layers of conductive indium-tin-oxide (ITO) stacks on a conventional TiO_2/SiO_2 DBR framework. These hybrid cDBRs are optimized to exhibit a high reflectance for wavelengths shorter than 500 nm to recycle photons for the front PEC cell and a high transmittance above 500 nm for the effective operation of the rear DSSC while maintaining conductivity to serve as the counter-electrode of the real cell. The PEC/DSSC tandem device with the hybrid cDBRs shows unassisted hydrogen evolution with an STH efficiency of 7.1%, which is the best performance obtained to date from an n-type oxide-based PEC cell.

Results

Concept and design of hybrid cDBRs. A schematic description of the PEC/DSSC tandem water splitting device with the bipolar electrodes is presented in Fig. 1. For the left part, one side is a $BiVO_4/WO_3$ photoanode on fluorine-doped tin oxide (FTO), and the other side is a hybrid cDBR with surface Pt coating (counter of DSSC). For the right part, one side is dye/TiO_2 (anode of DSSC) and the other side is Pt layer as the counter of the tandem device. The hybrid cDBR consists of alternating layers of dense- and porous-ITO layers on the TiO_2/SiO_2 DBR stack with the stopband edge near 500 nm. The fabrication process is described in detail in the Methods section. When the sunlight is incident to the tandem device, the $BiVO_4/WO_3$ photoanode mainly absorbs short-wavelength photons, both from the incident light and the recycled light reflected by the hybrid cDBR. The longer wavelength component of the solar irradiation transmits through the $BiVO_4/WO_3$ photoanode as well as the hybrid cDBR and is absorbed by the dye/TiO_2 electrode in the rear DSSC. The optical properties of the hybrid cDBR structures were optimized using a genetic algorithm (GA)[17,18] and experimentally realized using the oblique-angle deposition (OAD) technique to tune the refractive index of the conductive ITO on demand[19]. The measured refractive indices of the dense- and porous-ITO films were 1.92 and 1.45 at $\lambda = 550$ nm, respectively, indicating sufficient refractive index contrast for the DBR function. To identify an optimum hybrid cDBR structure, the absorptance spectra of the $BiVO_4/WO_3$ photoanode and the dye/TiO_2 electrode used in the DSSC (Fig. 2a) were taken into account in a figure of merit for the GA optimization as shown in equation (1),

$$\mathrm{FOM} = R_{<500\mathrm{nm}} + T_{>500\mathrm{nm}} = \frac{\int_{\lambda=300\mathrm{nm}}^{\lambda=500\mathrm{nm}} w(\lambda)R(\lambda)}{\int_{\lambda=300\mathrm{nm}}^{\lambda=500\mathrm{nm}} w(\lambda)} + \frac{\int_{\lambda=500\mathrm{nm}}^{\lambda=800\mathrm{nm}} w(\lambda)[1-R(\lambda)]}{\int_{\lambda=500\mathrm{nm}}^{\lambda=800\mathrm{nm}} w(\lambda)} \quad (1)$$

where $R(\lambda)$ is the reflectance of the hybrid cDBR and $w(\lambda)$ is a weighting factor based on the absorptance spectra of the $BiVO_4/WO_3$ photoanode and the dye/TiO_2 electrode in the DSSC. Note that instead of using ITO-only cDBRs, the hybrid structures consisting of ITO cDBRs stacked on the conventional three pairs of the TiO_2/SiO_2 DBR stack were designed to maximize the optical filtering property while maintaining the conductivity in the ITO cDBR. The GA-optimized single and double pairs of porous/dense ITO stacks (two-layer and four-layer ITO) on the TiO_2/SiO_2 DBR structure showed the optimal optical properties (Table 1); thus, these designs were chosen for the fabrication.

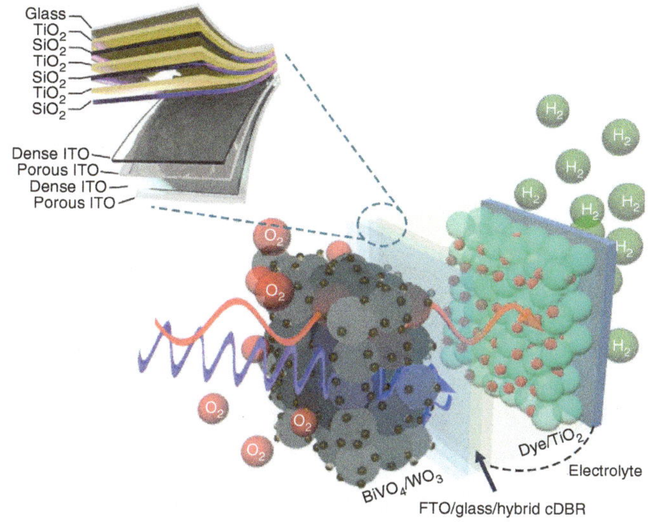

Figure 1 | Schematic illustration of the PEC/DSSC tandem device. For the left part, the back side of the photoanode is a hybrid cDBR consisting of an alternating dense and porous ITO layers on top of TiO_2/SiO_2 stack. The hybrid cDBR reflects short-wavelength light <500 nm (represented by the blue arrow) while transmitting longer-wavelength light (represented by the red arrow). Pt island is coated on its top. For the right part, dye/TiO_2 photoanode and Pt layer are placed at the two opposite sides.

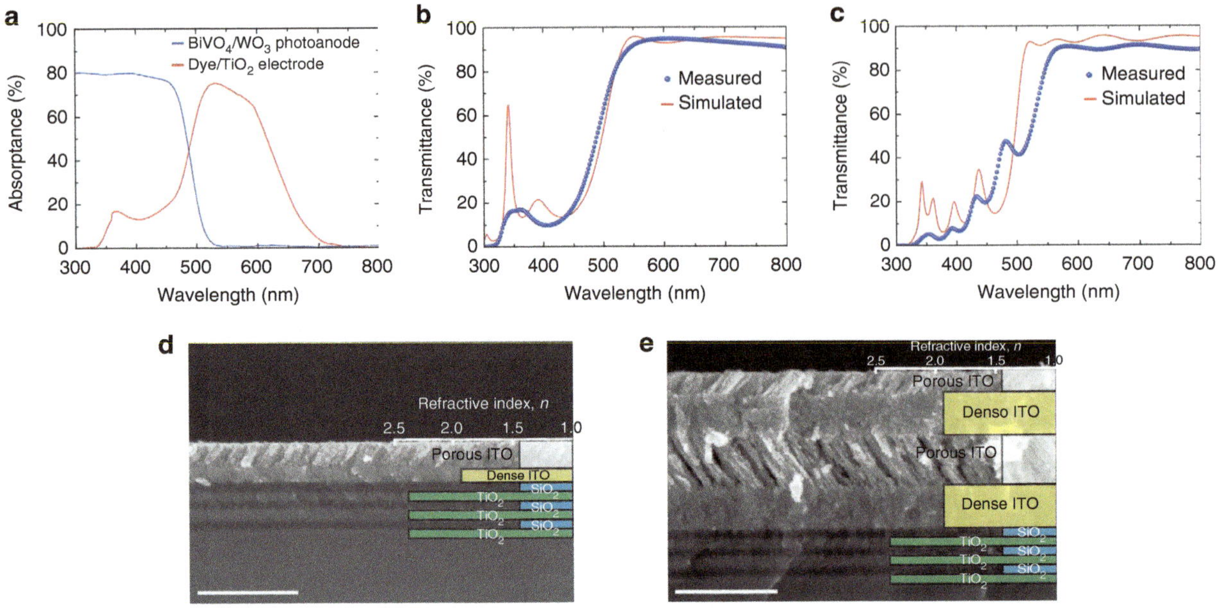

Figure 2 | The hybrid cDBR. (**a**) Absorptance spectra of the $BiVO_4/WO_3$ photoanode (blue solid line) and the dye/TiO_2 electrode measured behind the $BiVO_4/WO_3$ photoanode (red solid line). Measured (blue dotted lines) and simulated (red solid lines) transmittance spectra of the hybrid cDBR with (**b**) two- and (**c**) four-layer ITO. Cross-sectional SEM images of the hybrid cDBR with (**d**) two- and (**e**) four- layer ITO, showing clear interfaces between the layers. Scale bar, 500 nm.

Table 1 | Optimized structures of the hybrid cDBRs with 2- and 4-layer ITO.

	cDBR with 2-layer ITO	cDBR with 4-layer ITO
Structure		Porous ITO, 117 nm
		Dense ITO, 197 nm
	Porous ITO, 127 nm	Porous ITO, 249 nm
	Dense ITO, 94 nm	Dense ITO, 203 nm
	SiO_2, 65 nm	
	TiO_2, 32 nm	
	SiO_2, 60 nm	
	TiO_2, 50 nm	
	SiO_2, 64 nm	
	TiO_2, 25 nm	
	Glass substrate	
$R_{300\text{-}500nm}$	40.52%	40.61%
$R_{500\text{-}800nm}$	8.67%	8.09%
FOM	1.3186	1.3238

cDBR, conductive distributed Bragg reflector; FOM, figure of merit; ITO, indium-tin-oxide. The corresponding average reflectance R for the short-wavelength (300–500 nm) and long-wavelength (500–800 nm) regions.

Detailed GA optimization calculations and the experimental procedures for designing the hybrid cDBR stacks are provided in Supplementary Figs 1–3 and Supplementary Note 1.

Fabrication and characterization of hybrid cDBRs. The measured transmittance spectra (blue dotted lines) of the fabricated hybrid cDBR structures were compared with simulated values (red solid lines), as shown in Fig. 2b,c. The measured spectra were in good agreement with the simulated results, except for the region near 360 nm because of the absorption effect of the glass substrate. These results confirm that the fabricated hybrid cDBR structures with two- and four-layer ITO stacks predominantly reflect the short-wavelength component of the solar radiation to recycle high-energy photons for the PEC photoanode yet have a high transmittance in the long-wavelength component for effective absorption by the dye/TiO_2 electrode in the rear DSSC, as indicated by the figure of merit in the GA optimization. Figure 2d,e showed cross-sectional scanning-electron microscopy (SEM) images of the hybrid cDBR with two- and four-layer ITO stacks, respectively, with the refractive index profiles of the individual layers measured by ellipsometry. These SEM images demonstrated that the fabricated structures had clear interfaces between each layer, whose thickness was well controlled as designed by the GA optimization. The fabricated hybrid cDBRs with two- and four-layer ITO stacks showed high conductivity, with sheet resistances of 20.4 and $35.0\,\Omega\,sq^{-1}$, respectively, indicating a moderate electrical property as an electrode in addition to a superior optical property for the effective utilization of the solar radiation. Because both structures owned very similar optical results, the hybrid cDBR with two-layer ITO that showed a lower sheet resistance was chosen for the PEC tandem cell fabrication.

PEC performances. Bipolar conductive electrodes composed of conventional FTO and the hybrid cDBR were used in the tandem system[7,8], in which the photoanode materials mesoporous (W, Mo)-doped $BiVO_4/WO_3$ were placed on the FTO side and a thin Pt layer was coated on the hybrid cDBR stack. The typical cobalt electrolyte-based DSSC with a high open circuit potential (V_{oc}) was selected as the rear photovoltaic[20] to assemble the whole device. The PEC performances of photoanodes were measured under three-electrode system (Supplementary Fig. 4) and the performances of the tandem devices were evaluated using a two-electrode electrochemical configuration. For two-electrode measurement, before fabrication of the tandem device, the performances of the front PEC photoanode and the rear DSSC were evaluated independently. Notably, the J–V curves of the rear DSSC were obtained under the condition of 1 Sun illumination filtered by the semi-transparent front PEC photoanode with/without cDBR. The operating current density (J_{op}) and the operating potential (U_{op}) of the PEC tandem devices could be defined based on the individual two-electrode J–V curves (Fig. 3a)[21,22]. It can be seen that although the performance of

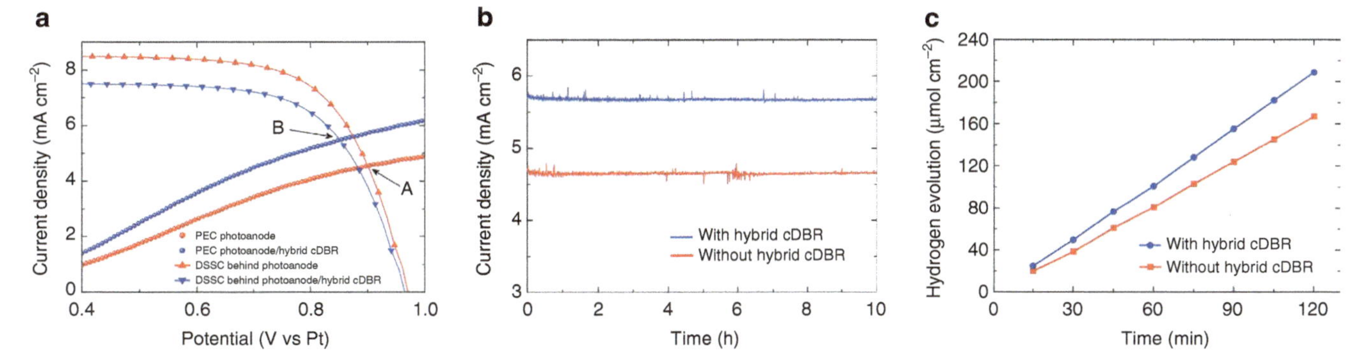

Figure 3 | PEC performances. (**a**) The measured PEC performance in a two-electrode system under AM1.5G illumination. The red dots correspond to the *J-V* curve of photoanode without the hybrid cDBR stack, and the red dotted line corresponds to that of DSSC (checked behind photoanode). The blue dots indicate the photoanode/hybrid cDBR data, and the blue dotted line indicates the DSSC data (determined behind the photoanode/hybrid cDBR). Points A and B define the operating current density (J_{op}) and the operating potential (U_{op}), respectively, for the tandem device without/with the hybrid cDBR interlayers; point B realizes a higher J_{op} under a lower U_{op}. (**b**) The long-term current density versus time (*J-t*) curve of the tandem device with (blue) and without (red) the hybrid cDBR stack. (**c**) The hydrogen evolutions of the tandem device with (blue) and without (red) the hybrid cDBR stack.

the rear DSSC was slightly lower under the filtered 1 Sun illumination condition due to the front PEC photoanode with the hybrid cDBR, the performance of the front photoanode was enhanced with the help of the hybrid cDBR. As a result, the higher operating current density J_{op} of the tandem system is achieved under a lower operating potential U_{op}, as illustrated by the points A and B. This result indicates that the restricted DSSC performance with the presence of the hybrid cDBR is a minor factor in the enhanced overall performance. Detailed information from theoretical studies about the effect of the hybrid cDBR on the light absorption of the front photoanode and rear DSSC, which further affects their respective performances, can be found in Supplementary Figs 5–7 and Supplementary Note 2.

The long-term current density versus time (*J–t*) performance of the PEC tandem devices with and without the hybrid cDBR were compared (Fig. 3b); the current density was well matched with the J_{op} (Fig. 3a), with highly stable behaviours for 10 h for both the tandem devices with or without cDBR. We confirmed the corresponding evolution of hydrogen (Fig. 3c) from the well-mixed gases from both anode and cathode chambers. Besides, in order to characterize the gas crossovers for our membrane-free system, we also sampled and analysed the gases from two chambers, respectively[23], as shown in Supplementary Fig. 8. During the measurement the potential drop due to the electrolyte resistive loss could be quite small, at 10^{-2} V level based on our tandem device operating current density, as well as the near-neutral pH electrolyte used[24]. In addition, the use of wireless tandem device for gas evolution in this work reduced the distance between the working and counter, and decreased the solution loss in further[25,26]. According to the results in Fig. 3, a steady photocurrent density of 5.75 mA cm^{-2} without any additional bias was achieved by the PEC tandem device with hybrid cDBR, which is 1.1 mA cm^{-2} greater than the value for the device without hybrid cDBR. The unassisted STH efficiency of the PEC tandem device with hybrid cDBR was estimated to be ~7.1% according to the equation $\eta_{STH} = J_{op} \times 1.23/P$ (*P* is the power of the illuminating light)[27,28].

Discussion

In summary, we demonstrated the enhanced STH conversion efficiency of an unassisted PEC solar water-splitting tandem cell incorporated with a hybrid cDBR, which serves as both an optical filter to effectively utilize photons as well as a counter-electrode of the rear DSSC. Through photon recycling by the hybrid cDBR, the issue of trade-off between the light absorption and the transmittance of the front PEC photoanodes could be solved. As a result, we demonstrated a 7.1% STH conversion efficiency without any external potential, which is the highest value reported to date for a PEC/solar cell tandem device. In this study, the optical properties of the hybrid cDBR were designed by GA optimization to be compatible with the photoanode ($BiVO_4/WO_3$) and the photovoltaics (DSSC). As a versatile technology exhibiting application-specific optical properties by GA optimization, the hybrid cDBR can be applied to many different PEC solar water-splitting systems and photovoltaics.

Methods

Preparation of WO_3 layer. The porous and highly transparent WO_3 host layer was prepared by surfactant-assisted synthesis. The precursor solution of H_2WO_4 (10:1 PEG:W weight ratio)[7] was prepared by slowly dissolving 0.9 g of tungsten powder in 5 ml of H_2O_2 (35%, Junsei) and stirring the solution for 6 h. Then, 10 ml of 2-propanol ($(CH_3)_2CHOH$, Junsei) was added to the solution, and stirring was continued for one additional day. Finally, 8 ml of polyethylene glycol (PEG, Aldrich) was added to the solution as a surfactant to provide a mesoporous film morphology. The substrate, FTO (TEC-8, Pilkilton) glass, was cleaned in a mixture of ethanol and acetone (1:1 by vol%) for 10 min with sonication and then immersed in a mixed H_2SO_4/H_2O_2 solution for another 10 min for surface hydrophilization. The prepared WO_3 precursor solution was dropped onto the FTO substrate, which was placed on an optic table. The sample was then dried at room temperature for ~20 min to obtain a uniform distribution of the solution on the top of the FTO substrate. The samples were annealed using a temperate gradient from 300 to 550 °C, in which the temperature was increased at a rate of 10 °C min^{-1} followed by annealing for 30 min at 550 °C. To obtain an optimal WO_3 layer, 20 μl of precursor solution was spread on the pre-defined FTO (1.5 × 1.5 cm^2) followed by the drying and annealing process for one cycle, and totally two cycles were done and used for the heterojunction fabrication.

Fabrication of (W, Mo)-doped $BiVO_4$ layer on WO_3 layer. The $BiVO_4$ precursor solution was prepared from a mixture of bismuth nitrate hexahydrate ($BiN_3O_9 \cdot 5H_2O$, 99.99% Aldrich), vanadyl acetylacetonate ($C_{10}H_{14}O_5V$, 98% Aldrich) and sodium molybdate (Na_2MoO_4, Aldrich) at a mole ratio of 100:96:4 that was added to a solution of 1:0.12 acetylacetone ($C_5H_8O_2$, Fluka) and acetic acid (CH_3COOH, 99.70%, DAE). After sonication for 10 min, a dark green solution was obtained. Care was taken such that visible floating matter was not present, and the solution was used within 1 day after preparation to avoid sedimentation and prevent void filling of the mesoporous WO_3 bottom layer. The final mole concentration of Bi was 0.09 M. A total of 100 μl of the $BiVO_4$ precursor solution was dropped on the as-prepared mesoporous WO_3 bottom layer and kept for 1 min for solution penetration with the following spin coating process (5 s, 500 r.p.m. plus 30 s, 1,500 r.p.m.). The samples were then dried at 100 °C for 10 min, and then annealing at 300, 400, 500 °C for 5 min, respectively. During the annealing processes, W could be naturally doped into the $BiVO_4$ because of the intimate contact of the two components in the embedded structure. This process was conducted in a dry and well-ventilated environment and repeated five times. Finally, the samples were annealed in a box furnace at 500 °C for 2 h.

FeOOH and NiOOH co-catalyst surface modification. The FeOOH/NiOOH layer was deposited as follows. Using photo-assisted deposition (1 Sun, AM 1.5G), FeOOH was deposited in a 0.1 M $FeSO_4$ (iron(II) sulphate heptahydrate, 99%, Aldrich) solution under 0.25 V versus Ag/AgCl for 13 min. NiOOH was then deposited in 0.1 M $NiSO_4$ (nickel(II) sulphate hexahydrate, 99%, Aldrich, adjusted pH to 6.7 with basic solution) under 0.11 V versus Ag/AgCl for 6 min. Finally, one additional NiOOH layer was formed by electrodeposition in 0.1 M $NiSO_4$ under 1.2 V versus Ag/AgCl for 1.5 min.

Optimization and fabrication of the hybrid cDBRs. The structures of the hybrid cDBR were designed by two-step GA optimization. First, the dielectric DBR structure composed of a TiO_2/SiO_2 multilayer stack was optimized. Wavelength-dependent refractive index profiles of sputter-deposited TiO_2 ($n = 2.40$ at $\lambda = 550$ nm) and SiO_2 ($n = 1.45$ at $\lambda = 550$ nm) measured by ellipsometry were used. Then, ITO cDBR stacks were optimized on the top of the dielectric DBR after fixing the previously optimized structure. The measured refractive indices and the extinction coefficients of the dense and the porous ITO layers were used for the optimization of the cDBRs (Supplementary Fig. 1a).

The hybrid cDBR stack was fabricated on the back side of the photoanode. First, a TiO_2/SiO_2 multilayer stack was deposited on the back side of the FTO glass by radio-frequency magnetron sputtering with targeted thicknesses. On the top of this dielectric DBR stack, the ITO cDBR stack was fabricated by OAD using electron-beam evaporation. The dense ITO layer was formed without substrate tilting ($\theta_{OAD} = 0°$), and the low-refractive index porous ITO layer was formed by OAD with a substrate tilt angle of 70° ($\theta_{OAD} = 70°$). To improve the optical transparency and conductivity of the ITO-based cDBR stack, the hybrid cDBR samples were annealed in ambient oxygen at 550 °C for 1 min using a rapid thermal annealing system.

Fabrication of dye-coated TiO_2 electrode of DSSC. A 40 mM titanium tetrachloride ($TiCl_4$, Aldrich) solution was prepared, and cleaned FTO substrates were immersed in this solution for 30 min (70 °C). Using the screen-print technique, nanocrystalline transparent TiO_2 electrodes (TiO_2 paste-Dyesol 18NR-T) were deposited onto the FTO glass substrates and then annealed at 500 °C for 30 min. The thickness of this layer was confirmed using an Alpha-step 250 surface profilometer (Tencor Instruments). The as-prepared TiO_2 electrodes were annealed at 500 °C for 30 min and then immersed in a THF/ethanol (v/v, 2:1) solution containing 0.3 mM JK-306 dye and 0.3 mM 4-[bis(9,9-dimethyl-9H-fluoren-2-yl)amino]benzoic acid coadsorbent for 12 h at room temperature.

Preparation of counter-electrode of the tandem device. After the fabrication of the DSSC photoanode on one side of FTO substrate, the other side of glass was coated with Pt layer by e-beam evaporation. This Pt layer is used as the hydrogen evolution reaction (HER) catalyst in the PEC tandem device (work as the counter of the whole tandem device). The dye-coated TiO_2 electrode side and the Pt side were electrically connected with silver paste.

Assembly of the PEC tandem device. After fabrication of hybrid cDBR stack at the back side of PEC photoanode, it was coated with an H_2PtCl_6 solution and heated at 300 °C for 30 min for its use as the counter-electrode of the DSSC. The two sides of glass (PEC photoanode on FTO and Pt islands on top of cDBR) were also electrically connected with silver paste. Two plates, the front PEC photoanode/Pt and the rear dye-adsorbed TiO_2/Pt (Pt for HER), were assembled using a 60-μm-thick Surlyn spacer (Surlyn-1702, Dupont). An electrolyte solution of 0.22 M $Co(bpy)_3(BCN_4)_2$, 0.05 M $Co(bpy)_3(BCN_4)_3$, 0.1 M $LiClO_4$ and 0.8 M 4-*tert*-butylpyridine in acetonitrile was then injected into the sandwiched cell through a hole drilled in the rear TiO_2 electrode. The hole was sealed with a Surlyn spacer and a cover glass. All of the exposed conductive glass or conductive material such as silver paste were totally covered by super glue with drying process before immersing into the electrolyte for measurement.

Characterizations and additional measurements. The evolution of hydrogen and oxygen was quantified with a gas chromatograph equipped with a thermal conductivity detector and a 5-Å molecular sieve column. A phosphate buffer solution (aqueous 0.1 M Na_2HPO_4, pH 9.2, and NaH_2PO_4, pH 4.0) after Ar-purging was used to adjust the pH of the electrolyte to 6.9. The membrane-free solar-driven water splitting system was used for gas evolutions. During the measurements, all cell compartments were completely sealed with rubber septum and glycerine, after then the gases generated from two chambers were intermittently sampled and analysed individually to calculate the gas crossovers. For more accurate value of the whole generated hydrogen, the evolved gases from two chambers were well mixed before analysed. The PEC performance of the photoanode was determined by using the potentiostat with different electrode system. For the two-electrode measurement, the Pt foil acted as both the reference and counter-electrode. For the three-electrode measurement, Ag/AgCl electrode was used as reference electrode and Pt foil was used as the counter-electrode. *J–V* curves were obtained under 1 Sun and AM 1.5 G using a solar simulator for irradiation (a 300-W xenon lamp (Newport)). The electrolyte was a 0.5 M Na_2SO_4 solution (adjusted with buffer solution). A silicon reference cell (Fraunhofer ISE, Certificate No. C-ISE269) was used to calibrate the exact 1 Sun condition. To measure the performances of the PEC tandem devices, the Pt (HER catalyst) on the back side of the rear electrode was removed. The working lead of the potentiostat was connected to the dye-coated TiO_2 electrode, and the counter and reference lines were connected to the separated platinum foil. The dimension of the used Pt foil was $1.5 \times 1.5\ cm^2$. The optical properties during the experiments were measured using a ultraviolet–visible spectrophotometer (UV-2401 PC, Shimadzu). Morphology analyses of the samples were performed using field-emission SEM (JSM-7000F).

References

1. Brillet, J. *et al.* Examining architectures of photoanode–photovoltaic tandem cells for solar water splitting. *J. Mater. Res.* **25,** 17–24 (2010).
2. Duret, A. & Gratzel, M. Visible light-induced water oxidation on mesoscopic alpha-Fe_2O_3 films made by ultrasonic spray pyrolysis. *J. Phys. Chem. B* **109,** 17184–17191 (2005).
3. Brillet, J. *et al.* Highly efficient water splitting by a dual-absorber tandem cell. *Nature Photon* **6,** 824–828 (2012).
4. Han, L. *et al.* Efficient water-splitting device based on a bismuth vanadate photoanode and thin-film silicon solar cells. *ChemSusChem* **7,** 2832–2838 (2014).
5. Ziegler, J. *et al.* Photoelectrochemical and photovoltaic characteristics of amorphous-silicon-based tandem cells as photocathodes for water splitting. *ChemPhysChem* **15,** 4026–4031 (2014).
6. Shina, K., Yoob, J.-B. & Park, J. H. Photoelectrochemical cell/dye-sensitized solar cell tandem water splitting systems with transparent and vertically aligned quantum dot sensitized TiO_2 nanorod arrays. *J. Power Sources* **225,** 263–268 (2013).
7. Kim, J. K., Shin, K., Cho, S. M., Lee, T.-W. & Park, J. H. Synthesis of transparent mesoporous tungsten trioxide films with enhanced photoelectrochemical response: application to unassisted solar water splitting. *Energy Environ. Sci* **4,** 1465–1470 (2011).
8. Shi, X. *et al.* Unassisted photoelectrochemical water splitting beyond 5.7% solar-to-hydrogen conversion efficiency by a wireless monolithic photoanode/dye-sensitised solar cell tandem device. *Nano Energy* **13,** 182–191 (2015).
9. Prévot, M. S. & Sivula, K. Photoelectrochemical tandem cells for solar water splitting. *J. Phys. Chem. C* **117,** 17879–17893 (2013).
10. Shi, X. *et al.* Efficient photoelectrochemical hydrogen production from bismuth vanadate-decorated tungsten trioxide helix nanostructures. *Nat. Commun* **5,** 4775 (2014).
11. Abdi, F. F. *et al.* Efficient solar water splitting by enhanced charge separation in a bismuth vanadate-silicon tandem photoelectrode. *Nat. Commun* **4,** 2195 (2013).
12. Reece, S. Y. *et al.* Wireless solar water splitting using silicon-based semiconductors and earth-abundant catalysts. *Science* **334,** 645–648 (2011).
13. Luo, J. *et al.* Water photolysis at 12.3% efficiency via perovskite photovoltaics and earth-abundant catalysts. *Science* **345,** 1593–1596 (2014).
14. O'Brien, P. G. *et al.* Selectively transparent and conducting photonic crystals. *Adv. Mater.* **22,** 611–616 (2010).
15. Calvo, M. E. *et al.* Photoconducting Bragg mirrors based on TiO2 nanoparticle multilayers. *Adv. Funct. Mater.* **18,** 2708–2715 (2008).
16. Park, J. T., Chi, W. S., Kim, S. J., Lee, D. & Kim, J. H. Mesoporous TiO_2 Bragg stack templated by graft copolymer for dye-sensitized solar cells. *Sci. Rep* **4,** 5505 (2014).
17. Poxson, D. J. *et al.* Demonstration of optical interference filters utilizing tunable refractive index layers. *Opt. Express* **18,** A594–A599 (2010).
18. Schubert, M. F. *et al.* Design of multilayer antireflection coatings made from co-sputtered and low-refractive-index materials by genetic algorithm. *Opt. Express* **16,** 5290–5298 (2008).
19. Schubert, M. F., Xi, J.-Q., Kim, J. K. & Schubert, E. F. Distributed Bragg reflector consisting of high- and low-refractive-index thin film layers made of the same material. *Appl. Phys. Lett.* **90,** 141115 (2007).
20. Yella, A. *et al.* Porphyrin-sensitized solar cells with cobalt (II/III)-based redox electrolyte exceed 12 percent efficiency. *Science* **334,** 629–634 (2011).
21. Weber, M. F. & Dignam, M. J. Efficiency of splitting water with semiconducting photoelectrodes. *J. Electrochem. Soc.* **131,** 1258–1265 (1984).
22. Chen, Y., Hu, S., Xiang, C. & Lewis, N. S. A sensitivity analysis to assess the relative importance of improvements in electrocatalysts, light absorbers, and system geometry on the efficiency of solar-fuels generators. *Energy Environ. Sci* **8,** 876–886 (2015).

23. Jin, J. *et al.* An experimental and modeling/simulation-based evaluation of the efficiency and operational performance characteristics of an integrated, membrane-free, neutral pH solar-driven water-splitting system. *Energy Environ. Sci* **7,** 3371–3380 (2014).
24. Singh, M. R., Papadantonakis, K., Xiang, C. & Lewis, N. S. An electrochemical engineering assessment of the operational conditions and constraints for solar-driven water-splitting systems at near-neutral pH. *Energy Environ. Sci* **8,** 2760–2767 (2015).
25. Haussener, S. *et al.* Modeling, simulation, and design criteria for photoelectrochemical water-splitting systems. *Energy Environ. Sci* **5,** 9922–9935 (2012).
26. Xiang, C., Meng, A. C. & Lewis, N. S. Evaluation and optimization of mass transport of redox species in silicon microwire-array photoelectrodes. *Proc. Natl Acad. Sci. USA* **109,** 15622–15627 (2012).
27. Liao, C.-H., Huang, C.-W. & Wu, J. C. S. Hydrogen production from semiconductor-based photocatalysis via water splitting. *Catalysts* **2,** 490–516 (2012).
28. Schoonman, J. & Van de Krol, R. Nanostructured materials for solar hydrogen production. *U.P.B. Sci. Bull. Series B* **73,** 31–44 (2011).

Acknowledgements
The authors gratefully acknowledge support from the NRF of Korea Grant funded by the Ministry of Science, ICT and Future Planning (NRF-2013R1A2A1A09014038, 2015M1A2A2074663, 2011-0030254) and Green Science project by POSCO.

Author contributions
J.H.P. and J.K.K. conceived the idea and designed the experiments. X.S. performed the main experiments and manuscript preparation. H.J, S.J.O. and I.Y.C performed the DBR design and experiments. J.K., I.T.C. and H.K.K. contributed materials/analysis for DSSCs. M.M. and K.Z. analysed the data. J.H.P. and J.K.K. finalized the paper.

Additional information

Competing financial interests: The authors declare no competing financial interests.

3

Optical modulation of nano-gap tunnelling junctions comprising self-assembled monolayers of hemicyanine dyes

Parisa Pourhossein[1], Ratheesh K. Vijayaraghavan[2,†], Stefan C.J. Meskers[2] & Ryan C. Chiechi[1]

Light-driven conductance switching in molecular tunnelling junctions that relies on photo-isomerization is constrained by the limitations of kinetic traps and either by the sterics of rearranging atoms in a densely packed monolayer or the small absorbance of individual molecules. Here we demonstrate light-driven conductance gating; devices comprising monolayers of hemicyanine dyes trapped between two metallic nanowires exhibit higher conductance under irradiation than in the dark. The modulation of the tunnelling current occurs faster than the timescale of the measurement ($\sim$1 min). We propose a mechanism in which a fraction of molecules enters an excited state that brings the conjugated portion of the monolayer into resonance with the electrodes. This mechanism is supported by calculations showing the delocalization of molecular orbitals near the Fermi energy in the excited and cationic states, but not the ground state and a reasonable change in conductance with respect to the effective barrier width.

[1] Stratingh Institute for Chemistry, and Zernike Institute for Advanced Materials, University of Groningen, Nijenborgh 4, Groningen 9747 AG, The Netherlands. [2] Molecular Materials and Nanosystems and Institute for Complex Molecular Systems, Eindhoven University of Technology, PO Box 513, Eindhoven 5600 MB, The Netherlands. † Present address: Department of Chemical Sciences, Indian Institute of Science Education and Research Kolkata, Mohanpur 741246, India. Correspondence and requests for materials should be addressed to R.C.C. (email: r.c.chiechi@rug.nl).

Devices in which the flow of electricity between two electrodes is modulated by organic compounds fall into two categories; bulk, thin-film devices and molecular, single-molecule. While both utilize synthetic chemistry to tune electrical properties, the latter combines molecular and electronic properties directly; rather than affecting charge-carriers en masse, tunnelling electrons directly utilize the orbital structure of individual molecules. Therefore, control over the composition, structure and conformation of molecules translates into device functionality. One of the most direct applications of this principle is photo-induced conductance switching, in which the tunnelling probability of a metal/molecule/metal junction is affected by photoisomerization *in situ*, typically leading to a change in tunnelling distance[1]. An immediate challenge to fabricating such devices is coupling light into the junction to affect the isomerization. The most common approach is to contact a self-assembled monolayer (SAM) with a (semi) transparent electrode such as graphene[2,3], MoS_2 (ref. 4), poly(3,4-ethylenedioxythiophene) polystyrene sulfonate[5] or sufficiently thin Au (ref. 6). Care must be taken to choose a SAM that is sufficiently robust and that in which isomerization is not sterically inhibited[7]. An insurmountable problem with this approach to conductance switching is that the SAMs (and single-molecule junctions[8]) require long soak times to affect switching, usually on the order of tens of minutes. Symmetric dithiols can be self-assembled into enables on pre-fabricated gaps instead of planar electrodes, but the soak times increase[9].

One approach to bring light-driven conductance switching into a useful time regime is to find systems that do not photoisomerize, but in which the tunnelling probability in the dark and light states differs. Such a system should exhibit photo-gating as opposed to reversible switching, the magnitude of tunnelling current should differ in the light and dark, and modulate on the timescale of a vertical excitation and relaxation. Battacharyya *et al.* measured the conductance of a fullerene-porphyrin dyad on indium tin oxide surfaces in break-junctions in the dark and under irradiation, observing a commensurate increase in tunnelling current that could not be attributed to photocurrent (the light-driven injection of carriers)[10]. They observed two distinct populations of conductivities, one low (dark) and one high (light). Under irradiation, the low-conductance population was depleted and the high-conductance population grew. This observation was only possible because of the long-lived state that results from the internal electron transfer from the porphyrin to the fullerene, allowing the coincidence of a dyad in the charge-separated excited state becoming trapped in the junction during an *I/V* sweep. The need for a transparent bottom electrode through which to irradiate the junction remains a challenge. In the aforementioned experiment, the authors used indium tin oxide, which is not suitable for forming high-quality SAMs for large-area measurements and devices. However, this result suggests that intramolecular charge-transfer may cause sufficiently significant electronic changes that can affect tunnelling charge transport in the excited state. And, while there are models that address the interaction of light with tunnelling junctions, theoretical development is hindered by experimental challenges[11].

This paper describes the reversible, light-induced gating of tunnelling currents through nano-gaps between two parallel Au nanowires separated by a SAM comprising hemicyanine dyes. The monolayer serves both as the active component, affecting the change in tunnelling current, and as a template for the fabrication of the nano-gap. Hemicyanine dyes couple donor and acceptor units through a conjugated bridge, leading to strong optical absorption and a significant charge-transfer character[12]. We designed a series of hemicyanine dyes that form well-defined SAMs that are capable of producing photocurrent under illumination[13]. The origin of the photocurrent is the charge transfer that occurs on excitation, which is sufficient to drive electrons into the Au electrode that supports the SAM, while a redox couple collects the resulting hole. This series of compounds is, therefore, a good candidate for photo-gating and this class of compounds has even been investigated in tunnelling junctions (in the ground state)[14,15]. Rather than utilizing a transparent electrode, we incorporated these dyes into SAM-templated addressable nano-gap (STAN) electrodes[16,17] in which a metal/SAM/metal sandwich structure is turned on its side and cut into thin sections by nanoskiving[18–20]. The resulting devices comprise two metallic nanowires in parallel that are separated by the cross-section of a SAM. The resulting structures are similar to the self-assembly of ensembles into pre-fabricated junctions[9], but the process is bottom–up and can, therefore, accommodate asymmetric chromophores. (And STANs are millimetres in length.) Thus, the molecules spanning the electrodes are exposed and can be irradiated directly during *I/V* sweeps. On vertical excitation, molecules incorporating hemicyanine dye moieties enter a state of high conductance in which the density of states (DOS) of the frontier orbitals delocalizes, effectively decreasing the width of the tunnelling barrier without undergoing any bond rearrangements or conformational changes. Control measurements on SAMs of alkanes show no change in conductance.

Results

Effects of irradiation. The device architecture and experimental design are shown in Fig. 1. The tunnelling current of STAN electrodes incorporating SAMs of the hemicyanine dye 1-(10-acetylsulfanyldecyl)-4-[2-(4-dimethylamino-phenyl)vinyl] quinolinium hexfluorophosphate (AminoPyr, shown in Fig. 1) increases during exposure to light. The current of STANs incorporating octadecanethiol (C18) is insensitive to light. We measured the photo-gating effect using lasers of three different wavelengths, 405, 532 and 650 nm. Due to the broad absorption of hemicyanines and the further broadening that occurs on adssoprtion to Au, all three wavelengths fall within the absorption spectrum of AminoPyr (Supplementary Fig. 1). We observed no significant difference between them and, therefore, present the light data as an aggregate data set of three lasers. (An important distinction between gating and switching is that switching requires two wavelengths to affect the two directions of switching; gating results in a high-conductance light state and a low-conductance dark state.) From these raw *J/V* data, we constructed histograms of $\log|J|$ for each value of V with the laser light switched on and off (which we refer to as 'light' and 'dark' respectively). The peaks of these histograms ($\mu_{\log}$) and the 95% confidence intervals (computed from $\sigma_{\log}$) are shown in the top left of Fig. 2. The histograms of all *J/V* data are shown in Supplementary Fig. 2. Every measurement under illumination did not lead to an increase in conductance nor did every measurement in the dark produce the same conductance. Rather, two populations emerged for AminoPyr, one centred at $\sim 10^{-5}$ and one at $\sim 10^{-3}$ A cm^{-2} for measurements in the light and dark, respectively. The population in the dark is approximately inverted in the light; under illumination the population at 10^{-5} depletes and the one at 10^{-3} grows. This inversion is visible in the bottom of Fig. 2, which shows conductance heat maps formed from histograms of $\log|\frac{dJ}{dV}|$ versus V in which the colours correspond to the frequencies of the bins, from dark blue (0) to yellow (~ 450). These plots show, in one picture, the entire data set; each point of the plots corresponds to a conductance (G as $\log|\frac{dJ}{dV}|$) at a particular voltage and the colour indicates the

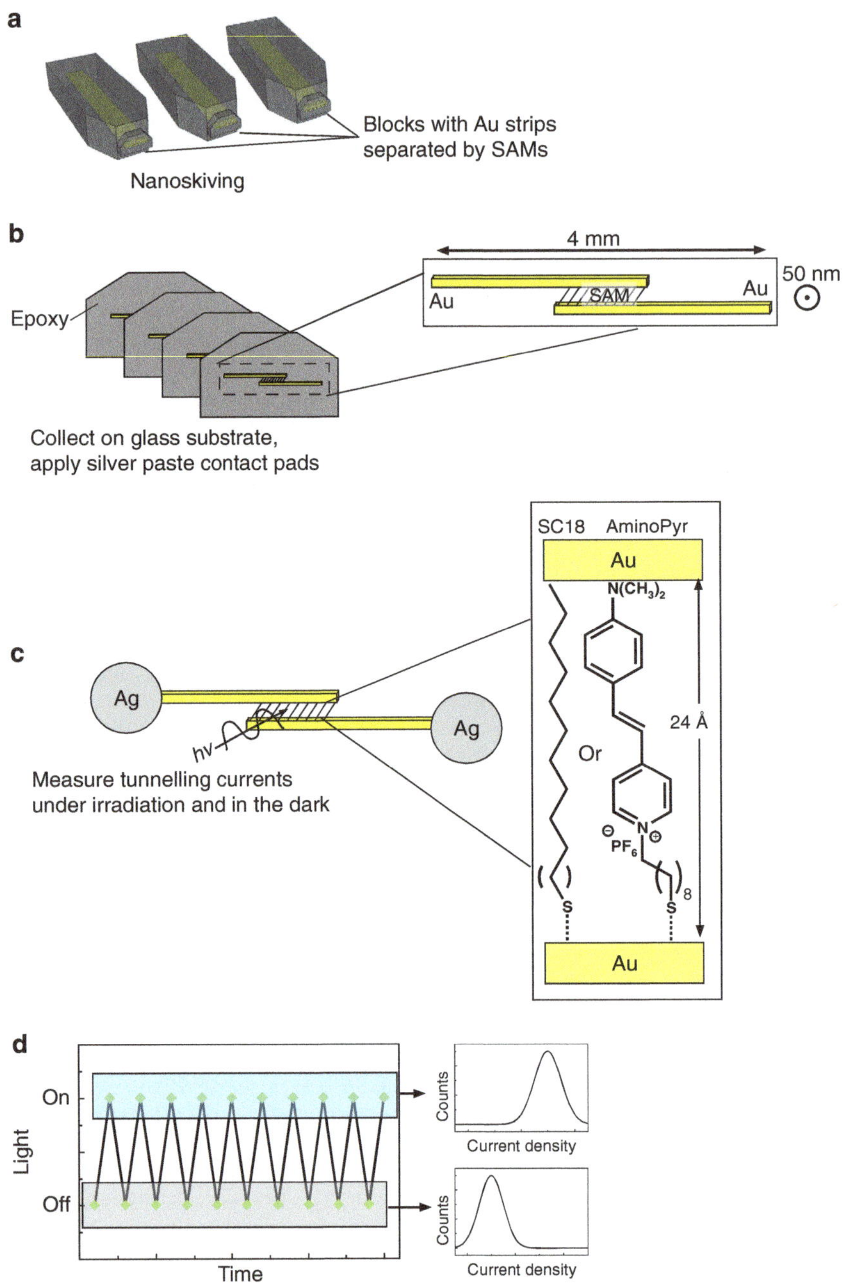

Figure 1 | Schematic of the fabrication of STAN devices comprising SAMs of AminoPyr and C18. (**a**) Au/SAM/Au sandwiches are embedded in epoxy. (**b**) These epoxy blocks are sectioned by nanoskiving to form parallel nanowires separated by SAMs as depicted in the inset. (**c**) The sections are transferred to glass substrates and contact pads are painted onto the nanowires using silver paste to form the STAN devices comprising either SAMs of AminoPyr or C18 as depicted in the inset. (**d**) Tunnelling currents are measured for each device in the light and dark sequentially (dark-measure-light-measure...). Histograms of *J* for each value of *V* are constructed from the light and dark data. Details for the fabrication of STANs (steps **a-c**) are given in refs 16 and 17.

frequency with which that *G*/*V* pair was observed. In the case of AminoPyr, the heatmap plots in Fig. 2 show two clear tranches of conductance with different frequencies that captures the entirety of the photo-gating phenomenon. The top right of Fig. 2 shows histograms and Gaussian fits of *J* at $+0.5$ V for AminoPyr in the light and dark showing the population inversion more clearly. The fits were made by selecting the peaks of the histograms, that is, with the assumption that the histograms in the light and dark each describe a bi-modal distribution.

Figure 3 shows conductance heatmap plots of STANs comprising C18 (bottom) with a population centred at $\sim 10^{-4.2}$ and a tail extending to $\sim 10^{-2}$ A cm^{-2} in the dark, however, the population does not change under illumination. The top right plot shows Gaussian fits made with the assumption that the histograms in the light and dark each describe a bi-modal distribution for the sake of comparison to AminoPyr; the conductance heat maps of C18 clearly show only one tranche for each state. Unlike AminoPyr, the magnitude of the two Gaussian fits does not change; fits around -4.5 are almost identical in the light and dark. From the plots in Figs 2 and 3 we can draw three important conclusions: first, the magnitude of the conductance in the dark is roughly length-dependent because

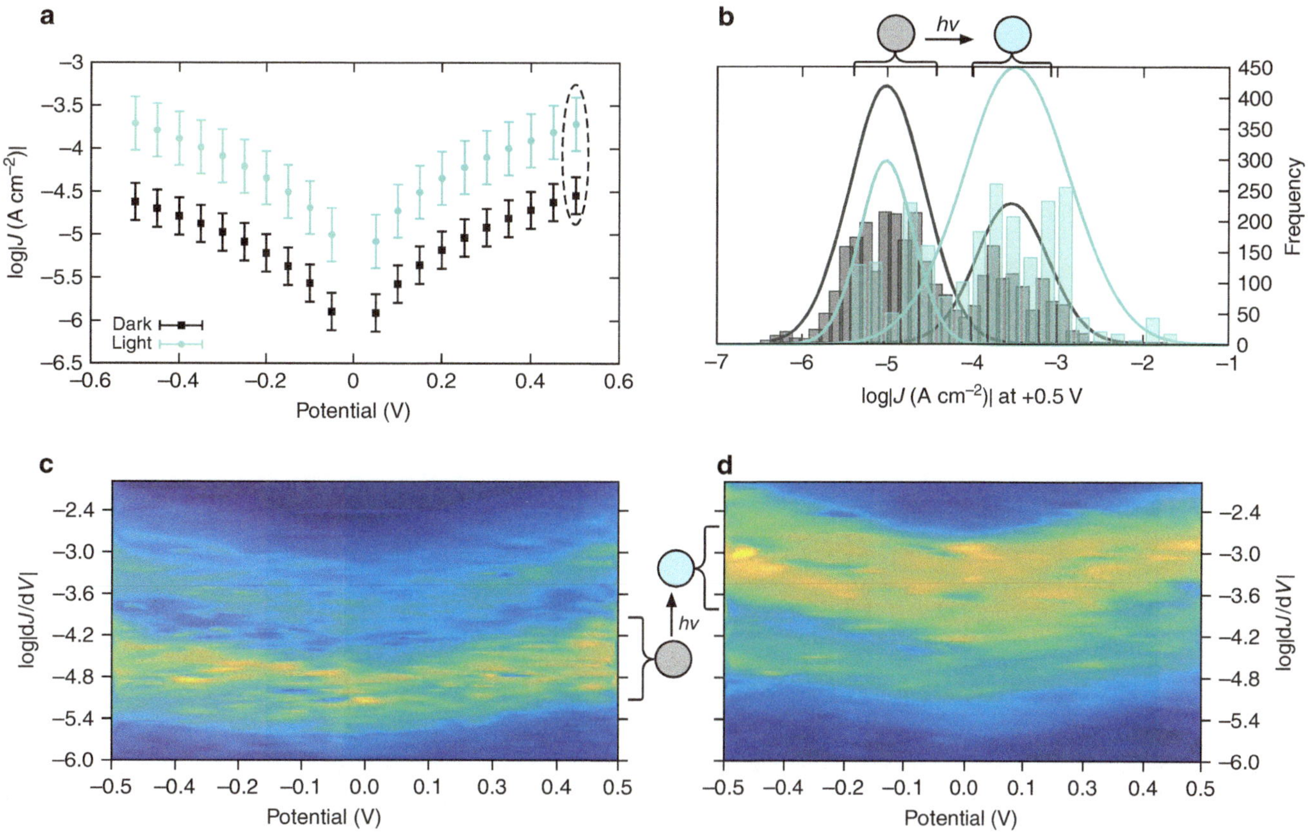

Figure 2 | Summary of conductance data for AminoPyr in the dark and under irradiation. (**a**) Current-density versus voltage plots of AminoPyr-dark (back squares) and AminoPyr-light (cyan circles). The points correspond to Gaussian means from fits of histograms of log|J| for each value of V. The error bars are the 95% confidence intervals (see Supplementary Methods for details). (**b**) The histograms and Gaussian fits of log|J| at +0.5 V (corresponding to the data in the dashed oval in the top-left plot). (**c**) Conductance heatmap plot for AminoPyr-dark showing histograms binned to log|$\frac{dJ}{dV}$| (conductance, y axis) versus potential (in V, x axis). The colours correspond to the frequencies of the histograms; lighter colours indicate higher frequencies (yellow ~450.) (**d**) An identical heatmap plot for AminoPyr-light . The coloured circles correspond the the data in the dark (grey) and under illumination (cyan). A schematic of the junction is shown above the top-left plot and the structure of AminoPyr is shown between the top two plots. These data show a clear, statistical difference between AminoPyr-dark and AminoPyr-light.

both C18 and AminoPyr are approximately the same length and exhibit the same conductance (in the dark); Second, only STANs comprising AminoPyr show two distinct tranches of conductance and the populations of these tranches approximately inverts between the dark and light states; Finally, all of the STAN devices investigated in the light and the dark exhibit approximately symmetric, bowl-shaped conductance plots with positive curvature, which indicates that the dominant mechanism of conduction is likely tunnelling (which we previously established for STANs of alkanes[16]). Although the curvature of the plots of AminoPyr and C18 is slightly different, the shape of AminoPyr-dark and AminoPyr-light remains unchanged, which is a strong indication that the absorption of light increases the tunnelling probability and is not the result of photocurrent or the appearance of a hopping mechanism.

Conductance gating. While there are no established methodologies for characterizing light-induced changes in conductance in tunnelling junctions, a figure of merit in any device configuration is reversibility. Typically J at some voltage is plotted against either the number of cycles (or time at a constant rate of switching). Since virtually all conductance switching is based on (photo)isomerization, reversibility is generally limited by fatigue of the SAM; irreversible pathways, disorder, degradation and so on. Typical devices last for fewer than 10 cycles[2,4–6], with the notable exception of a graphene/monolayer/graphene system based on azobenene isomerization that switches for at least 50 cycles[3]. Direct comparisons with our photo-gating measurements are difficult because of the statistical nature of the observation of switching and the completely different mechanism by which the change in conductance occurs. Battacharyya *et al.* were only able to report conductance histograms in the light and dark (which are sufficient to demonstrate the switching phenomenon) because of the nature of break-junction measurements[10]. Since each observation is sampling the junction in one, specific state (a particular binding mode or the absence of a molecule), the peak conductance—and the gating effect—can only be discerned by assembling a histogram of events (plateaux of G) after many thousands of observations. Our STAN junctions, however, are monolithic devices and conductance switching occurs in-place, thus we report the statistics of switching events rather than the formation of junctions.

We acquire data sequentially and build up conductance histograms by collecting many values of J at each value of V for a device in a particular state that is constant with the exception of the changing variable of illumination. That is, we can construct a plot of J in time similar to those reported for conductance switching based on photoisomerization. However, we do not always observe switching either because it did not occur (due to a subtle change with the laser, a 'bad' SAM, or some other uncontrollable variable) or because that particular device was

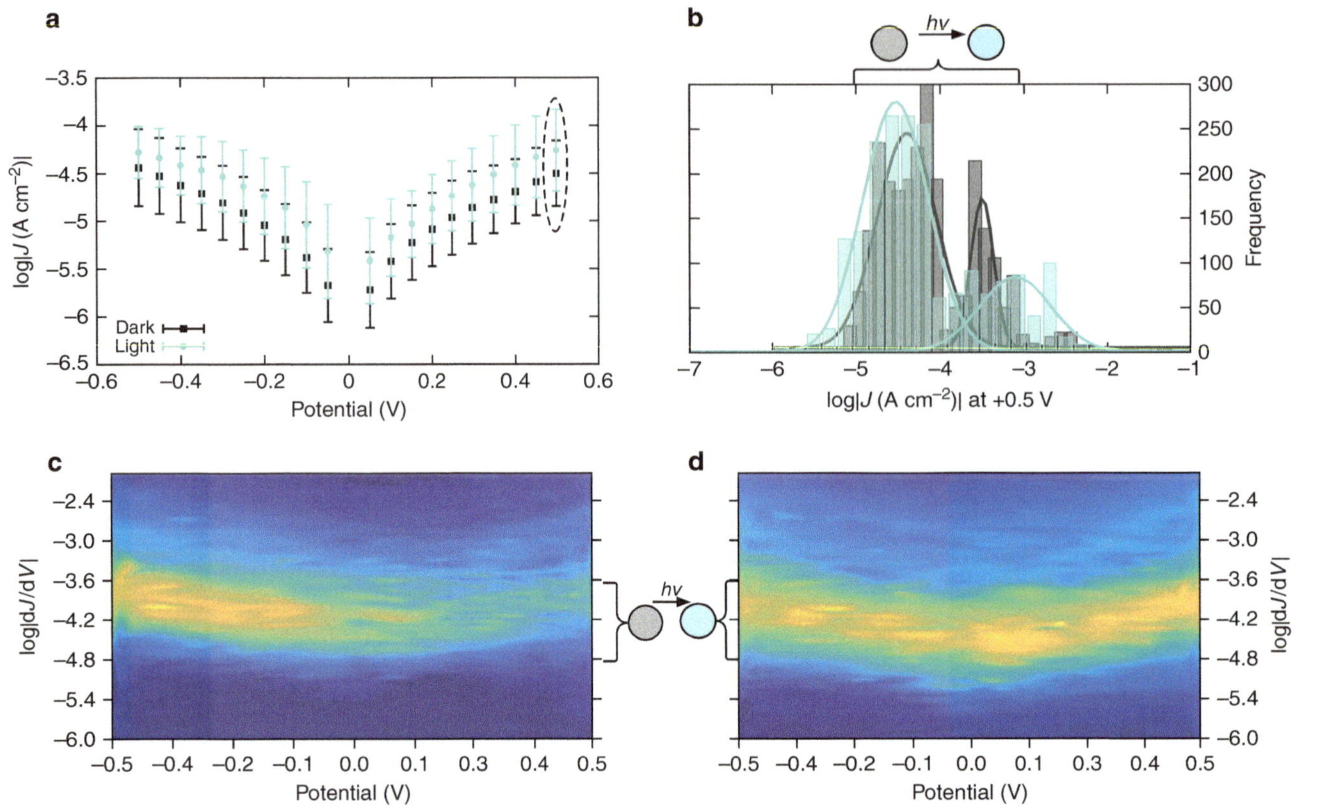

Figure 3 | Summary of conductance data for C18 in the dark and under irradiation. Plots of C18 constructed identically to those shown for AminoPyr in Fig. 2. (**a**) *J/V* curves in the light and dark. (**b**) histograms of *J* at +0.5 V in the light and dark. Conductance heatmap plots in the dark (**c**,**d**). These plots show that there is no statistical difference between C18-dark and C18-light.

under-sampled when it failed; as Fig. 2 shows, the histograms in the light and dark overlap slightly. Thus, what we show in Fig. 4 is 100 consecutive cycles in which switching was observed for each 'on/off' event. Sequences of 100 cycles can span at most two devices (because each was cycled ~100 times), but the average time between cycles is ~60 s, which is limited by the integration time of the *J/V* sweeps. The full range of 1,000 and a plot showing that the apparent gating of C18 was smaller in magnitude and not in phase with irradiation are shown in the Supplementary Figs 3 and 4 and discussed in Supplementary Note 1. (The statistical proof of photo-gating is captured in Fig. 2.) The failure mode of our devices is the collapse of the STAN electrodes due to electrostatic stress (and/or the migration of Au atoms) because there is no fatigue associated with exciting a dye molecule on the order of 100 times. Isomerization causes fatigue of the SAM, which typically leads to dampening of the *J/t* plots, while our devices work for a finite number of *J/V* cycles irrespective of switching events and then short abruptly. The yields of working STAN devices are not straightforward to quantify because they oscillate between 100 and 0% as the knife cuts through a defect in the SAM. Typically, when we encounter a shorted device, cleaving ~1 μm from the block results in non-shorted devices, but the duration of spans of working devices depends on the identity of the SAM. Typical average yields for long-chain SAMs are ~70% (ref. 16).

The microscopic mechanism of the photo-gating phenomenon is not possible to derive from conductance data alone, but we can reasonably exclude some possibilities. We were unsuccessful in resolving *J* in time with a lock-in amplifier and thus can only state that the experimentally observed switching time is <60 s (the time resolution of our measurement). And the only other data point we were able to obtain from the series of hemicyanines

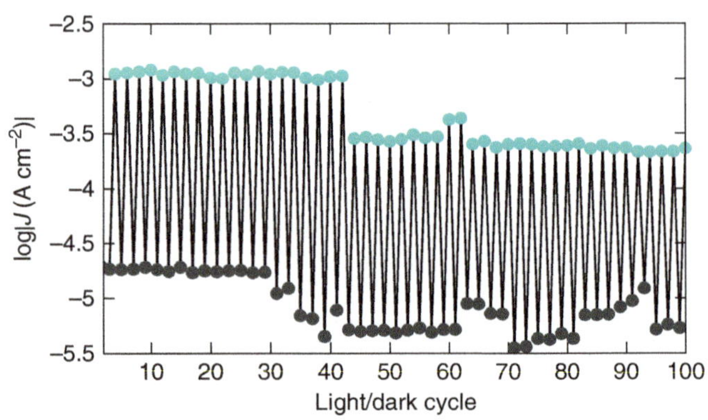

Figure 4 | Plot of conductance cycling of AminoPyr. The values of log|*J*| at +0.5 V as a function of light/dark cycles for a range of 100 consecutive cycles. The cyan circles correspond to log|*J*| under irradiation and the black circles to log|*J*| in the dark. The time between each cycle is on the order of 60 s. The entire range (1,000 cycles) is shown in Supplementary Fig. 3.

(OMePyr , Supplementary Note 2) did not provide statistically significant differences in *J* (Supplementary Figs 5–7). The lack of wavelength dependence is not surprising if one considers the cross-section of the SAM compared with the electrodes; ~1 nm of SAM compared with 100 nm for each electrode. Thus, the absorption of light may take place at the electrodes, which transfer energy to the SAM. Surface-enhanced Raman spectroscopy of molecules adsorbed to Au STAN electrodes shows a strong dependance on the width of the gap with a maximum at 5 nm (ref. 21). Figure 5 shows the angular

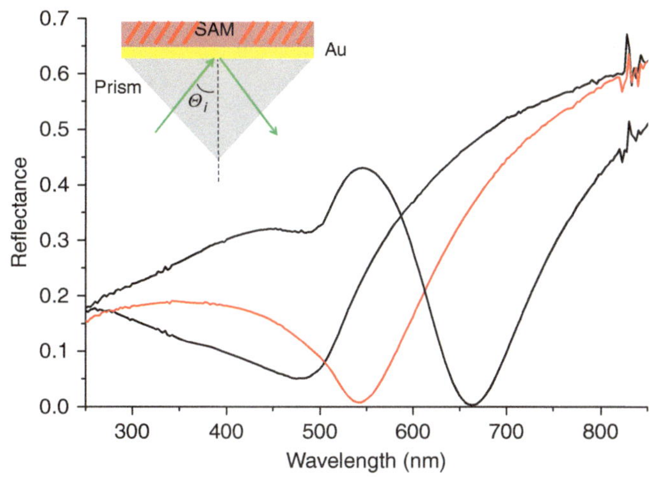

Figure 5 | Angular dependence of reflectance for SAMs of AminoPyr on Au. Decreases in reflectance indicate Plasmon absorption, showing that Plasmon absorptions occur across all the wavelengths used to excite the STAN devices. The different curves correspond to different values of Θ_i: 80° (blue), 55° (red) and 47° (black). The absorbance becomes more pronounced and red-shifted at 47°, which is comparable to the angles used to irradiate the STAN devices.

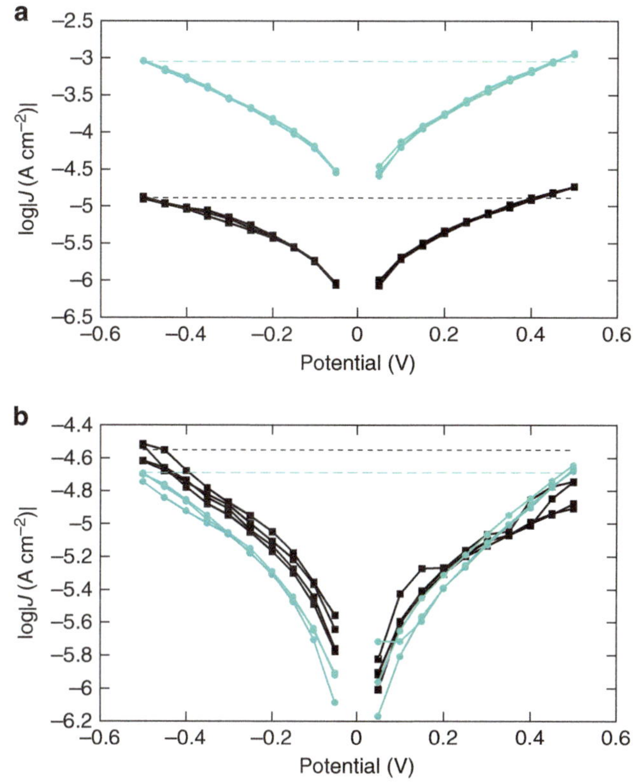

Figure 6 | Individual *J/V* plots of AminoPyr and C18. Plots of log|*J*| versus *V* for AminoPyr (**a**) and C18 (**b**) from a single device showing three traces each from a single dark-light-dark-light-dark-light cycle. The dashed lines show the slight asymmetry of the device in both the dark (black squares) and light (cyan circles) states.

dependence of reflectance of SAMs of AminoPyr on planar Au. There is a strong dip centred between ~450 and ~650 nm corresponding to Plasmon absorption that shifts to lower energies at 47°, which is within the range used to irradiate the STAN electrodes (the surfaces of the two electrodes parallel to the substrate). At 47° the normalized reflectance begins to drop rapidly at ~700 nm, reaching a minimum at 650, mirroring the absorbance spectrum shown in Supplementary Fig. 1. Given that photo-gating requires only the absorption of a photon to induce a high-conductance (excited) state, it is reasonable to expect that irradiation of a Au substrate support a SAM of AminoPyr will lead to a fraction of molecules populating this state. In the absence of any dramatic shifts in absorbance induced by the Van der Waals contact to the 'top' electrode in the assembled STAN devices irradiation at wavelengths below ~700 nm would, therefore, induce a high-conductance state by energy-transfer from the Plasmon to the SAM. And Plasmon energy-transfer processes have been observed for SAMs of AminoPyr in a liquid-cell photovoltaic device[22].

Finite-element analysis of STANs shows a significant field enhancement in the gap as compared with the edges, suggesting that, in addition to the Plasmon modes present in planar substrates, STANs can act like antennas, directing energy into the gap containing the SAMs (ref. 23). This field enhancement alone could modulate the conductivity of the gap; tunnelling current can be generated directly from plasmon resonances across metallic nano-gaps bridged by dithiols at $0.6-1.0$ eV in the terahertz regime[24]. That mechanism is unlikely in our case given the difference in energies and the fact that it would be present in C18, for which we observe no increase in J on irradiation. Likewise any thermal effects arising from the irradiation of the STAN electrodes would have to arise from the SAM of AminoPyr. Although we cannot rule out the thermal excitation of AminoPyr, it would still constitute a photo-gating phenomenon; only the mechanism of energy-transfer from the light would differ.

The lack of any obvious wavelength dependence could be due to the broad absorption of the Plasmon mode; given the proximity of the molecules to the electrode (the π-system is in direct contact), sufficient near-field modes would be active to conceal a weak-wavelength dependance from the direct absorption of chromophores. However, a difference in absorption would correspond to a difference in the fraction of molecules in the light state and J saturates very quickly as a function of the fraction of excited chromophores (Supplementary Fig. 8 and Supplementary Note 3). Changes in $\log|J|$ of less than $\sigma_{\log}$, if present, would require testing hundreds of devices to resolve statistically. Thus, the expected behaviour irrespective of the details of light absorption and energy-transfer is a threshold whereby J jumps from a low-conductance to a high-conductance state at a critical intensity of light independent of wavelength below ~700 nm. Such behaviour is fully consistent with our observations.

Another possibility is that, regardless of the mechanism, once promoted to an excited state, the chromophores simply inject an electron (or hole) into the adjacent electrode, giving rise to an increase in current. If this current were of sufficient magnitude to observe as an increase in J, it would almost certainly show a difference in the symmetry of the J/V curves in the light and dark states and/or a measurable photo-voltage; neither of these are present. Figure 6 shows no difference in the J/V curves of AminoPyr-dark and AminoPyr-light, and very little difference in symmetry (ratio of $\log|J|$ at ± 0.5 V) between AminoPyr and C18. In fact, C18 is slightly more asymmetric and in the opposite direction as AminoPyr. (The asymmetry in both cases probably arises from the disparate interfaces.) We propose instead that the change in conductance is due to the excited-state singly occupied molecular orbital (SOMO) of AminoPyr-light coming into resonance with the electrodes and reducing the effective tunnelling distance to that of the alkyl tail as depicted in Supplementary Fig. 9.

A simple way to model a metal/SAM/metal tunnelling junction in which the SAM contains molecules in different conductive states is to treat these molecules as 'thin-area' defects[25] or, in our case as 'high-*J* defects.' In the dark, the expectation is that the entire population of molecules is in a nominally low-conductance state defined by the thickness of the SAM and the details of the tunnelling barrier. When a fraction of molecules populates a high-conductance state, they can be considered high-*J* defects because, although they do not change in length, the width of the tunnelling barrier decreases and they exhibit an increase in conductance analogous to the decrease in distance that occurs in, for example, photo-induced *cis/trans* isomerization of an azo-benzene. Thus, in the light, the expectation is that a significant fraction of the population is in a high-*J* state and carries most of the tunnelling current. This model predicts that high-*J* defects have a pronounced influence on the total magnitude of *J* that scales with the decrease in effective tunnelling distance (but not light intensity), while the remainder of the SAM—in the ground state—remains in the low-*J* state. According to this model, if the conductance of AminoPyr-light (high-*J*) corresponds to an effective distance of 55% of AminoPyr-dark (low-*J*), only 2% of the population of molecules in the SAM need be in the AminoPyr-light state for them to carry >85% of the total current through the junction. Thus, a steady-state of population consisting of 98% dark-state and 2% light-state (excited or charge-separated) molecules of AminoPyr could cause the higher value of *J* that we attribute to AminoPyr-light. The threshold population necessary to affect a measurable difference in *J* varies inversely with the difference in conductivity between the dark and light states (the change in effective distance), but it only reaches 10% at a difference in conductivity that corresponds to an effective distance of 50%. To put this estimate into context, an azo-benzene switch (based on photoisomerization) showed a change in *J* of $10^{1.4}$ A cm^{-2} with a change in distance of 35% (ref. 6) and a change in *J* of 10^{1} A cm^{-2} with a change in distance of 21% (ref. 3). We observe a maximum difference in *J* of $\sim 10^{2}$ A cm^{-2} for AminoPyr , for which the chromophore comprises ~50% of the total molecular length (these data are plotted in the Supplementary Figs 8 and 10). This model is discussed in detail in Supplementary Note 3; using experimentally derived parameters, it correctly predicts the observed change in *J* via the mechanism described above.

The high-*J* defect model explains how photo-excitation can lead to an overall increase in *J*, but it is silent on the underlying mechanism by which a molecule in the excited state is more conductive in a tunnelling junction. To elucidate that mechanism, we performed density functional theory (DFT) calculations on AminoPyr in the ground and (triplet) excited state as well as the formally charged state in which an electron is transferred to the electrode on excitation as has been observed for SAMs of AminoPyr in dye-sensitized photovoltaic devices[13]. For computational simplicity (as discussed in Supplementary Note 4) we simulated the electrodes as six-atom Au clusters and truncated the alkane spacers to two carbons, which is sufficient to decouple the thiol-electrode interface from the hemicyanine-electrode interface. These data are summarized in the top panel of Fig. 7 and show that the density of states (DOS) derived from the frontier orbitals is completely localized to the electrodes and the alkane spacer in the ground state in agreement with the model proposed above; tunnelling is off-resonant and dominated by molecular length in the ground state. In the excited state, this DOS delocalizes across the entire molecule, effectively hybridizing the hemicyanine chromophore with the electrode. The DOS of the cation (not pictured) is identical. These observations further support the proposed model; in the excited state, the hemicyanine (55% of the total tunnelling distance) presents a significantly lower or absent tunnelling barrier as its frontier orbitals are engaged. The bottom panel of Fig. 7 are simulated transmission curves (with E_F set to -5.0 eV) in the ground, excited and cation states. These data show two main resonance peaks, both of which are outside of or on the edge of the bias window in the ground state. In the excited state, the feature below E_F shifts almost 1 eV towards the centre of the bias window (E_F). This shift is also visible in the cation state together with an additional resonance just above E_F. Thus, these transmission curves support the hypothesis that AminoPyr-light is more conductive than AminoPyr-dark whether or not AminoPyr becomes oxidized.

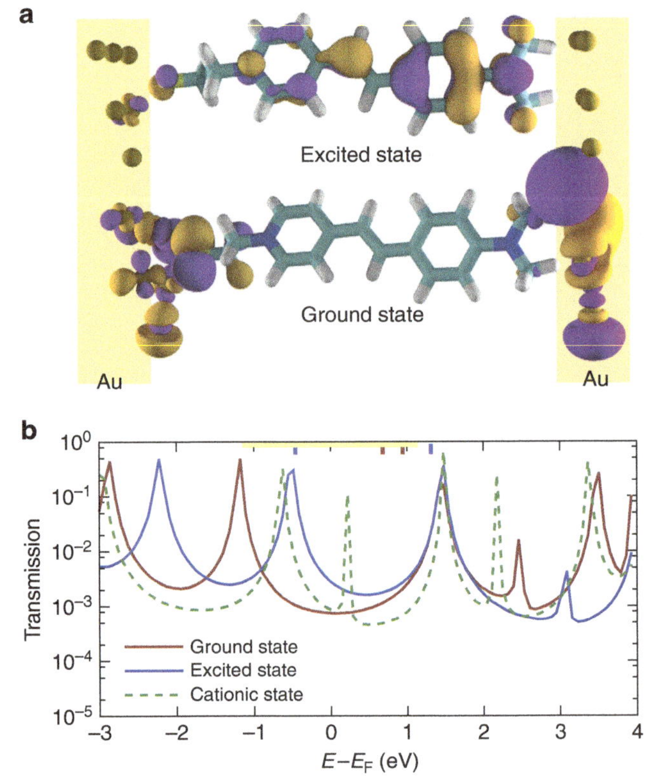

Figure 7 | Calculated molecular orbitals and transmission of AminoPyr between Au electrodes. (**a**) The frontier orbitals of AminoPyr with truncated alkyl chains between two six-atom Au clusters in the ground and photo-excited states (Supplementary Note 2). The frontier orbitals of the cationic state are identical to the excited state. (**b**): The transmission spectra of the same structures in the ground, excited and cationic states with E_F set to -5.0. The positions of the frontier orbitals for the calculated structures are indicated with thick lines across the top of the plot. The yellow bar spans a 2 eV window centered at 0. Because the frontier orbitals only span the electrodes in the excited and cationic states, exciting AminoPyr or formally transferring and electron to the electrode shift a highly transmissive feature from outside to inside the bias window in agreement with experimental observations. In the cationic state an additional feature appears very near E_F.

Discussion

We observed the photo-gating of molecular tunnelling junctions comprising hemicyanine dyes in a STAN electrode configuration. These are monolithic devices that can be fabricated on-demand and the entire life cycle from fabrication to wiring to measurement occurs under ambient conditions; no clean rooms, high vacuum set-ups and so on. (except for the initial deposition of Au). This configuration also obviates the need for a transparent electrode by exposing the 'side' of a SAM between two parallel nanowires. While there are some top–down techniques that similarly expose the sides of molecules[26], they are limited to symmetric molecules, are not compatible with SAMs and are gated with electric fields instead of light. Thus, STAN electrodes are unique in their ability to allow the interrogation of tunnelling junctions comprising SAMs (and asymmetric molecules) with light. Control experiments on junctions incorporating alkanedithiols show no evidence of photo-gating, thus we can rule out the electrodes as the source of the change in conductance. We can therefore ascribe the photo-gating effect to the SAM and propose a mechanism whereby the fraction of molecules in the light-state modulate the conductance of the entire junction. This model correctly predicts the observed magnitude of J in the light and agrees with literature values of changes in J on photoisomerization.

While control experiments on SAMS of C18 rule out effects arising from the electrodes, the lack of the ability to characterize the SAMs in an assembled STAN device hinders proving the proposed mechanism directly. Plasmons probably play a role, as modes are active across the entire absorption window of SAMs of AminoPyr and surface-enhanced Raman effects have been observed in STAN electrodes. The lack of an obvious wavelength dependence supports the hypothesis that Plasmons are involved in absorption and energy-transfer processes, but further experiments on sets of molecules in which the identity and dimensions of the electrodes are varied are needed to elucidate the mechanism fully. The lack of an obvious dependence of the magnitude of the change in conductance on the intensity of excitation (and/or absorbance as a function of wavelength) is consistent with the proposed mechanism of high- and low-J 'defects' modulating the conductance of the entire junction, which is in turn supported by DFT calculations. Properly testing this hypothesis will, however, require synthesizing series of dyes in which the relative lengths of the chromophore and alkyl spacer are varied. Relating the exact mechanism of light absorption and energy-transfer to the magnitude of conductance switching is a matter for future investigation, as this paper only presents robust experimental evidence of a photo-gating phenomenon in solid-state, bottom devices fabricated by nanoskiving.

The proposed photo-gating mechanism does not involve isomerization or the rearrangement of atoms, thus the gating time is much shorter than ordinary switching and does not show rapid fatigue. It therefore differs from previous reports of conductance switching for which the mechanism is generally attributed to a change in effective tunnelling distance following a *cis*/*trans* isomerization or a change in conjugation patterns with ring opening/closure. Thus, while switching requires several minutes of soak time to affect a chance in conductance, photo-gating is limited only by the timescale of a vertical excitation and relaxation back to the ground state. Simulated transmission spectra and calculated DOS derived from the frontier orbtials show that tunnelling probability (conductivity) of AminoPyr is higher in both the excited and cationic states. A vertical excitation and/or quenching by charge-transfer to the electrode can, therefore, increase the conductivity and we can reasonably assert that the change in conductivity occurs on the order of picoseconds. This work represents a step towards realizing functional devices based on molecular tunnelling junctions, in this case devices that can respond to light in real-time.

Methods

STAN electrodes. The nano-gap devices were fabricated by sectioning sandwich structures of Au/SAM/Au where SAM is a SAM of either C18 or AminoPyr using an ultramicrotome (nanoskiving). A representative scanning electron micrograph is shown in Supplementary Fig. 11. The SAM was grown on the buried interface of a thin film of Au that was template-stripped from a technical grade Si wafer. The second Au layer was deposited by thermal vapour deposition directly onto the Au/SAM structure. The fabrication and electrical measurement of STAN electrodes incorporating C18 is also described in detail elsewhere[16,17]. A representative s.e.m. image of a STAN of AminoPyr is shown in Supplementary Fig. 10 The limiting step in the fabrication is the deposition of Au directly onto the SAM that serves a template for the gap. We were able to fabricate STANs of AminoPyr and OMePyr successfully, but SAMs of the other hemicyanine compounds in the series[13] proved too fragile to serve as gap templates. We collected the STAN electrodes on 1 cm glass substrates in 'ribbons' of 5–10 devices from the water bath of the diamond knife that we used for nanoskiving. After drying the ribbons at 60 °C for 1 h, we painted silver paste contacts onto the ends of the wires (as depicted in Fig. 1), which we contacted inside a home-built Faraday cage equipped with an electrometer and a moveable stage. We investigated the STAN devices by sweeping the potential between ± 0.5 V and collecting J/V curves while alternating the light conditions between dark and irradiation with a diode laser. After a sufficient number of J/V sweeps or when a device failed, we continued the measurement on the next device by advancing the stage and contacting the next device (large data sets across multiple devices were acquired in rapid succession). A more detailed description of the experimental procedure is provided in the Supplementary Note 2.

***J*/*V* measurements.** We collected J/V data in four different situations: (1) irradiation with green light ($\lambda = 532$ nm), (2) irradiation with purple light ($\lambda = 405$ nm), (3) irradiation with red light ($\lambda = 650$ nm), (4) no irradiation (dark). All lasers are continuous wave (CW), < 1 mW. The angle at which the light struck the substrates on which the devices were mounted varied between 30 and 60°. In static photo-switching studies each device is interrogated first in a stable ground state and then, after several minutes of irradiation, interrogated again in a (kinetically trapped) switched state. Since our devices do not rely on photoisomerization, we characterized their J/V properties by interrogating them sequentially in the dark and under illumination. That is, for each device we acquired several J/V traces under ambient conditions, switched on the laser and acquired several more, switched if off and measured several more and so on, until we reached ~ 100 switching events or the device shorted. The magnitude of the currents were in the pA-nA regime and required both amplification and integration, which takes place in a Keithley 6430 electrometer. As a result, we were unable to resolve the current in time with greater resolution than one J/V sweep, which is on the order of one minute.

References

1. Kudernac, T., Katsonis, N., Browne, W. R. & Feringa, B. L. Nano-electronic switches: Light-induced switching of the conductance of molecular systems. *J. Mater. Chem.* **19**, 7168–7177 (2009).
2. Li, T. *et al.* Ultrathin reduced graphene oxide films as transparent top-contacts for light switchable solid-state molecular junctions. *Adv. Mater.* **25**, 4164–4170 (2013).
3. Seo, S., Min, M., Lee, S. M. & Lee, H. Photo-switchable molecular monolayer anchored between highly transparent and flexible graphene electrodes. *Nat. Commun.* **4**, 1920 (2013).
4. Margapoti, E. *et al.* A 2D semiconductor-self-assembled monolayer photoswitchable diode. *Adv. Mater.* **27**, 1426–1431 (2015).
5. Kronemeijer, A. J. *et al.* Reversible conductance switching in molecular devices. *Adv. Mater.* **20**, 1467–1473 (2008).
6. Ferri, V. *et al.* Light-powered electrical switch based on cargo-lifting azobenzene monolayers. *Angew. Chem. Int. Ed.* **47**, 3407–3409 (2008).
7. Pace, G. *et al.* Cooperative light-induced molecular movements of highly ordered azobenzene self-assembled monolayers. *Proc. Natl Acad. Sci. USA* **104**, 9937–9942 (2007).
8. Lara-Avila, S. *et al.* Light-triggered conductance switching in single-molecule dihydroazulene/vinylheptafulvene junctions. *J. Phys. Chem. C* **115**, 18372–18377 (2011).
9. Meng, F. *et al.* Orthogonally modulated molecular transport junctions for resettable electronic logic gates. *Nat. Commun.* **5**, 3023 (2014).
10. Battacharyya, S. *et al.* Optical modulation of molecular conductance. *Nano Lett.* **11**, 2709–2714 (2011).
11. Galperin, M. & Nitzan, A. Molecular optoelectronics: the interaction of molecular conduction junctions with light. *Phys. Chem. Chem. Phys.* **14**, 9421–18 (2012).
12. Mishra, A., Behera, R. K., Behera, P. K., Mishra, B. K. & Behera, G. B. Cyanines during the 1990s:? a review. *Chem. Rev.* **100**, 1973–2012 (2000).
13. Vijayaraghavan, R. K., Gholamrezaie, F. & Meskers, S. C. J. Photovoltaic effect in self-assembled molecular monolayers on gold: influence of orbital energy level alignment on short-circuit current generation. *J. Phys. Chem. C* **117**, 16820–16829 (2013).
14. Ashwell, G., Tyrrell, W. D. & Whittam, A. J. Molecular rectification: self-assembled monolayers of a donor-(π-bridge)-acceptor chromophore connected via a truncated Au-S-(CH 2) 3 bridge. *J. Mater. Chem.* **13**, 2855–2857 (2003).

15. Ashwell, G., Tyrrell, W. D. & Whittam, A. J. Molecular rectification:? self-assembled monolayers in which donor-(π-bridge)-acceptor moieties are centrally located and symmetrically coupled to both gold electrodes. *J. Am. Chem. Soc.* **126,** 7102–7110 (2004).
16. Pourhossein, P. & Chiechi, R. C. Directly addressable sub-3 nm gold nanogaps fabricated by nanoskiving using self-assembled monolayers as templates. *ACS Nano* **6,** 5566–5573 (2012).
17. Pourhossein, P. & Chiechi, R. C. Fabricating nanogaps by nanoskiving. *J. Vis. Exp.* **13,** e50406–e50406 (2013).
18. Lipomi, D. J. *et al.* Survey of materials for nanoskiving and influence of the cutting process on the nanostructures produced. *ACS Appl. Mater. Interfaces* **2,** 2503–2514 (2010).
19. Xu, Q., Rioux, R. M. & Whitesides, G. M. Fabrication of complex metallic nanostructures by nanoskiving. *ACS Nano* **1,** 215–227 (2007).
20. Lipomi, D. J., Martinez, R. V. & Whitesides, G. M. Use of thin sectioning (nanoskiving) to fabricate nanostructures for electronic and optical applications. *Angew. Chem. Int. Ed.* **50,** 8566–8583 (2011).
21. Zhou, Z. *et al.* From 1D to 3D: tunable sub-10nm gaps in large area devices. *Adv. Mater.* **28,** 2956–2963 (2016).
22. Gholamrezaie, F., Vijayaraghavan, R. K. & Meskers, S. C. J. Photovoltaic action in a self-assembled monolayer of hemicyanine dyes on gold from dissociation of surface plasmons. *Appl. Phys. Lett.* **106,** 183303 (2015).
23. Zaretski, A. V. *et al.* Using the thickness of graphene to template lateral subnanometer gaps between gold nanostructures. *Nano Lett.* **15,** 635–640 (2015).
24. Tan, S. F. *et al.* Quantum plasmon resonances controlled by molecular tunnel junctions. *Science* **343,** 1496–1499 (2014).
25. Weiss, E. A. *et al.* Influence of defects on the electrical characteristics of Mercury-Drop junctions: self-assembled monolayers of n-Alkanethiolates on rough and smooth silver. *J. Am. Chem. Soc.* **129,** 4336–4349 (2007).
26. Cui, A., Dong, H. & Hu, W. Nanogap electrodes towards solid state single-molecule transistors. *Small* **11,** 6115–6141 (2015).

Acknowledgements

R.C.C. and P.P. were supported by the European Research Council for the ERC Starting Grant 335473 (MOLECSYNCON). We thank the Center for Information Technology of the University of Groningen for their support of the Peregrine high performance computing cluster.

Author contributions

R.K.V. synthesized and characterized the the compounds. P.P. fabricated the devices and performed the *J*/*V* measurements. S.C.J.M. supervised the synthesis and characterization and R.C.C. supervised the project. All the authors contributed to preparing figures and writing the manuscript.

Additional information

Competing financial interests: The authors declare no competing financial interests.

4

Surface engineering of hierarchical platinum-cobalt nanowires for efficient electrocatalysis

Lingzheng Bu[1], Shaojun Guo[2,3], Xu Zhang[4], Xuan Shen[5], Dong Su[5], Gang Lu[4], Xing Zhu[6], Jianlin Yao[1], Jun Guo[6] & Xiaoqing Huang[1]

Despite intense research in past decades, the lack of high-performance catalysts for fuel cell reactions remains a challenge in realizing fuel cell technologies for transportation applications. Here we report a facile strategy for synthesizing hierarchical platinum-cobalt nanowires with high-index, platinum-rich facets and ordered intermetallic structure. These structural features enable unprecedented performance for the oxygen reduction and alcohol oxidation reactions. The specific/mass activities of the platinum-cobalt nanowires for oxygen reduction reaction are 39.6/33.7 times higher than commercial Pt/C catalyst, respectively. Density functional theory simulations reveal that the active threefold hollow sites on the platinum-rich high-index facets provide an additional factor in enhancing oxygen reduction reaction activities. The nanowires are stable in the electrochemical conditions and also thermally stable. This work may represent a key step towards scalable production of high-performance platinum-based nanowires for applications in catalysis and energy conversion.

[1] College of Chemistry, Chemical Engineering and Materials Science, Soochow University, Jiangsu 215123, China. [2] Department of Materials Science and Engineering, College of Engineering, Peking University, Beijing 100871, China. [3] Department of Energy and Resources Engineering, College of Engineering, Peking University, Beijing 100871, China. [4] Department of Physics and Astronomy, California State University, Northridge 91330, California, USA. [5] Center for Functional Nanomaterials, Brookhaven National Laboratory, Upton, New York 11973, USA. [6] Testing and Analysis Center, Soochow University, Jiangsu 215123, China. Correspondence and requests for materials should be addressed to S.-J.G. (email: guosj@pku.edu.cn) or X.-Q.H. (email: hxq006@suda.edu.cn)

Fuel cells are highly desired devices for advanced portable electronic devices and electrical vehicles as they are promising candidates for providing a sustainable, clean and efficient energy supply[1–5]. The fuel cell catalyses the reactions between a fuel (methanol, ethanol and molecular hydrogen) at the anode and molecular oxygen at the cathode, converting chemical energy into electrical power via the electrochemical process[6–8]. Catalysts are the central components of fuel cell technologies and largely dictate their ultimate performance. To date, Pt-based nanomaterials are recognized as the most effective electrocatalysts for both the anodic oxidation reaction and the cathodic oxygen reduction reaction (ORR) in the fuel cells[9–11]. Currently, the state-of-the-art electrocatalysts are almost exclusively small platinum (Pt) nanoparticles (NPs) supported on carbon black (Pt/C). However, the practical large-scale commercialization of fuel cells is still a challenge because of the high cost, poor reaction kinetics, loss of electrochemical surface area and catalytic activity over long-term duration of commercial Pt catalysts[12–16].

Precise control over material structures at the nanoscale allows for command over both surface structure and composition, thereby imparting enhanced and new functionality[17–20]. In the search for high-performance electrocatalysts[9–10,21–24], one-dimensional (1D) Pt alloy nanostructures represent a promising direction for advanced catalysis because they possess multiple advantages, such as inherent anisotropic morphology, high flexibility, high surface area and high conductivity, compared with their 0D counterparts, making them work more effectively for ORR activity and durability enhancement than traditional Pt alloy NPs[25,26]. This is because during the catalysis process, individual nanowires (NWs) can have a higher surface area interacting with carbon support than NPs, which not only enhances electron transfer between oxygen and the Pt surface, but also facilitates bonding between the NWs and the carbon support, rendering high stability. These advantages have stimulated explorations of new synthetic routes to create various Pt-*M* (*M* = iron (Fe)-, cobalt (Co)-, nickel (Ni))-based NWs, but unfortunately usually leading to conventional large-diameter Pt*M*-based NWs with smooth surfaces (lacking the strong electron effect), which are, however, not beneficial for creating more active sites for maximizing electrocatalytic performance, and ultrathin Pt-*M* NWs with limited durability because of their relatively small diameters[27,28]. Therefore, the biggest issue in using Pt-*M* NWs for ORR is that there is always a trade-off between activity and durability, making them impractical for fuel cell devices. In this regard, the creation of a new class of Pt*M* NWs with improved utilization of Pt atoms for simultaneous high catalytic activity and durability is highly desirable but remains a significant challenge.

Herein, we report the synthesis of a class of hierarchical Pt-Co NWs enclosed with high-density high-index facts through a robust large-scalable wet-chemical approach as electrocatalysts for fuel cell reactions. Unlike conventional Pt-*M* NWs, these hierarchical Pt-Co NWs with highly uneven surfaces are tailored to have an ordered intermetallic structure, high-index facets and a Pt-rich surface, which can meet several critical design criteria for efficient nanoelectrocatalysts[29]. They are fully functional for both high-performance anodic and cathodic fuel cell reactions, such as methanol oxidation reaction (MOR), ethanol oxidation reaction (EOR) and ORR that largely outperform those based on the commercial Pt/C catalyst. In particular, they deliver unprecedented ORR performance with the promising specific and mass activities of 7.12 mA cm^{-2} and 3.71 A mg$^{-1}_{Pt}$ at 0.9 V versus reversible hydrogen electrode (RHE), which are 39.6 and 33.7 times higher than those of the state-of-the-art Pt/C catalyst (Johnson Matthey, 20 wt% Pt, 0.18 mA cm^{-2} and 0.11 A mg$^{-1}_{Pt}$), making the hierarchical Pt-Co NWs the most active electrocatalyst among the Pt-Co-based nanomaterials for ORR reported to date, to the best of our knowledge[30]. Moreover, the hierarchical Pt-Co NWs show superior cycling stability in the ORR condition with negligible activity decay over the course of 20,000 cycles and are also thermally stable with no morphology change after annealing up to 500 °C. Their high thermal stability may provide a catalytic platform for high-temperature catalytic reactions, which is not easily achieved by particulate nanostructures because of their deformation or aggregation during the annealing treatment[31,32]. These hierarchical Pt-Co NWs enclosed with high-index facets enabled by our colloidal chemistry technique reported herein are promising catalyst candidates with unprecedented catalytic activity and stability for practical proton exchange membrane fuel cells.

Results

Synthesis and characterization of hierarchical Pt-Co NWs. The wet-chemical approach to produce the hierarchical Pt-Co NWs involves the use of platinum(II) acetylacetonate (Pt(acac)$_2$) and cobalt(III) acetylacetonate (Co(acac)$_3$) as the metal precursors, oleylamine as solvent and surfactant, cetyltrimethylammonium chloride (CTAC) as the structure-directing agent, and glucose as a reducing agent. Briefly, Pt(acac)$_2$, Co(acac)$_3$, oleylamine, glucose and CTAC in the controlled molar ratios were added into a 35-ml glass vial. After the glass vial had been capped, the mixture was fully ultrasonicated for around 1.5 h, heated from room temperature to 160 °C in 0.5 h, then kept at 160 °C for 8 h in an oil bath, and finally naturally cooled to room temperature (see the Methods section for details). The resulting black colloidal products were subjected to centrifugation and washed three times with ethanol/cyclohexane (v:v, 9:1) mixture. A variety of synthetic parameters, such as precursor, reducing agent and surfactant, were carried out for studying the optimal conditions for synthesizing well-defined Pt-Co NWs. The results show that the successful preparation of hierarchical Pt-Co NWs is highly controlled by the amount of CTAC, the select of Co(acac)$_3$ and the use of glucose (Supplementary Figs 1–4).

Transmission electron microscopy (TEM) and high-angle annular dark-field scanning TEM (HAADF-STEM) were used to characterize the morphology of the as-prepared hierarchical Pt-Co NWs. The TEM and STEM images show that the product consists of abundant nanostructure with 1D contour at the first glance (Fig. 1 and Supplementary Fig. 5). The NWs were the dominant product with the typical yield over 90%. The length of the NWs is in the range of several micrometres and the average aspect ratio is ~50. Strikingly, high-magnification HAADF-STEM (inset in Fig. 1a) and TEM images (Fig. 1b) reveal that each NW has a highly uneven diameter along their entire length, exhibiting highly uneven and the crenel-like hierarchical nanostructure. The overall Pt/Co composition is quantitatively determined to be 74.8/25.2 (Pt_3Co), measured by inductively coupled plasma atomic emission spectroscopy (ICP-AES) and TEM energy-dispersive X-ray spectroscopy (TEM-EDS; Fig. 1c). The surface molar ratio of Pt to Co obtained from X-ray photoelectron spectroscopy (XPS) is 95.2/4.8, much higher than that of overall Pt/Co composition, revealing the Pt-rich surface structure (Supplementary Fig. 6). The powder X-ray diffraction (PXRD) pattern of the hierarchical Pt-Co NWs (Fig. 1d) displays the distinct facet-centred cubic structure associated with the alloyed Pt_3Co (Pt_3Co, JCPDS No. 29-0499). The alloyed structure was also confirmed by the STEM-EDS elemental mapping analysis, where the distributions of Pt and Co are even through the whole NW, despite the hierarchical Pt_3Co NWs having very uneven surfaces (Fig. 1e and Supplementary Fig. 7).

The hierarchical Pt_3Co NWs are single crystalline, revealed by both selected-area electron diffraction and atomic resolution

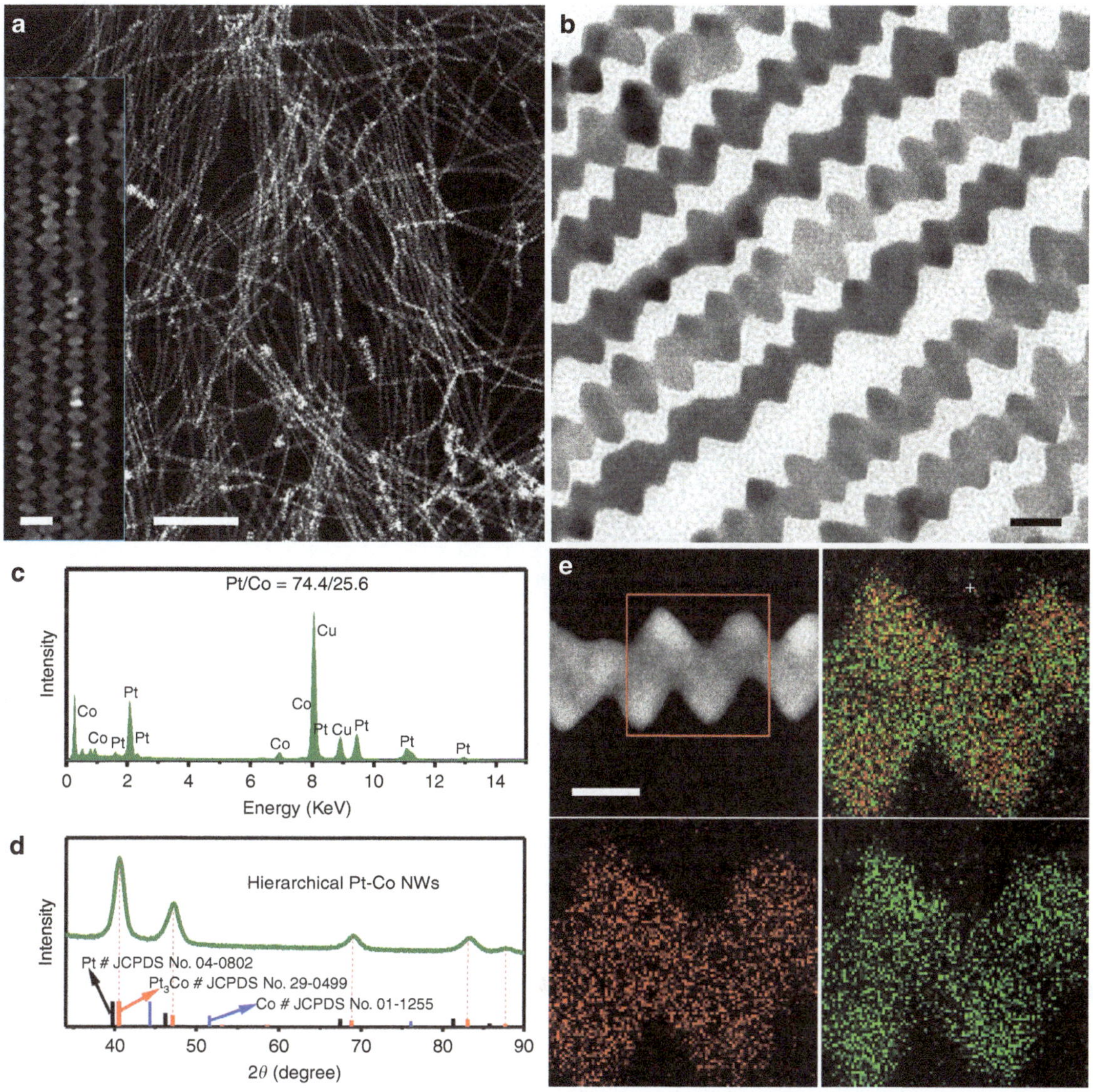

Figure 1 | Morphology and structure characterization of hierarchical Pt-Co nanowires. Representative (**a**) STEM image, (**b**) TEM image, (**c**) TEM-EDS, (**d**) PXRD pattern and (**e**) STEM-ADF image and EDS elemental mappings of the hierarchical Pt_3Co NWs. Inset in **a** is an enlarged STEM image. The composition is Pt/Co = 74.8/25.2, as revealed by ICP-AES. The scale bars in **a**, inset of **a**, **b** and **e** are 200, 20, 10 and 10 nm, respectively.

HAADF-STEM measurements (Fig. 2a,b and Supplementary Fig. 7). To identify the unconventional NWs, a number of techniques was utilized to further analyse the structure of hierarchical Pt_3Co NWs. Figure 2c is an atomic-resolution HAADF-STEM image along the [001] zone axis. The local ordering is observed, corresponding to the $L1_2$-ordered intermetallic structure of Pt_3Co. The projected $L1_2$ unit cell is composed of a periodic square array of pure Co columns surrounded by Pt columns at the face centres and corners of each unit cell (Fig. 2g). As the contrast of HAADF-STEM image is proportional to the average atomic numbers of the columns (Z-contrast),[33] the Pt columns have higher brightness than that of the Co columns in Fig. 2d–f. This is different from the homogeneous alloyed phase, which has all columns with equal intensity and without the super periodicity. From the structural perspective, the ordered phases with definite composition and structure can provide the predictable control over structural, geometric and electronic effects for catalysis optimization, which cannot be afforded by the widely created conventional alloys.[30] In order to identify the crystal indices of the flat planes in the hierarchical Pt_3Co NWs, we applied a HAADF-STEM electron tomography technique to visualize the three-dimensional (3D) structure of the hierarchical Pt_3Co NWs. The HAADF-STEM images were acquired from -65° to 65° with the intervals of 1°. Figure 2h–j and the inset show the 3D reconstructed tomograms of an individual hierarchical Pt_3Co NW at -60°, 0° and 60° and corresponding HAADF-STEM images. Combined the results from electron tomography with those from the atomic resolution HAADF-STEM image shown in Fig. 2k, we found the [001] plane of the NWs aligns very close to the grid plane. Furthermore, by analysis of the cross-sectional structure along <101> direction of Pt_3Co NWs, we could deduce the indexes of the flat planes as [310] and [110] (Fig. 2l). As the hierarchical Pt_3Co NWs contain many uneven surfaces, high-density [310], high-index facets will be introduced onto individual NWs, which is the key to achieve high electrocatalyic activity.

Growth mechanism. The ability to make hierarchical Pt_3Co NWs with unconventional surface structures is the most striking

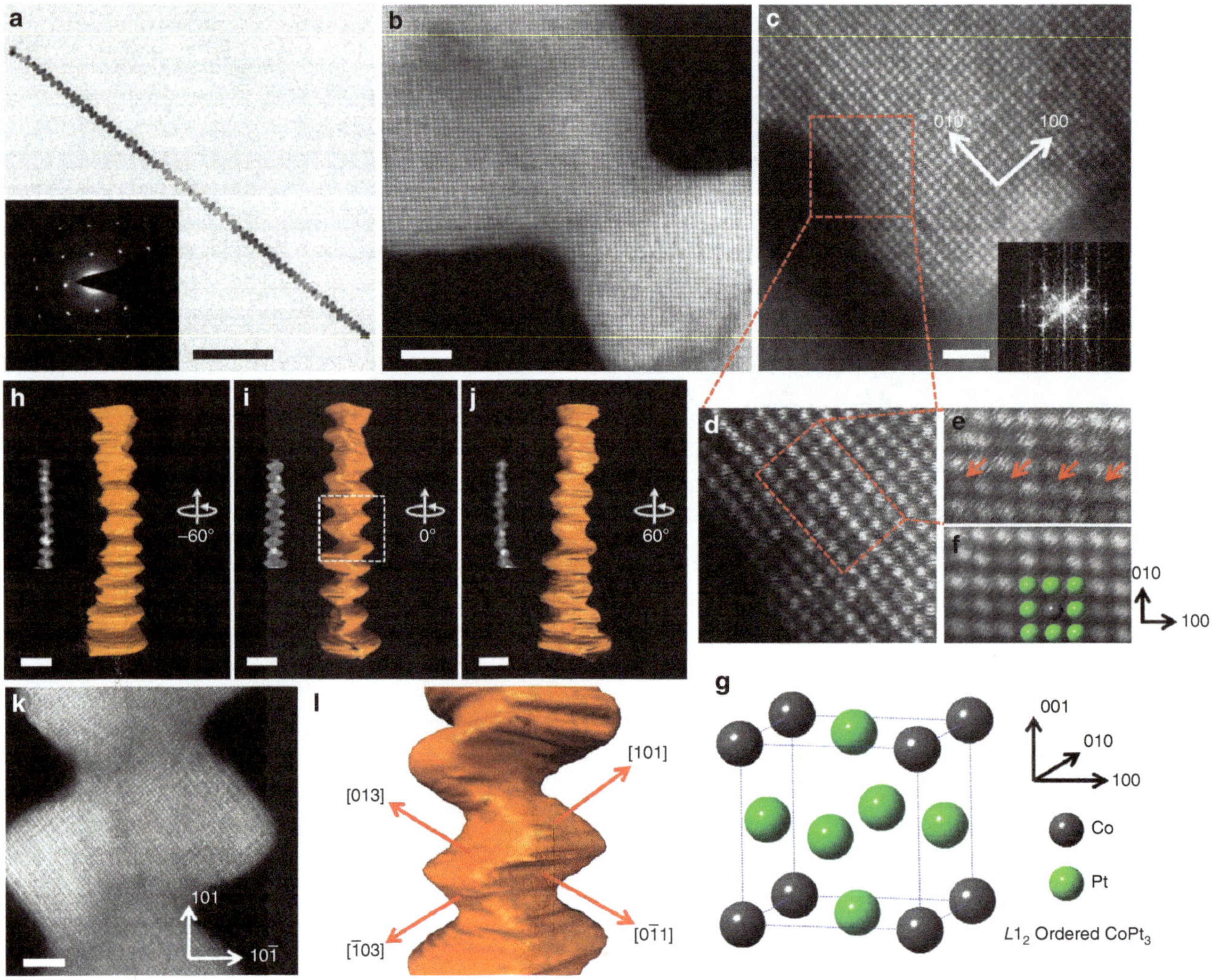

Figure 2 | Structure analysis of hierarchical Pt-Co nanowires. (**a**) TEM image of an individual hierarchical Pt_3Co NW. The inset shows the corresponding selected-area electron diffraction pattern. (**b**,**c**) Atomic-resolution HAADF-STEM images of Pt_3Co NWs along [001] zone axis. The inset in (**c**) shows the corresponding Fast Fourier Transform (FFT) pattern. (**d**,**e**)The enlarged HAADF-STEM images from the region with $L1_2$-ordered intermetallic structure. (**f**) FFT-filtered HAADF-STEM image from the same area of **e**. (**g**) Schematic diagram of the $L1_2$-ordered intermetallic structure of Pt_3Co. (**h**-**j**) A series of reconstructed 3D tomograms at viewing angles of $-60°$ (**h**), 0° (**i**) and 60° (**j**), the as-acquired HAADF-STEM images are shown in the insets. (**k**) An atomic resolution HAADF-STEM image at the viewing angles of 0°, and (**l**) indexes of the flat planes of the Pt_3Co NWs. The scale bars in **a**-**c**, **h**-**k** are 100, 2, 1, 10, 10, 10 and 2 nm, respectively.

feature of our approach. 1D Pt-*M* nanostructures with such features have not been achieved previously, to the best of our knowledge. To understand their growth mechanism, the intermediates collected at different growth durations were carefully investigated (Fig. 3a–h). The TEM results show that ultrathin NWs were already formed at the early stage of the synthesis (30 min, Fig. 3a). These ultrathin NWs are of essentially pure Pt nanomaterials, as revealed by the PXRD and energy dispersive X-ray spectroscopy analyses (Fig. 3e,f). The diameter of ultrathin NWs is measured to be around 2 nm. As the reaction time was prolonged to 1 h, the surface of the NWs became more uneven (Fig. 3b), along with the average diameter increased from 2 to 5 nm. 8.9% Co was detected in the NWs of 5 nm (Fig. 3f). Together with the increase of the diameter in the NWs, further increase in the Co content and the presence of obvious humps in the intermediates were observed after 3.0 h (Fig. 3c,f). When the reaction time was 5.0 h, typical hierarchical PtCo NWs were obtained (Fig. 3d) with the Pt/Co ratio of about 3/1 (Fig. 3f). When the reaction time was increased to 8 h, there was no significant change in the diameter and morphology of NWs (Fig. 3g,h). These time-dependent structure and composition changes reveal that the formation of such hierarchical Pt_3Co NWs relies upon the initial formation of ultrathin Pt NWs, the reduction of Co species onto the preformed NWs, and hereafter the inter-diffusion to form alloyed Pt_3Co NWs (Fig. 3e,f). As time goes on, the diffraction peaks shift towards higher 2θ values with the increase of Co content (Fig. 3e). During the reaction time from 1.0 to 3.0 h, both the peak positions in the PXRD patterns and the structure of the NWs in unevenness change obviously. High-resolution TEM (HRTEM) and elemental mappings of the hierarchical Pt-Co NW intermediates at different growth times were also carried out. As shown in the HRTEM images (Supplementary Fig. 8), as the reaction proceeds, the humps on the surface of the preformed Pt NWs gradually evolve. As revealed by the HAADF-STEM elemental mappings, the Pt and Co distributions of each hierarchical Pt-Co NW intermediates are even throughout the whole growth process, despite the contents of Co increase as the reaction time prolonged

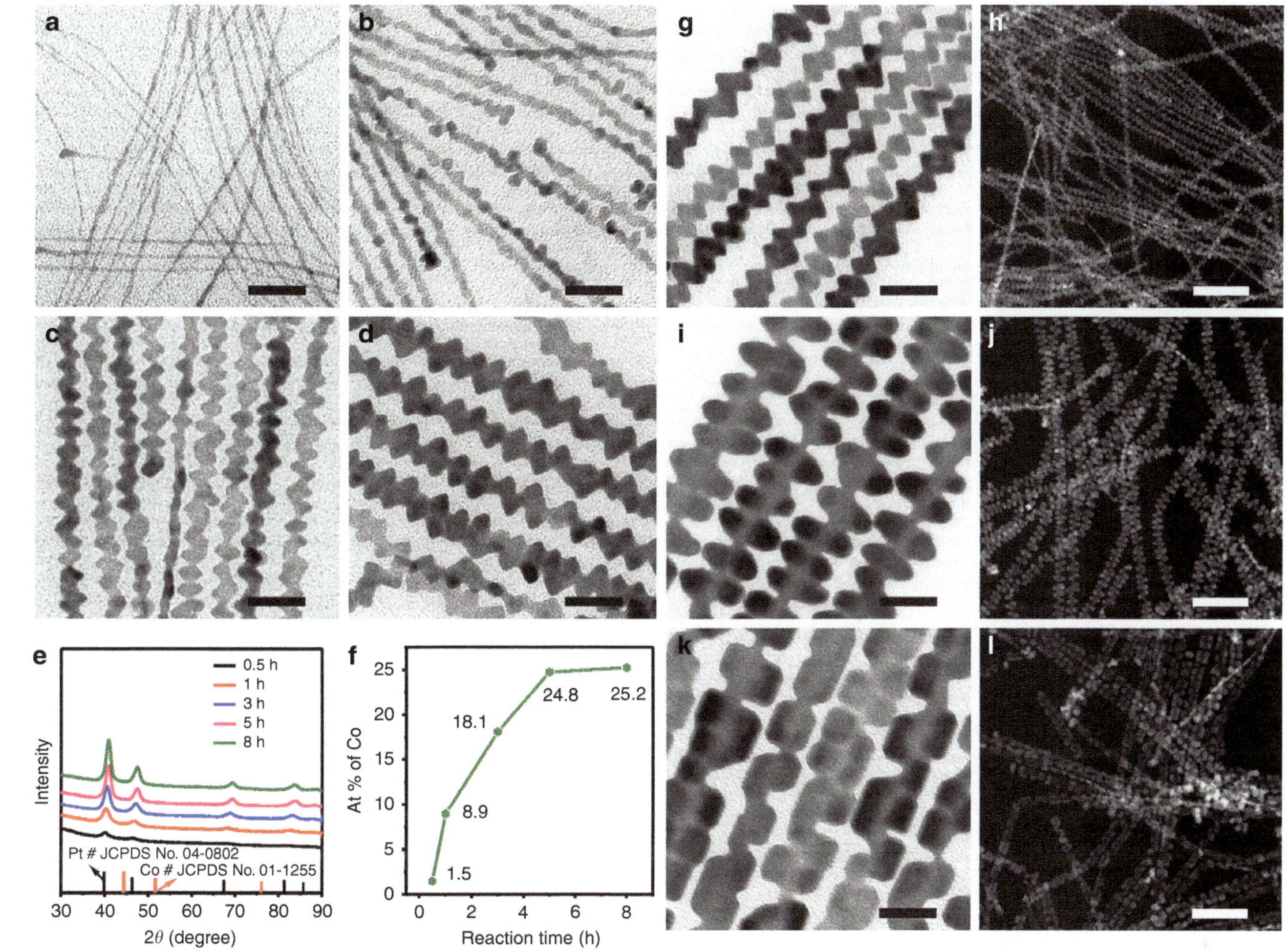

Figure 3 | Growth mechanism and structural tuning. Typical TEM images of hierarchical Pt_3Co NWs intermediates collected from (**a**) 0.5, (**b**) 1, (**c**) 3 and (**d**) 5 h. (**e**) PXRD patterns of hierarchical Pt_3Co NWs intermediates collected from the reactions at different reaction times. (**f**) The composition ratio changes of Co to Pt for the hierarchical Pt_3Co NWs intermediates, as determined by ICP-AES measurements. Typical TEM images and HAADF-STEM images of (**g-h**) hierarchical Pt_3Co NWs, (**i,j**) hierarchical Pt_7Co NWs and (**k,l**) hierarchical $Pt_{10}Co$ NWs collected from 8 h reactions. The scale bars in **a-d**, **g**, **i** and **k** are 20 nm. The scale bars in **h,j** and **l** are 100 nm.

(Supplementary Fig. 9). As the presence of very clear humps is coincident with the large increase of Co content in the Pt_3Co NWs, we infer that the different reduction potentials and diffusions of Pt and Co should be the major driving force for the creation of such hierarchical NWs in the present condition, being consistent with the synthesis of the unconventional silicon NWs with periodic shell, where the diffusion of silicon adatoms is believed to be the most significant kinetic factor in controlling the growth of periodic shell.[34]

These TEM and composition studies as a function of reaction time indicate that the amount of $Co(acac)_3$ introduced can significantly affect the growth of the hierarchical Pt-Co NWs. As shown in Supplementary Fig. 4a,b, the conventional NWs rather than hierarchical ones were obtained when the synthesis was carried out in the absence of $Co(acac)_3$, whereas the introduction of $Co(acac)_3$ could greatly impact the growth of NWs (Fig. 3i–l). If the amount of $Co(acac)_3$ was reduced from 9.0 to 3.0 and 1.5 mg, the diameter of the obtained hierarchical Pt-Co NWs changed from 12 to 18 and 21 nm, associated with the reduced density of bumps (Fig. 3i–l and Supplementary Figs 10–16). When the amount of $Co(acac)_3$ was increased to 12.0 mg, the diameter and structure of hierarchical Pt-Co NWs were almost unchanged (Supplementary Fig. 4c,d). The hierarchical Pt-Co NWs were yielded once $Co(acac)_3$ was fed, although the hierarchical feature was highly depended on the amount of $Co(acac)_3$ introduced. These results might be caused by the amount of $Co(acac)_3$, which could dictate the reduction/diffusion behaviour of Co during the growth, and thus control the hierarchical degree of the Pt-Co NWs.[34]

MOR and EOR performance. The hierarchical Pt-Co NWs were loaded on a commercial carbon (C, Vulcan) support by simply sonicating the Pt-Co NWs and C in cyclohexane, then washed with an ethanol/cyclohexane mixture and an ethanol/acetic acid mixture. Through such treatments, the hierarchical Pt-Co NWs were uniformly deposited onto carbon support with slightly composition changes from $Pt_{74.8}Co_{25.2}$ to $Pt_{75.5}Co_{24.5}$, while retaining their original nanostructures (Fig. 4a and Supplementary Figs 17 and 18). The limited loss of the Co was mainly caused by the formation of Pt-rich surface after this treatment, as confirmed by XPS characterization (Supplementary Fig. 19). Once the Pt-rich surface is present as a thin shell in a core/shell structure, both its activity and durability can be greatly enhanced as a result of the simultaneous downshift of its d-band centre and surface strain induced in the Pt surface[25,27].

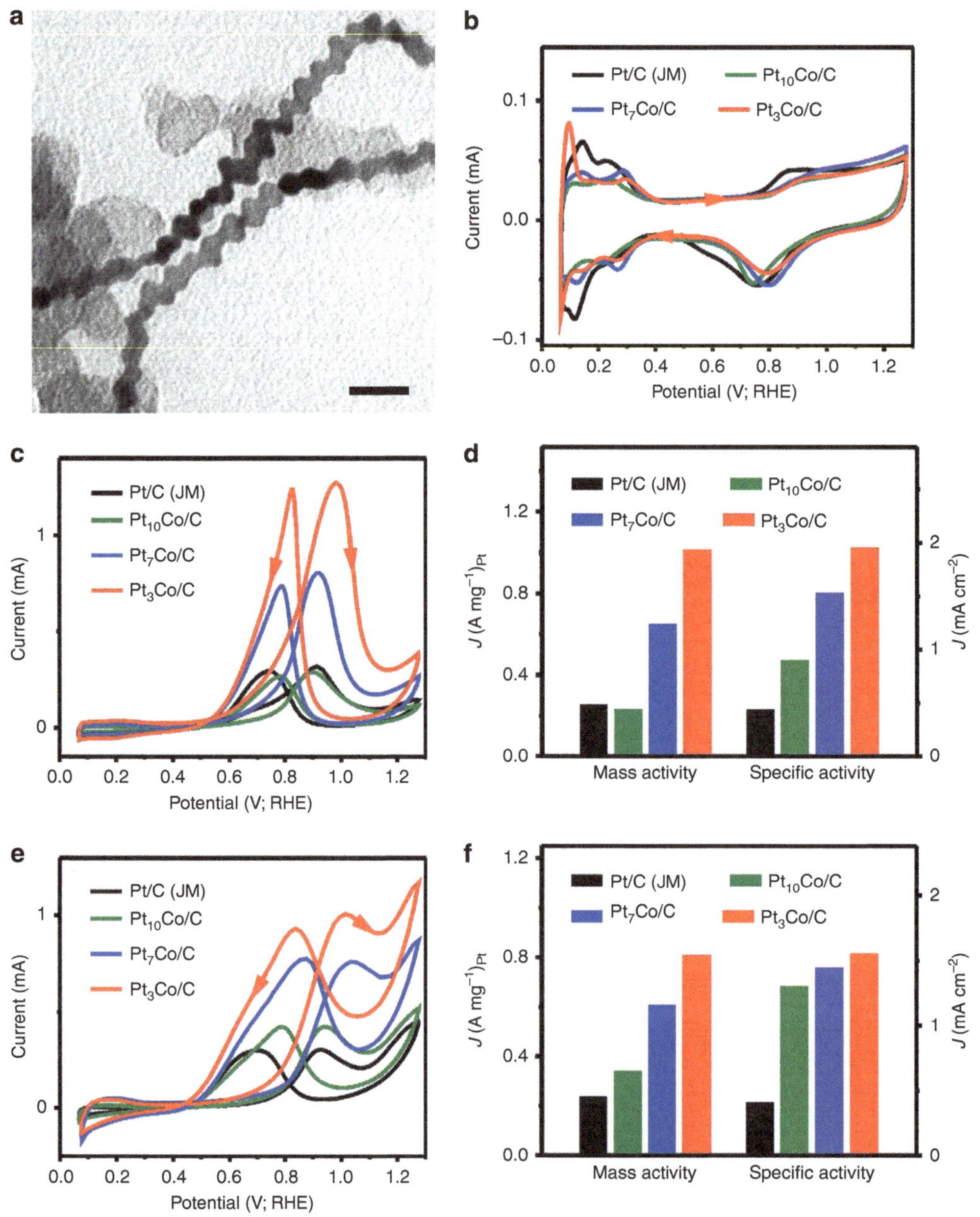

Figure 4 | Alcohol electro-oxidation performance comparison. (**a**) Representative TEM image and (**b**) CVs of different catalysts in 0.1 M $HClO_4$ solution. (**c**) CVs and (**d**) histogram of mass and specific activities of different catalysts for MOR in 0.1 M $HClO_4$ + 0.2 M methanol solution. (**e**) CVs and (**f**) histogram of mass and specific activities of different catalysts for EOR in 0.1 M $HClO_4$ + 0.2 M ethanol solution. The scale bar in **a** is 20 nm.

The resulting hierarchical Pt-Co NWs catalysts (Pt-Co NWs/C) were initially evaluated as anodic oxidation electrocatalysts for both MOR and EOR. To this end, three hierarchical PtCo NWs/C with distinct compositions, namely Pt_3Co NWs/C (Supplementary Figs 5–7 and 17), Pt_7Co NWs/C (Supplementary Figs 11–13 and 17) and $Pt_{10}Co$ NWs/C (Supplementary Figs 14–16 and 17), were selected, and benchmarked against commercial Pt/C from JM (Pt/C, 20 wt% Pt on Vulcan XC72R carbon, Pt particle size: 2–5 nm, Supplementary Fig. 20). The catalysts were dispersed in a mixture of isopropanol/water/Nafion, and then drop onto glassy carbon electrode. The loading amount of Pt was kept to be 1.25 µg for all the catalysts. The cyclic voltammograms (CVs) of different electrocatalysts in N_2-purged 0.1 M perchloric acid solution at a sweep rate of 50 mV s^{-1} are shown in Fig. 4b. The current responses from hydrogen adsorption/desorption processes appear in the potential range of 0.05–0.35 V. The electrochemically active surface areas (ECSAs) of Pt_3Co NWs/C, Pt_7Co NWs/C, $Pt_{10}Co$ NWs/C, Pt_3Co NPs/C and Pt/C catalysts were calculated to be 52.1, 42.3, 25.8, 38.7 and 58.8 m^2 g^{-1}, respectively. Among all the Pt-Co NWs/C catalysts, the Pt_3Co NWs/C catalysts display the highest ECSA of 52.1 m^2 g^{-1} because of the most high-density dendritic structure, comparable to that of the commercial Pt/C (58.8 m^2 g^{-1}) catalyst.

The MOR measurements were carried out in 0.1 M $HClO_4$ + 0.1 M CH_3OH solutions at a sweep rate of 50 mV s^{-1}. Figure 4c compares the CVs of three different Pt-Co NWs/C and commercial Pt/C for methanol electrooxidation. The mass and specific activities of all the catalysts are presented in Fig. 4d, in which the Pt_3Co NWs/C shows the highest mass activity and specific activity. The peak current densities in the forward (positive) potential scan were 1.95, 1.53, 0.90 and 0.44 mA cm^{-2} on the Pt_3Co NWs/C, Pt_7Co NWs/C, $Pt_{10}Co$ NWs/C and Pt/C catalyst, respectively. The maximum mass activity of Pt_3Co NWs/C is measured to be 1.02 A mg^{-1}_{Pt}, which is 1.59, 4.43 and 4.08 times higher than those of the Pt_7Co NWs/C, $Pt_{10}Co$ NWs/C and Pt/C catalysts, respectively (Fig. 4d). We also evaluated the electrocatalytic activities of the Pt-Co NWs/C catalysts for EOR. As expected, all the Pt-Co NWs/C catalysts exhibit excellent electrocatalytic activity towards EOR (Fig. 4e), in which Pt_3Co NWs/C has the highest activities by displaying excellent specific activity of 1.55 mA cm^{-2} and high mass activity of 0.81 A mg^{-1}_{Pt}, which are 3.78 and 3.37 times higher than those of commercial Pt/C, respectively (Fig. 4f), and higher than those of a home-made Pt_3Co NPs/C (Supplementary Figs 21 and 22 and Supplementary Tables 1 and 2).

ORR performance. With their unique structural features, the Pt-Co NWs/C catalysts were further evaluated as cathodic electrocatalysts for ORR. Figure 5a shows the ORR polarization curves of the different catalysts recorded at room temperature in an O_2-saturated 0.1 M $HClO_4$ solution. The kinetic currents at 0.9 V versus RHE were normalized to ECSA and the mass loading of Pt to obtain the specific and mass activities, respectively. Among three kinds of Pt-Co NWs/C investigated, Pt_3Co NWs/C has the highest specific activity of 7.12 mA cm^{-2} at 0.9 V versus RHE (Fig. 5b) with an impressive improvement factor of 39.6 over commercial Pt/C catalyst. Its mass activity reaches to 3.71 A mg^{-1}_{Pt}, 33.7-fold higher than that of commercial Pt/C catalyst. Although the Pt_3Co NPs/C catalyst has the same composition with the Pt_3Co NWs/C catalyst, the ORR activities of the Pt_3Co NPs/C catalyst are much lower than the Pt_3Co NWs/C catalyst (Fig. 5 and Supplementary Fig. 22), further displaying the super-high performance for ORR of the Pt_3Co NWs/C catalyst (Supplementary Table 3). Those results make the Pt_3Co NWs/C the most efficient catalyst ever achieved in the Pt-Co-based ORR catalysts to date to the best of our knowledge[30], whose specific and mass activities are higher than those of the recent Pt-based nanocatalysts.[35-38]

The hierarchical Pt-Co NWs/C catalyst also shows superior electrochemical stability, accessed by using an accelerated durability test between 0.6 and 1.1 V (versus RHE) in 0.1 M $HClO_4$ at a scan rate of 100 mV s^{-1}. Figure 5c,d shows the ORR polarization curves of the hierarchical Pt_3Co NWs/C catalyst before and after 10,000, 15,000 and 20,000 potential cycles. After 20,000 potential cycles, the activity of the hierarchical Pt_3Co NWs/C catalyst is still as high as 3.41 A mg^{-1}_{Pt} (91.9% of the initial value), which is 31-fold higher than that of the commercial Pt/C catalyst. The morphology and composition of hierarchical Pt_3Co NWs/C catalyst after 20,000 electrochemical cycles was thoroughly studied (Fig. 5g,h and Supplementary Figs 23 and 24), showing there is no morphology (Fig. 5g and Supplementary Figs 23 and 24), alloyed structure (Fig. 5h and Supplementary Fig. 24) and composition changes (Supplementary Fig. 23) on the hierarchical Pt-Co NWs after 20,000 potential cycles. For a comparison, only ~60% of initial activity could be retained (Fig. 5e,f), and serious NP aggregation was observed for the commercial Pt/C catalyst after the identical durability tests (Supplementary Fig. 20c,d). All these data indicate that hierarchical Pt-Co NWs are stable for ORR in strong acid solution.

Theoretical investigation. To shed light on the high-ORR performance on the hierarchical PtCo NWs, we performed density functional theory (DFT) calculations for the oxygen adsorption energy (E_O) on the same high-index NW facets as observed in the HAADF-STEM measurements. E_O is widely used as a descriptor for ORR activities, and there exists an optimal value of Eo, for which the ORR activity reaches the maximum[39,40]. For convenience, this optimal E_O was shift to 0 eV, thus ΔE_O represented the difference of a given E_O value relative to this optimal reference. It is known that the surface Co atoms in the PtCo NWs could dissolve under the electrochemical conditions, and be etched away, leading to core/shell-like structures with pure Pt as the shell. For such core/shell structures, in general, both surface strain and ligand effect could influence their catalytic activities, which are the subject of our first-principles simulations. Two typical adsorption sites, a bridge site at the edge of the NW and a hollow site on the facet of the NW, were examined on three surfaces, including [110], [310] and (2×1) missing-row reconstructed (MRR)-[310]. To examine the ligand effect on each of the three surfaces, we compared the pure Pt surface and the Pt_3Co alloy surface with the top-layer Co atoms completely etched away (Pt_3Co/Pt). To examine the strain effect, we calculated ΔE_O as a function of a compressive strain, ranging from 0 to 3%. The results of ΔE_O are shown in Fig. 6a,b and Supplementary Fig. 25. Moreover, we also calculated ΔE_O on the [111] facet of Pt NPs with the particle diameter ranging from 3 to 8 nm using a combined quantum mechanics-molecular mechanics method,[41,42] and the results are shown in Fig. 6c. First, we found that the threefold hollow sites on both [110] and MRR-[310] surfaces were highly active for ORR. Their ΔE_O values were much closer to zero than that of Pt NPs, suggesting that the threefold hollow sites on the NWs could offer superior ORR activities to the Pt NPs. This similar enhancement in ORR activities has also been observed on the threefold hollow sites of high-index [211] surfaces calculated by Norskov *et al.*[43] Hence, the additionally enhanced ORR activities in our Pt_3Co NWs can be attributed to the threefold hollow sites on the high-index facets. Second, ΔE_O was found to increase linearly as a function of the compressive strain on both [110] and MRR-[310] surfaces, as shown in Fig. 6a,b. The compressive strain could relieve the over-binding of oxygen to the surfaces, thus enhance the ORR activities. In contrast, the ligand effect was negligible as the ΔE_O values on the pure Pt and the Pt_3Co/Pt surfaces were almost the same. Therefore, the further improved ORR activities on the high-index faceted PtCo NWs stem from: (i) the active threefold hollow sites and (ii) the compressive strain on the Pt surface owing to the smaller atomic size of Co.

Anti-aggregation properties. Another significant feature of the hierarchical Pt-Co NWs is that they can exhibit excellent anti-aggregation properties upon the high-temperature annealing treatment. As shown in Supplementary Fig. 26, the hierarchical Pt_3Co NWs can be thermally stable up to 500 °C, as the 1D hierarchical structure was largely maintained after they were annealed at 500 °C for 1 h in N_2 atmosphere. The high thermal stability of the hierarchical Pt_3Co NWs is beneficial for the design of catalytic systems for high-temperature reactions or chemical processes, which is not easily achieved by the particulate Pt nanocrystals because of their poor anti-deformation ability.

Discussion

To summarize, we demonstrate a large-scalable, wet-chemical approach to synthesize a class of Pt-Co NWs consisting of large-density high-index facets and Pt-rich surfaces. These features enable them to exhibit impressive activity and durability towards both the alcohol oxidations and ORR, to satisfy requirements for efficient electrocatalytic applications. They show high specific and mass activities for both MOR and EOR, largely outperforming those based on the commercial Pt/C catalyst. For the cathodic ORR, they can deliver unprecedented activities that are higher than the best Pt-Co-based catalysts for ORR, representing the highest activities ever achieved in the

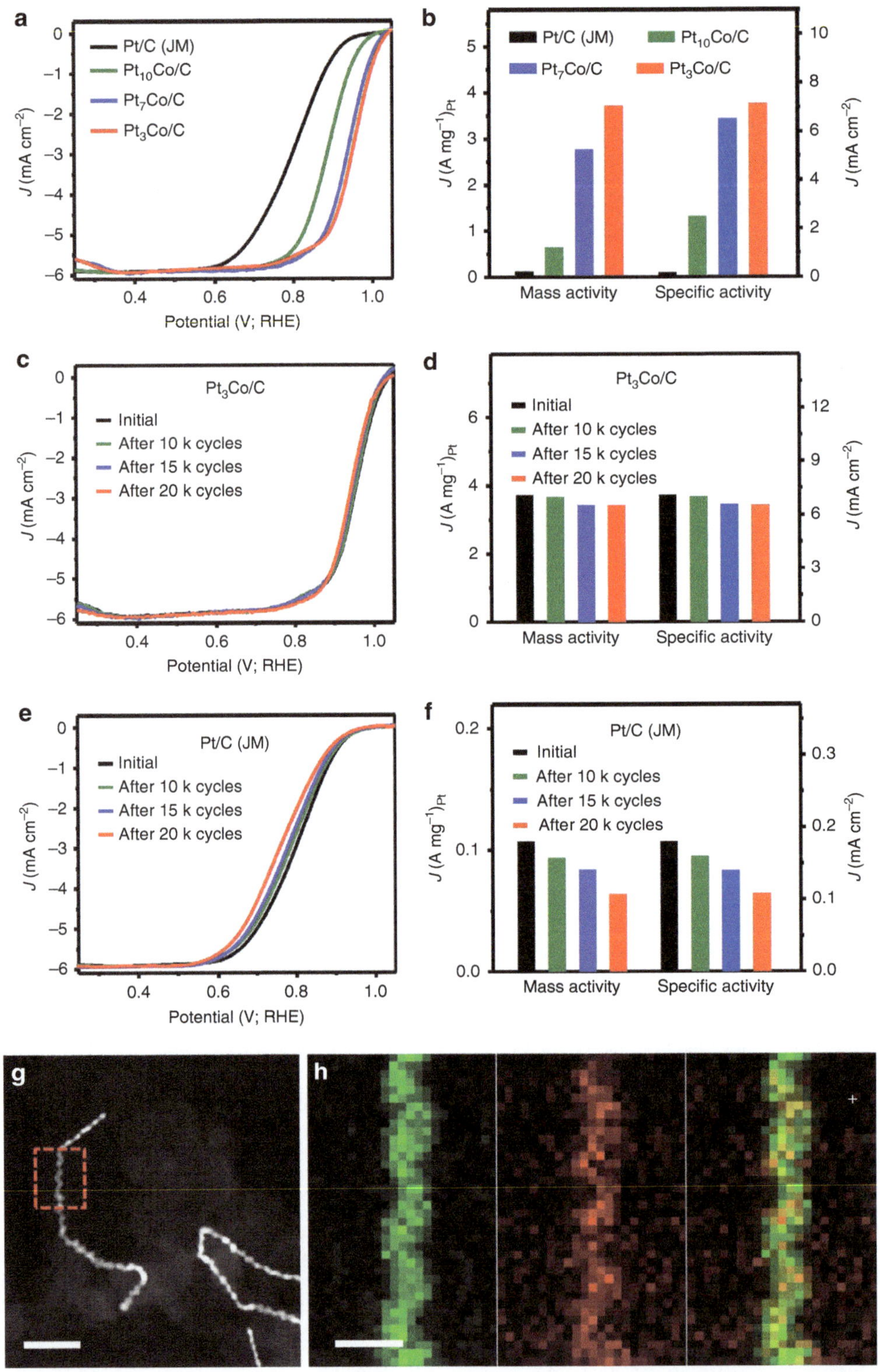

Figure 5 | ORR performance comparison. (**a**) ORR polarization curves and (**b**) ORR-specific activities and mass activities of different catalysts. The ORR polarization curves were recorded at room temperature in an O_2-saturated 0.1 M $HClO_4$ aqueous solution at a sweep rate of 10 mV s^{-1} and a rotation rate of 1,600 r.p.m. (**c**) ORR polarization curves of the hierarchical Pt_3Co NWs/C catalyst before and after 10,000, 15,000 and 20,000 potential cycles between 0.6 and 1.1 V versus RHE. (**d**) The changes on specific activities and mass activities of the hierarchical Pt_3Co NWs/C catalyst before and after 10,000, 15,000 and 20,000 potential cycles. (**e**) ORR polarization curves of the commercial Pt/C catalyst before and after 10,000, 15,000 and 20,000 potential cycles between 0.6 and 1.1 V versus RHE. (**f**) The changes on specific activities and mass activities of the commercial Pt/C catalyst before and after 10,000, 15,000 and 20,000 potential cycles. The durability tests were carried out at room temperature in 0.1 M $HClO_4$ at a scan rate of 100 mV s^{-1}. (**g**) HAADF-STEM image and (**h**) elemental mappings of the hierarchical Pt_3Co NWs/C after 20,000 cycles. The scale bars in **g** and **h** are 50 and 10 nm, respectively. The elemental mapping in **h** comes from the red-dotted line-marked area in **g**.

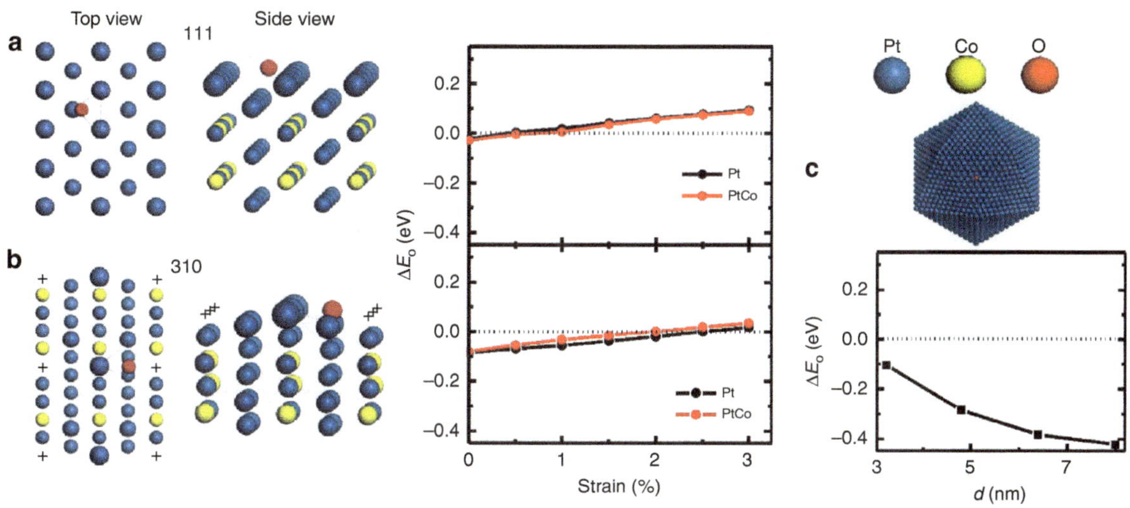

Figure 6 | DFT calculations of oxygen adsorption energy. (**a**) Atomistic model of [110] surface and ΔE_O as a function of compressive strain. (**b**) Atomistic model of the MRR-[310] surface and ΔE_O as a function of compressive strain. Blue cross represents the missing atomic row on the top layer of [310] surface. The black and red curves correspond to ΔE_O values on the pure Pt and Pt_3Co/Pt surface. (**c**) ΔE_O on [111] facet of the Pt NPs as a function of the particle size. The NP was modelled by an icosahedron with 20 [111] facets. The horizontal dashed line indicates the optimal ΔE_O value.

Pt-Co-based ORR catalysts, to the best of our knowledge. DFT studies indicate that the high ORR activities on the PtCo NWs mainly originate from the hollow sites on the [110] and [310] high-index facets of NWs. Most significantly, without sacrificing the ORR activities, they show good stability for ORR over the long-term course of 20,000 cycles. Therefore, our finding may open up opportunities for the rational design of practically relevant catalysts with excellent activity and superior durability, which could impact broad areas such as fine chemical productions, fuel cells, batteries and beyond.

Methods

Preparation of hierarchical Pt_3Co NWs. In a typical preparation of hierarchical Pt_3Co NWs, platinum(II) acetylacetonate ($Pt(acac)_2$, 10 mg), cobalt (III) acetylacetonate ($Co(acac)_3$, 9.0 mg), CTAC (32.0 mg), glucose (60.0 mg) and 5 ml oleylamine were added into a vial (volume: 35 ml). After the vial had been capped, the mixture was ultrasonicated for 90 min. The resulting homogeneous dark green mixture was then heated from room temperature to 160 °C in 30 min and maintained at 160 °C for 8 h in an oil bath, before it was cooled to room temperature. The resulting colloidal products were collected by centrifugation and washed three times with an ethanol/cyclohexane mixture. The synthesis conditions for hierarchical Pt_7Co and $Pt_{10}Co$ NWs were similar to that of hierarchical Pt_3Co NWs except changing the amount of $Co(acac)_3$ to 3.0 and 1.5 mg, respectively.

Characterization. PXRD patterns were collected using an X'Pert-Pro X-ray powder diffractometer equipped with a Cu radiation source ($\lambda = 0.15406$ nm). The morphology and size of the NWs were determined by TEM (Hitachi, HT7700) at 120 kV. Selected-area electron diffraction, HRTEM, HAADF-STEM and the 3D visualization of tomographic reconstruction tests were conducted on a JEOL-2100F transmission electron microscope at an acceleration voltage of 200 kV. High-resolution STEM imaging and elemental mapping analysis were carried out on a Hitachi HD2700C (200 kV) STEM with a probe aberration corrector, at the Center for Functional Nanomaterials, Brookhaven National Lab. Low-resolution energy dispersive X-ray spectroscopy was performed on a scanning electron microscope (Hitachi, S-4700). All the XPS spectra of the hierarchical Pt-Co NWs were collected by XPS (Thermo Scientific, ESCALAB 250 XI). The concentration of all the catalysts was determined by the ICP-AES (710-ES, Varian). The catalysts after the stability tests were collected and scratched off the electrode with the aid of sonication in ethanol for further TEM characterization.

Electrochemical measurements. The catalysts were redispersed in a mixture solvent containing isopropanol and Nafion (5%; v-v = 1:0.005) to form a homogeneous catalyst ink by sonicating for 30 min. The concentration of Pt was fixed to be 0.25 mg ml^{-1} based on ICP-AES measurement. Five microlitres of the dispersion was transferred onto the GC electrode. The Pt/C catalyst (20 wt%, 2–5 nm Pt NPs on Vulcan XC-72 carbon support, JM), was prepared by sonicating its ink in an aqueous dispersion (1.25 mg ml^{-1}). Five microlitres of the dispersion was transferred onto the GC electrode with the loading amount of metal at 1.25 µg.

A CHI660E electrochemical analyzer (CHI Instruments) was used for testing the MOR and EOR of different catalysts. The MOR and EOR measurements were carried out in 0.1 M $HClO_4$ + 0.2 M methanol and 0.1 M $HClO_4$ + 0.2 M ethanol at room temperature, respectively. The CV technique and conventional three-electrode system were applied to evaluate the electrochemical properties of various catalysts. A leak-free saturated calomel electrode was used as the reference electrode and a Pt wire was used as the counter electrode. The ECSA was determined by integrating the hydrogen adsorption charge on the CVs at the rate of 50 mV s^{-1} at room temperature in N_2-saturated 0.1 M $HClO_4$ solution. In the CV measurements for MOR and EOR, the electrode potential was scanned at a rate of 50 mV s^{-1}. All the catalyst electrodes were cleaned before data collection with a steady-state CV. The current was normalized to Pt amount and ECSA to get mass and specific activities, respectively.

The electrochemical measurements for ORR were performed using a glassy-carbon Rotating Disk Electrode (Pine Research Instrumentation, diameter: 5 mm, area: 0.196 cm^2) connected to an installation of rotating electrode speed control (Pine Research Instrumentation, model: AFMSRCE). A leak-free saturated calomel electrode was used as the reference electrode and a Pt wire was used as the counter electrode. The electrolyte was 0.1 M $HClO_4$. The loading amount of Pt for the hierarchical $Pt_{10}Co$ NWs/C, Pt_7Co NWs/C, Pt_3Co NWs/C, Pt_3Co NPs/C and the commercial Pt/C (JM) catalysts was all kept at 6.4 µg cm^{-2}. ORR measurements were conducted in 0.1 M $HClO_4$ solutions purged with the saturated O_2 during the measurements. The scan and rotation rates for ORR measurements were 10 mV s^{-1} and 1,600 r.p.m., respectively. In the ORR polarization curves, the current densities were normalized in reference to the geometric area of the glassy carbon electrode (0.196 cm^2). For each catalyst, the kinetic current was normalized to the loading amount of Pt and ECSA in order to generate mass and specific activities, respectively. The accelerated durability tests were performed at room temperature in 0.1 M $HClO_4$ solutions by applying the cyclic potential sweeps between 0.6 and 1.1 V versus RHE at a sweep rate of 100 mV s^{-1} for 20,000 cycles.

DFT calculations. The oxygen adsorption energy E_O was defined as $E_O = E[\text{surf} + O] - E[\text{surf}] - E[O_2]/2$, where $E[\text{surf} + O]$ and $E[\text{surf}]$ are the total energies of the [110] or [310] surface with and without the O adsorbate, respectively. $E[O_2]$ is the total energy of an oxygen molecule. A four-layer slab was used in the calculations. The atoms in the top two layers were fully relaxed while the rest of the atoms were fixed in their equilibrium positions. The DFT calculations were carried out using the VASP package[44,45] with the projector augmented wave pseudopotentials[46] and Perdew–Burke–Ernzerh of generalized gradient approximation.[47] An energy cutoff of 400 eV was used for the plane-wave basis set. The Brillouin zone was sampled on the basis of the Monkhorst-Pack scheme[48] with a 7 × 5 × 1 k-point mesh. The force convergence criterion for atomic relaxation was 0.02 eV Å^{-1}.

References

1. Debe, M. K. Electrocatalyst approaches and challenges for automotive fuel cells. *Nature* **486,** 43–51 (2012).
2. Wang, Y. J., Wilkinson, D. P. & Zhang, J. J. Noncarbon support materials for polymer electrolyte membrane fuel cell electrocatalysts. *Chem. Rev.* **111,** 7625–7651 (2011).
3. Winter, M. & Brodd, R. J. What are batteries, fuel cells, and supercapacitors? *Chem. Rev.* **104,** 4245–4269 (2004).
4. Dresselhaus, M. S. & Thomas, I. L. Alternative energy technologies. *Nature* **414,** 332–337 (2001).
5. Bacon, F. T. Fuel cells: Will they soon become a major source of electrical energy? *Nature* **186,** 589–592 (1960).
6. Su, D. S. & Sun, G. Q. Nonprecious-metal catalysts for low-cost fuel cells. *Angew. Chem. Int. Ed.* **50,** 11570–11572 (2011).
7. Chen, Z. W., Higgins, D., Yu, A. P., Zhang, L. & Zhang, J. J. A review on non-precious metal electrocatalysts for PEM fuel cells. *Energy Environ. Sci.* **4,** 3167–3192 (2011).
8. Bing, Y. H., Liu, H. S., Zhang, L., Ghosh, D. & Zhang, J. J. Nanostructured Pt-Alloy eectrocatalysts for PEM fuel cell oxygen reduction reaction. *Chem. Soc. Rev.* **39,** 2184–2202 (2010).
9. Porter, N. S., Wu, H., Quan, Z. W. & Fang, J. Y. Shape-control and electrocatalytic activity-enhancement of Pt-based bimetallic nanocrystals. *Acc. Chem. Res.* **46,** 1867–1877 (2013).
10. Wu, J. B. & Yang, H. Platinum-based oxygen reduction electrocatalysts. *Acc. Chem. Res.* **46,** 1848–1857 (2013).
11. Chen, J. Y., Lim, B., Lee, E. P. & Xia, Y. N. Shape-controlled synthesis of platinum nanocrystals for catalytic and electrocatalytic applications. *Nano Today* **4,** 81–95 (2009).
12. Gasteiger, H. A. & Markovic, N. M. Just a Dream - or Future Reality? *Science* **324,** 48–49 (2009).
13. Nørskov, J. K., Bligaard, T., Rossmeisl, J. & Christensen, C. H. Towards the computational design of solid catalysts. *Nat. Chem* **1,** 37–46 (2009).
14. Greeley, J. *et al.* Alloys of platinum and early transition metals as oxygen reduction electrocatalysts. *Nat. Chem* **1,** 552–556 (2009).
15. de Bruijn, F. A., Dam, V. A. T. & Janssen, G. J. M. Review: durability and degradation issues of PEM fuel cell components. *Fuel Cells* **8,** 3–22 (2008).
16. Gasteiger, H. A., Kocha, S. S., Sompalli, B. & Wagner, F. T. Activity benchmarks and requirements for Pt, Pt-alloy, and non-Pt oxygen reduction catalysts for PEMFCs. *Appl. Catal. B Environ* **56,** 9–35 (2005).
17. Lim, B. *et al.* Pd-Pt bimetallic nanodendrites with high activity for oxygen reduction. *Science* **324,** 1302–1305 (2009).
18. Tian, N., Zhou, Z. Y., Sun, S. G., Ding, Y. & Wang, Z. L. Synthesis of tetrahexahedral platinum nanocrystals with high-index facets and high electro-oxidation activity. *Science* **316,** 732–735 (2007).
19. Tang, Y. & Ouyang, M. Tailoring properties and functionalities of metal nanoparticles through crystallinity engineering. *Nat. Mater.* **6,** 754–759 (2007).
20. Shevchenko, E. V., Talapin, D. V., Kotov, N. A., O'Brien, S. & Murray, C. B. Structural diversity in binary nanoparticle superlattices. *Nature* **439,** 55–59 (2006).
21. Chen, C. *et al.* Highly crystalline multimetallic nanoframes with three-dimensional electrocatalytic surfaces. *Science* **343,** 1339–1343 (2014).
22. Strasser, P. Catalysts by platonic design. *Science* **349,** 379–380 (2015).
23. Zhang, H., Jin, M. & Xia, Y. Enhancing the catalytic and electrocatalytic properties of Pt-based catalysts by forming bimetallic nanocrystals with Pd. *Chem. Soc. Rev.* **41,** 8035–8049 (2012).
24. Stamenkovic, V. R. *et al.* Improved oxygen reduction activity on $Pt_3Ni(111)$ via increased surface site availability. *Science* **315,** 493–497 (2007).
25. Guo, S., Zhang, S., Su, D. & Sun, S. Seed-mediated synthesis of core/shell FePtM/FePt (M = Pd, Au) nanowires and their electrocatalysis for oxygen reduction reaction. *J. Am. Chem. Soc.* **135,** 13879–13884 (2013).
26. Xia, B. Y., Ng, W. T., Wu, H. B., Wang, X. & Lou, X. W. Self-Supported interconnected Pt nanoassemblies as highly stable electrocatalysts for low-temperature fuel cells. *Angew. Chem. Int. Ed.* **51,** 7213–7216 (2012).
27. Guo, S. J., Zhang, S. & Sun, S. H. Tuning nanoparticle catalysis for the oxygen reduction reaction. *Angew. Chem. Int. Ed.* **52,** 8526–8544 (2013).
28. Chen, A. C. & Holt-Hindle, P. Platinum-based nanostructured materials: synthesis, properties, and applications. *Chem. Rev.* **110,** 3767–3804 (2010).
29. Bu, L. *et al.* A general method for multimetallic platinum alloy nanowires as highly active and stable oxygen reduction catalysts. *Adv. Mater.* **27,** 7204–7212 (2015).
30. Wang, D. *et al.* Structurally ordered intermetallic platinum-cobalt core-shell nanoparticles with enhanced activity and stability as oxygen reduction electrocatalysts. *Nat. Mater.* **12,** 81–87 (2013).
31. Carpenter, M. K., Moylan, T. E., Kukreja, R. S., Atwan, M. H. & Tessema, M. M. Solvothermal synthesis of platinum alloy nanoparticles for oxygen reduction electrocatalysis. *J. Am. Chem. Soc.* **134,** 8535–8542 (2012).
32. Snyder, J., McCue, I., Livi, K. & Erlebacher, J. Structure/processing/properties relationships in nanoporous nanoparticles as applied to catalysis of the cathodic oxygen reduction reaction. *J. Am. Chem. Soc.* **134,** 8633–8645 (2012).
33. Jesson, D. E. & Pennycook, S. J. Incoherent imaging of crystals using thermally scattered electrons. *Proc. R. Soc. Lond Ser. A* **449,** 273–293 (1995).
34. Day, R. W. *et al.* Plateau-Rayleigh crystal growth of periodic shells on one-dimensional substrates. *Nat. Nanotechnol* **10,** 345–352 (2015).
35. Huang, X. *et al.* A facile strategy to Pt_3Ni nanocrystals with highly porous features as an enhanced oxygen reduction reaction catalyst. *Adv. Mater.* **25,** 2974–2979 (2013).
36. Guo, S. *et al.* FePt and CoPt nanowires as efficient catalysts for the oxygen reduction reaction. *Angew. Chem. Int. Ed.* **52,** 3465–3468 (2013).
37. Cui, C., Gan, L., Heggen, M., Rudi, S. & Strasser, P. Compositional segregation in shaped Pt alloy nanoparticles and their structural behaviour during electrocatalysis. *Nat. Mater.* **12,** 765–771 (2013).
38. Choi, S. I. *et al.* Synthesis and characterization of 9 nm Pt-Ni octahedra with a record high activity of $3.3A/mg_{Pt}$ for the oxygen reduction reaction. *Nano Lett.* **13,** 3420–3425 (2013).
39. Norskov, J. K., Rossmeisl, J., Logadottir, A. & Lindqvist, L. Origin of the overpotential for oxygen reduction at a fuel-cell cathode. *J. Phys. Chem. B* **108,** 17886–17892 (2004).
40. Stamenkovic, V. *et al.* Changing the activity of electrocatalysts for oxygen reduction by tuning the surface electronic structure. *Angew. Chem. Int. Ed.* **45,** 2897–2901 (2006).
41. Zhang, X. & Lu, G. Computational design of core/shell nanoparticles for oxygen reduction reactions. *J. Phys. Chem. Lett.* **5,** 292–297 (2014).
42. Zhang, X., Lu, G. & Curtin, W. A. Multiscale quantum/atomistic coupling using constrained density functional theory. *Phys. Rev. B* **87,** 054113 (2013).
43. Peterson, A. A. *et al.* Fininte-size effects in O and CO adsorption for the late transition metals. *Top. Catal* **55,** 1276–1282 (2012).
44. Kresse, G. & Hafner, J. *Ab Initio* molecular dynamics for liquid metals. *Phys. Rev. B* **47,** 558–561 (1993).
45. Kresse, G. & Furthmüller, J. Efficient iterative schemes for *ab initio* total-energy calculations using a plane-wave basis set. *Phys. Rev. B* **54,** 11169–11186 (1996).
46. Blöchl, P. E. Projector augmented-wave method. *Phys. Rev. B* **50,** 17953–17979 (1994).
47. Perdew, J. P., Burke, K. & Ernzerhof, M. Generalized gradient approximation made simple. *Phys. Rev. Lett.* **77,** 3865–3868 (1996).
48. Monkhorst, H. J. & Pack, J. D. Special points for brillouin-zone integrations. *Phys. Rev. B* **13,** 5188–5192 (1976).

Acknowledgements

This work was financially supported by the start-up supports from the Soochow University and Peking University, Young Thousand Talented Program, the National Natural Science Foundation of China (21571135), the National Basic Research Program of China (Battery) and the Priority Academic Program Development of Jiangsu Higher Education Institutions (PAPD). The work at California State University Northridge was supported by the US Army Research Office via the MURI grant W911NF-11-1-0353. TEM work carried out in part at the Center for Functional Nanomaterials, Brookhaven National Laboratory, which is supported by the US Department of Energy, Office of Basic Energy Sciences, under Contract No. DE-SC0012704.

Author contributions

X.H. and S.G. conceived and supervised the research. X.H., S.G. and L.B. designed the experiments. X.H., S.G., L.B., X.S., D.S., X.Z. and J.G. performed most of the experiments and data analysis. X.H., S.G., L.B., X.Z., D.S., J.Y. and J.G. participated in various aspects of the experiments and discussions. X.Z. and G.L. performed the DFT simulations. X.H., S.G. and L.B. wrote the paper. All authors discussed the results and commented on the manuscript.

Additional information

Competing financial interests: The authors declare no competing financial interests.

Strain-relief by single dislocation loops in calcite crystals grown on self-assembled monolayers

Johannes Ihli[1,*,†], Jesse N. Clark[2,3,*], Alexander S. Côté[4,†], Yi-Yeoun Kim[1], Anna S. Schenk[1], Alexander N. Kulak[1], Timothy P. Comyn[5], Oliver Chammas[6], Ross J. Harder[7], Dorothy M. Duffy[4], Ian K. Robinson[8] & Fiona C. Meldrum[1]

Most of our knowledge of dislocation-mediated stress relaxation during epitaxial crystal growth comes from the study of inorganic heterostructures. Here we use Bragg coherent diffraction imaging to investigate a contrasting system, the epitaxial growth of calcite ($CaCO_3$) crystals on organic self-assembled monolayers, where these are widely used as a model for biomineralization processes. The calcite crystals are imaged to simultaneously visualize the crystal morphology and internal strain fields. Our data reveal that each crystal possesses a single dislocation loop that occupies a common position in every crystal. The loops exhibit entirely different geometries to misfit dislocations generated in conventional epitaxial thin films and are suggested to form in response to the stress field, arising from interfacial defects and the nanoscale roughness of the substrate. This work provides unique insight into how self-assembled monolayers control the growth of inorganic crystals and demonstrates important differences as compared with inorganic substrates.

[1] School of Chemistry, University of Leeds, Woodhouse Lane, Leeds LS2 9JT, UK. [2] Stanford PULSE Institute, SLAC National Accelerator Laboratory, 2575 Sand Hill Road, Menlo Park, California 94025, USA. [3] Center for Free-Electron Laser Science (CFEL), Deutsches Elektronensynchrotron (DESY) Notkestrasse 85, 22607 Hamburg, Germany. [4] Department of Physics and Astronomy, University College London, Gower Street, London WC1E 6BT, UK. [5] Institute for Materials Research, University of Leeds, Leeds LS2 9JT, UK. [6] School of Physics and Astronomy, University of Leeds, Leeds LS2 9JT, UK. [7] Advanced Photon Source, Argonne, Illinois 60439, USA. [8] London Centre for Nanotechnology, University College London, 17-19 Gordon Street, London WC1H 0AH, UK. * These authors contributed equally to this work. † Present addresses: Paul Scherrer Institute, 5232 Villigen PSI, Switzerland. (J.I.); London Centre for Nanotechnology, University College London, 17-19 Gordon Street, London WC1H 0AH, UK (A.S.C.). Correspondence and requests for materials should be addressed to J.I. (email: johannes.ihli@psi.ch) or to I.K.R. (email: i.robinson@ucl.ac.uk) or to F.C.M. (email: F.Meldrum@leeds.ac.uk.).

The control of crystal growth at interfaces is fundamental to a wide range of processes of scientific, environmental and technological importance. The epitaxial growth of crystals, where a single crystal of one compound grows with a unique orientation on the surface of a second[1], attracts particular interest. This phenomenon has received considerable attention for the fabrication of thin film and nanoparticulate semiconductor, ferroelectric and superconducting devices, where the properties of these heterostructures can be tuned according to the degree of strain introduced into the supported thin film[2–4]. The strain, in turn, arises from the lattice mismatch present between the substrate and crystalline thin film, where this increases in value until a critical film thickness is reached[5]. Beyond this point, the strain energy can be relieved through the introduction of misfit dislocations, where these must have edge character parallel to the substrate/crystal interface to reduce strain[1,6,7]. Due to their technological importance and suitability for study with techniques such as transmission electron microscopy (TEM), these systems have provided the vast majority of our current knowledge about stress relaxation and dislocation formation during epitaxial crystal growth.

In the work described here, we profit from recent advances in imaging methods[8] to investigate strain and associated dislocation formation in a quite different example of epitaxial crystal growth—the growth of calcite ($CaCO_3$) crystals on organothiol self-assembled monolayers (SAMs). The ability of organized organic matrices such as Langmuir monolayers[9], Langmuir–Shaeffer films[10] and SAMs to direct the orientation, and sometimes even the polymorph, of inorganic crystals has received significant attention, where these systems provide excellent models for biomineralization processes such as the formation of mollusc shell nacre. Of these studies, the precipitation of calcite on organothiol SAMs on coinage metals is the best-characterized, where the nucleation plane can be selected according to the SAM chain length[11], packing geometry/tilt[12], the terminal group[13] and its degree of ionization[14,15], and the type of metal substrate[13,16]. That the precipitated calcite crystals are co-aligned within monocrystalline Au {111} domains demonstrates an epitaxial relationship between the SAM and the crystal lattice[17], and selection of the nucleation face has been proposed to arise from a stereochemical match between the orientation of the SAM headgroups and the ions in the crystal nucleation face[13,18]. Subsequent studies have refined this view and have suggested that templating of inorganic crystals by the organic interface is a cooperative process[12,19,20], in which structural feedback between the crystal and monolayer ensures selection of the most favourable combination of the orientation of the SAM and crystal[21–25].

However, many questions remain regarding the mechanisms by which organic matrices can control crystallization. Indeed, while Langmuir–Shaeffer films[10] and SAMs[13,19] have been observed to change conformation when directing the growth of oriented calcite, little is known about whether the organic matrix can cause parallel changes in the crystal lattice. The work described in this article uses Bragg coherent diffraction imaging (BCDI)[8] to address this question. BCDI provides a unique method for simultaneously visualizing both the morphology of a crystal, and the strain fields within it at a spatial resolution of ~100 nm, without the requirement for sample preparation methods. Our data demonstrate how a 'soft' organic matrix can inform the structure of a 'hard' inorganic crystal and show that heterogeneous nucleation of a calcite single crystal on a SAM results in deformation of the calcite crystal lattice. As a key finding, we show that the stresses originating at the crystal/SAM interface give rise to the formation of a single dislocation loop within each crystal, where the geometry of this dislocation is entirely different to that of the misfit dislocation loops, characteristically seen in purely inorganic epitaxial heterostructures. The roughness of the substrate—where this is intrinsic to all SAMs prepared on evaporated metal films—is considered to lie at the heart of this effect. These results, therefore, provide a unique insight into the mechanisms by which SAMs control both the growth and defect structures of crystals, where the latter is intimately linked to the mechanical properties.

Results

Morphological development of calcite crystals on SAMs. Oriented calcite crystals were precipitated on 11-mercaptoundecanoic acid SAMs on Au (111)/Si (001) using a hanging drop set-up, in which 200 µl drops of 5 mM $CaCl_2$ solution were suspended from a SAM (Fig. 1)[26]. These substrates were then placed in a closed desiccator containing solid ammonium carbonate, and crystallization was allowed to proceed for up to 30 min. Characterization of the crystal orientations using acquired pole measurements showed that the majority were oriented with the {012} or the {113} plane parallel to the SAM (Supplementary Fig. 1). Analysis of the morphological development of the crystals demonstrated that the initial form was roughly pyramidal and that it appears to comprise an aggregate of smaller particles (Fig. 1). This is indicative of crystallization via an amorphous calcium carbonate precursor phase, as is expected under these reaction conditions[12,27]. These particles then convert to irregular tetrahedra with three, well-defined {104} faces directed into the solution, while further growth leads to the truncation of the vertex, where the longest sides meet. This generates an additional {104} face. Finally, as growth normal to the substrate begins to dominate over growth adjacent to the substrate, the crystals undergo a morphological transition to full rhombohedra.

This sequence of morphologies can be readily explained in terms of surface free energy minimization, where the surface-to-volume ratio of the calcite crystals and the relative contributions of the crystal/solution and the crystal/SAM interfaces change during growth. Winterbottom constructions, which provide a phenomenological prediction of the equilibrium shapes of crystals located on solid substrates, were performed on an interface for a fixed crystal orientation and volume, under variation of the SAM/crystal interfacial energy. A comparable morphological transition to that seen experimentally was observed under the increasing interfacial energy, as reflects the decreasing influence of the substrate/crystal interface on the crystal morphology during growth (Fig. 1). Significantly, a larger variation between the experimental and predicted morphologies was observed at later growth stages, where the Winterbottom constructions terminate with a regular rhombohedron, as compared with the elongated rhombohedron seen experimentally[28].

Bragg coherent diffraction imaging. Calcite crystals of sizes 1–4 µm were characterized at different stages of the morphological development using BCDI, where this provides a simultaneous visualization of the crystal morphology and its internal strain. BCDI is achieved by illumination of a crystal with a coherent X-ray beam, whose coherence volume is larger than that of the crystal. A series of two-dimensional (2D) diffraction patterns is collected around a selected Bragg reflection for different points on the rocking curve, and is used to generate a coherent, three-dimensional (3D) X-ray diffraction pattern, which is formed by scattering from all parts of the crystal. In this lens-free form of microscopy, analysis of the coherent, 3D X-ray diffraction pattern through application of iterative phase-retrieval algorithms[29] generates a complex-valued reconstructed crystal

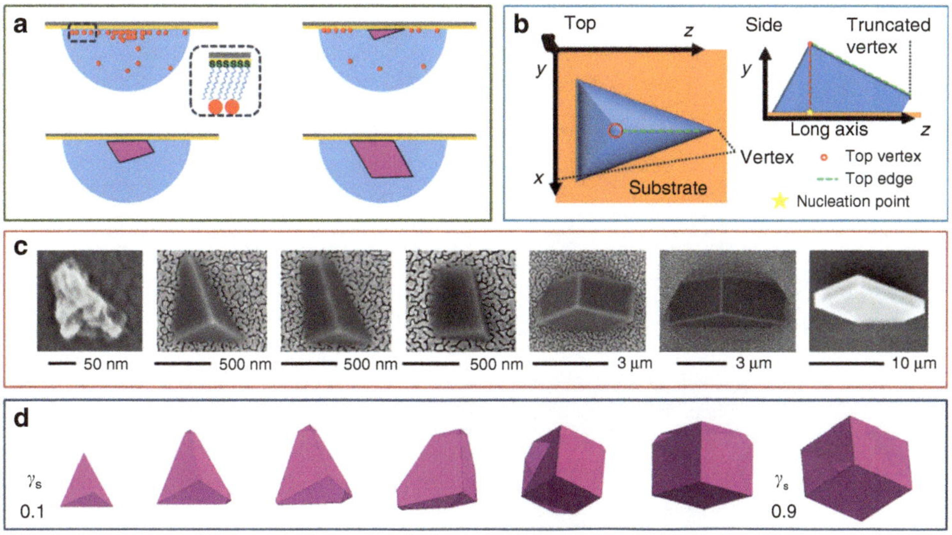

Figure 1 | Morphological progression of calcite nucleated on COOH terminated SAMs. (**a**) A schematic of the experimental set-up, where CO_2 and NH_3 diffuse into hanging droplets of $CaCl_{2(aq)}$, causing amorphous calcium carbonate formation (red dot). An oriented tetrahedron of calcite bounded by planar {104} faces then forms, whose growth leads to the development of an additional facet as a truncation of the long axis. Further growth then results in a transformation to rhombohedral calcite. (**b**) Schematic images of the tetrahedral growth form, showing the location of the truncation face. (**c**) Scanning electron microscope images showing the morphological development of the calcite crystals. (**d**) Winterbottom reconstructions predicting the morphological development of the calcite crystals, where these are of identical volume with stepwise increasing relative interfacial energy of the crystal/SAM (γ_s; 0.1–0.9). Crystal/ water interfacial energy values used were taken from Duffy and colleagues[36].

$\rho(r)$. The reconstructed amplitude provides a 3D representation of the specimen's electron density distribution $|\rho(r)|$, where this is sensitive to the crystallinity of the specimen. The phase shifts in the reconstructed complex amplitude, $\arg[\rho(r)] = \phi(r)$, are sensitive to small variations in lattice deformation and are proportional to a projection of the vector displacement field $u(r)$ of the atoms (from the ideal lattice points) and the scattering vector Q via $\phi(r) = u(r)\ Q$ (ref. 30). Here, a series of 2D diffraction patterns were collected from oriented calcite crystals at an angle corresponding to the off-specular {104} reflection. Stacks of 2D diffraction patterns were then inverted using an approach based on guided phase retrieval[31,32]. Detailed descriptions of the BCDI experiments and image reconstructions are provided in the Method section and the Supplementary Methods[8].

BCDI and TEM analysis of calcite crystals on SAMs. Representative BCDI reconstructions from two randomly selected calcite crystals are shown in Fig. 2. The images shown are top-down and bottom-up projections of the iso-surface renderings of (Fig. 2a) the reconstructed electron densities (amplitudes), where this provides a visualization of the crystal morphologies and (Fig. 2b) the projected displacements (phase), which correspond to lattice strains. The displacements are represented by a cyclic colour map projected onto the recorded electron density. A colour shift towards red ($+d/2$) corresponds to a lattice contraction, while a shift towards blue ($-d/2$) equates to lattice dilation, where d is equal to the spacing between adjacent lattice planes. A colour change across the whole scale corresponds to a displacement of one unit cell in a particular direction. The crystals labelled (i) and (ii) correspond to different stages of growth. Crystal (i) is 1.4 μm in size, approximately tetrahedral in shape, and shows the beginning of a new truncation face. At 2 μm in size, crystal (ii) appears to be at a later stage of development, where it shows smoother faces and a well-defined {104} truncation. In addition, the 'top edge' of this crystal is now almost parallel to the substrate (as is also seen at later stages in the Winterbottom constructions).

The BCDI reconstructions also provide a unique opportunity to examine the influence of the SAM on the crystal structure. In both crystals, the face adjacent to the SAM exhibits a degree of roughness that is consistent with atomic force microscopy measurements of the SAM functionalized gold substrate (Supplementary Fig. 2)[33]. That the nucleation face of the calcite crystals are themselves roughened suggests that the crystal grows, so as to preserve interfacial contact. Interestingly, examination of the nucleation faces of crystals (i) and (ii) also reveals the presence of two adjacent surface cusps of sizes ~70–100 nm on each face. These intriguing features can be seen more clearly in cross-sections of the crystals (Fig. 2c), which show that they lie in a plane approximately parallel to the truncated vertex. Confirmation that these cusps correspond to physical features in the crystals was obtained by TEM of thin sections prepared by focused ion beam (FIB) milling. Figure 3 shows electron micrographs of a prepared section, and its location with respect to the original crystal. These reveal a linear feature of length 85 nm and width 10–20 nm (arrowed), whose location is commensurate with the surface cusps observed using BCDI (Fig. 3c).

The strain present within the calcite crystals (as inferred from the projected displacements; Fig. 2b,d) shows that lattice deformation is concentrated at the edges and corners of each of the crystals. This strain becomes more localized around the corners as the crystal grows in size. These distributions of lattice displacements can be attributed to their elastic anisotropy, as supported by finite element (FE) modelling (Supplementary Fig. 3). Placing a {012} oriented, tetrahedral calcite crystal under a uniform surface stress of 1.5 N m^{-1} resulted in a comparable displacement profile within the crystal to that observed experimentally.

Returning to the surface cusps visualized within crystals (i) and (ii), each of these are associated with localized strain fields that radiate from the substrate into the crystal. Importantly, these regions possess both a hollow core and a spiral phase/displacement (turquoise circle in Fig. 2d), where this combination of features identifies them as dislocations[8,32].

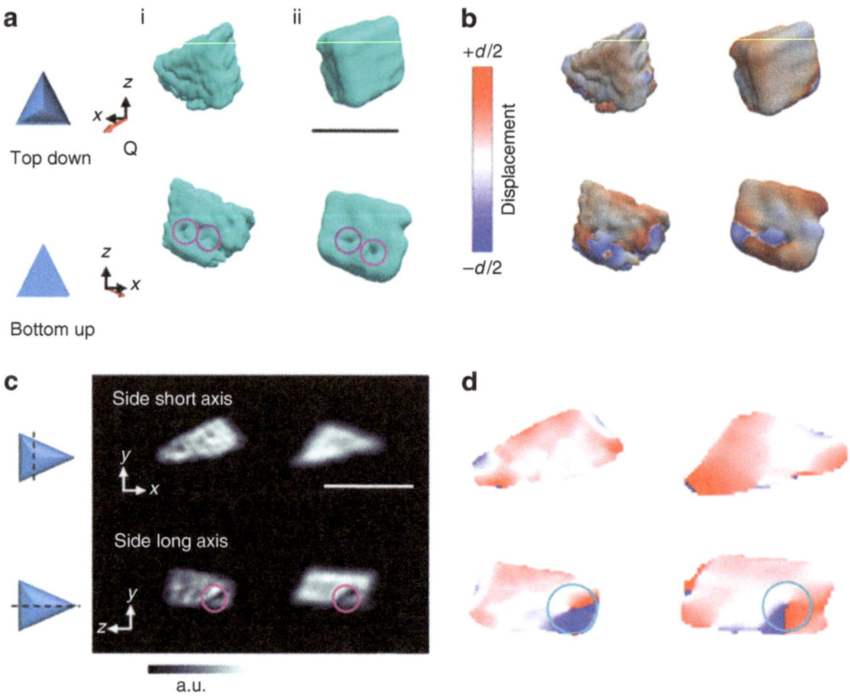

Figure 2 | Bragg coherent diffraction imaging (BCDI) reconstructions. Reconstructions of two different calcite crystals (i) and (ii), which were nucleated on carboxylate-terminated SAMs, are shown. The reconstructed crystal shapes from BCDI amplitude measurements are shown in **a**, where these are viewed from the directions indicated. The surface cusps, which appear on the bases of both crystals, are circled. (**b**) The projected displacements ($-d/2$ blue lattice dilation and $+d/2$ red lattice contraction) of the crystals. (**c**) Sections through the electron density maps and (**d**) sections cut through the displacement maps, where these made through the centres of the crystals, normal to the substrate. The surface cusps (magenta circle) and areas of spiral displacement (cyan circle) are highlighted. The beam direction is along the *z* axis, with the *y* axis vertical, while the sample/substrate is located at a set scattering angle towards the beam direction (*z*) with **Q** the scattering vector. Scale bar, 1.8 μm.

Further examination of the strain fields associated with these cusps then demonstrated that each pair of surface cusps actually form part/are the surface expressions of a single dislocation loop (Fig. 4i,ii; Supplementary Movies 1 and 2). As shown in Fig. 2d, the planes of these dislocation loops are approximately parallel to the new 'truncation' faces, which suggests that they are not randomly located. Notably, with the exception of these single dislocation loops, crystals (i) and (ii) appear mostly dislocation free.

Discussion

Crystallization at interfaces has been well studied for the epitaxial growth of crystalline thin films on solid substrates, and misfit dislocations often form in the growing layer to relieve the elastic strain associated with lattice mismatch with the substrate[1,5,6]. This strain energy increases, as the volume of the crystal increases, and when the stored elastic energy exceeds the energy cost of making a dislocation, dislocation loops may be nucleated to relieve the stress. In the absence of existing dislocations, such loops will generally nucleate at the free surface of the growing crystal, where the nucleation barriers are lower than in the bulk (Fig. 5a). The loops then grow under the influence of the misfit stress until they reach the interface[6]. The final loop has edge character parallel to the interface, to relieve misfit strain, and some screw character on the sections perpendicular to the interface.

Our BCDI study of calcite crystals grown on SAMs shows that, while dislocations also form in this system, they are entirely different from conventional misfit dislocations in that the ends of the loop lie at the crystal/substrate interface rather than the free surface of the crystal. The observed dislocation loops must either have been created during crystal growth or have nucleated and grown in response to stress. Although we are not currently in a position to conclusively identify their origins, their location and orientation is strongly indicative of a mechanism in which the dislocation loop nucleates at the crystal/SAM interface and then expands on a slip plane into the interior of the crystal during crystal growth. That they terminate at the crystal/SAM interface, as opposed to a growth surface, appears to rule out a mechanism, where they form as a result of a growth defect. For that to be the case, two independent dislocations would have to grow in from the substrate, before curving round to form a continuous loop at a later stage of the growth. While it is possible that two dislocations that were sufficiently close could combine due to the interaction of their stress fields, the end of the ones formed here are ≈ 1 μm apart. Further, the two separated dislocations arms could only unite if their ends have the same Burgers vectors. A far more credible explanation is that the dislocations nucleate and grow under the internal stress field of the crystal to generate a loop. Further, their configuration is consistent with their nucleation at the base of the (012) oriented calcite tetrahedron, where the stress—which reduces the barrier to dislocation nucleation—is concentrated (Fig. 5b).

Both modelling and experimental studies have shown that the innate flexibility of SAMs is fundamental to their ability to support the epitaxial growth of calcite crystals. There is a very large lattice mismatch between an idealized 11-mercaptoundecanoic acid SAM on Au and the (012) calcite face, which can be compensated by a high density of surface vacancies or surface steps in the calcite[34]. If it is assumed that the SAM is commensurate with the gold substrate, then the lattice mismatch is 0.2% in the calcite <001> direction and 26% in the perpendicular $<12\bar{1}>$ direction[14]. Unlike SAMs, which are

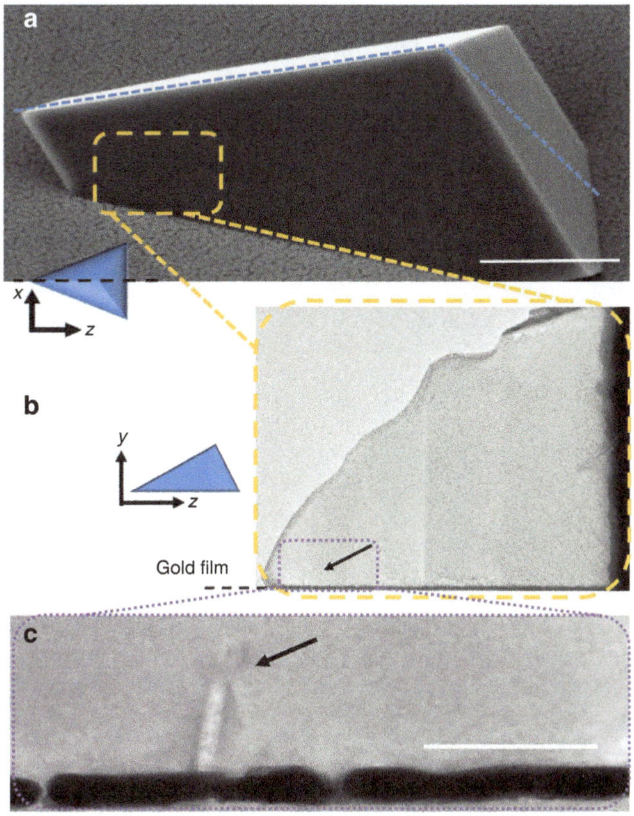

Figure 3 | Electron micrographs of oriented sections of precipitated calcite crystals. (**a**) Scanning electron microscope image of the sample crystal, where the dotted blue line shows the direction of the cut. Scale bar, 1 μm. (**b**) The selectively thinned tip of the prepared lamella, whose location with respect to the original crystal is indicated in the yellow box shown in **a**. (**c**) Shows a higher-magnification image of the front end of the tip (arrowed in **b**). Arrowed in **c** is a feature of size ~85 × 15 nm, which corresponds to the 'surface cusps' imaged using BCDI. Scale bar, 100 nm.

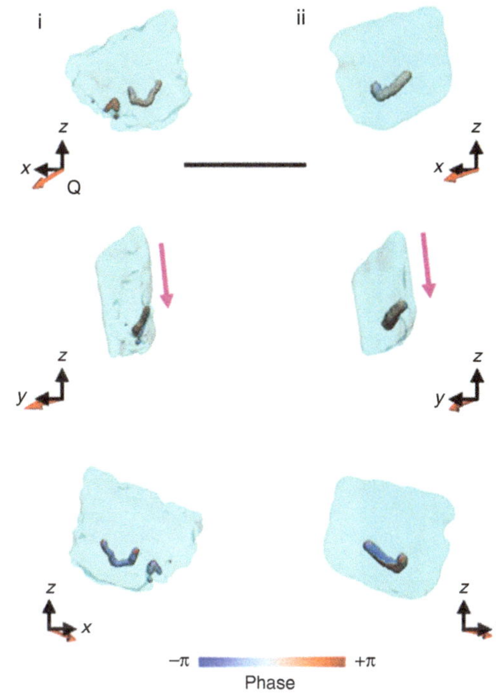

Figure 4 | Iso-surface rendering of the defects present within the oriented calcite crystals. The defects shown have a low electron density core that is surrounded by a spiral deformation field/phase, and correspond to dislocation loops. In both crystals (i and ii) the plane of the loop is directed towards the truncation facet (pink arrow). Defects are given in respect to their location within the crystal, as given by semi-transparent projections of the electron density. **Q** is the scattering vector. Scale bar, 1.8 μm.

easily deformable, calcite has a high elastic modulus and so it will not readily deform to match the SAM. A cycle of mutual control, in which both the organic and mineral components induce complementary local order across the interface then leads to the formation of a critical crystalline region, which defines the nucleation face.

Although stress due to interfacial defects undoubtedly contributes to the nucleation of the dislocation loops viewed here, the magnitude of the stress is unlikely to be high enough to overcome the barrier for dislocation growth. An additional source of internal stress is the roughness of the Au/SAM substrate, which has surface features of ≈50 × 10 nm. If the calcite crystal grows so as to maintain contact with the curved substrate, as suggested by our data, this will result in bending of the lattice planes close to the substrate, which will induce a complex stress field in the crystal. This, in turn, could be partially relieved by the formation of a dislocation. The total stored elastic energy in a crystal due to bending can be calculated using FE modelling by applying a stress to the base of the tetrahedron (Fig. 5c), while the elastic energy of a dislocation loop can be estimated ($E_{dislocation} \approx Gb^2l/2$). Given typical values for calcite of the Burgers vector $b \approx 0.8$ nm, shear modulus $G = 35$ GPa, and the dislocation length, $l = 1$ μm, the dislocation energy can be estimated as $\sim 10^{-14}$ J. Therefore, it can be seen from Fig. 5c that for an interfacial stress of 6 N m^{-1} on the base of the tetrahedron, a crystal of volume 0.2 μm^3 (comparable to the crystals analysed here) will have more stored elastic energy than it costs to create a 1-μm dislocation loop. This lends support to the argument that the dislocation relieves interface induced stress.

After nucleation, growth of the dislocation loop is governed by the preferential calcite slip system of {104} and {012} planes, and $<\bar{4}\,\bar{2}\,1>$ slip directions[35], and the growing dislocation loop is constrained to lie in one of these planes. As the plane of the observed dislocation loops is approximately parallel to the {104} truncation face, we can conclude that the {104} planes are the preferred slip planes for stress relief on SAMs. A schematic representation of the dislocation configuration is illustrated in Fig. 5d. This dislocation is consistent with the configurations of both of the observed dislocations. It is noted that the {104} $<\bar{4}\,\bar{2}\,1>$ slip system generally requires high temperature and/or pressure for the activation[35]. However, the dislocation loops are unambiguously observed in our crystals, and nucleation and growth appears to be the only plausible explanation for their origin. Rough surfaces have been shown to have very high local stresses (up to 4 GPa in TiO_2 nanoparticles[36]) that would make the nucleation of dislocation loops entirely feasible. It is also possible that dislocation motion is easier in micron-sized crystals than in their bulk counterparts, which may also favour their growth in our small crystals.

It remains striking, however, that each calcite crystal imaged here possesses a single dislocation loop, where each are of similar size and in an identical location with respect to the final crystal morphology. This suggests that the control over calcite nucleation and growth exerted by the monolayer is highly reproducible, and that the developments of the dislocation loop and crystal morphology are intimately linked. Finally, it is interesting to

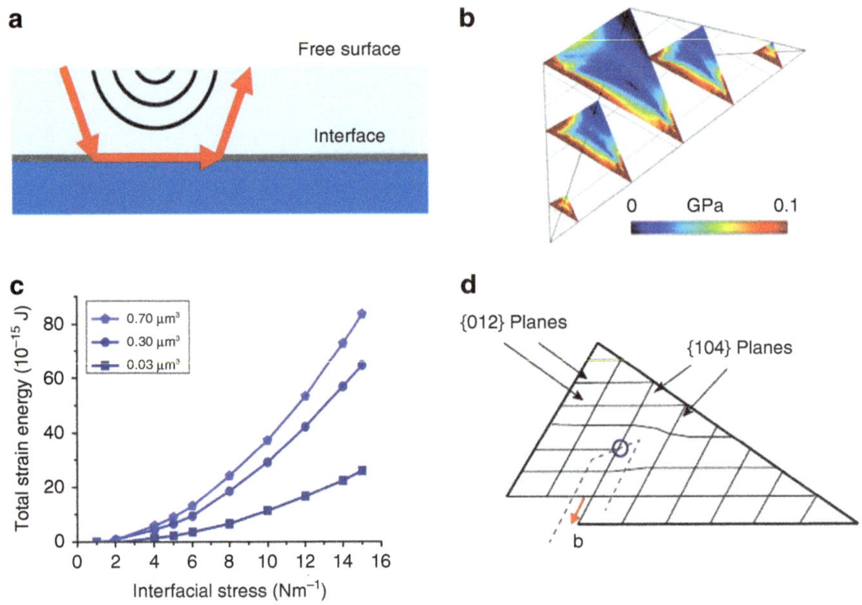

Figure 5 | Dislocation geometry. (**a**) Schematic of a classical 'epitaxy' misfit dislocations (red arrow), between two materials with a close lattice match. (**b**) Calculated stress distribution inside a calcite tetrahedron caused by an interfacial stress. The von Mises stress distribution is shown in vertical slices through the crystal. The interfacial stress was set to 10 Nm^{-1}. (**c**) Total elastic energy, calculated using FE, due to interfacial stress for three different sized crystals. (**d**) Lattice planes on a cross-section of a (012) tetrahedron, showing the configuration of a dislocation on a (104) slip plane, with a $[\bar{4}\bar{2}1]$ Burgers vector. The blue circle shows where the dislocation line cuts the plane and the dashed line shows the dislocation line, which has screw character on the vertical segments and edge character on the horizontal segment (perpendicular to the plane). The red arrow shows the Burgers vector (**b**). This configuration is consistent with the experimental observations.

note that the tetrahedral calcite crystals grown at the gas/liquid interface frequently reach diameters of $\approx 20\,\mu m$ (refs 37,38), as compared with the maximal value of $\approx 2\,\mu m$ observed on SAMs here. Again, the energy cost associated with the interfacial stress and dislocation loops would be expected to give rise to this reduced footprint on the substrate.

In summary, we have used X-ray BCDI to gain unique insight into the epitaxial growth of calcite crystals on organic SAMs, where this topic has received enormous interest in the literature, thanks to its relevance to biomineralization processes. High-resolution images of the internal strain present within micron-scale calcite crystals precipitated at SAMs reveal the presence of single dislocation loops within each crystal, where their configurations suggest that they play a role in the morphological development of the crystal. This provides an alternative explanation to the elastic strain model[24], which proposes that elastic deformation of both the SAM and calcite lattice gives rise to anisotropic growth. Importantly, the observed dislocation loops also exhibit entirely different geometries to the misfit dislocations generated during conventional epitaxial thin film formation. Our data strongly suggest that the nanoscale roughness present in SAM/evaporated metal film systems is fundamental to the formation of these defects, where this provides new insight into the factors that govern crystallization on these substrates. The results presented, therefore, provide an important contribution to our knowledge about stress relaxation and dislocation formation during epitaxial crystal growth, and demonstrate that the roughness of the substrate—be it organic or inorganic in nature—should be considered when controlling interfacial strain and defects during epitaxial crystal growth.

Methods

Materials. Analytical grade $(NH_4)_2CO_3$ and $CaCl_2 \cdot 2H_2O$, were used as received. Solutions were prepared using Milli-Q standard 18.2 MΩcm and experiments were performed at a temperature of 21 °C. Glassware was soaked overnight in 10% w/v NaOH, followed by rinsing with dilute HCl and washing with Milli-Q water.

Substrate preparation. Functionalized SAMs were prepared on freshly deposited noble metal films. Thin films were deposited either on silicon wafer or cleaned glass slides using a Mantis Qprep 250 deposition system at a base pressure $< 10^{-6}$ mbar. A 2-nm Cr was initially deposited to promote substrate adhesion, followed by the evaporation of 25–50 nm of Au at ≤ 0.1 nm s^{-1}. Monolayer formation on metal substrate was initiated by immersion in a 1 mM solution of 11-mercaptoundecanoic acid in ethanol. The prepared SAMs were then thoroughly rinsed with ethanol and Milli-Q water, and were subsequently dried under nitrogen.

Mineral deposition. Preferentially oriented calcite was obtained by diffusion methods[26]. A measure of 200-µl droplets of 5 mM $CaCl_2$ solution were hung from an inverted substrate, which was in turn placed in a sealed container (2 l) in the presence of $(NH_4)_2CO_{3(s)}$ (2 g). This geometry prevented 'homogenously' formed calcite from settling onto the surface. $(NH_4)_2CO_{3(s)}$ decomposition into $CO_{2(g)}$ and $NH_{3(g)}$ created the required supersaturation for $CaCO_3$ precipitation, and provided a gradual increase in supersaturation that ensured a sufficient number density of single crystals. Samples were removed after 30 min of incubation, giving rise to crystals 1–4 µm in diameter, which are suitable for coherent imaging. They were washed in ethanol and left to dry.

CDI set-up. Bragg CDI experiments were performed at beamline 34-ID-C of the Advanced Photon Source, Argonne National Laboratory, USA. An undulator produced X-rays that were monochromatized using a silicon (111) double-crystal monochromator to an energy of 9 keV. Calcite crystals on a substrate were placed on a diffractometer that had its rotation centre aligned with the X-ray beam. Slits were used to aperture the X-rays to reduce the illuminated area. An X-ray sensitive charge-coupled device (Princeton instruments) with 1300 × 1300 square pixels of side length 22.5 µm was positioned at the desired diffraction angle for an off-specular (104) reflection 2.5 m from the sample. To measure its full 3D diffraction patterns, the crystal was rotated by 0.3 degree with 0.003 degree step size. At each rotation angle, a 2D slice of the 3D far-field diffraction pattern was recorded. By stacking all 2D diffraction frames together, a complete 3D diffraction pattern is obtained, from which real-space images can be reconstructed. Due to the small size of the crystals (1–4 µm), the illumination can be considered almost completely coherent.

Reconstruction algorithm. Images were obtained by performing iterative phase retrieval[29] on 3D coherent diffraction patterns. Complete knowledge (both amplitude and phase) of the diffracted wavefield allows an image to be obtained via an inverse Fourier transform. Provided the diffraction data is oversampled, that is the sample has its Fourier transform sampled at least twice the Nyquist frequency (or alternatively its auto-correlation is sampled at least at the Nyquist frequency) and the crystal is isolated, phase retrieval can be performed. The basic phase-retrieval process begins with a guess for the diffracted phase before applying an inverse Fourier transform to yield a first estimate of the crystal. After enforcing the constraint that the crystal is isolated, this new crystal iterate is Fourier transformed to yield an estimate for the 3D-diffracted wavefield. Consistency with the measured intensity is enforced, while retaining the current estimate of the phase. This process is repeated until a self-consistent solution is reached using combinations of current and previous estimates for the crystal. For this work an approach was used that combined guided phase retrieval[31], with low- to high-resolution (or multi-resolution) reconstructions[32]. A detailed description is provided in the Supplementary Methods.

Finite element calculation. The calculations were performed with COMSOL Multiphysics 4.2a. Each element of the elasticity tensor of calcite was rotated to the crystallographic orientation of a crystal with a (012) nucleation plane. Surface stress was modelled by applying a 1-nm thick membrane, which was then uniformly contracted on each crystal facet (Supplementary Fig. 3).

Characterization. Scanning electron micrographs of uncoated specimen were obtained using an FEI Nova NanoSEM 650. Crystal growth was followed using an inverted Olympus IX-70 confocal microscope. Crystal orientation was determined using a Bruker D8 Advanced diffractometer equipped with a CuKα_1 X-ray source in pole configuration, using a step size 1.5° at 2.5 s (Psi 0–90, Phi 0–360). Substrates were characterized using atomic force microscopy (Bruker dimensions 3100 AFM) in tapping mode (Brucker Tespa; resonance frequency 345–385 kHz, K 20–80 Nm^{-1}) at a scan rate of 1.98 Hz with pixel dimension of 512 × 512. Images of the internal structure of the oriented calcite crystals were obtained using high-resolution TEM imaging of thin sections prepared by FIB milling. Sample preparation was performed using an FEI Nova200 Dual Beam FIB/scanning electron microscopy. The ion beam was operated at 30 kV and at beam currents between 0.1 and 5 nA. Lift-out was performed *in situ* using a Kleindiek micromanipulator. The samples were then analysed with a FEI Tecnai F20 200 kV field emission gun–TEM fitted with an Oxford Instruments INCA 350 EDX system/80 mm X-Max SDD detector and a Gatan Orius SC600A Charge Coupled Device (CCD) camera.

References

1. Van der Merwe, J. H. Misfit dislocation generation in epitaxial layers. *Crit. Rev. Solid State Mater. Sci.* **17,** 187–209 (1991).
2. Shiraki, Y. & Sakai, A. Fabrication technology of SiGe hetero-structures and their properties. *Surf. Sci. Rep.* **59,** 153–207 (2005).
3. O'Reilly, E. P. Valence band engineering in strained-layer structures. *Semicond. Sci. Technol.* **4,** 121–137 (1989).
4. Oh, S. H. & Park, C. G. Misfit strain relaxation by dislocations in $SrRuO_3/SrTiO_3$ (001) heteroepitaxy. *J. Appl. Phys.* **95,** 4691–4704 (2004).
5. Jain, S. C., Harker, A. H. & Cowley, R. A. Misfit strain and misfit dislocations in lattice mismatched epitaxial layers and other systems. *Phil. Mag. A* **75,** 1461–1515 (1997).
6. Hull, R. & Bean, J. C. Misfit dislocations in lattice-mismatched epitaxial films. *Crit. Rev. Solid State Mater. Sci.* **17,** 507–546 (1992).
7. Chen, Y. & Washburn, J. Structural transition in large-lattice-mismatch heteroepitaxy. *Phys. Rev. Letts.* **77,** 4046–4049 (1996).
8. Clark, J. N. *et al.* Three-dimensional imaging of dislocation propagation during crystal growth and dissolution. *Nat. Mater.* **14,** 780–784 (2015).
9. Heywood, B. R. & Mann, S. Template-directed nuclatiosn and growth of inorganic materials. *Adv. Mater.* **6,** 9–20 (1994).
10. Berman, A. *et al.* Total alignment of calcite at acidic polydiacetylene films: cooperativity at the organic-inorganic interface. *Science* **269,** 515–518 (1995).
11. DiMasi, E. *et al.* Complementary control by additives of the kinetics of amorphous $CaCO_3$ mineralization at an organic interface: in-situ synchrotron X-ray observations. *Phys. Rev. Letts.* **97,** 045503 (2006).
12. Lee, J. R. I. *et al.* Structural development of mercaptophenol self-assembled monolayers and the overlying mineral phase during templated $CaCO_3$ crystallization from a transient amorphous film. *J. Am. Chem. Soc.* **129,** 10370–10381 (2007).
13. Aizenberg, J., Black, A. J. & Whitesides, G. M. Control of crystal nucleation by patterned self-assembled monolayers. *Nature* **398,** 495–498 (1999).
14. Travaille, A. M. *et al.* Highly oriented self-assembled monolayers as templates for epitaxial calcite growth. *J. Am. Chem. Soc.* **125,** 11571–11577 (2003).
15. Darkins, R., Côté, A. S., Freeman, C. L. & Duffy, D. M. Crystallization rates of calcite from an amorphous precursor in confinement. *J. Cryst. Growth* **367,** 110–114 (2013).
16. Nielsen, M. H. & Lee, J. R.I in *Methods in Enzymology* Vol. 532 (ed. De Yoreo, J. J.) 209–224 (Academic Press, 2013).
17. Travaille, A. M. *et al.* Aligned growth of calcite crystals on a self-assembled monolayer. *Adv. Mater.* **14,** 492–495 (2002).
18. Aizenberg, J., Black, A. J. & Whitesides, G. M. Oriented growth of calcite controlled by self-assembled monolayers of functionalized alkanethiols supported on gold and silver. *J. Am. Chem. Soc.* **121,** 4500–4509 (1999).
19. Ahn, D. J., Berman, A. & Charych, D. Probing the dynamics of template-directed calcite crystallization with *in situ* FTIR. *J. Phys. Chem.* **100,** 12455–12461 (1996).
20. Lee, J. R. I. *et al.* Cooperative reorganization of mineral and template during directed nucleation of calcium carbonate. *J. Phys. Chem. C* **117,** 11076–11085 (2013).
21. Quigley, D., Rodger, P. M., Freeman, C. L., Harding, J. H. & Duffy, D. M. Metadynamics simulations of calcite crystallization on self-assembled monolayers. *J. Chem. Phys.* **131,** 094703 (2009).
22. Volkmer, D., Fricke, M., Vollhardt, D. & Siegel, S. Crystallization of (012) oriented calcite single crystals underneath monolayers of tetra(carboxymethoxy)calix 4 arenes. *Dalton. Trans.* **24,** 4547–4554 (2002).
23. Freeman, C. L., Harding, J. H. & Duffy, D. M. Simulations of calcite crystallization on self-assembled monolayers. *Langmuir* **24,** 9607–9615 (2008).
24. Pokroy, B. & Aizenberg, J. Calcite shape modulation through the lattice mismatch between the self-assembled monolayer template and the nucleated crystal face. *CrystEngComm.* **9,** 1219–1225 (2007).
25. Harding, J. H., Freeman, C. L. & Duffy, D. M. Oriented crystal growth on organic monolayers. *CrystEngComm.* **16,** 1430–1438 (2014).
26. Ihli, J., Bots, P., Kulak, A., Benning, L. G. & Meldrum, F. C. Elucidating mechanisms of diffusion-based calcium carbonate synthesis leads to controlled mesocrystal formation. *Adv. Funct. Mater.* **23,** 1965–1973 (2013).
27. Stephens, C. J., Kim, Y.-Y., Evans, S. D., Meldrum, F. C. & Christenson, H. K. Early Stages of crystallization of calcium carbonate revealed in picoliter droplets. *J. Am. Chem. Soc.* **133,** 5210–5213 (2011).
28. Zucker, R., Chatain, D., Dahmen, U., Hagège, S. & Carter, W. C. New software tools for the calculation and display of isolated and attached interfacial-energy minimizing particle shapes. *J. Mater. Sci.* **47,** 8290–8302 (2012).
29. Fienup, J. R. Phase retrieval algorithms: a comparison. *Appl. Opt.* **21,** 2758–2769 (1982).
30. Clark, J. N. *et al.* Ultrafast three-dimensional imaging of lattice dynamics in individual gold nanocrystals. *Science* **341,** 56–59 (2013).
31. Chen, C.-C., Miao, J., Wang, C. W. & Lee, T. K. Application of optimization technique to noncrystalline x-ray diffraction microscopy: guided hybrid input-output method. *Phys. Rev. B* **76,** 064113 (2007).
32. McCallum, B. C. & Bates, R. H. T. Towards a strategy for automatic phase retrieval from noisy fourier intensities. *J. Mod. Opt.* **36,** 619–648 (1989).
33. Borukhin, S. & Pokroy, B. Formation and elimination of surface nanodefects on ultraflat metal surfaces produced by template stripping. *Langmuir* **27,** 13415–13419 (2011).
34. Côté, A. S., Darkins, R. & Duffy, D. M. Modeling calcite crystallization on self-assembled carboxylate-terminated alkanethiols. *J. Phys. Chem. C* **118,** 19188–19193 (2014).
35. DeBresser, J. H. P. & Spiers, C. J. Strength characteristics of the r, f, and c slip systems in calcite. *Tectonophysics* **272,** 1–23 (1997).
36. Darkins, R., Sushko, M. L., Liu, J. & Duffy, D. M. Stress in titania nanoparticles: an atomistic study. *Phys. Chem. Chem. Phys.* **16,** 9441–9447 (2014).
37. Hashmi, S. M., Wickman, H. H. & Weitz, D. A. Tetrahedral calcite crystals facilitate self-assembly at the air-water interface. *Phys. Rev. E* **72,** 041605 (2005).
38. Loste, E., Díaz-Martí, E., Zarbakhsh, A. & Meldrum, F. C. Study of calcium carbonate precipitation under a series of fatty acid langmuir monolayers using brewster angle microscopy. *Langmuir* **19,** 2830–2837 (2003).
39. Ihli, J. *et al.* Dataset for BCDI study of Calcite Crystals Grown on Self-Assembled Monolayers, Research Data Leeds Repository. Available at http://doi.org/10.5518/53.

Acknowledgements

This work was supported by FP7-advanced grant from the European Research Council (J.N.C. and I.K.R.), by an Engineering and Physical Sciences Research Council Leadership Fellowship (F.C.M., J.I. and Y.Y.K.; EP/H005374/1) and EPSRC grant EP/K006304/1 (A.N.K.). F.C.M., A.S.S., D.M.D. and A.S.C. are also supported by an EPSRC Programme Grant (grant EP/I001514/1) that funds the Materials Interface with Biology (MIB) consortium. Some of the experimental work was carried out at Advanced Photon Source Beamline 34-ID-C, built with funds from the US National Science Foundation under Grant DMR-9724294 and operated by the US Department of Energy, Office of Science, Office of Basic Energy Sciences, under contract DE-AC02-06CH11357.

Author contributions

J.I. and J.N.C. designed the project; J.I. prepared the samples; J.N.C., J.I. and R.J.H., performed the BCDI experiments; J.N.K. performed the image reconstructions; J.I. and T.P.C. acquired the pole figures; A.N.K., Y.Y.K. and J.I. acquired the FIB images; O.C. and J.I. acquired the AFM data; J.N.C., J.I. and A.S.S. analysed the data, D.M.D. and A.S.C. performed the FE studies; and J.I., J.N.C., F.C.M., D.M.D. and I.K.R. wrote the paper. All the authors read and commented on the manuscript.

Additional information

Competing financial interests: The authors declare no competing financial interests.

6

Structure sensitivity in the nonscalable regime explored via catalysed ethylene hydrogenation on supported platinum nanoclusters

Andrew S. Crampton[1], Marian D. Rötzer[1], Claron J. Ridge[1,2], Florian F. Schweinberger[1], Ueli Heiz[1], Bokwon Yoon[3] & Uzi Landman[3]

The sensitivity, or insensitivity, of catalysed reactions to catalyst structure is a commonly employed fundamental concept. Here we report on the nature of nano-catalysed ethylene hydrogenation, investigated through experiments on size-selected Pt_n ($n = 8$–15) clusters soft-landed on magnesia and first-principles simulations, yielding benchmark information about the validity of structure sensitivity/insensitivity at the bottom of the catalyst size range. Both ethylene-hydrogenation-to-ethane and the parallel hydrogenation-dehydrogenation ethylidyne-producing route are considered, uncovering that at the <1 nm size-scale the reaction exhibits characteristics consistent with structure sensitivity, in contrast to structure insensitivity found for larger particles. The onset of catalysed hydrogenation occurs for Pt_n ($n \geq 10$) clusters at $T > 150$ K, with maximum room temperature reactivity observed for Pt_{13}. Structure insensitivity, inherent for specific cluster sizes, is induced in the more active Pt_{13} by a temperature increase up to 400 K leading to ethylidyne formation. Control of sub-nanometre particle size may be used for tuning catalysed hydrogenation activity and selectivity.

[1] Chair of Physical Chemistry, Catalysis Research Center, Chemistry Department, Technische Universität München, Lichtenbergstraße 4, Garching 85748, Germany. [2] Air Force Research Laboratory, Energetic Materials Branch, 2306 Perimeter Road, Eglin AFB, Florida 32542, USA. [3] School of Physics, Georgia Institute of Technology, Atlanta, Georgia 30332-0430, USA. Correspondence and requests for materials should be addressed to U.L. (email: uzi@gatech.edu).

The classification and systematization[1–3] of heterogeneously catalysed reactions as 'demanding' or 'facile', with the former term describing dependence on catalyst particle size and the latter associated with reactions that are size invariant, has been a central guiding theme of research in heterogeneous catalysis for over half a century[4–7]. The early focus on particle size came from the realization that the fraction of atoms situated at a metal particle's surface to the total number of metal atoms, sometimes termed the 'fraction of exposed atoms'[6] or 'degree of dispersion'[1], varies with particle size, approaching unity for very small particles (diameter ≤1 nm) and taking small values for large particles. It was recognized early, based on simple geometric (polyhedral) models (tetrahedral, octahedral, cuboctahedral and so on), that the above-mentioned size-related classification may be related to the size-dependent abundance of atomic configurations characterized by surface atoms with low coordination numbers (for example, edges or corners). These under-coordinated atoms were expected to express varying proclivities for engaging in chemical reactions depending on the nature of the reactants and the ambient conditions[1,4–6].

The advent and proliferation of atomic and molecular scale surface science preparation and characterization techniques in the 1970s focused on reactions catalysed on well-defined single-crystal model systems. It then became customary (extending up to date) to classify heterogeneously catalysed reactions as structure sensitive or structure insensitive, corresponding to the aforementioned classifications of demanding and facile, in order to better describe the primary difference between these single-crystal model systems. An example of a structure-sensitive reaction is ammonia synthesis through the reaction of nitrogen and hydrogen on iron single-crystal surfaces where the Fe(111) surface exhibits a higher turnover frequency (TOF) than Fe(100) or Fe(110)[8]. Ethylene hydrogenation catalysed by platinum serves as an example of a structure-insensitive reaction, that is, no change in the TOF has been observed on different single-crystal planes[9]. The latter fundamental and technological important reaction stands out as the first reaction for which a catalytic mechanistic pathway has been formulated (the Horiuti-Polanyi mechanism, HP), and it forms the focus of our investigation in this paper[10].

The classification of the ethylene hydrogenation as structure insensitive is supported by decades of experimental work employing mostly surface science techniques for reactions taking place on well-characterized extended transition metal surface planes (see Supplementary Note 1 for details), or investigations on samples made of surface-supported distributions of relatively large metal particles with sizes ranging from 1.5 to 10 nm diameter. Although full understanding pertaining to the microscopic origins of the observed structure insensitivity is still lacking[5,7,11,12], a commonly cited view holds that formation of carbonaceous species (in particular ethylidyne) in the course of the hydrogenation reaction, leads to saturation coverage (poisoning) of the metal surface that results in masking of structural features that could have otherwise brought about different reactivities, and/or alternative reactive outcomes, for surfaces showing different crystallography; that is, if not for the formation of poisonous carbonaceous species, metal surfaces with different crystallography (that is, different Miller indices) would have exhibited structure sensitivity.

Metal particles with sizes reduced to the <1 nm in 'diameter' range (containing 10–20 atoms) may be viewed as quasi-zero-dimensional (0D) quantum dots, characterized by electronic quantum size effects[13,14] and specific atomic organization motifs, that lead to the emergence of unique size-dependent physical and chemical properties. In this regime, extrapolations from larger sizes using arguments relying on surface-to-volume ratios and scaling relations based on the enumeration of special sites (for example, corner and edge atoms) as a function of the size are not operative[15–17]. In this size scale—that is, in the limit of ultimate dispersion—where almost all of the atoms of the metal cluster are essentially surface atoms and where almost all the cluster atoms can be classified as under-coordinated (compared with the bulk), a new materials' size regime is entered[13,14,18,19]. Explorations of materials' properties in this limiting reduced-size scale requires experiments and first-principles theoretical treatments with atom-by-atom resolution. This is indeed the approach that we have taken in this study.

In quest of appraising the applicability of the fundamental concept of structure sensitivity/insensitivity to nanoscale catalysts, founded on the basis of observations made at larger size scales, we focus here on the hydrogenation of ethylene catalysed by size-selected platinum nanoclusters containing between 7 and 40 Pt atoms supported on an MgO(100)/Mo(100) surface, which is inactive for ethylene hydrogenation. As previously mentioned, ethylene hydrogenation serves often as a generic example of a structure insensitive reaction, obeying the parallel reactions selectivity criterion (PRSC)[1], with the full-hydrogenation-to-ethane channel and the alternative half-hydrogenation-to-ethyl/dehydrogenation-to-ethylidyne pathway serving as the operative two parallel reactions. The PRSC stipulates that: 'The best way to proceed is to choose a molecule reacting along two parallel paths and measure the selectivity defined as the rate of the two parallel reactions. If the two products come from different adsorbed states requiring different surface structures, a change of selectivity with dispersion or mode of preparation of the metal may be found. The most unequivocal case is when the specific activity for one of the parallel reactions changes from one catalyst to the next, whereas the specific activity for the other remains unchanged". Thus, our present experimental and theoretical studies of the ethylene hydrogenation reaction catalysed by nanoscale platinum clusters with atom-by-atom size-controlled resolution, joined with the first adaptation of the PRSC to nanoscale catalysts, provides, as elaborated below, benchmark results pertaining to the reaction's structure sensitivity/insensitivity at the bottom of the catalyst particle size range. As in the case of larger catalyst particles, both the ethylene-hydrogenation-to-ethane channel and the parallel ethylidyne-producing, poisoning, route must be considered, culminating in the fundamental uncovering that at the ultimate <1 nm size scale, the ethylene hydrogenation reaction catalysed by supported platinum particles exhibits characteristics consistent with being a structure-sensitive reaction, in contrast to the accepted structure insensitivity of the reaction found for particles with a lower dispersion.

Results

Research plan. We begin with temperature programmed reaction (TPR) experiments, starting from the low temperature adsorption of the two reactants (H_2 and C_2H_4) where all reaction pathways are quenched (except H_2 dissociation which may occur spontaneously even at low temperatures, see Methods for details). Subsequent measurement of the thermally desorbed species (reaction products and/or unreacted molecules), made at increased temperature, yields an effective hierarchical scan of the spectrum of activation energies operative in the catalysed reaction under study; for technical detail see Methods. It should be noted that the MgO(100) thin film used in this study has been determined to be defect poor[19–21] by a combined metastable impact electron spectroscopy (MIES) and temperature programmed desorption (TPD) study (see Methods and Supplementary Figs 2–4 for the spectroscopic and chemical characterization of the thin film). Analysis of the results of the above-mentioned characterizations lead us to conclude that the films prepared for the investigations described herein are essentially free of surface defects such as F-centres or other under-coordinated electron trapping sites, and that effects because of roughness, disorder effects, island formation and grain boundaries/corners/steps, are of no consequence (or below the detection limit) for the interpretation of the measurements and analysis presented in this paper.

The information obtained from these measurements is directly amenable to analysis and comparison with *ab initio*, first-principles, theoretical calculations (see Methods for details) that provide microscopic insights about the reaction mechanisms. To go beyond the above one-catalytic-cycle experiments, the reactivity of the size-selected Pt clusters is subsequently studied under isothermal, quasi-steady-state, conditions with the use of a pulsed molecular beam of ethylene. Under such circumstances, all the reaction channels that open at temperatures up to the selected temperature of the experiment are operative, including parallel reaction paths that are potentially detrimental to the full hydrogenation of ethylene to ethane, namely the aforementioned coking reactions such as the one resulting in the formation of ethylidyne.

One-catalytic-cycle experiments. Representative cluster-TPR spectra depicted in Fig. 1a and compared with one measured for a Pt(111) surface show negligible reactivity of the Pt_9 cluster and ethane production for Pt_{10} and Pt_{13} peaking at ~ 150 K, that is ~ 100 K below the temperature measured for the extended Pt(111) surface. This indicates similar activation energies for clusters in this size range that are considerably lower than the one found for the reaction catalysed on the Pt(111) surface. The integrated areas under the measured TPR curves, normalized to the number of atoms per cluster (Fig. 1b) illustrate that the supported platinum clusters catalyse the formation of more ethane per Pt atom than the corresponding amount generated from the Pt(111) surface, with a maximum found for Pt_{13}. Most importantly, the temperature-dependent variation of the ethylene hydrogenation reactivity of the Pt_n ($n = 9$, 10 and 13) clusters as a function of their size and, in particular, the appearance of a threshold size ($n = 10$) for onset of the catalytic activity, suggest that ethylene hydrogenation catalysed by Pt nanoclusters in the 1 nm size range falls under the classification of a size (structure)-sensitive reaction, in contrast to the commonly accepted structure insensitivity when catalysed by larger particles and extended surfaces. As demonstrated below, this conjectured new classification in the sub-nanometre size range is indeed supported by measurement of TOFs manifesting size (structure) sensitivity under steady-state conditions at room temperature (300 K).

Structures and reaction mechanisms from *ab-initio* calculations. In light of the above-observed size dependencies, it is evident that uncovering the atomic-arrangement motifs of the sub-nanometre-size supported Pt cluster catalysts is imperative to advancing our understanding of the physical principles underlying the above-noted apparent structural sensitivity of the catalysed reactions. The development and implementation of efficient and practical optimization strategies (global optimization in particular[22]) is a vexing problem of great importance in diverse fields, including the physical sciences, where we mention continuing efforts in the areas of protein folding, and the atomic arrangements in biomolecules, clusters and crystals[23]. A main difficulty in this area originates from the fact that the global extremum of a real multivariate function (for example, the total potential energy of an atomic cluster as a function of the locations of the atoms) is actually a local property, thus requiring an exhaustive search.

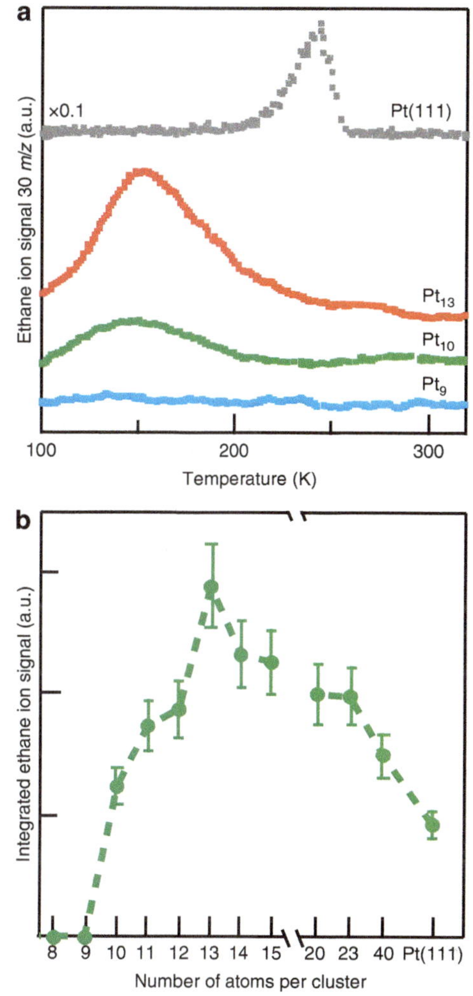

Figure 1 | TPR of ethylene hydrogenation on Pt_n clusters and Pt(111). (**a**) The measured ethane ion signals ($m/z = 30$) for Pt_9, Pt_{10}, Pt_{13} and Pt(111) as a function of substrate temperature are shown. (**b**) The integrated signal from the curves in **a** as well as for cluster sizes up to Pt_{40} normalized to the number of Pt atoms, are displayed. The TPR experiments were performed by first dosing 0.4 H_2 molecules per MgO surface atom ($1\,cm^2 = 2.25 \times 10^{15}$ surface atoms) followed by 0.4 C_2H_4 ethylene molecules per MgO surface atom at 100 K followed by applying a temperature ramp of $2\,K\,s^{-1}$. The error bars represent a 16% error, which was determined from multiple measurements on a single cluster size.

Consequently, a large number of diverse methods for global structure optimizations have been proposed (see, for example, refs 22,24 and citations therein). Although significant progress has been achieved, proving that the global extremum has indeed been found seems to be a rather unattainable task for most systems of interest[22].

The optimal (lowest-energy) structures of the surface-supported Pt_9, Pt_{10} and Pt_{13} clusters determined through first-principles calculations (based on extensive searches, guided by cluster structural motifs gathered for other systems and by previously proposed search methodologies[24]) are displayed in Fig. 2 as well as in Supplementary Figs 7–15. It is evident that these clusters prefer a three-dimensional (3D) prismatic two-layer structure. In addition, we observe that these adsorbed clusters possess a large number of higher-energy isomeric structures (with the preferred ones having 3D geometries), reflecting the complex nature of their potential energy surface; see in particular the extensive rich spectra of structural isomers in Supplementary

Figs 7–13. Although we focus in the following on the lowest-energy clusters, it is possible that some higher-energy isomers are also present on the supporting surface.

It is pertinent to remark here that the optimal (minimum total energy) configurations, as well as those of the higher-energy isomers, were determined through unconstrained relaxation of the atomic positions in conjunction with density functional calculations of the total energy (see Methods for details). The bare clusters were positioned on the MgO(100) surface, and their configurations, together with the positions of the atoms of the underlying Mg(100) substrate, were optimized via unconstrained relaxations. Once the optimal cluster structures have been determined, the binding (adsorption) energies and atomic positions of the reactants (that is, the individual H_2 and C_2H_4 molecules, as well as multiply adsorbed, and co-adsorbed, systems) were determined through total energy minimization; similarly, in determination of the reaction path, the degrees of freedom of all atoms of the system (substrate, cluster and reactants) have been relaxed except for the very few describing the chosen reaction coordinate (for example, the distance, or combination of distances, between atoms of the reactants), which is varied in a stepwise manner (recording the total energy change along the change in the reaction coordinate yields the reaction energy profile and the activation barrier for the reaction (see Methods for details)).

The relaxation procedure described above allows for adsorbate/reactant-induced structural relaxations of the underlying metal cluster and metal-oxide (MgO(100)) substrate; this form of adsorbate-induced cluster-catalyst relaxation (Supplementary Fig. 16 where inter-isomer structural crossover is shown) is a manifestation of a more general class of cluster relaxation processes termed fluxionality[14,18], which includes also cluster relaxation in the course of a chemical reaction, named dynamical reaction-induced cluster fluxionality[18]. By considering systematically only relaxed configurations, the size-dependent fluxional propensity of clusters in the size regime studied here becomes one of the characteristics that determine the response of the clusters to the presence of reactants and intermediates (for example, ethylidyne) and the consequent size-dependent reactivities and the course of reactions catalysed by these clusters. Finally, in addition to being of physical significance in the description of the mechanisms of cluster-catalysed reactions, the inter-isomer structural conversion shown in Supplementary Fig. 16 illustrates the use of the dynamical reaction-induced cluster fluxionality as a method for uncovering structural isomers; in fact, it has been used by us to discover the ground-state structure of the metal cluster in Pt_{13}/MgO (see Fig. 2e,f and Supplementary Figs 11 and 16).

The spin state of the bare-adsorbed Pt clusters is $N_{\uparrow}$-$N_{\downarrow} = 2$ (where N_s is the number of electrons with spin s = ↑ or ↓), and they are found, through total charge differences (calculated before and after adsorption and shown in Fig. 2a,c,e and Bader charge analysis (BCA, see Fig. 2b,d,f and Supplementary Figs 14 and 15)), to be negatively charged with the total excess Bader electronic charges on the adsorbed cluster being: $\delta\rho_B[Pt_n/MgO] = 1.68$ e, 1.78 e and 1.92 e for $n = 9$, 10 and 13, respectively. We find that whereas the total excess electronic charges on the $n = 9$ and 10 clusters are quite similar, close to a

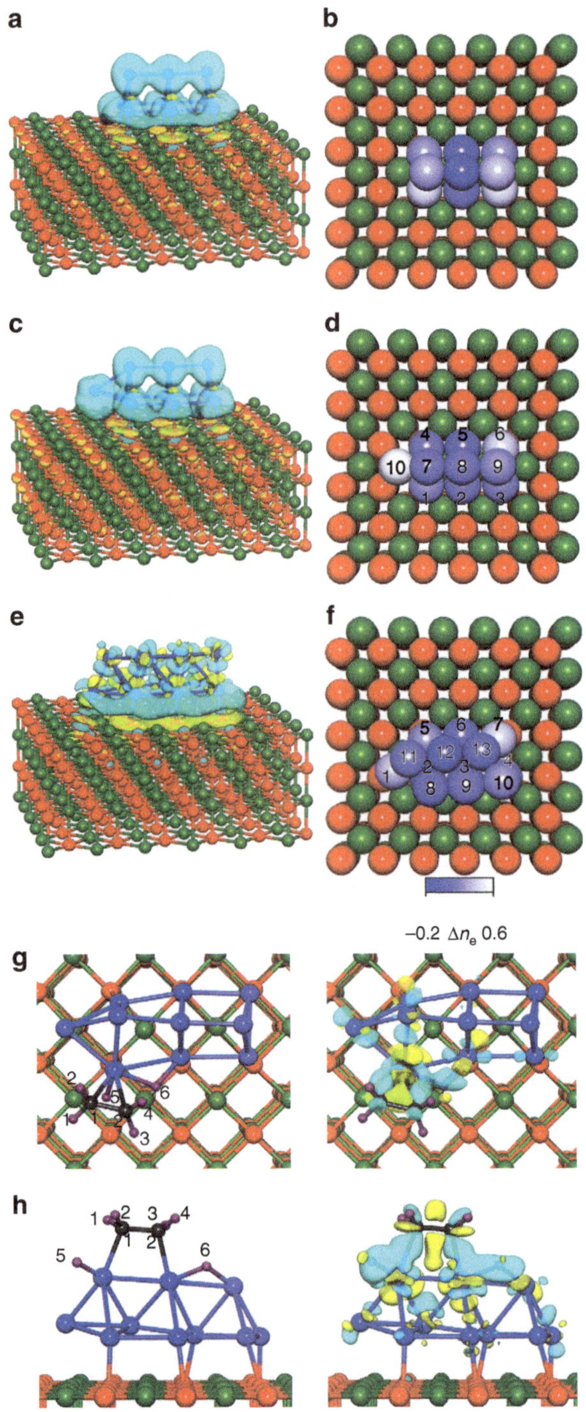

Figure 2 | Optimal configurations and charge distribution of bare Pt_n/MgO and co-adsorbed $C_2H_4 + H_2$. **a**, **c** and **e**, corresponding to Pt_9, Pt_{10} and Pt_{13} clusters, the blue and yellow contour hyper-surfaces correspond to excess (light blue) and deficient (yellow) charge distributions obtained as the difference between the total charges before and after adsorption of the clusters; these hypersurfaces are drawn such that the excess electronic (negative) charge inside the light blue hypersurface is 50% of the total electronic charge and the same for the positive charge inside the yellow hypersurfaces. Bader charge analysis is given in **b**,**d** and **f**, with lighter colour corresponding to excess number of electrons (that is, excess negative charge on the corresponding atom); for the values of the Bader charges, see Supplementary Figs 14 and 15. Co-adsorption of C_2H_4 and H_2 on Pt_{10}/MgO is shown for the π (**g**) and di-σ (**h**) bonding modes. The adsorption geometries are shown on the left, and on the right, we depict the bonding frontier orbitals of the adsorption system (the light blue and yellow denote different signs of the wave function); the σ-type Pt-C bonds are clearly seen in **h** (note the directed wave function contours on the right). In both **g** and **h**, atoms 5 and 6 are the proximal dissociated co-adsorbed H atoms.

third (0.51e) of $\delta\rho_B[Pt_{10}/MgO]$ is located for the Pt_{10} cluster on the capping atom (the 10th atom in Fig. 2d); similarly, a large contribution to $\delta\rho_B[Pt_{13}/MgO]$ is also found on the capping atom

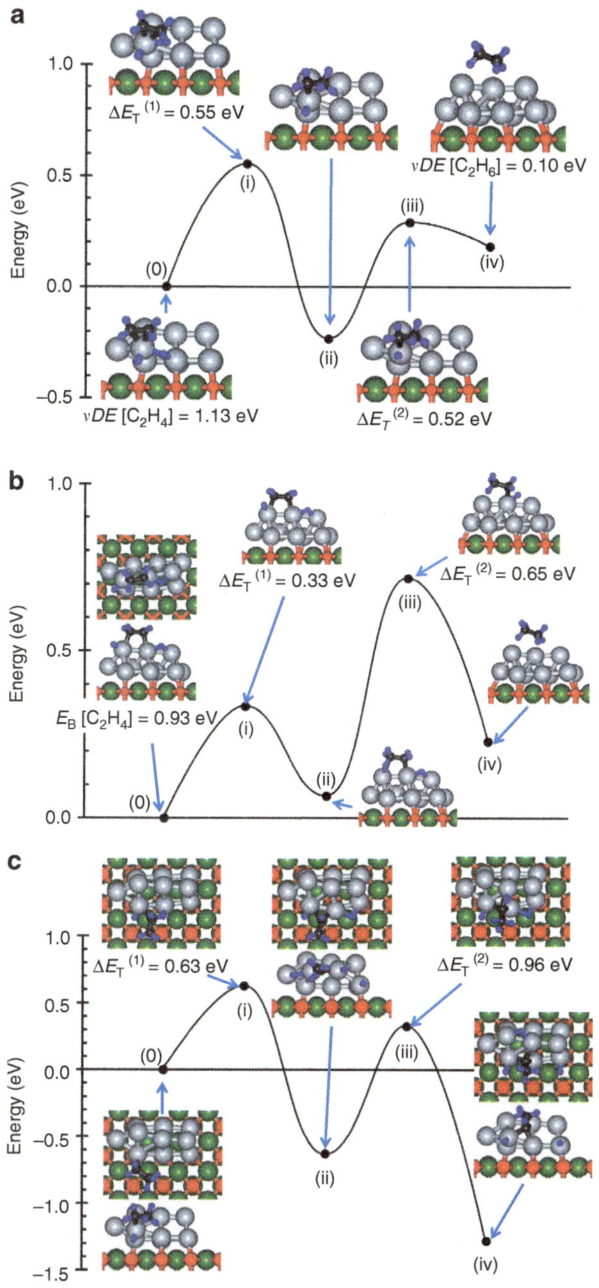

Figure 3 | Calculated first-principles steered reaction pathways catalysed by Pt_{10}/MgO. (a,b) Low-activation-barrier reaction pathways for the π **(a)** and di-σ **(b)** bonded ethylene (configuration on the left, marked 0), co-adsorbed with dissociated hydrogen. The activation energy barriers are denoted as $\Delta E_T^{(k)}$, $k=1,2$. The reaction proceeds through the two successive hydrogenation steps described in the Horiuti-Polanyi mechanism. **(c)** SRP for the low-temperature generation of ethylidyne ($\equiv CCH_3$) on Pt_{10}/MgO, starting from the ethyl ($-CH_2CH_3$) intermediate (configuration (0), on the left) generated in the first step of the reaction for the π-bonded ethylene (depicted as configuration (ii) in **a**). The two activation barriers correspond to dehydrogenation processes, resulting in a strongly adsorbed ethylidyne molecule ($\equiv CCH_3$, configuration (iv)). Surmounting the barrier for the second dehydrogenation process is assisted by the highly exothermic ($\sim$1.35 eV) formation of the $-CHCH_3$ intermediate (iii) resulting from the first dehydrogenation step.

(the first atom (marked 1) of Pt_{13}, shown in Fig. 2f, which is found to have an excess local charge of 0.34 e out of a total of 1.92 e, with the atoms marked 7 having also a large excess charge of 0.36 (Supplementary Fig. 15). For the Pt_{10} cluster, this excess charge distribution leaves a smaller amount of excess charge (1.78 e − 0.51 e = 1.27 e) to be distributed over the remaining nine atoms of the Pt_{10} cluster in comparison with the total excess charge (1.68 e) available for distribution over the compact (no capping atom) Pt_9 triangular prism cluster. This smaller amount of negative excess charge on most of the atoms of the Pt_{10} cluster compared with the Pt_9 case, are found to influence binding sites, adsorption energies and consequent reaction pathways on the adsorbed clusters as well as demonstrating that, in this size-range, even the addition of a single atom, while not changing the basic geometric structure, can drastically alter the electronic properties of a cluster.

Previous investigations (see review in ref. 7) identified two adsorption modes of ethylene on the (111) surface of platinum, the π and di-σ modes, where in the former the ethylene is thought to coordinate to a single Pt atom through a π bond, with the ethylene maintaining (at least to a large extent) sp^2 hybridization, and in the latter the adsorbed molecule is attached to two adjacent Pt atoms (in a η^2 manner) through two σ bonds showing a significant extent of sp^3 hybridization. On the extended Pt(111) surface, the π-bonded ethylene was found to be the preferred adsorption configuration only at very low temperatures, whereas for $T = 100$–240 K di-σ ethylene is preferred under ultra-high vacuum (UHV) conditions, with ethylidyne forming at higher temperatures. All the theoretical computational studies[25] corroborated the experimentally observed preference for di-σ over the π ethylene absorption mode, whereas the weakly adsorbed π-bonded species is experimentally believed to be the main reactant during hydrogenation of ethylene[9].

Contrary to extended surfaces, for the magnesia-supported Pt_n ($n \geq 9$) clusters, we find essentially the same propensity for adsorption in the π or di-σ mode, with the corresponding vertical desorption energies (VDEs) of the C_2H_4 molecules (co-adsorbed with a dissociated H_2 molecule in neighbouring sites) calculated to be VDE = 1.13 eV and 0.93 eV for the π and di-σ modes, respectively (see Figs 2g,h and 3a,b for details); note, in particular, the much higher value of the desorption energy for the π-bonded molecule compared with the corresponding calculated (0.87 eV (ref. 25) and measured (0.41 ± 0.10 eV (ref. 26) values on the Pt(111) surface. Furthermore, the π-bonded molecule (see Figs 2g and 3a for adsorption on Pt_{10}/MgO)) has a C–C bond length d(C–C) = 1.425 Å, which is close to the experimental value of 1.41 Å (ref. 27) this value is somewhat larger than d(C–C) = 1.334 Å value found in the isolated C_2H_4 molecule, indicating a certain degree of sp^3 rehybridization. In the di-σ adsorbed ethylene molecule, we find evidence for a larger degree of sp^3 hybridization, with d(C–C) = 1.469 Å, which is closer to the value found in the isolated C_2H_6 molecule (d(C–C) = 1.528 Å). Similar results were found by us for the two adsorption modes of ethylene on the supported Pt_{13} cluster (see the π-bonded adsorption configuration, marked 0, with VDE = 1.54 eV in Fig. 4a). The charge redistributions for both ethylene adsorption modes reflect the frontier orbital interactions leading to ethylene chemisorption, described by the Dewar–Chatt–Duncanson model[28,29] (see also Supplementary Note 2), involving electron donation from the highest occupied (π) ethylene orbital into an empty d orbital of the metal, and back-donation from the filled Pt d-orbitals into the ethylene lowest unoccupied (antibonding) π^* orbital. The resulting electron charge isosurfaces (see right panels in Fig. 2g,h) portray the resulting redistribution of the electron density. Both the above bonding processes as well as the repulsive interactions

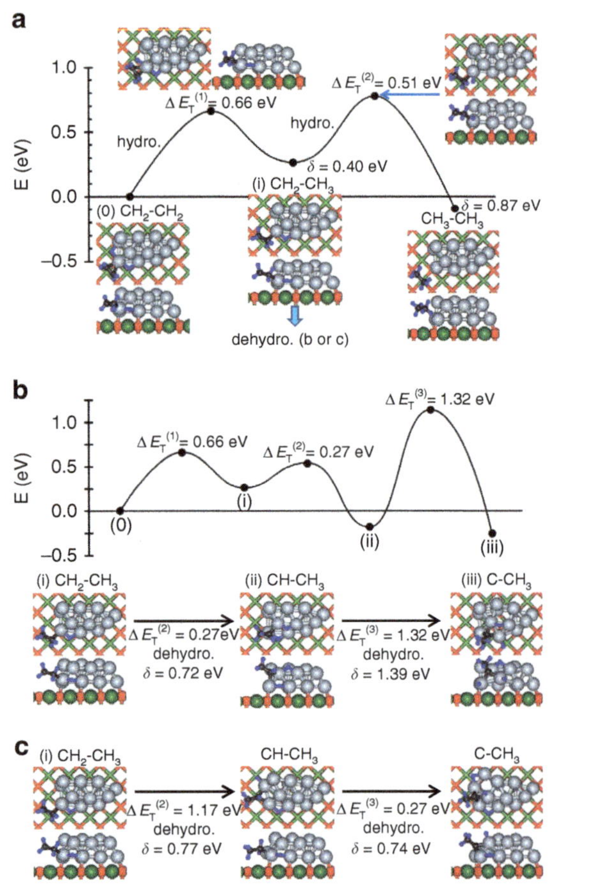

Figure 4 | Calculated first-principles steered reaction pathways catalysed by Pt_{13}/MgO. (**a**) Low-activation-barrier reaction pathway for hydrogenation of π-bonded ethylene configuration co-adsorbed with dissociated hydrogen on Pt_{13}/MgO. The reaction proceeds through the two successive hydrogenation steps described in the HP mechanism. The blue downward arrow near configuration (i) indicated the adsorbed ethyl intermediate (i) is the starting configuration for the **b** and **c** pathways. (**b**,**c**) Alternative reaction channels yielding ethylidyne (≡C-CH_3), starting from the π-bonded ethylene configuration co-adsorbed with dissociated hydrogen (marked (0) in **a**). The ethylene hydrogenation reaction producing ethane (see pathways in **a**) entails step (0)→(i), with $\Delta E_T{}^{(1)} = 0.66$ eV, followed by the activated second hydrogenation ($\Delta E_T{}^{(2)} = 0.51$ eV) that yields a weakly adsorbed C_2H_6 molecule. Two alternative dehydrogenation reaction channels that lead to formation of ethylidyne (starting from the ethyl, CH_2CH_3, intermediate, marked (i) in **a**, see blue downward arrow in **a**) are shown in **b** and **c**. The **b** and **c** pathways entail higher energy barriers (1.32 eV and 1.17 eV, in **b** and **c**, respectively) and thus their activation requires heating above room temperature, resulting in marked poisoning of the ethane-producing hydrogenation channel (see Fig. 5). The graph (profile) of the reaction path is shown only for channel **b**. The energy difference between the top of an activation barrier and the subsequent local minimum on the reaction pathway is given (in eV) by δ; this energy gives the height of the reverse reaction step, or it can assist (in part) passage over the subsequent energy barrier.

between the filled frontier orbitals of ethylene and occupied d-states of the metal catalyst particle (termed as Pauli[30,31] or four-electron repulsion[32]) depend on: (i) coordination of atoms making up the Pt particle, with reduced coordination, such as at edge and corner atoms, resulting in decreased repulsion, and (ii) occupation of metal d-states (with the Pauli repulsive interaction between filled orbitals of the metal cluster and the adsorbate increasing at cluster sites having larger excess electron charge); site-coordination and charging patterns can be ascertained from the calculated information given in Fig. 2a–f, and similar considerations apply also to the catalysed ethylene hydrogenation mechanism discussed below. In this context, it is pertinent to recall here our systematic study[33] of the effect of charging by cluster–substrate interactions on the adsorption and low-temperature combustion reaction of carbon monoxide and dioxygen, catalysed by MgO-supported gold clusters of sizes similar to those considered in the current investigation.

To gain insights into the microscopic ethylene hydrogenation mechanism, we carried out extensive first-principles steered reaction-pathway (SRP) simulations. For Pt_9, no low-activation-energy hydrogenation channel was found in the simulations (see Supplementary Figs 17–20 for details), in agreement with the TPR results (Fig. 1). On the other hand, multiple reaction pathways, characterized by relatively low activation barriers, were found in SRP simulations for the larger clusters, as illustrated in Figs 3 and 4 for Pt_{10} and Pt_{13}, respectively.

The simulated pathways starting from a π-bonded ethylene molecule co-adsorbed with a dissociated H_2 on the Pt_{10} and Pt_{13} clusters are shown in Figs 3a and 4a, respectively. For both cluster sizes, we observe two successive energy barriers (denoted as $\Delta E_T(i)$, $i = 1,2$) corresponding to the two hydrogenation steps of the HP mechanism, whose heights are sufficiently low to permit the experimentally observed low temperature reactivity, see Fig. 1 for details. In the first reaction step (marked (0)-(i) in Fig. 3a for Pt_{10}/MgO), one of the adsorbed hydrogen atoms approaches the nearest carbon atom (initially $d(C^{(1)}\text{–}H^{(1)}) = 2.432$ Å). The first activation barrier $\Delta E_T^{(1)} = 0.55$ eV, is characterized by $d(C^{(1)}\text{–}H^{(1)}) = 1.25$ Å and $d(C^{(1)}\text{–}C^{(2)};\ \Delta E_T^{(1)}) = 1.487$ Å (compared with $d(C\text{–}C) = 1.487$ Å calculated for gaseous ethyl C_2H_5) portraying formation of an adsorbed ethyl-like top-of-the-barrier intermediate. In the local minimum (marked (ii)) that follows the first barrier $d(C^{(1)}\text{–}H^{(1)}) = 1.105$ Å, $d(C^{(1)}\text{–}C^{(2)};\ \text{(ii)}) = 1.529$ Å, and the adsorbed ethyl intermediate is inclined with respect to the cluster ($d(C^{(1)}\text{–}Pt^{(1)}) = 3.065$ Å and $d(C^{(2)}\text{–}Pt^{(1)}) = 2.080$ Å). The closer value at the barrier-top of $d(C^{(1)}\text{–}C^{(2)};\ \Delta E_T^{(1)})$ to $d(C^{(1)}\text{–}C^{(2)};\ \text{(ii)})$ then to the initial state of the adsorbed ethylene, $d(C^{(1)}\text{–}C^{(2)};(0)) = 1.425$ Å, qualifies $\Delta E_T^{(1)}$ as a 'late transition state barrier'. Full hydrogenation occurs in the second activation process—(ii) C_2H_5→(iii) $\Delta E_T^{(2)} = 0.52$ eV→(iv) adsorbed C_2H_6 with VDE = 0.1 eV—resulting in the product ethane molecule. The pathways of the hydrogenation reaction of the di-σ-adsorbed ethylene molecule (shown for Pt_{10}/MgO in Fig. 3b) resemble the one described above, with the first activation barrier being lower (0.33 eV); for geometric details of the reaction pathways, see Supplementary Figs 21–27. Similar results were found in our SRP simulations for the Pt_{13}/MgO cluster (see Fig. 4a for the π-bonded ethylene molecule); for details, see Supplementary Figs 28–36).

The principal result of our analysis to this point is the discovery and explanation of the onset of low-temperature ethylene hydrogenation processes (peaking at temperatures of ~150 K) on magnesia-supported 3D Pt_n clusters for $n \geq 10$, with the calculated activation barriers for *both* the π and di-σ C_2H_4 adsorption modes on the Pt cluster catalysts being significantly smaller than those obtained for the Pt(111) surface; compared with values calculated for the lowest coverage (1/9) on Pt(111)[25]. This reflects significant enhancement of the hydrogenation reaction, achieved through the use of supported platinum clusters of <1 nm diameter (on the order of 10–15 Pt atoms); note, however, the absence of catalytic activity of 3D Pt_n clusters with $n \leq 9$. The measured larger hydrogenation reactivity of the Pt_{13} cluster correlates with the calculated larger number of active sites on this cluster for both H_2 dissociative adsorption (see Supplementary Fig. 1 and Methods for details), and low-barrier hydrogenation of co-adsorbed ethylene and hydrogen.

In addition to the low-activation-energy pathways found on the Pt_n/MgO ($n \geq 10$), we have found on these clusters (particularly on the larger cluster, Pt_{13}/MgO) several reaction sites characterized by higher ethylene-hydrogenation activation energy barriers; typically, $\Delta E_T > 0.8–1.0$ eV (see Supplementary Figs 21–36 for details). Most significantly, along with the low-temperature hydrogenation pathways (Fig. 3a,b) we found for the Pt_{10}/MgO catalyst a relativelyc low-barrier dehydrogenation reaction channel resulting in the formation of ethylidyne (Fig. 3c), whereas for the larger Pt_{13}/MgO cluster we found multiple high-energy ethylidyne formation pathways; see Fig. 4c,d manifesting energy-barriers of $\Delta E_T \sim 1.2–1.3$ eV. This finding implies for Pt_{13} an onset of significant ethylidyne formation upon heating to higher temperatures ($T \sim 350–400$ K), which acts as a blocking, or poisonous, agent for ethylene hydrogenation; for the Pt_{10} clusters such alternative reaction channels are open already at 300 K, see, for example, Fig. 3c.

Isothermal multi-catalytic-cycle experiments. The pulsed molecular beam technique was applied to determine TOFs under quasi-steady-state isothermal conditions (see Methods for experimental details). This method allows probing and verification, under actual catalytic conditions, of the trends deduced from the TPR experiments, as well as confirmation of the predictions obtained from first-principles calculations regarding ethylidyne formation (see above). At these elevated temperatures, processes such as cluster sintering become more probable, but our data did not show any indication that this was occurring during the time scale of our experiments and studies performed on size-selected clusters also cast strong doubt that sintering occurs already at 400 K, see Methods for further discussion.

The measured ethane TOFs at 300 K displayed in Fig. 5a (blue) show that at this temperature Pt_n clusters with $n < 10$ are active, as expected from the calculated activation energies on Pt_9. Furthermore, at this temperature, the distinct onset of reactivity for n>9 observed in the TPR, one-catalytic-cycle, experiments (see Fig. 1 for details) is not operative, and reactivity variations are observed only in a narrow size window for Pt_{11}–Pt_{14}, with a pronounced maximum activity for Pt_{13}. This demonstrates the emergence of a size window where the reaction is *size* (structure) sensitive, in contrast with the size (structure) insensitivity observed for other cluster sizes, nanoparticles and single-crystal surfaces[9,34–36]. This serves to illustrate that clusters (of the same chemical identity but different sizes) can exhibit both structure sensitivity and insensitivity and that a single atom can change this designation.

To investigate the influence on the hydrogenation of ethylene brought about by surface species generated by alternative reactions, for example, the aforementioned ethylidyne production channel, we have repeated the measurement of the TOF at 300 K, but this time after the clusters were exposed to 10 pulses of ethylene at 400 K in the presence of deuterium (Fig. 5a, red curve). The absolute change of the TOF for the sizes measured in Fig. 5a, depicted in Fig. 5b, clearly shows that Pt_{13}, as well as the other clusters within the enhanced reactivity size window shown in Fig. 5a, deactivate as a result of the 400 K heating step, whereas the other sizes show no significant change in activity. After the 400 K step, all cluster sizes show TOFs similar to the one measured on Pt(111), that is, the reaction has turned size insensitive. This observation is consistent with the results of the calculations (Figs 3c and 4c,d), where the activation barrier for the dehydrogenation pathway to the (hydrogenation blocking) ethylidyne species was found to be thermally accessible already at temperatures as low as $T < 300$ K for Pt_n ($n \leq 10$), whereas for Pt_{13} ethylidyne production was predicted to require heating to $T > 350$ K. Consequently, the size-insensitive clusters, as well as Pt(111), are all passivated already at 300 K, and only with a temperature increase do the more active sizes follow. To optimize reactivity for the hydrogenation of ethylene, it is thus not only essential to choose the right size window between Pt_{11} and Pt_{14}, but also the right temperature window $160\,K \leq T \leq 350$ K, where the hydrogenation of ethylene is thermally accessible but not the formation of poisoning species. This size-dependent behaviour of two competing reaction channels offers the potential for controlling and tuning more applied hydrogenation and dehydrogenation reactions on an atom-by-atom basis.

Further evidence for the formation of inhibiting carbon species has been obtained through the use of CO adsorption and infrared reflection absorption spectroscopy to probe the cluster catalysts before and after the reaction, as co-adsorption of carbonaceous species on platinum is known to induce a redshift in the CO stretch frequency because of electron donation to the metal[37–40]. The attribution of the redshifts reported here to carbon was strengthened by the fact that after the TPR experiment in Fig. 1a, no deuterium adsorption was observed on Pt_9, Pt_{10} and Pt_{13} indicating a blocking effect of an adsorbate on the clusters. The only plausible explanation would be a dehydrogenated product from ethylene. Although a change in cluster shape could also possibly induce such a small redshift, the inability of this hypothesis to reconcile the TPD and infrared data led to the assignment of co-adsorbed carbonaceous species as the origin of the redshift.

After running the reaction under conditions of highest activity (300 K), ten Langmuir of CO were adsorbed at 100 K and an infrared spectrum recorded (Fig. 5c, blue curves). The CO stretch is red-shifted by 13 cm^{-1} compared with spectra measured from pure metallic sites on all cluster sizes (see the shifted spectra in Fig. 5c and Fig. 5d for both the shifted and unshifted values, respectively). Importantly, the cluster of highest reactivity, Pt_{13}, also exhibits a shoulder (marked with an arrow in Fig. 5c) where the CO stretch on a clean cluster is found. In comparison, the Pt(111) crystal shows a much larger redshift (83 cm^{-1}, compare the frequencies given by the triangle and filled blue dot on the right in Fig. 5d), which is attributed to adsorbed ethylidyne[26,41,42]. Performing the same experiment after the 400 K heating steps brings about an even larger redshift of the CO stretch (34 cm^{-1}, compare the red and blue curves in the spectra displayed in Fig. 5c, and summarized in Fig. 5d), indicating a larger influence of inhibiting dehydrogenation products as well as complete site blocking on Pt_9. On Pt(111), the absorption peak decreases slightly in intensity but remains at the same position, which correlates to the expected stability of ethylidyne up to 400 K (ref. 43).

The finding that a CO stretch peak position comparable to that measured for CO on a clean (bare) cluster is found after reaction at 300 K only for the most active size, Pt_{13}, reflects the intrinsic resistance of ethylene on Pt_{13} to follow the dehydrogenation pathways leading to ethylidyne formation, as predicted by our theoretical results. The formation of more carbonaceous species on Pt_{13} (portrayed by a larger redshift after pulsing at 400 K) induces the subsequent structure insensitivity, previously established for other cluster sizes, nanoparticles and single crystals.

This combined pulsed molecular beam and infrared reflection absorption spectroscopy study demonstrates for the first time that the formation of carbon species is indeed the key factor underlying structure insensitivity of ethylene hydrogenation on larger particles and extended surfaces of platinum.

Discussion

The investigations discussed in this paper demonstrate that classification of a reaction as structure sensitive, or insensitive,

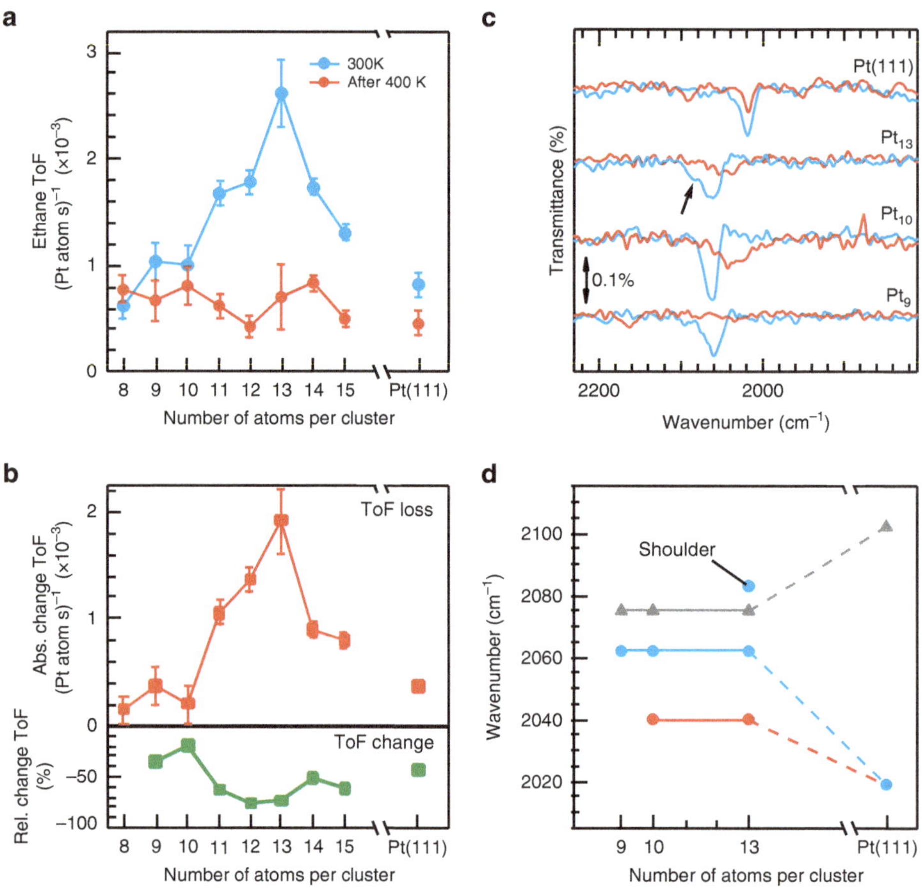

Figure 5 | Pulsed molecular beam of $C_2H_4 + H_2$ reacting on Pt_n/MgO along with infrared spectra. (**a**) Measured TOF at 300 K of the ethylene hydrogenation on Pt_{8-15} clusters and Pt(111) single-crystal surface, before (blue) and after (red) running the identical reaction at 400 K. The measured TOF has been normalized to the number of platinum atoms for Pt_{8-15} and the Pt(111) surface. A background deuterium pressure of 2×10^{-6} mbar was used for all experiments. (**b**) Decrease of the turnover frequency between the blue and red measurements as well as this value represented as a percent loss. (**c**) Infrared reflection absorption spectra of 10 L CO after the experiments in **a** measured at 100 K. Blue spectra were taken after pulsing at 300 K and red after the clusters had been exposed to the same reaction conditions at 400 K and again at 300 K. (**d**) Position of the CO stretch on clean cluster samples (grey data points) as well as after the blue and red data points were acquired. The error bars represent the standard deviation of multiple activity measurements on the same cluster size.

must be reassessed for materials with sizes in the non-scalable sub-nanometre size regime. This size range has been shown to exhibit surface chemical properties, which can be modulated by a single atom, and where each particle size displays its own mode of behaviour. The previously reported structure insensitivity of ethylene hydrogenation on Pt is shown here to be untenable in the cluster size range of Pt_7–Pt_{40}, where we find structure sensitivity of ethylene hydrogenation to emerge, with maximum reactivity for Pt_{13}. Structure-insensitive behaviour was observed to be inherent for specific cluster sizes at ambient temperatures and can be induced in the more active sizes (for example, Pt_{13}) by an increase in reaction temperature, which opens dehydrogenation reaction channels leading to the formation of carbonaceous species evidenced in the infrared spectra and predicted by first-principles calculations. Aside from addressing the applicability, at this size range, of common catalyst classification as structure sensitive or insensitive, our findings point to the possibility of controlling the activity and selectivity of hydrogenation and dehydrogenation reactions catalysed by clusters of these sizes. We remark that rather than treating an industrial hydrogenation catalyst under industrial conditions, we focus here, with the use of model catalysts of nanoscale dimensions, on a basic science question of great interest in the field of heterogeneous catalysis pertaining to the concept of structure sensitivity/insensitivity at the bottom of the catalyst size scale. For the catalyst cluster sizes considered here, the reaction is investigated in a temperature range lying mostly below, and up to, room temperature (with certain thermal treatment involving temporal heating to 400 K), which is lower than the temperatures used in typical industrial catalytic hydrogenation applications; the finding that nanocluster catalysts (particularly made of noble metals) catalyse reactions at temperatures significantly lower than larger clusters and extended surfaces made of these metals is rather common now (see, for example, refs 13,14,19,20,44,45). Correspondingly, the temperature-window selectivity and thermal-tuning that we discuss and demonstrate are operative, and are kinetically relevant, for these nanocatalysts under the above-specified temperature conditions. This also suggests that in conjunction with future employment of nanocatalysts in industrial-type processes, the concept of temperature selectivity/tuning windows, discussed in this work from a basic science perspective, may emerge also as one of practical relevance.

In the non-scalable size regime, the above findings make element-specific generalizations of catalytic properties, based on single crystals and nanoparticles, rather questionable. The first-principles simulations show that even similar atomic geometric arrangements can lead to widely differing physical and chemical properties. Consequently, each reaction must be individually

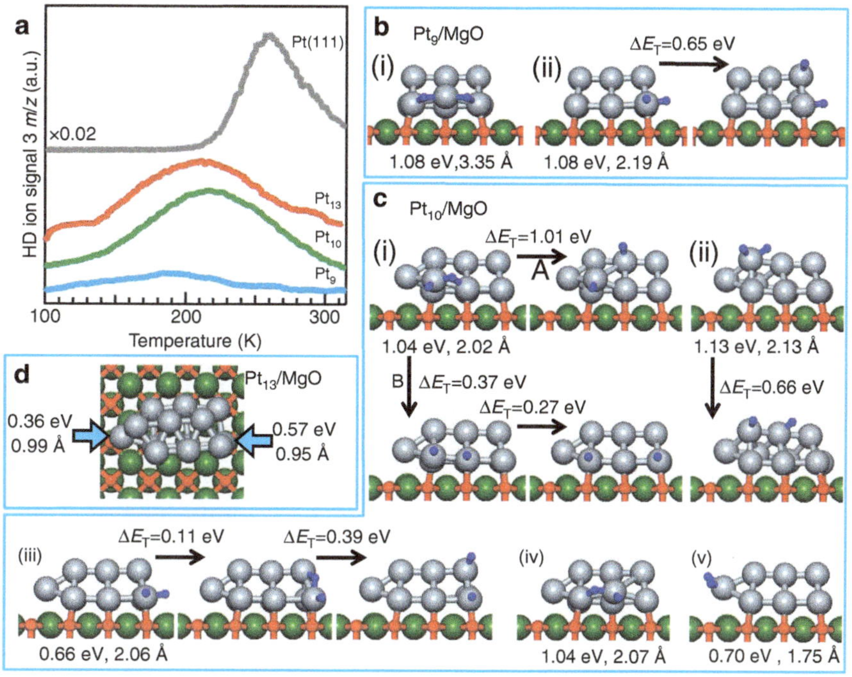

Figure 6 | Hydrogen adsorption on supported platinum clusters. (**a**) TPR measurements of the H/D exchange on Pt_n/MgO, $n = 9$, 10, 13 and on Pt(111). (**b**,**c**) Calculated hydrogen dissociative adsorption at various sites, and diffusion on Pt_9/MgO and Pt_{10}/MgO, respectively. The pair of values under an adsorption configuration give the H_2 adsorption energy and distance between the dissociated H atoms. ΔE_T denotes the activation barrier for diffusion between the corresponding H-atom configurations on the Pt cluster. (**d**) On Pt_{13}/MgO, H_2 dissociates at all sites except the ones marked by an arrow. Spheres colored green denote magnesium atoms, oxygen atoms are colored red, platinum atoms are coloured red, Pt atoms in grey, and hydrogen atoms are depicted by small blue spheres. The cluster configurations correspond to the lowest energy cluster-adsorption configuration on MgO(100).

tested in an atom-by-atom, size-selective, manner, in order to disentangle and reveal the catalytic properties of such systems. The use of structure (in)sensitivity has been a most useful concept for the description and systematization of catalytic processes on single crystals and particles possessing well-defined single-crystal facets. However, with decreasing particle sizes (that is, for higher dispersion), and in particular in the sub-nanometre size regime, extrapolations from systems of larger size are likely to become inadequate, requiring instead an atom-by-atom reassessment.

In closing, the results that we reported here demonstrate that hydrogenation and dehydrogenation reactions can be controlled and tuned by controlling the precise atomic size of the catalyst particle. This has implications for a broad range of chemical processes[46] ranging from chemical synthesis, to food chemistry, and energy applications, where the insight garnered here, as well as future investigations along these lines, could facilitate a deeper understanding of applied reaction systems, and may offer design strategies for catalysts with enhanced activity and/or selectivity.

Methods

Sample preparation and characterization. All experiments were perform in an ultra-high vacuum (UHV) chamber with a base pressure of 1×10^{-10} mbar[47]. The Mo(100) (MaTeck, Germany, 0.785 cm^2) single crystal was cleaned by heating to 2,000 K and subsequent oxidation at 900 K in a 5×10^{-7} mbar O_2 background (5.5 purity, Air Liquide Germany). The purity of the crystal was checked with Auger electron spectroscopy (AES) and ultraviolet photoelectron spectroscopy (UPS)[48]. The MgO(100) film was then grown at 600 K on the Mo(100) single crystal by evaporating a magnesium (≥99.95%, Merck Germany) ribbon in front of the crystal in a 5×10^{-7} mbar background of O_2. After annealing the film at 800 K for $t > 10$ min, the film was characterized with AES and is estimated to have a thickness of ten atomic layers, the purity was further confirmed with AES, UPS and MIES.

The presence of surface defects has also been investigated and the films have been determined to be defect poor and exhibit a high degree of order. Oxygen vacancies (F-centres) have been shown to be detectable via MIES as a clear band beginning at a binding energy (BE) of 1 eV and peaking at 2 eV (refs 20,49). The MIES spectrum in Supplementary Fig. 2 shows this region from the MgO film prepared as described above. No characteristic emission peak below 4 eV is observed, clearly showing the absence of F-centre type defects or other under-coordinated electron-trapping sites.

Surface roughness can also play a factor in thin films and can also influence catalytic activity. Previous surface characterization of thin MgO films concluded that above 7 ML thin film thickness, global surface roughness decreases and (1 × 1) low energy electron diffraction (LEED) spots are visible[50]. As the films grown here are thicker than 7 ML and we have previously published this same LEED pattern from our film growth procedure[21], the films grown for this study possess low surface roughness.

This same property has also been characterized with the use of MIES, where a narrowing of the O2p band indicates surface ordering[51]. Their spectrum showed a full-width half-maximum O2p peak of 3.2 eV for a disordered film and 2.5 eV for an ordered film. The MIES spectrum in Supplementary Fig. 3 (complete spectrum of Supplementary Fig. 2) shows the full-width half-maximum of the O2p peak (BE = 5.5 eV) to be 2.13 eV for the films grown for this study. Again, this is further evidence that the MgO film is highly ordered.

Grain boundaries and islands can be directly detected using a CO TPD. The CO TPD from the thin film used in this study is displayed in Supplementary Fig. 4. The sample was dosed at 100K with 1 CO molecule per MgO surface atom, and a temperature ramp of 2 K/s was applied (see below for technical details regarding the TPD experiment). The spectrum shows only a single, narrow desorption peak at 105 K, which is concordant with a defect poor film. The absence of island formation and the absence of exposed Mo domains in our films was confirmed by the nonappearance of a characteristic desorption peak at about 290 K in our CO TPD measurement (see ref. 52).

In conclusion, the above data clearly show that our films are defect poor, in that they do not exhibit the established spectral and chemical properties expected from F-centres (or other under-coordinated electron trapping sites), rough/disordered films, island-like films and grain boundaries/corners/steps and so on.

The Pt(111) (MaTeck) single crystal was cleaned by cycles of argon ion sputtering at 900 K, followed by oxidation for 5 min at 650 K in a 5×10^{-7} O_2 background pressure and annealing at 1,300 K for 1 min. The purity was checked with AES, UPS and MIES.

The clusters were generated using a laser evaporation source; a detailed description can be found elsewhere[47]. In brief, the second harmonic of a Nd:YAG laser (InnoLas DPSS, 532 nm, 100 Hz) is focused onto a rotating platinum target (99.95% purity, Alfa-Aesar), with each pulse being thermalized and extracted into the vacuum with a delayed pulse of helium gas (6.0 purity, Westfalen). Electrostatic lenses guide the clusters to a bender where the positively charged species are guided into a quadrupole mass spectrometer (QMS; Extrel, 16,000 a.m.u.). The cluster beam is then size selected and a single cluster size is deposited onto the MgO(100)/Mo(100) substrate. Retarding field analyses ensure that the clusters have a kinetic energy of no more than 1 eV per atom for soft-landing conditions. The clusters are neutralized by electron tunnelling through the MgO film and the resulting current is recorded and integrated in order to calculate a cluster coverage. It is assumed that each cluster has unit charge. All TPR curves shown were performed on 9×10^{12} clusters. The majority of the pulsed valve experiments were also performed with this coverage and extrapolating data measured on lower coverages to 9×10^{12} did not produce any significant variations in the TOF.

Cluster stability. Cluster stability at the relevant temperatures was ensured using a variety of techniques. For platinum clusters of various sizes, we have done repeated, consecutive, HD exchange or deuterium TPD experiments up to ~320–350 K and observe no change in the desorption peak temperature or area. From this observation, it was concluded that the TPD results (Fig. 1 in the manuscript) indeed reflect true size effects. In addition, the pulse-to-pulse variations, seen in Supplementary Fig. 6B at 300 and 400 K for Pt_9 and Pt_{10} (Supplementary Fig. 5), also reflect the stability of the clusters (against sintering, coalescence or fragmentation) under the reaction conditions of the isothermal (multi-cycle) experiments presented in this work. Any appreciable degree of sintering would be coupled with a very clear change in activity in both cases, which was not observed for the cluster sizes determined as being structure insensitive in a TPR run or at isothermal conditions.

Size-selected clusters have also been extracted from the experimental setup for stability studies on a silica support using nanoplasmonic sensing and transmission electron microscopy[53]. It was shown that after 21 h at 453 K, no sintering of size-selected samples (Pt_{22} and Pt_{68}) is observed. In addition, Pt_{68} resists sintering at 533 K during the hydrogen oxidation reaction. STM results from size-selected Pd clusters show that on relatively strongly, supported binding boronitride film surfaces, annealing to 500 K for 5 min does not change the cluster size distribution (height histogram)[54]. The underlying principle of this stability is on the one hand the suppression of the Ostwald ripening through the relatively weak interaction of the transporting species (Pt-atoms) with the oxide support[54] and the monodispersity of the cluster samples with the lack of the driving force for Ostwald ripening[53]. In fact, in these experiments, Ostwald ripening is only observed when reaching the Hüttig temperature of the cluster material, which for platinum is ~650 K. The observed absence of Smoluchowski ripening can be understood when considering the relative large binding energies of the clusters with the support material. A reliable estimate of the energy that anchors the cluster to the underlying MgO(100) supporting surface is given by the calculated vertical BE (vBE), given as

$$E_{\mathrm{vBE}}^{(n)} = E[\mathrm{Pt}_n]_{\mathrm{Pt}n/\mathrm{MgO}} + E[\mathrm{MgO}]_{\mathrm{Pt}n/\mathrm{MgO}} - E[\mathrm{Pt}_n/\mathrm{MgO}], \quad (1)$$

where $E[\mathrm{Pt}_n/\mathrm{MgO}]$ is the total energy of the structurally relaxed (energy optimized) ground state (lowest energy isomer) of the Pt_n cluster adsorbed on the MgO surface, and $E[X] \times_{/\mathrm{MgO}}$ is the total energy of the isolated component ($X = Pt_n$ or MgO) of the combined adsorption system, each calculated in the geometry that it assumes in the relaxed (optimized) configuration. The calculated values for $E_{\mathrm{vBE}}^{(n)}$ for the three cluster sizes considered in this work, $n = 9$, 10 and 13, are: $E_{\mathrm{vBE}}{}^{(n)} = 6.96$, 7.65 and 9.09 eV, respectively. These values reflect strong bonding of the clusters to the MgO(100) surface, inhibiting Smoluchowski ripening processes.

TPR and H/D exchange experiments. The temperature programmed reaction (TPR) experiments were performed by dosing 0.4 H_2 (5.0 purity, Air Liquide) molecules per MgO surface atom ($1\ \mathrm{cm}^2 = 2.25 \times 10^{15}$ surface atoms) followed by 0.4 C_2H_4 (3.5 purity, Westfalen) ethylene molecules per MgO surface atom, using a calibrated molecular beam doser at a crystal temperature of 100 K. The crystal was then positioned ~5 mm away from a skimmer leading to a differentially pumped QMS (Balzers QMA 430, Liechtenstein) chamber. A temperature ramp was then applied (Eurotherm 2408) and the ion signal of 30 m/z (ethane parent peak) was recorded.

Before executing the TPR measurements displayed in Fig. 1, we performed measurements and theoretical calculations pertaining to a key step of the HP ethylene hydrogenation mechanism, namely the dissociative adsorption of H_2. To this end, we utilize isotopic H/D exchange (scrambling) TPR measurements. These HD-exchange experiments were performed in the same manner as described above for the other TPR measurements, except that a gas mixture of H_2 and D_2 (100.00%, Westfalen) was used where each partial pressure equaled 0.4 molecules per surface atom. A mass of 3 m/z was measured for these experiments. H/D exchange spectra have been recorded for a range of cluster sizes and the results for Pt_9, Pt_{10} and Pt_{13} are displayed in Fig. 6a showing a limited activity for the Pt_9 cluster compared with that of the larger clusters (peaking at $T \sim 200$ K), along with the measured record for the Pt(111) surface that shows onset of H/D exchange at a higher temperature and a peak centred at 260 K.

Hydrogen molecules are predicted from our DFT calculations to dissociatively adsorb on all Pt atoms of the bottom layer of Pt_9 (Fig. 6b), but do not bind to top layer atoms of the cluster; the only way that an H atom may adsorb on the top layer is by diffusing (with an activation energy of 0.65 eV) from a bottom atom after dissociation, see Fig. 6b(ii) for details). The limited adsorption and dissociation of hydrogen on the Pt_9 cluster correlate with the measured H/D exchange (Fig. 6a).

On the other hand, dissociative H_2 adsorption occurs readily at all sites of the bottom layer of the supported Pt_{10} and Pt_{13} clusters, with diffusion of the dissociated adsorbed hydrogen atoms between bottom layer sites entailing small barriers of the order of 0.1–0.3 eV, and interlayer diffusion requiring barriers that are as low as 0.40 eV. Unlike the case of the smallest cluster (Pt_9), H_2 dissociatively adsorbs on one of the second layer atoms of the Pt_{10} clusters (see Fig. 6c(ii), diffusing to a neighbouring upper layer atom with an activation energy of 0.66 eV), and the Pt_{13} cluster dissociatively adsorbs H_2 at all sites (except one, see right arrow, Fig. 6d and Supplementary Fig. 1). The abundant sites for dissociative adsorption on both the bottom and top layers of the Pt_{10} and Pt_{13} clusters and the facile inter-site H-diffusion on these clusters correlate with the measured low-temperature H/D exchange signal (Fig. 6a).

The pulsed molecular beam technique. The pulsed molecular beams technique used in our laboratory has been previously described[55], and a brief overview will be given here. A piezo pulser, which allows for highly reproducible gas pulses, was filled with a 2-Torr background pressure of ethylene. From the time-resolved QMS response to a single ethylene pulse, it was determined that there are $\sim 10^{14}$ gas molecules per pulse giving a local pressure of $\sim 5 \times 10^{-7}$ mbar. A deuterium background pressure of 2×10^{-6} mbar was established with a leak valve and the temperature of the crystal was increased to 300 K. The crystal was again placed ~5 mm in front of the skimmer (same QMS as for the TPD experiments) and ethylene was pulsed onto the surface at a rate of 0.1 Hz. The ethane production was measured by monitoring the mass signal at 31 m/z with an oscilloscope (LeCroy Wave-runner 44Xi-A). Within the signal wave, we define a quasi-steady-state regime, where the ethane production maintains a constant value over the course of 80 ms. Approximately 20 pulses are then recorded and the quasi-steady-state signal is averaged and used to calculate the TOF. Supplementary Fig. 6A shows a representative average of pulses for Pt_{11}. A region displaying steady-state ethane production (in grey) is then defined and the TOF calculated from this signal. Supplementary Fig. 6B shows the pulse to pulse signal for the same experiment displayed in Supplementary Fig. 6A. The calibration of the mass spectrometer was performed by determining the QMS signal of a saturated monolayer of CO on Pt(111) using the TPD technique and then determining the sensitivity factor between ethane and CO. It should be noted that although we measure ethane at $m/z = 31$, the factor between the signal from equal amounts of pure ethane ($m/z = 30$) and deuterated ethane, measured on the peak $m/z = 31$, is roughly[56] 2. As this error is systematic, no influence on the trend observed, or even the order of magnitude of the signal, occurs.

Infrared reflection absorption spectroscopy. (Thermo Electron Corp. Nicolet FT-6700) was performed in single reflection mode with an external MCT-detector (Thermo Electron Corp., MCTA-TRS). At the experimental steps, a background spectrum was acquired by averaging 256 scans at a resolution of $4\ \mathrm{cm}^{-1}$ with the sample at ~100 K. Ten Langmuir of CO were dosed and a sample spectra recorded with the same parameters as the background.

First-principles computational methodology. To model the Pt_n/MgO systems, with $n = 8$, 9, 10, 13, we employed a four-layer MgO(100) slab with a calculational cell consisting of 7×6 unit cells; each layer consisted of 42 Mg and 42 oxygen atoms with the atoms in the bottom layer held stationary (at the experimental lattice constant of 4.21 Å) and the atoms in the other three layers allowed to relax to the optimal atomic arrangement. The experiments in this paper were performed for MgO films with thickness of ten layers or more. Under these conditions, there is no effect of the underlying Mo(100) substrate on the behaviour at the top of the film. Indeed, the effect of the Mo substrate has been predicted early on theoretically[57] to be limited to MgO films with a thickness of four to five layers; see also predicted thickness-dependent reactivities, where the catalytic activity of deposited metal clusters was predicted to depend on the thickness of the underlying metal-oxide film adsorbed on a metal substrate[58,59]. This has been verified experimentally in later reactivity studies of gold nanoclusters supported on MgO films of variable thicknesses[20] and most convincingly through work function measurements[60]. The above-mentioned prediction and experimental verifications justify the procedure used in our calculations, where no effect of the underlying Mo substrate on the charging behaviour at the MgO/Pt nanocluster interface is considered.

The optimal (minimum total energy) configurations were determined (through unconstrained relaxation of the atomic positions, using a conjugate gradient search) when the calculated energy converged within 0.001 eV. The bare clusters were positioned on the MgO(100) surface, and their configuration, together with the positions of the atoms of the underlying substrate, was optimized via unconstrained relaxations as described above. The calculational supercell, which included the MgO(100) slab and a 24-Å-thick vacuum region, was periodically replicated. Once the optimal cluster structures have been determined the binding (adsorption) energies and atomic positions of the reactants (that is, the individual H_2 and C_2H_4 molecules, as well as multiply adsorbed, and co-adsorbed, systems) were determined through total energy minimization. This relaxation procedure allows for adsorbate/reactant-induced structural relaxations of the underlying metal cluster and metal-oxide (MgO(100) substrate; this form of adsorbate-induced cluster-catalyst relaxation (illustrated in Supplementary Fig. 16) is a manifestation of a more general class of cluster relaxation processes termed fluxionality[18], which includes also dynamical reaction-induced cluster fluxionality.

In the first-principles calculations of the reaction profiles (pathways), a reaction coordinate was judiciously chosen; typically, the reaction coordinate consists of the distance between two atoms of the reactant molecules (for example, an adsorbed H atom of a dissociated hydrogen molecule and the C atom of a reacting ethylene molecule); the reaction coordinate can in general be a combination of geometrical parameters (interatomic distances and/or angles). For each value of the reaction coordinate, the total energy of the system, calculated with density functional theory (DFT) was optimized through unconstrained relaxation of all of the other degrees of freedom of the system (reactants, other adsorbents, Pt cluster atoms and MgO slab atoms atoms). The reaction profiles (reaction paths) were obtained via repeating such calculations for a sequence of values of the chosen reaction coordinate. These calculations yield results that are the same as, or very close to, those obtained by other methods, for example, the nudged elastic band and variants thereof; see the ref. 13.

It is pertinent to note here that the structural relaxation following reactant adsorption and subsequent to each reaction step (using a conjugate gradient method) allows for distortions of the combined system (underlying support surface, adsorbed cluster and reactants) and leads to certain adsorption and reaction—induced structural rearrangements; for example, in some cases it can bring about transformation of the cluster geometry to a different structural isomeric form, as in the aforementioned example of dynamical reaction-induced cluster fluxionality (Supplementary Fig. 16). Global modifications of the structural motif of the adsorbed cluster/adsorbate system involve prohibitive activation energy barriers and are unlikely to occur as these transformations would be kinetically non-viable, particularly for the adsorption and reaction conditions considered in our studies that involve relatively low temperatures, ranging from below room temperature up to 400 K. Consequently, for kinetic reasons, it appears that relaxations involving local minima of the free-energy landscape of the combined catalytic system (rather than the global ones) are more effective here in steering the reaction pathway and determining the catalysed reaction barriers.

All the aforementioned first-principles electronic structure calculations employed the DFT method as implemented in the VASP-DFT package, using a plane-wave basis with a kinetic energy cutoff of 400 eV, projector augmented wave (PAW) pseudopotentials[61] and the Perdew, Burke & Ernzerhof (PBE) generalized gradient approximation for the exchange-correlation potential[62–66]. Γ- point sampling of the Brillouin zone was used.

References

1. Boudart, M. in *Advances in Catalysis* Vol. 20 (eds Herman, P., Weisz, P. B. & Eley, D. D.) 153–166 (Academic, 1969).
2. Boudart, M., Aldag, A., Benson, J. E., Dougharty, N. A. & Girvin Harkins, C. On the specific activity of platinum catalysts. *J. Catal.* **6,** 92–99 (1966).
3. Boudart, M., Aldag, A. W., Ptak, L. D. & Benson, J. E. On the selectivity of platinum catalysts. *J. Catal.* **11,** 35–45 (1968).
4. Bond, G. C. The origins of particle size effects in heterogeneous catalysis. *Surf. Sci* **156** Part 2 966–981 (1985).
5. Bond, G. C. Supported metal catalysts: some unsolved problems. *Chem. Soc. Rev.* **20,** 441–475 CS9912000441 (1991).
6. Che, M. & Bennett, C. O. in *Advances in Catalysis* Vol. 36 (eds Herman, P., Weisz, P. B. & Eley, D. D.) 55–172 (Academic, 1989).
7. Somorjai, G. A. & McCrea, K. Roadmap for catalysis science in the 21st century: a personal view of building the future on past and present accomplishments. *Appl. Catal. A* **222,** 3–18 (2001).
8. Boudart, M. & Djéga-Mariadassou, G. *Kinetics Of Heterogeneous Catalytic Reactions* (Princeton Univ., 1984).
9. Cremer, P. S., Su, X., Shen, Y. R. & Somorjai, G. A. Ethylene hydrogenation on Pt(111) monitored *in situ* at high pressures using sum frequency generation. *J. Am. Chem. Soc.* **118,** 2942–2949 (1996).
10. Horiuti, J. & Polanyi, M. Mechanism of the ethylene hydrogenation. *Trans. Faraday Soc* **30,** 1164–1172 (1934).
11. Boudart, M. Heterogeneous catalysis by metals. *J. Mol. Catal* **30,** 27–38 (1985).
12. Zaera, F. Key unanswered questions about the mechanism of olefin hydrogenation catalysis by transition-metal surfaces: a surface-science perspective. *Phys. Chem. Chem. Phys.* **15,** 11988–12003 (2013).
13. Heiz, U. & Landman, U. *Nanocatalysis* (Springer, 2007).
14. Landman, U., Yoon, B., Zhang, C., Heiz, U. & Arenz, M. Factors in gold nanocatalysis: oxidation of CO in the non-scalable size regime. *Top. Catal* **44,** 145–158 (2007).
15. Falsig, H. *et al.* Trends in the catalytic CO oxidation activity of nanoparticles. *Angew. Chem. Int. Ed.* **47,** 4835–4839 (2008).
16. Hvolbæk, B. *et al.* Catalytic activity of Au nanoparticles. *Nano Today* **2,** 14–18 (2007).
17. Janssens, T. V. W. *et al.* Insights into the reactivity of supported Au nanoparticles: combining theory and experiments. *Top. Catal.* **44,** 15–26 (2007).
18. Häkkinen, H., Abbet, S., Sanchez, A., Heiz, U. & Landman, U. Structural, electronic, and impurity-doping effects in nanoscale chemistry: supported gold nanoclusters. *Angew. Chem. Int. Ed.* **42,** 1297–1300 (2003).
19. Sanchez, A. *et al.* When gold is not noble: nanoscale gold catalysts. *J. Phys. Chem. A* **103,** 9573–9578 (1999).
20. Harding, C. *et al.* Control and manipulation of gold nanocatalysis: effects of metal oxide support thickness and composition. *J. Am. Chem. Soc.* **131,** 538–548 (2009).
21. Heiz, U. & Schneider, W.-D. Size-selected clusters on solid surfaces. *Crit. Rev. Solid State Mater. Sci* **26,** 251–290 (2001).
22. Barhen, J., Protopopescu, V. & Reister, D. TRUST: a deterministic algorithm for global optimization. *Science* **276,** 1094–1097 (1997).
23. Wales, D. J. & Scheraga, H. A. Global optimization of clusters, crystals, and biomolecules. *Science* **285,** 1368–1372 (1999).
24. Wolf, M. D. & Landman, U. Genetic algorithms for structural cluster optimization. *J. Phys. Chem. A* **102,** 6129–6137 (1998).
25. Zhao, Z.-J., Moskaleva, L. V., Aleksandrov, H. A., Basaran, D. & Rösch, N. Ethylidyne formation from ethylene over Pt(111): a mechanistic study from first-principle calculations. *J. Phys. Chem. C* **114,** 12190–12201 (2010).
26. Kubota, J., Ichihara, S., Kondo, J. N., Domen, K. & Hirose, C. π-Bonded ethene on Pt(111) surface studied by IRAS. *Surf. Sci.* **357–358,** 634–638 (1996).
27. Cassuto, A., Mane, M. & Jupille, J. Ethylene monolayer and multilayer on Pt(111) below 52 K: determination of bond lengths by near-edge X-ray fine structure. *Surf. Sci.* **249,** 8–14 (1991).
28. Chatt, J. & Duncanson, L. A. Olefin co-ordination compounds. Part III. Infra-red spectra and structure: attempted preparation of acetylene complexes. *J. Chem. Soc.* 2939–2947 (1953).
29. Dewar, M. A review of π Complex Theory. *Bull. Soc. Chim* **18,** C 71–C 79 (1951).
30. Neurock, M. & van Santen, R. A. A first principles analysis of C-H bond formation in ethylene hydrogenation. *J. Phys. Chem. B* **104,** 11127–11145 (2000).
31. van Santen, R. A. & Neurock, M. *Molecular Heterogeneous Catalysis: A Conceptual and Computational Approach* (Wiley-VCH, 2006).
32. Sautet, P. & Paul, J.-F. Low temperature adsorption of ethylene and butadiene on platinum and palladium surfaces: a theoretical study of the diσ/π competition. *Catal. Lett.* **9,** 245–260 (1991).
33. Yoon, B. *et al.* Probing of charging and reactivity of supported gold nanoclusters: Bonding of CO to gold octamers adsorbed on magnesia. *Science* **307,** 403–407 (2005).
34. Cortright, R. D., Goddard, S. A., Rekoske, J. E. & Dumesic, J. A. Kinetic study of ethylene hydrogenation. *J. Catal.* **127,** 342–353 (1991).
35. Rioux, R. M., Song, H., Hoefelmeyer, J. D., Yang, P. & Somorjai, G. A. High-surface-area catalyst design: synthesis, characterization, and reaction studies of platinum nanoparticles in mesoporous SBA-15 silica. *J. Phys. Chem. B* **109,** 2192–2202 (2005).
36. Song, H. *et al.* Hydrothermal growth of mesoporous SBA-15 silica in the presence of PVP-stabilized Pt nanoparticles: synthesis, characterization, and catalytic properties. *J. Am. Chem. Soc.* **128,** 3027–3037 (2006).
37. Beebe, Jr. T. P. & Yates, Jr. J. T. Spectroscopic detection of (111) facets on supported Pd crystallites: Site blocking by ethylidyne on Pd/Al2O3. *Surf. Sci* **173,** L606–L612 (1986).
38. Chen, P., Kung, K. Y., Shen, Y. R. & Somorjai, G. A. Sum frequency generation spectroscopic study of CO/ethylene coadsorption on the Pt(1 1 1) surface and CO poisoning of catalytic ethylene hydrogenation. *Surf. Sci* **494,** 289–297 (2001).
39. Lundwall, M. J., McClure, S. M. & Goodman, D. W. Probing Terrace and Step Sites on Pt Nanoparticles Using CO and Ethylene. *J. Phys. Chem. C* **114,** 7904–7912 (2010).
40. Rioux, R. M. *et al.* Adsorption and co-adsorption of ethylene and carbon monoxide on silica-supported monodisperse pt nanoparticles: volumetric adsorption and infrared spectroscopy studies. *Langmuir* **24,** 198–207 (2008).
41. Kesmodel, L. L., Dubois, L. H. & Somorjai, G. A. LEED analysis of acetylene and ethylene chemisorption on the Pt(111) surface: Evidence for ethylidyne formation. *J. Chem. Phys.* **70,** 2180–2188 (1979).
42. Mohsin, S. B., Trenary, M. & Robota, H. J. Infrared identification of the low-temperature forms of ethylene adsorbed on platinum/alumina. *J. Phys. Chem.* **92,** 5229–5233 (1988).
43. Land, T. A., Michely, T., Behm, R. J., Hemminger, J. C. & Comsa, G. Direct observation of surface reactions by scanning tunneling microscopy: Ethylene→ethylidyne→carbon particles→graphite on Pt(111). *J. Chem. Phys.* **97,** 6774–6783 (1992).
44. Lang, S. M., Bernhardt, T. B., R.N. Barnett, R. N. & Landman, U. Methane activation and catalytic ethylene formation on free Au_2^+. *Angew. Chem. Int. Ed.* **49,** 980–983 (2010).
45. Lang, S. M., Bernhardt, T. B., R.N. Barnett, R. N. & Landman, U. Temperature-tunable selective methane catalysis on Au_2^+: from cryogenicpartial oxidation yielding formaldehyde to cold ethylene production. *J. Phys. Chem. C* **115,** 6788–6795 (2011).

46. Ertl, G., Knözinger, H., Schüth, F. & Weitkamp, J. *Handbook of Heterogeneous Catalysis* 2 edn (Wiley-VCH, 2008).
47. Heiz, U., Vanolli, F., Trento, L. & Schneider, W.-D. Chemical Reactivity of size-selected supported clusters: an experimental setup. *Rev. Sci. Inst* **68,** 1986–1994 (1997).
48. Schweinberger, F. F. *et al.* Submonolayer sensitive adsorption study of trichloroethene on single crystal surfaces by means of MIES, UPS and TPD. *Surf. Sci.* **609,** 18–29 (2013).
49. Kolmakov, A., Stultz, J. & Goodman, D. W. Characterization of surface defects on MgO thin films by ultraviolet photoelectron and metastable impact electron spectroscopies. *J. Chem. Phys.* **113,** 7564–7570 (2000).
50. Benedetti, S., Benia, H. M., Nilius, N., Valeri, S. & Freund, H. J. Morphology and optical properties of MgO thin films on Mo(0 0 1). *Chem. Phys. Lett.* **430,** 330–335 (2006).
51. Kim, Y. D., Stultz, J. & Goodman, D. W. Characterization of MgO(1 0 0) thin film growth on Mo(1 0 0). *Surf. Sci* **506,** 228–234 (2002).
52. Zaera, F., Kollin, E. & Gland, J. L. Observation of an unusually low CO stretching frequency: CO chemisorption on a Mo(100) surface. *Chem. Phys. Lett.* **121,** 464–468 (1985).
53. Wettergren, K. *et al.* High sintering resistance of size-selected platinum cluster catalysts by suppressed Ostwald ripening. *Nano Lett.* **14,** 5803–5809 (2014).
54. Fukamori, Y. *et al.* Fundamental insight into the substrate-dependent ripening of monodisperse clusters. *ChemCatChem* **5,** 3330–3341 (2013).
55. Harding, C. J. *et al.* Dual pulsed-beam controlled mole fraction studies of the catalytic oxidation of CO on supported Pd nanocatalysts. *J. Catal.* **255,** 234–240 (2008).
56. Amenomiya, Y. & Pottie, R. F. Mass spectra of some deuterated ethanes. I. The effect of ionizing voltage. *Can. J. Chem.* **46,** 1735–1739 (1968).
57. Ricci, D., Bongiorno, A., Pacchioni, G. & Landman, U. Bonding trends and dimensionality crossover of gold nanoclusters on metal-supported MgO thin films. *Phys. Rev. Lett.* **97,** 036106 (2006).
58. Yoon, B. & Landman, U. Electric field control of structure, dimensionality, and reactivity of gold nanoclusters on metal-supported MgO films. *Phys. Rev. Lett.* **100,** 056102 (2008).
59. Zhang, C., Yoon, B. & Landman, U. Predicted oxidation of CO catalyzed by Au nanoclusters on a thin defect-free MgO film supported on a Mo(100) surface. *J. Am. Chem. Soc.* **129,** 2228–2229 (2007).
60. Vaida, M. E. *et al.* Ultrathin magnesia films as support for molecules and metal clusters: Tuning reactivity by thickness and composition. *Phys. Status Solidi B* **247,** 1001–1015 (2010).
61. Kresse, G. & Joubert, D. From ultrasoft pseudopotentials to the projector augmented-wave method. *Phys. Rev. B* **59,** 1758–1775 (1999).
62. Levy, M. & Perdew, J. P. Tightly bound and convexity constraint on the exchange-correlation-energy functional in the low-density limit, and other formal tests of generalized-gradient approximations. Phys. Rev. B **48,** 11638 (1993). *Erratum* **55,** 13321 (1997).
63. Levy, M. & Perdew, J. P. Tight bound and convexity constraint on the exchange-correlation-energy functional in the low-density limit, and other formal tests of generalized-gradient approximations. *Phys. Rev. B* **48,** 11638–11645 (1993).
64. Perdew, J. P. *Electronic Structure of Solids.* (eds Ziesche, P. & Eschrig, H.) 11–20 (Akademie, 1991).
65. Perdew, J. P., Burke, K. & Ernzerhof, M. Generalized gradient approximation made simple. *Phys. Rev. Lett.* **77,** 3865–3868 (1996).
66. Perdew, J. P. *et al.* Atoms, molecules, solids, and surfaces: Applications of the generalized gradient approximation for exchange and correlation. *Phys. Rev. B* **46,** 6671–6687 (1992).

Acknowledgements

The experimental work has been supported by the European Research Council (ERC) through the advanced research grant (246645-ASC3), and by the DFG through project HE3454/23-1. B.Y. was supported by the Air Force Office for Scientific Research (AFOSR) and U.L. by grant No. FG05-86ER45234 from the Office of Basic Energy Sciences of the US Department of energy (DOE). Computations were carried out at the Georgia Tech Center for Computational Materials science.

Author contributions

A.S.C., U.L. and U.H. conceived the project, A.S.C., M.D.R., C.J.R. and F.F.S. contributed to the experimental work, A.S.C., M.D.R, C.J.R., F.F.S. and U.H. analysed the experimental results, B.Y. executed the theoretical computations, B.Y. and U.L. analysed the computational results, A.S.C, U.L. and U.H. wrote the manuscript.

Additional information

Competing financial interests: The authors declare no competing financial interests.

7

First-principles quantum dynamical theory for the dissociative chemisorption of H_2O on rigid Cu(111)

Zhaojun Zhang[1,*], Tianhui Liu[1,2,*], Bina Fu[1], Xueming Yang[1,2,3] & Dong H. Zhang[1,3]

Despite significant progress made in the past decades, it remains extremely challenging to investigate the dissociative chemisorption dynamics of molecular species on surfaces at a full-dimensional quantum mechanical level, in particular for polyatomic-surface reactions. Here we report, to the best of our knowledge, the first full-dimensional quantum dynamics study for the dissociative chemisorption of H_2O on rigid Cu(111) with all the nine molecular degrees of freedom fully coupled, based on an accurate full-dimensional potential energy surface. The full-dimensional quantum mechanical reactivity provides the dynamics features with the highest accuracy, revealing that the excitations in vibrational modes of H_2O are more efficacious than increasing the translational energy in promoting the reaction. The enhancement of the excitation in asymmetric stretch is the largest, but that of symmetric stretch becomes comparable at very low energies. The full-dimensional characterization also allows the investigation of the validity of previous reduced-dimensional and approximate dynamical models.

[1] State Key Laboratory of Molecular Reaction Dynamics and Center for Theoretical and Computational Chemistry, Dalian Institute of Chemical Physics, Chinese Academy of Sciences, Dalian 116023, China. [2] Department of Chemical Physics, University of Science and Technology of China, Hefei 230026, China. [3] Center for Advanced Chemical Physics and 2011 Frontier Center for Quantum Science and Technology, University of Science and Technology of China, Hefei 230026, China. * These authors contributed equally to this work. Correspondence and requests for materials should be addressed to B.F. (email: bina@dicp.ac.cn) or to D.H.Z. (email: zhangdh@dicp.ac.cn).

The molecular interactions and chemical reactions at metal surfaces are of great importance to many industrial applications. Dissociative chemisorption is one of the simplest surface chemical reactions, which undergoes the dissociation of a gas-phase molecule upon its collision with a surface, leading to chemisorbed products. It is a fundamental and often the rate-limiting step in many industrial heterogeneous catalytic processes, such as steam reforming, ammonia synthesis and the water–gas shift reaction[1]. Enormous progress has been obtained for the dissociative chemisorption dynamics in the past decades[2–18], due to its significance in surface chemistry and catalysis.

Thanks to the molecular beam and laser techniques, many pioneering experimental studies with quantum state resolution have been carried out, in which rich dynamics information, including mode specificity and bond selectivity in dissociative chemisorption processes was uncovered[3–7,12–16]. Owing to the involvement of a surface, the theoretical simulation for a gas-surface reaction is much more complicated than for a chemical reaction that occurs exclusively in the gas phase, even with small molecules. There has been intense effort to develop theory for reliable dynamics calculations of dissociative chemisorption of molecular species on surfaces. The purpose of this quest for quantitatively accurate theory is to provide a definitive understanding of the dynamics of dissociative chemisorption, which ultimately provides the basis for the control of chemical reactions at gas-surface interfaces and confident prediction of a model of heterogenous catalysis. It is gratifying that the system of diatomic molecules, dissociating on metal surfaces is sufficiently small to be amenable to full-dimensional (six-dimensional, 6D) quantum dynamics calculations on an accurate potential energy surface (PES)[8,19–32]. In particular for the dissociation of hydrogen molecule on metal surfaces, the state-of-the-art dynamics calculations, as well as the underlying chemical theory have been tested at a level close to that achieved for simple reactions in the gas phase[8,33].

The dissociative chemisorption of polyatomic molecules with multiple vibrational modes offers much richer complexity and mode specificity, thus much more attention has been attracted to them[34–36]. Despite significant progress made in the past two decades, it remains extremely challenging to investigate the dissociative chemisorption of polyatomic molecules on surfaces at a full-dimensional quantum mechanical level, due to the difficulties in constructing reliable high-dimensional PESs and developing practical quantum mechanical methodologies.

The classical approximation for nuclear motion has been an alternative to a full quantum characterization of dissociative chemisorption of polyatomic molecules. Although the *ab initio* molecular dynamics (AIMD) with energies computed on the fly, and quasiclassical trajectory calculations based on global PESs have shed valuable light on the dissociative chemisorption dynamics[37–41], both the two approaches neglect quantum effects and the former is also computationally expensive. It is worth noting that the effect of surface atomic motion can be taken into account with the *ab initio* molecular dynamics. An accurate full quantum characterization of dissociative chemisorption of polyatomic molecules is highly desirable in view of potentially important quantum effects, such as tunnelling, zero-point energy (ZPE) and resonances.

The dissociative chemisorption of water on transition-metal surfaces is of fundamental importance in understanding many industrial heterogeneous catalytic processes, such as steam reforming, and represents the rate-limiting step in low-temperature water–gas shift reaction on copper catalysts[42,43]. A total of nine degrees of freedom should be considered on a rigid surface, rendering it difficult to develop an accurate, global PES and formidable to carry out a fully coupled nine-dimensional (9D) quantum dynamics calculations. Previous studies of water on surfaces resorted to approximate or reduced-dimensional quantum models.

An approximate approach relying on a reaction path Hamiltonian[44], markedly simplifies the polyatomic problem. Farjamnia and Jackson[45] employed the reaction path Hamiltonian approach to investigate the dissociative chemisorption of water on Ni(111), which treats all the internal motions of water using a vibrationally adiabatic basis set, and produces the minimum-barrier-site reaction probability from wave packet propagations. Sudden models were used to average over the impact sites and lattice motion[46].

Development of the reduced-dimensional quantum approaches is prevalent in understanding the dissociation of water on metal surfaces, starting from the simplest pseudo-diatomic molecule model by Tiwari and co-workers[47]. Guo and co-workers performed 6D quantum dynamics calculations to study the mode specificity of H_2O and bond selectivity of HOD on a rigid flat Cu(111) surface[48–51], employing their 6D PESs developed by permutationally invariant polynomials (PIPs)[52,53]. The 6D model neglects the effects of impact sites and surface corrugation. The obtained strong mode specificity can be rationalized by recently proposed sudden vector projection model[54,55].

The first quantum-state-resolved molecular beam experiment on the dissociative chemisorption of D_2O on Ni(111) demonstrated a large enhancement in reactivity upon excitation of the asymmetric stretch of D_2O (ref. 16). The observed mode specificity was semiquantitatively understood by Guo and Beck and co-workers[16] using the 6D quantum dynamical approach, where the effects of impact sites and lattice motion were approximated using the sudden models proposed by Jackson and co-workers[39,46,56–58]. Recently, Jiang and Guo[51] investigated 6D site-specific dissociation probabilities of this system, but based on a new 9D PES developed by PIP-neural network (PIP-NN) method.

Very recently, we reported the first seven-dimensional (7D) quantum dynamics study for the dissociative chemisorption of H_2O on Cu(111), based on an accurate 9D PES developed by NN approach[18,59]. The dissociation probabilities exhibit strong azimuthal angle dependence, indicating the 6D quantum model neglecting the azimuthal angle can introduce substantial errors.

The new 7D quantum dynamics study for the title gas-surface reaction has provided new insight into the dissociative chemisorption dynamics[18,59]. Nevertheless, it is still a provisional model with reduced-dimensional approximation, in which the surface lateral coordinates are fixed in a specific site of the impact. A quantitatively accurate description of the title gas-surface reaction can only be verified by comparison with a fully coupled 9D quantum mechanical approach based on an accurate global PES.

In this work, we report the full-dimensional quantum dynamics study of the dissociative chemisorption of H_2O on a rigid Cu(111) surface, allowing all the nine molecular degrees of freedom to be fully coupled. The challenging full-dimensional quantum mechanical calculations not only offer the dynamics features with the highest accuracy, but also allow a conclusive examination of previous dynamical approximations.

Results

9D dissociation probabilities. The full-dimensional (9D) dissociation probability for H_2O initially in the ground vibrational state, together with the 7D results on the fixed top, bridge, fcc, hcp and transition-state (TS) sites are illustrated in Fig. 1 in the kinetic energy region of (0.9, 1.8) eV. Due to the non-reactive of

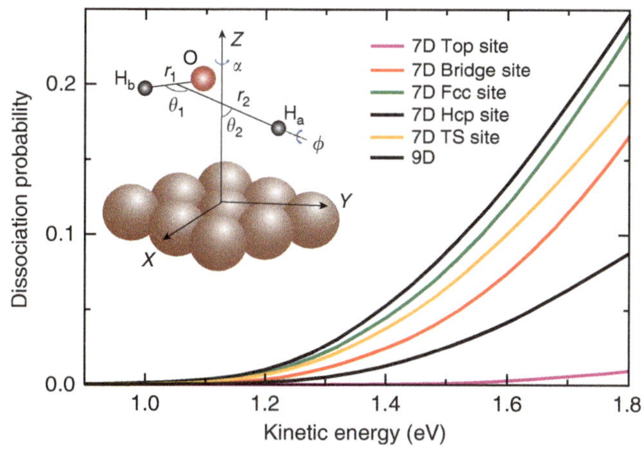

Figure 1 | Full-dimensional (9D) and seven-dimensional (7D) site-specific results. The 9D dissociation probability and 7D site-specific (top, bridge, fcc, hcp and TS sites) probabilities with H_2O initially in the ground rovibrational state (000). The 9D molecular coordinates of the H_2O/Cu(111) system are shown in the inset.

one OH bond in current theory, the computed dissociation probabilities were multiplied by a factor of 2 that correspond to the results for two reactive bonds. On the whole, all the dissociation probabilities are smooth functions of the kinetic energy, and increase monotonically with increasing kinetic energy. As has been discussed before[29,30,41,51,59], the dissociation probability for H_2O fixed at the top site is extremely small, due to the topography of the PES and dynamical process, but not the static barrier height. Significant differences between the 9D dissociation probability and the results calculated by reduced-dimensional (7D) fixed-site approach are seen. The 9D probability is substantially smaller or larger than the 7D fixed-site probabilities in the entire energy region. At the kinetic energy of 1.8 eV, the 9D probability is larger than the probability on the fixed top site by a factor of over 8, but smaller than the rest of four fixed-site results by a factor of roughly 1–2. In addition, these factors turn to much larger with the decreasing kinetic energy. Obviously, the 7D quantum model neglecting the surface lateral coordinates cannot produce quantitatively accurate dissociation probabilities and the full-dimensional calculations are remarkably important to characterize the dynamics.

Vibrational excitations. As the water reactant has three vibrational modes, including the symmetric stretching (ν_1), bending (ν_2) and asymmetric stretching (ν_3) modes, respectively, various vibrational excitations of $H_2O(\nu_1\nu_2\nu_3)$ should have different influences on its interaction with metal surfaces. We are now capable of gaining insight into the mode specificity of the title gas-surface reaction at a full-dimensional quantum mechanical level, which is significant in the reactivity control and an in-depth understanding of the dynamics.

Figure 2 shows the 9D dissociation probabilities for H_2O initially in six lowest non-rotating vibrational states in both linear and log scales. The results are compared as a function of total energy, which is measured with respect to the asymptotic energy of ground rovibrational state of $H_2O(000)$. As seen in Fig. 2a, the dissociation probabilities of five vibrationally excited states are all larger than that of the ground state in the entire energy region, implying that excitations in vibrational modes of H_2O enhance the reactivity for the dissociative chemisorption of H_2O on Cu(111). The reactivity enhancement by exciting the stretching modes is consistent with the Polanyi's rules for a late-barrier reaction, and the enhancement due to the bending mode is

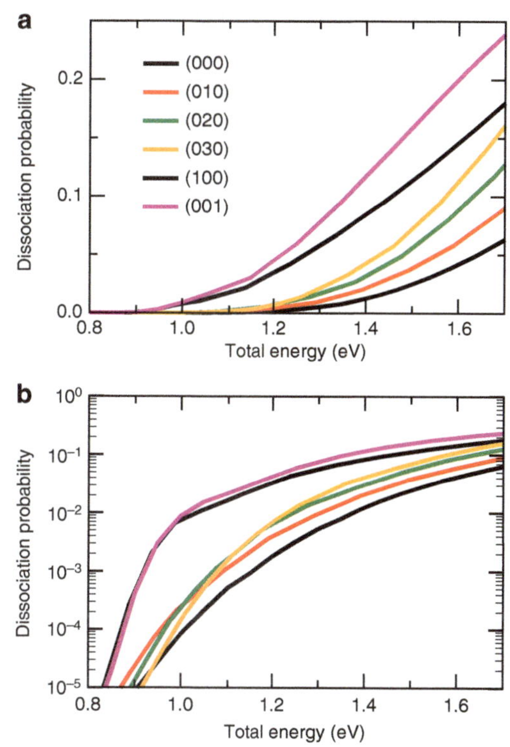

Figure 2 | The reactivity enhancement by vibrational excitations of reactant. (**a**) The full-dimensional dissociation probabilities with H_2O in six non-rotating vibrational states as a function of the total energy. (**b**) Same as **a**, except in a logarithmic scale.

evidenced by the fact that the H–O–H angle of the saddle point (119.21°) is larger than that in isolated H_2O (104.41°). The three bending overtones (010), (020) and (030) enhance the dissociation probability, but with smaller degrees of enhancement as compared with the stretching modes, although the excitation energy of the (030) state is even larger than that of the asymmetric stretch (001). Vibrational excitations in the stretching modes are more efficient than excitations in the bending modes, due to the stronger couplings of H_2O stretch with the reaction coordinate. It is interesting to find that the enhancement of excitation in the asymmetric stretching mode is larger than that of the symmetric stretching mode (100), consistent with the performance in 7D[59], but not in the 6D calculations[50], where the larger enhancement of symmetric stretching mode is observed.

From the corresponding dissociation probabilities shown in a logarithmic scale in Fig. 2b, we can obviously see that in the very low-energy regions below the classical barrier (1.08 eV) there exist considerable dissociation probabilities for the ground vibrational state of H_2O. The ZPEs of reactant asymptote and TS are 0.57 and 0.43 eV, respectively, so that the ZPE-corrected barrier height is 0.94 eV. Thus, the reactivity below the classical barrier (down to roughly 0.9 eV) is due to both the ZPE and tunnelling effects. In addition, the overall reactivity enhancement from vibrational excitations increases as the total energy decreases. The enhancement factors of the reactivity relative to the ground state at the total energy of 1.1 eV are 2.98, 5.52, 4.47, 34.26 and 53.99 for the (010), (020), (030), (100) and (001) states, respectively, which are much larger than those of 1.42, 2.02, 2.57, 3.04 and 3.81 at the total energy of 1.7 eV, respectively. The largest enhancement of excitation in the asymmetric stretching mode is presumably due to the high excitation energy of this mode and its strong couplings with the reaction coordinate. We also note that the enhancement of excitation in the symmetric stretching mode

Table 1 | Vibrational efficacies for five non-rotating excited vibrational states in dissociative chemisorption of H_2O on Cu(111) calculated by full-dimensional quantum dynamical approach.

Vibrational state	Vibrational energy (eV)	Vibrational efficacy, η					
		$P_0=10^{-5}$		$P_0=10^{-4}$	$P_0=10^{-3}$	$P_0=10^{-2}$	$P_0=0.05$
(010)	0.192	1.18	(1.05)*	1.29	1.34	1.39	1.43
(020)	0.378	1.04	(0.884)*	1.12	1.22	1.35	1.42
(100)	0.437	1.17	(0.999)*	1.32	1.54	1.79	1.84
(001)	0.449	1.16	(0.932)*	1.30	1.52	1.81	1.93
(030)	0.560	0.98		1.04	1.13	1.26	1.37

*The numbers in parentheses are the 6D results for the vibrational efficacy in ref. 50.

become comparable to the asymmetric stretching mode at very low energies (< 1.0 eV).

The relative enhancement is often quantified in terms of the vibrational efficacy, which is described as $\eta = [E_i(0, P_0) - E_i(v, P_0)]/(E_v - E_0)$, where $E_i(0, P_0)$ and $E_i(v, P_0)$ are the kinetic energy for the vibrational ground (0) and excited (v) states to give the same dissociation probability P_0, while E_0 and E_v are the corresponding vibrational energies[45,50]. The computed results are shown in Table 1. The values of η are always larger than unity, indicating excitations in all the vibrational modes are more efficacious than the same amount of translational energy in promoting the reaction. The vibrational efficacy increases with the increase of dissociation probabilities and collision energies. Besides, we see much larger values of vibrational efficacy, that is, much stronger mode specificity in the present 9D work than those obtained by the 6D quantum model[50].

Validity of site-averaging approximations. A full-dimensional quantum dynamical characterization of a polyatomic molecule-surface reaction is ideal, which is often many theoreticians strive for, due to the fact that it can provide more reliable results and a quantitatively accurate understanding of the reaction dynamics. Nevertheless, such calculations for polyatomic dissociations, such as an explicit treatment of the 15 degrees of freedom for rigid CH_4, are still formidable. Employing the reduced-dimensional dynamical model, as well as the impact site approximation to accurately approximate the full-dimensional quantum mechanical reactivity should be a reliable and promising approach. The current full-dimensional quantum dynamics study for H_2O/Cu(111) gives a good opportunity to validate two existing implementations of site-averaging approximations for polyatomic molecule-surface reactions.

The site-sudden approximation is proposed by Jackson and Nave[46] and also extensively used[16,39,56,60], which characterizes the dynamics with the assumption that dissociation probabilities at different impact sites have the same energy dependence as that at one single site, for example, TS site, but varying with the static barrier height. With that approximation, the potentials at various impact sites are approximated by a harmonic expansion around one specific site.

Recent dynamics calculations demonstrated that the reactivity mainly depends on the topography of the PES rather than the static barrier height[29,30,32,41,51], indicating the site-averaging approximation with harmonic potential (SAHP) employed in polyatomic molecule-surface reactions might not be quite reliable[16,39,46,56,60]. We recently proposed and validated a new site-averaging approximation with exact potential (SAEP) in diatomic molecule-surface reactions[29–31], where the full-dimensional reactivity was well reproduced by averaging reduced-dimensional fixed-site results over multiple impact sites with appropriate relative weights. As in the following, both the two implementations of site-averaging approximations are first rigorously examined in the dissociative chemisorption of water on metal surfaces.

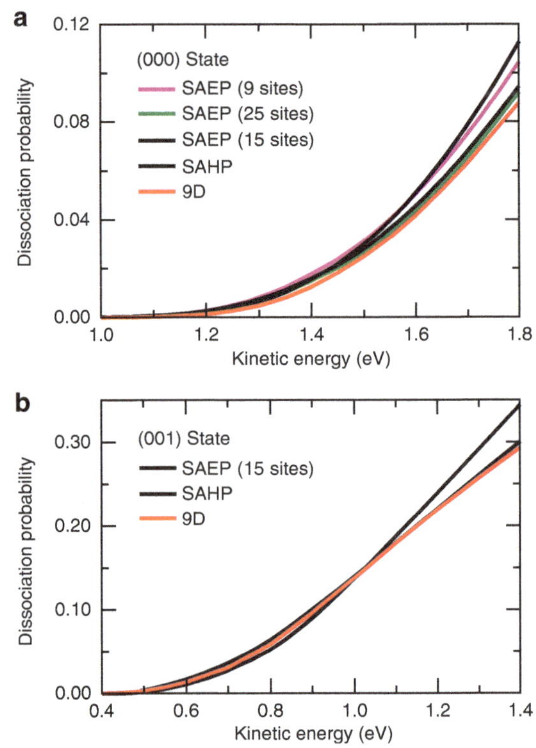

Figure 3 | 9D and site-averaging 7D results. (**a**) Comparisons of dissociation probabilities obtained by averaging the 7D results over 9, 25 and 15 impact sites with appropriate relative weights (SAEP), the probability obtained by averaging over the results for many (*X,Y*) sites using SAHP, and the full-dimensional dissociation probability, for H_2O initially in the (000) state. (**b**) The dissociation probabilities obtained by the SAEP and SAHP approximations, together with the full-dimensional dissociation probability, for H_2O initially in the (001) state.

The dissociation probabilities from the two implementations of site-averaging approximations (SAEP and SAHP) are compared with the 9D dissociation probability in Fig. 3. First, the SAEP results by averaging 7D dissociation probabilities over 9 sites and 25 sites in the C_{3v} unit cell of Cu(111), with appropriate relative weights were calculated for H_2O initially in the ground vibrational state (000; see Supplementary Fig. 1 and Supplementary Discussion for the schematic of the distribution of 9 sites and 25 sites, and the detailed site-averaging approach). As shown in Fig. 3a, the overall behaviour of the nine site-averaging dissociation probability resembles that of the 9D dissociation probability, although the magnitude of the former is larger than the latter particularly at high kinetic energies. The agreement between the 25 site-averaging probability using the SAEP and 9D probability is much better and the differences between them are negligible, indicating the validity of SAEP is confirmed again in the title polyatomic molecule-surface reaction.

As the static barrier heights and 7D dissociation probabilities of the hcp and fcc sites (Fig. 1) are quite similar, we assume the C_{6v} symmetry of the Cu(111) surface, where the fcc and hcp sites are indistinguishable on the PES. As a result, the 25 sites included in the averaging can be reduced to 15 sites further (one-half of the triangle), as shown in Fig. 4. The 15 site-averaging probability agree quite well with the 25 site-averaging probability, as well as the 9D result, indicating the full-dimensional probability of (000) state can be accurately reproduced by averaging 7D dissociation probabilities over 15 sites.

Following Jackson and Nave[46], as detailed in the Supplementary Discussion, we obtained the dissociation probability averaging over all important impact sites with SAHP. We can see in Fig. 3a that the dissociation probability with SAHP is close to the 15/25 site-averaging probability with SAEP and 9D probability in the low-energy region, but larger than them at kinetic energies >1.5 eV.

To further validate the two implementations of site-averaging approximations, we also compare the SAEP, SAHP and 9D probability curves for the vibrational excitation in asymmetric stretching mode of H_2O (001) in Fig. 3b. The agreement between the 15 site-averaging probability with SAEP and 9D probability is excellent and impressive, which further confirms the validity of our SAEP. However, the probability with SAHP is slightly smaller at low kinetic energies, but much larger at kinetic energies >1.0 eV than the 9D result for the (001) state. Those discrepancies imply the SAHP might introduce some errors in describing the full-dimensional quantum mechanical reactivity.

To give a more explicit comparison between the two implementations of site-averaging approximations, the exact static barrier heights on the PES and those obtained using the HP approximation at 15 impact sites mapped in the irreducible triangle within the approximate C_{6v} symmetry are illustrated in Fig. 4a,b, respectively. Obviously, the site-specific barrier height obtained by HP is always larger than the exact barrier height on the PES. More significant differences are seen when the distance between a specific site (X,Y) and the TS site increases. For example, the barrier height with HP at the hcp site (1.24 eV) is 0.15 eV larger than the exact barrier height (1.09 eV) on the PES,

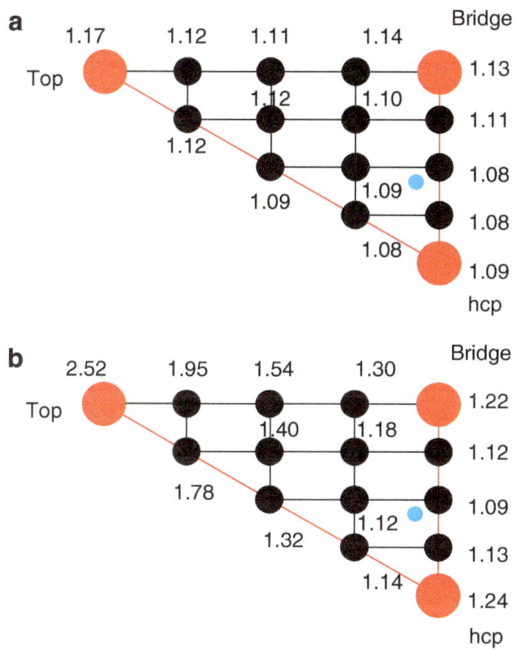

Figure 4 | Static barrier heights. (**a**) Static barrier heights (in eV) obtained on the exact fitted PES for 15 impact sites in the irreducible triangle of Cu(111) within the approximate C_{6v} symmetry. The site numbers are also labelled and the TS site is indicated by a blue circle. (**b**) Same as **a**, except by the HP approximation.

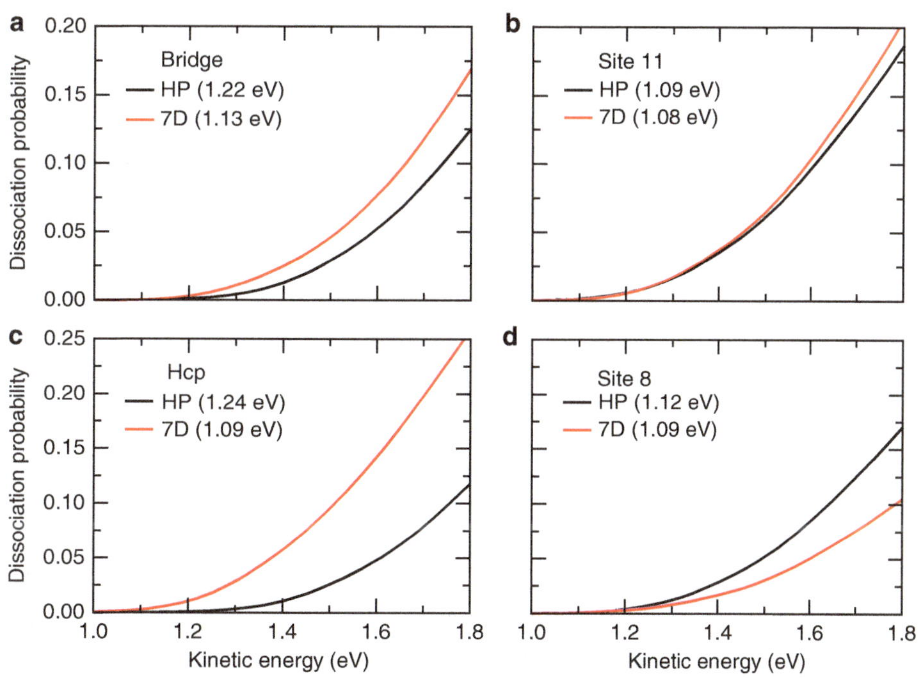

Figure 5 | Site-specific dissociation probabilities. (a-d) The site-specific dissociation probabilities obtained by the HP approximation and those calculated by 7D quantum dynamics calculations at the fixed bridge site, site 11, hcp site and site 8, respectively, with H_2O initially in the ground rovibrational state. The barrier heights obtained from the HP approximation and the exact PES are also indicated in the parentheses.

while at the top site the former (2.52 eV) is 1.35 eV larger than the latter (1.17 eV). These results indicate that the HP expansion cannot be validated for the barrier corrugation. The potential expanded in a harmonic approximation around a fixed site can be too high, which is expected in general and was discussed by Jackson *et al.*[39]

Furthermore, the site-specific dissociation probabilities obtained by HP and by the 7D quantum dynamics calculations for the (000) state of H_2O at 15 impact sites discussed above are calculated and compared. The results at four impact sites, that is, the bridge site, site 11, hcp site and site 8, as shown in Fig. 4a are illustrated in Fig. 5. Those results at rest of 10 sites are presented in Supplementary Figs 2–4, except for the top site, where the probability obtained by HP in the energy region considered are completely zero, due to the very high barrier height (2.52 eV) from HP. Overall, significant differences are observed between the two batches of results in the entire energy region, indicating the HP approximation can give rise to substantial errors in attaining the 7D site-specific dissociation probabilities. The farther this specific site is away from the TS site, the more remarkable differences are seen between the two batches of results. Although we did not see such significant differences in Fig. 3 after averaging many site-specific dissociation probabilities, it can be fortunate because the positive and negative errors induced by the HP approximation balanced out in the averaging.

Discussion

We have carried out the full-dimensional quantum dynamics study for the dissociative chemisorption of H_2O on a rigid Cu(111) surface, which rigorously treats and fully couples all the nine molecular degrees of freedom, to achieve a quantitatively accurate understanding of this centrally important gas-surface reaction. An accurate, full-dimensional PES constructed recently by the NN fitting to density functional theory (DFT) energy points was employed. The full-dimensional model allows the surface corrugation, as well as the translational movement of the H_2O moiety. Large differences between the full-dimensional (9D) and site-specific 7D reactivity are observed, indicating the significance of the challenging full-dimensional quantum mechanical calculations. The excitations in five lowest vibrational modes greatly promotes the reaction, which are more efficient than increasing the translational energy. The enhancement of the excitation in the asymmetric stretching mode (001) is largest, to which that of the symmetric stretch (100) become comparable at very low energies, whereas that of the bending mode (010) is smallest, presumably due to the strong couplings of H_2O stretch with the reaction coordinate. The bending excited state (030) has the highest vibrational energy, but gives rise to the smallest vibrational efficacy. The full-dimensional quantum dynamics study also allows the investigation of the validity of two implementations of site-averaging approximations. It is found that the full-dimensional reactivity can be accurately reproduced by averaging site-specific 7D results from quantum mechanical calculations over 15 impact sites, validating our SAEP in this polyatomic molecule-surface reaction. In contrast, the SAHP can introduce substantial errors in obtaining the site-specific dissociation probabilities. The resulting reactivity from SAHP may not accurately approximate the full-dimensional reactivity. The validity of SAEP observed in diatomic molecule-surface reactions and the H_2O/Cu(111) reaction should generally hold in many gas-surface reactions. It probably provides a good opportunity to investigate the dissociative chemisorption dynamics of polyatomic molecules on metal surfaces at a full-dimensional quantum mechanical level by site-averaging the reduced-dimensional results over multiple impact sites. The surface lattice motion should also be considered, so that a direct comparison with the experimental data can be achieved.

Methods

9D PES. The 9D PES was constructed by using the NN fitting to a total of 81102 DFT energy points[18]. This fit results in an overall very small root mean square error of only 9.0 meV, but is significantly smaller (6.0 meV) for energy points <2.0 eV relative to the $H_2O + Cu(111)$ asymptote, representing the unprecedented fitting accuracy for PESs of polyatomic-surface reactions. Readers can refer to Supplementary Methods for the details of 9D PES.

9D time-dependent wave packet approach. A total of nine degrees of freedom should be considered for the fully coupled quantum dynamics calculations of the dissociative chemisorption of H_2O on a corrugated, rigid metal surface. The 9D Hamiltonian and time-dependent wave function for the title reaction is expressed in terms of molecule coordinates (x, y, Z, r_1, r_2, θ_1, θ_2, ϕ, α ; shown in Supplementary Fig. 5). Details of quantum dynamical methodology are given in the Supplementary Methods section and the convergence of quantum results with respect to numerical parameters is shown in Supplementary Figs 6 and 7.

References

1. Chorkendorff, I. & Niemantsverdriet, J. W. *Concepts of Modern Catalysis and Kinetics* (Wiley-VCH, 2003).
2. Kroes, G. J. Frontiers in surface scattering simulations. *Science* **321,** 794–797 (2008).
3. Beck, R. D. *et al.* Vibrational mode-specific reaction of methane on a nickel surface. *Science* **302,** 98–100 (2003).
4. Luntz, A. C. Toward vibrational mode control in catalysis. *Science* **302,** 70–71 (2003).
5. Juurlink, L. B. F., McCabe, P. R., Smith, R. R., DiCologero, C. L. & Utz, A. L. Eigenstate-resolved studies of gas-surface reactivity: $CH_4(\nu_3)$ dissociation on Ni(100). *Phys. Rev. Lett.* **83,** 868–871 (1999).
6. Maroni, P. *et al.* State-resolved gas-surface reactivity of methane in the symmetric C–H stretch vibration on Ni(100). *Phys. Rev. Lett.* **94,** 246104 (2005).
7. Schmid, M. P., Maroni, P., Beck, R. D. & Rizzo, T. R. Surface reactivity of highly vibrationally excited molecules prepared by pulsed laser excitation: $CH_4(2\nu_3)$ on Ni(100). *J. Chem. Phys.* **117,** 8603–8606 (2002).
8. Díaz, C. *et al.* Chemically accurate simulation of a prototypical surface reaction: H_2 dissociation on Cu(111). *Science* **326,** 832–834 (2009).
9. Kroes, G. J. Towards chemically accurate simulation of molecule-surface reactions. *Phys. Chem. Chem. Phys.* **14,** 14966–14981 (2012).
10. Ran, Q., Matsiev, D., Auerbach, D. & Wodtke, A. Observation of a Change of Vibrational excitation mechanism with surface temperature: HCl collisions with Au(111). *Phys. Rev. Lett.* **98,** 237601 (2007).
11. Nave, S. & Jackson, B. Vibrational mode-selective chemistry: methane dissociation on Ni(100). *Phys. Rev. B* **81,** 233408 (2010).
12. Smith, R. R., Killelea, D. R., DelSesto, D. F. & Utz, A. L. Preference for vibrational over translational energy in a gas-surface reaction. *Science* **304,** 992–995 (2004).
13. Killelea, D. R., Campbell, V. L., Shuman, N. S. & Utz, A. L. Bond-selective control of a heterogeneously catalyzed reaction. *Science* **319,** 790–793 (2008).
14. Yoder, B. L., Bisson, R. & Beck, R. D. Steric effects in the chemisorption of vibrationally excited methane on Ni(100). *Science* **329,** 553–556 (2010).
15. Chen, L., Ueta, H., Bisson, R. & Beck, R. D. Vibrationally bond-selected chemisorption of methane isotopologues on Pt(111) studied by reflection absorption infrared spectroscopy. *Faraday Discuss* **157,** 285–295 (2012).
16. Hundt, P. M., Jiang, B., van Reijzen, M. E., Guo, H. & Beck, R. D. Vibrationally promoted dissociation of water on Ni(111). *Science* **344,** 504–507 (2014).
17. Golibrzuch, K., Bartels, N., Auerbach, D. J. & Wodtke, A. M. The dynamics of molecular interactions and chemical reactions at metal surfaces: testing the foundations of theory. *Annu. Rev. Phys. Chem.* **66,** 399–425 (2015).
18. Liu, T., Zhang, Z., Fu, B., Yang, X. & Zhang, D. H. A seven-dimensional quantum dynamics study of the dissociative chemisorption of H_2O on Cu(111): effects of azimuthal angles and azimuthal angle-averaging. *Chem. Sci.* **7,** 1840–1845 (2016).
19. Groß, A. Reactions at surfaces studied by ab initio dynamics calculations. *Surf. Sci. Rep.* **32,** 291–340 (1998).
20. Kroes, G. J., Gross, A., Baerends, E.-J., Scheffler, M. & McCormack, D. A. Quantum theory of dissociative chemisorption on metal surfaces. *Acc. Chem. Res.* **35,** 193–200 (2002).

21. Kroes, G. J., Baerends, E. J. & Mowrey, R. C. Six-dimensional quantum dynamics of dissociative chemisorption of $(v=0, j=0)H_2$ on Cu(100). *Phys. Rev. Lett.* **78,** 3583–3586 (1997).
22. Dai, J. & Light, J. C. Six dimensional quantum dynamics study for dissociative adsorption of H_2 on Cu(111) surface. *J. Chem. Phys.* **107,** 1676–1679 (1997).
23. Kroes, G. J., Baerends, E. J. & Mowrey, R. C. Six-dimensional quantum dynamics of dissociative chemisorption of H_2 on Cu(100). *J. Chem. Phys.* **107,** 3309–3323 (1997).
24. Gross, A., Wilke, S. & Scheffler, M. Six-dimensional quantum dynamics of adsorption and desorption of H_2 at Pd(100): steering and steric effects. *Phys. Rev. Lett.* **75,** 2718–2721 (1995).
25. Nave, S., Lemoine, D., Somers, M. F., Kingma, S. M. & Kroes, G. J. Six-dimensional quantum dynamics of $(v=0,j=0)D_2$ and of $(v=1,j=0)H_2$ scattering from Cu(111). *J. Chem. Phys.* **122,** 214709 (2005).
26. Nieto, P. *et al.* Diffractive and reactive scattering of H_2 from Ru(0001): experimental and theoretical study. *Phys. Chem. Chem. Phys.* **13,** 8583–8597 (2011).
27. Sementa, L. *et al.* Reactive scattering of H_2 from Cu(100): comparison of dynamics calculations based on the specific reaction parameter approach to density functional theory with experiment. *J. Chem. Phys.* **138,** 044708 (2013).
28. Liu, T., Fu, B. & Zhang, D. H. Six-dimensional potential energy surface of the dissociative chemisorption of HCl on Au(111) using neural networks. *Sci. China Chem.* **57,** 147–155 (2013).
29. Liu, T., Fu, B. & Zhang, D. H. Six-dimensional quantum dynamics study for the dissociative adsorption of DCl on Au(111) surface. *J. Chem. Phys.* **140,** 144701 (2014).
30. Liu, T., Fu, B. & Zhang, D. H. Six-dimensional quantum dynamics study for the dissociative adsorption of HCl on Au(111) surface. *J. Chem. Phys.* **139,** 184705 (2013).
31. Liu, T., Fu, B. & Zhang, D. H. Validity of the site-averaging approximation for modeling the dissociative chemisorption of H_2 on Cu(111) surface: a quantum dynamics study on two potential energy surfaces. *J. Chem. Phys.* **141,** 194302 (2014).
32. Hu, X., Jiang, B., Xie, D. & Guo, H. Site-specific dissociation dynamics of H_2/D_2 on Ag(111) and Co(0001) and the validity of the site-averaging model. *J. Chem. Phys.* **143,** 114706 (2015).
33. Kroes, G. J. & Díaz, C. Quantum and classical dynamics of reactive scattering of H_2 from metal surfaces. *Chem. Soc. Rev.* doi: 10.1039/C5CS00336A (2016).
34. Nave, S., Tiwari, A. & Jackson, B. Dissociative chemisorption of methane on ni and pt surfaces: mode-specific chemistry and the effects of lattice motion. *J. Phys. Chem. A* **118,** 9615–9631 (2014).
35. Jiang, B., Yang, M., Xie, D. & Guo, H. Quantum dynamics of polyatomic dissociative chemisorption on transition metal surfaces: mode specificity and bond selectivity. *Chem. Soc. Rev.* doi: 10.1039/C5CS00360A (2016).
36. Jiang, B., Alducin, M. & Guo, H. Electron-hole pair effects in polyatomic dissociative chemisorption: water on Ni(111). *J. Phys. Chem. Lett.* **7,** 327–331 (2016).
37. Sacchi, M., Wales, D. J. & Jenkins, S. J. Mode-specificity and transition state-specific energy redistribution in the chemisorption of CH_4 on Ni(100). *Phys. Chem. Chem. Phys.* **14,** 15879–15887 (2012).
38. Nattino, F. *et al.* Ab initio molecular dynamics calculations versus quantum-state-resolved experiments on $CHD_3 + Pt(111)$: new insights into a prototypical gas-surface reaction. *J. Phys. Chem. Lett.* **5,** 1294–1299 (2014).
39. Jackson, B., Nattino, F. & Kroes, G. J. Dissociative chemisorption of methane on metal surfaces: tests of dynamical assumptions using quantum models and ab initio molecular dynamics. *J. Chem. Phys.* **141,** 054102 (2014).
40. Shen, X. J., Lozano, A., Dong, W., Busnengo, H. F. & Yan, X. H. Towards bond selective chemistry from first principles: methane on metal surfaces. *Phys. Rev. Lett.* **112,** 046101 (2014).
41. Jiang, B. & Guo, H. Dynamics of water dissociative chemisorption on ni(111): effects of impact sites and incident angles. *Phys. Rev. Lett.* **114,** 166101 (2015).
42. Ovesen, C. V., Stoltze, P., Norskov, J. K. & Campbell, C. T. A kinetic model of the water gas shift reaction. *J. Catal.* **134,** 445–468 (1992).
43. Henderson, M. A. The interaction of water with solid surfaces: fundamental aspects revisited. *Surf. Sci. Rep.* **46,** 1–308 (2002).
44. Miller, W. H., Handy, N. C. & Adams, J. E. Reaction path Hamiltonian for polyatomic molecules. *J. Chem. Phys.* **72,** 99–112 (1980).
45. Farjamnia, A. & Jackson, B. The dissociative chemisorption of water on Ni(111): mode- and bond-selective chemistry on metal surfaces. *J. Chem. Phys.* **142,** 234705 (2015).
46. Jackson, B. & Nave, S. The dissociative chemisorption of methane on Ni(100): reaction path description of mode-selective chemistry. *J. Chem. Phys.* **135,** 114701 (2011).
47. Mondal, A., Seenivasan, H. & Tiwari, A. K. Water dissociation on Cu(111): effects of molecular orientation, rotation, and vibration on reactivity. *J. Chem. Phys.* **137,** 094708 (2012).
48. Jiang, B., Xie, D. & Guo, H. Vibrationally mediated bond selective dissociative chemisorption of HOD on Cu(111). *Chem. Sci.* **4,** 503–508 (2013).
49. Jiang, B., Li, J., Xie, D. & Guo, H. Effects of reactant internal excitation and orientation on dissociative chemisorption of H_2O on Cu(111): quasi-seven-dimensional quantum dynamics on a refined potential energy surface. *J. Chem. Phys.* **138,** 044704 (2013).
50. Jiang, B., Ren, X., Xie, D. & Guo, H. Enhancing dissociative chemisorption of H_2O on Cu(111) via vibrational excitation. *Proc. Natl Acad. Sci. USA* **109,** 10224–10227 (2012).
51. Jiang, B. & Guo, H. Quantum and classical dynamics of water dissociation on Ni(111): a test of the site-averaging model in dissociative chemisorption of polyatomic molecules. *J. Chem. Phys.* **143,** 164705 (2015).
52. Braams, B. J. & Bowman, J. M. Permutationally invariant potential energy surfaces in high dimensionality. *Int. Rev. Phys. Chem.* **28,** 577–606 (2009).
53. Bowman, J. M., Czakó, G. & Fu, B. High-dimensional ab initio potential energy surfaces for reaction dynamics calculations. *Phys. Chem. Phys. Chem.* **13,** 8094–8111 (2011).
54. Jiang, B. & Guo, H. Relative efficacy of vibrational versus translational excitation in promoting atom-diatom reactivity: rigorous examination of Polanyi's rules and proposition of sudden vector projection (SVP) model. *J. Chem. Phys.* **138,** 234104 (2013).
55. Jiang, B. & Guo, H. Prediction of Mode Specificity, Bond Selectivity, Normal Scaling, and Surface Lattice Effects in water dissociative chemisorption on several metal surfaces using the sudden vector projection model. *J. Phys. Chem. C* **118,** 26851–26858 (2014).
56. Jackson, B. & Nave, S. The dissociative chemisorption of methane on Ni(111): the effects of molecular vibration and lattice motion. *J. Chem. Phys.* **138,** 174705 (2013).
57. Tiwari, A. K., Nave, S. & Jackson, B. Methane dissociation on Ni(111): a new understanding of the lattice effect. *Phys. Rev. Lett.* **103,** 253201 (2009).
58. Tiwari, A. K., Nave, S. & Jackson, B. The temperature dependence of methane dissociation on Ni(111) and Pt(111): mixed quantum-classical studies of the lattice response. *J. Chem. Phys.* **132,** 134702 (2010).
59. Liu, T., Zhang, Z., Fu, B., Yang, X. & Zhang, D. H. Mode specificity for the dissociative chemisorption of H_2O on Cu(111): A quantum dynamics study on an accurately fitted potential energy surface. *Phys. Chem. Chem. Phys.* **18,** 8537–8544 (2016).
60. Jiang, B. *et al.* Mode selectivity in methane dissociative chemisorption on Ni(111). *Chem. Sci.* **4,** 3249–3254 (2013).

Acknowledgements

This work was supported by the National Natural Science Foundation of China (grant nos 91421315, 21433009, 21590804 and 21303197), the Ministry of Science and Technology of China (2013CB834601), the Youth Innovation Promotion Association (2015143) and the Chinese Academy of Sciences.

Author contributions

D.H.Z., X.Y. and B.F. conceived and supervised the research; Z.Z. and T.L. performed the research; Z.Z., T.L., B.F. and D.H.Z. analysed the data; and B.F. and D.H.Z. wrote the manuscript.

Additional information

Competing financial interests: The authors declare no competing financial interests.

8

Polymorphism in magic-sized $Au_{144}(SR)_{60}$ clusters

Kirsten M.Ø. Jensen[1,*], Pavol Juhas[2,*], Marcus A. Tofanelli[3], Christine L. Heinecke[3], Gavin Vaughan[4], Christopher J. Ackerson[3] & Simon J.L. Billinge[1,2]

Ultra-small, magic-sized metal nanoclusters represent an important new class of materials with properties between molecules and particles. However, their small size challenges the conventional methods for structure characterization. Here we present the structure of ultra-stable $Au_{144}(SR)_{60}$ magic-sized nanoclusters obtained from atomic pair distribution function analysis of X-ray powder diffraction data. The study reveals structural polymorphism in these archetypal nanoclusters. In addition to confirming the theoretically predicted icosahedral-cored cluster, we also find samples with a truncated decahedral core structure, with some samples exhibiting a coexistence of both cluster structures. Although the clusters are monodisperse in size, structural diversity is apparent. The discovery of polymorphism may open up a new dimension in nanoscale engineering.

[1] Department of Applied Physics and Applied Mathematics, Columbia University, New York, New York 10027, USA. [2] Condensed Matter Physics and Materials Science Department, Brookhaven National Laboratory, Upton, New York 11973, USA. [3] Department of Chemistry, Colorado State University, Fort Collins, Colorado 80523, USA. [4] European Synchrotron Radiation Facility, 38043 Grenoble, France. * These authors contributed equally to this work. Correspondence and requests for materials should be addressed to C.J.A. (email: chris.ackerson@colostate.edu) or to S.J.L.B. (email: sb2896@columbia.edu).

The promise of nanotechnology, to engineer materials at the nanoscale with improved properties, is predicated on the idea that material structure and properties are fundamentally modified on this scale. Gold clusters are prototypical inorganic materials that exemplify this[1–8]. In addition to being technologically important in their own right[9], they are a model system for studying this paradigm, as they form ultra-stable 'magic number' molecule-like clusters of different sizes[10,11]. A major challenge, in the majority of cases where the clusters cannot be crystallized, is to determine their structure. We overcome this 'nanostructure problem'[12] by using atomic pair distribution function (PDF) analysis of X-ray diffraction (XRD) data to study the structure of $Au_{144}(SR)_{60}$ (where R is the organic part of the thiol), one of the largest of the ultra-stable magic-sized clusters with known composition[13,14]. The PDF data successfully yield the core structure, with the surprising result that these clusters exhibit polymorphism. In very recent studies, single crystal structure determination illustrated that the much smaller $Au_{38}(SR)_{24}$ interconverts reversibly between two forms, depending on temperature[15]. Here we use PDF to show that polymorphism exists also in the large $Au_{144}(SR)_{60}$ cluster, representing the size regime in the transition between clusters forming non-bulk geometric structures and bulk face-centred cubic (fcc) nanoparticles[8]. The discovery of polymorphism brings an additional dimension to the phase space for nanoscale engineering.

The $Au_{144}(SR)_{60}$ structure has already been subject to many studies. Initially described as a ubiquitous 29-kDa core-mass compound[14,16,17], more recently the composition was determined by mass spectrometry as $Au_{144}(SR)_{60}$ (refs 18,19). Lopez-Acevedo *et al.*[20] developed a detailed structural model, tested by density functional theory (DFT), where the cluster consists of an icosahedral gold core surrounded by a gold/thiol surface layer. NMR (nuclear magnetic resonance) studies later suggested that all ligands are in symmetry equivalent positions[21]. Scanning transmission electron microscopy (STEM) studies by Bahena *et al.*[22] were consistent with the icosahedral core and by introducing the NMR symmetry requirement in theoretical calculations they proposed a symmetrized structure model featuring an equivalent ligand arrangement[22]. This model consists of a gold core of 54 atoms arranged as two Mackay icosahedral shells (Fig. 1a), whereas a 60-atom layer covers the 55-site inner core in an 'anti-Mackay' manner (Fig. 1b). The surface of the cluster structures consist of -SR-Au-SR-type structures (Fig. 1d), referred to as 'staples'[23], and in combination this gives the full proposed structure as illustrated in Fig. 1c. The structure is closely related to that of $Pd_{145}(CO)(PEt_3)$ determined by single-crystal XRD[24].

In this study, we apply atomic PDF analysis to $Au_{144}(SR)_{60}$. PDF analysis (described in Supplementary Note 1) has become widely used for nanostructure analysis[25–29] and is a potential tool for nanostructure solution[30–32]. In recent times, PDF has also been applied to the fingerprinting of gold nanocluster structure[33,34]. PDF goes beyond conventional X-ray powder diffraction, which typically covers only a narrow range of reciprocal space[35] and neglects diffuse scattering. The total scattering approach contains significantly more structural information, allowing a quantitative assessment of the structure that is impossible with conventional data from such small particles. We apply PDF nanostructure analysis to $Au_{144}(SR)_{60}$ clusters prepared with different ligands (Fig. 1e): phenylethane thiol (PET), *para*-mercaptobenzoic acid (*p*-MBA), butane thiol (SC4), hexanethiol (SC6) and dodecanethiol (SC12). Sample homogeneity is characterized by electrospray ionization–mass spectrometry (ESI–MS) and electrochemical methods. The approach results in full quantitative refinements of the structure of the gold core, with a semi-quantitative assessment of the surface structure. Surprisingly, we find two distinct structural forms for this cluster's core, one based on icosahedra seen in smaller clusters, proposed earlier for this 144 gold atom cluster[20–22], and one based on close packed decahedra that resemble larger gold clusters and bulk gold. The discovery of polymorphism in gold nanoclusters opens up a new dimension in nanoparticle engineering, presenting the

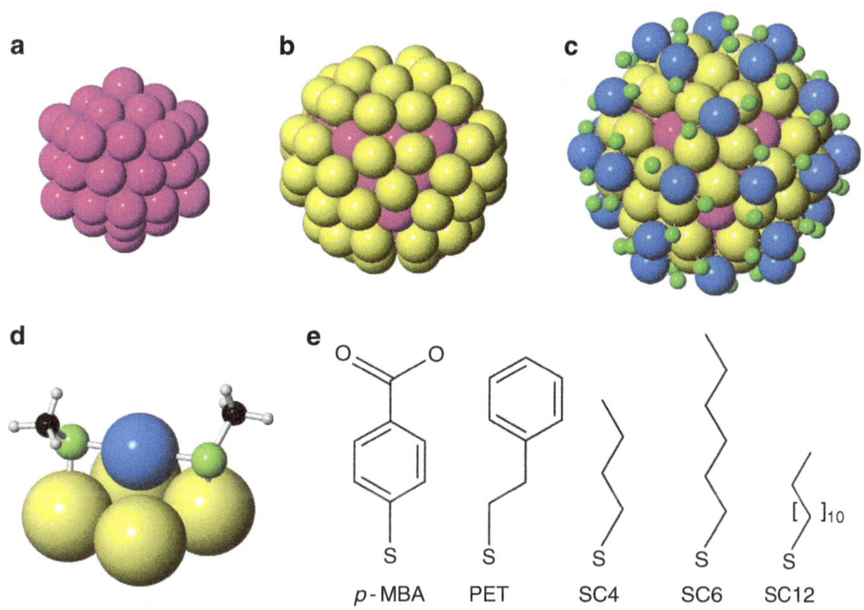

Figure 1 | Structure of the icosahedral $Au_{144}(SR)_{60}$ cluster. (**a**) Fifty-four atom gold core consisting of two Mackay icosahedron shells. (**b**) The icosahedral gold core (pink) is covered by 60 gold atoms (yellow) making up the grand core. (**c**) Total structure, where the grand core is covered in 'staples'—green atoms represent sulfur, whereas blue atoms represent gold in the staple structure. The organic carbon chains have been left out for clarity. (**d**) Illustration of staple structure on gold surface. (**e**) Thiolate ligands used in the study. From left: *p*-MBA, PET, SC4, SC6 and SC12.

possibility of engineering nanoparticle structure, as well as size and morphology.

Results

PDFs from $Au_{144}(SR)_{60}$. We first investigate the sample prepared with SC6 ligands. ESI-MS data (see Supplementary Fig. 1 and Supplementary Table 1) confirmed homogeneity of this sample, with at least 90% of the sample being $Au_{144}(SC6)_{60}$ and a small byproduct (<10%) with ESI-MS peaks, which can be assigned to $Au_{137}(SR)_{56}$ (ref. 36). Previously in the literature, this impurity signal has been assigned to $Au_{144}(SR)_{60}$ fragments[37,38]. No other cluster sizes, such as $Au_{130}(SR)_{50}$ or $Au_{133}(SR)_{52}$ were detected. The low Q scattering signal (where $Q = 4\pi \sin(\theta)/\lambda$ is the magnitude of the scattering vector), corresponding to conventional XRD data, the total scattering structure function $F(Q)$ and the PDF, $G(r)$, from this sample at 100 K are shown in Fig. 2. Owing to the small size of the gold clusters, only very broad scattering peaks are present in the low Q signal (Fig. 2a), resulting in too little information to attempt a total structure solution by crystallographic means. In the total scattering structure function $F(Q)$ (Fig. 2b) we see that the diffuse scattering extends over a wide range of reciprocal space containing scattering features with rich information that cannot be resolved when just the low Q conventional XRD data are used. The PDF, plotted in blue in Fig. 2c, is the Fourier transform of the data in Fig. 2b. This real-space function contains peaks at distances separating pairs of atoms in the structure. The observation of sharp peaks in real-space indicates that the gold clusters have a well-defined structure. The peaks in $G(r)$ disappear above 12.5 Å, which puts a lower bound on the diameter of the gold core of the clusters. The first large peak at *ca.* 2.9 Å is the nearest-neighbour gold–gold distance, r_{nn}, and is sharp. The strength and sharpness of the low r peaks suggests a high multiplicity for these distances, indicating a rather well-packed structure. In Fig. 2c, we also show the experimental PDFs from the $Au_{144}(PET)_{60}$ sample plotted in green. The similarity of the PDFs from $Au_{144}(SC6)_{60}$ and $Au_{144}(PET)_{60}$ indicates that these clusters have identical core structures and also establishes the reproducibility of the PDF measurements.

The red line in Fig. 2c shows the PDF from $Au_{144}(p\text{-MBA})_{60}$. There is a remarkable difference between this PDF and those of $Au_{144}(SC6)_{60}$ and $Au_{144}(PET)$. These clusters have the same size, as evident from the disappearance of sharp features in the PDF, by the characteristic, well-defined differential pulse voltammetry and from mobility in polyacrylamide gel electrophoresis consistent with $Au_{144}(SR)_{60}$. Both the SC6, PET- and p-MBA-protected preparations, formed the poorly diffracting hexagonal plate crystals previously observed for these compounds[39,40]. The sharp PDF peaks indicate that the $Au_{144}(p\text{-MBA})_{60}$ clusters also have a well-defined ordered structure. However, their structure is remarkably different from that of the SC6- and PET-terminated clusters: the $Au_{144}(SR)_{60}$ clusters are exhibiting polymorphism. We now explore quantitatively the two structural polymorphs, Form I and Form II, of these clusters.

Form I. We begin by calculating PDFs from candidate structures suggested in the literature to compare with the data. The relative atomic positions are highly constrained in the modelling (described in detail in Supplementary Note 2), with only five parameters allowed to vary: a scale factor accounting for the overall PDF intensity, a uniform cluster expansion factor that allows the cluster structure to contract or expand, two isotropic atomic displacement parameter applied separately to the core and surface atoms, as well as a parameter accounting for correlated atomic motion[41]. Therefore, good fits to the data are a

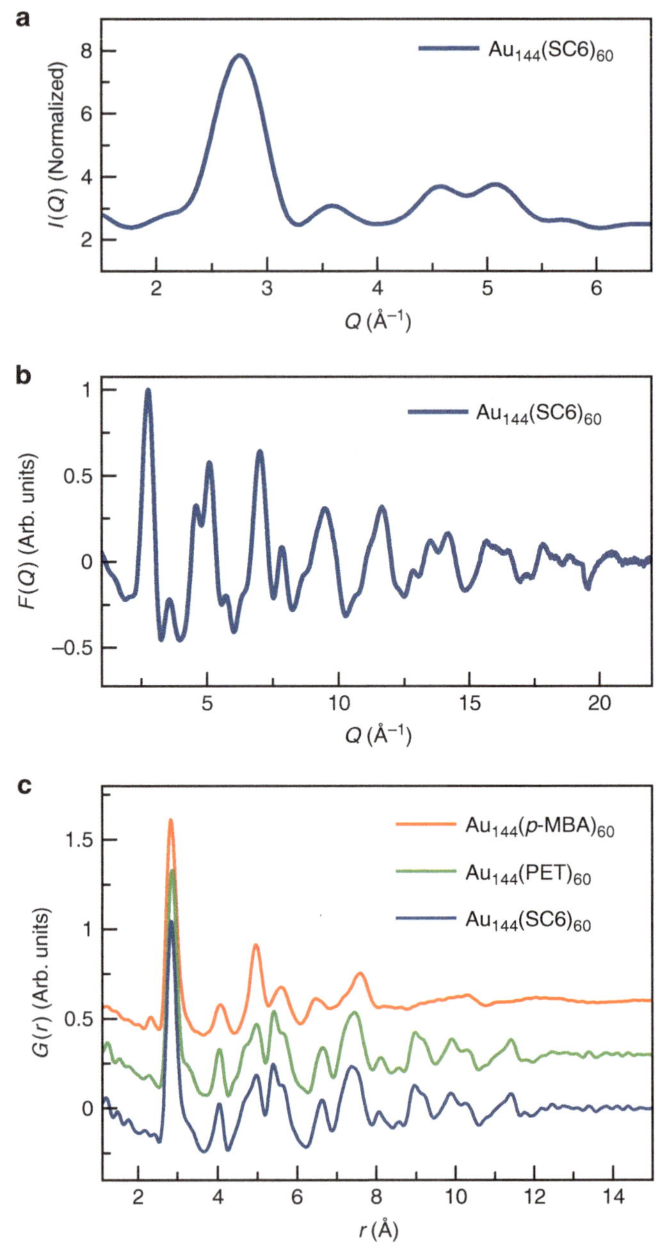

Figure 2 | Collected scattering data for $Au_{144}(SR)_{60}$ clusters. (**a**) Low Q scattering data, corresponding to the conventional XRD signal for the $Au_{144}(SC6)_{60}$ sample. (**b**) Total scattering structure function $F(Q)$ for $Au_{144}(SC6)_{60}$. (**c**) PDFs obtained from $Au_{144}(SC6)_{60}$, $Au_{144}(PET)_{60}$ and $Au_{144}(p\text{-MBA})_{60}$.

strong indicator that the model has captured the correct geometry of the core. To simplify the models, only the Au and S atoms were included in the refinements, as the scattering signal from the organic ligands is negligible (see Supplementary Note 3). Figure 3a shows the calculated PDF from the model suggested by Bahena *et al.*[22], fitted to the SC6 data[20]. The result of a refinement to the same data, but using the structure reported by Lopez-Acevedo *et al.*[20] is shown in Supplementary Fig. 2. Both models describe the main features of the PDF of the $Au_{144}(SC6)_{60}$ nanoclusters very well, with the Bahena structure giving a slightly better fit to the PDF with agreement factor $R_W = 16.3\%$. This confirms the proposed structures of previous theoretical and STEM studies[21,22].

Similar fits to the $Au_{144}(PET)_{60}$ sample are given in Supplementary Fig. 3 and Supplementary Table 2, also showing good agreement with the icosahedral model ($R_W = 15.8\%$). Furthermore, a direct comparison of the experimental data from the PET and SC6 data show that the two samples give rise to practically identical PDFs as illustrated in Supplementary Fig. 4, where the difference curve between the two PDFs is essentially a flat line. Interestingly, the ESI–MS data indicated ~16% $Au_{137}(SR)_{56}$ in the PET sample, that is, a higher fraction than seen in the SC6 sample. The flat difference curve between the two PDFs would not be expected if the byproduct signal in ESI–MS is coming from a different cluster, that is, $Au_{137}(SR)_{56}$. Thus, the PDFs either indicate that the byproduct signal is coming from fragments of $Au_{144}(SR)_{60}$ created during the ESI–MS measurement, or that the core structure of $Au_{137}(SR)_{56}$ is indistinguishable to that of $Au_{144}(SR)_{60}$ Form I. As we see later, the PDF is quite sensitive to small changes in core structure and, although the latter scenario cannot be ruled out, the former is more probable, indicating that our samples are pure $Au_{144}(SR)_{60}$. If the latter scenario is correct, it establishes that the core of $Au_{137}(SR)_{56}$ is highly similar to that of $Au_{144}(SR)_{60}$.

The data shown in Fig. 2 are obtained at 100 K. Scattering data from $Au_{144}(PET)_{60}$ were also taken at 300 K, showing no structural changes between the two temperatures (Supplementary Fig. 5). Furthermore, we measured data using three different X-ray energies, ranging from 39 to 87 keV, and all PDFs (Supplementary Fig. 5) showed the same structure.

Form II. We now turn to the structure of the $Au_{144}(p\text{-MBA})_{60}$ cluster, which has the very different PDF evident in Fig. 2c. The homogeneity of this cluster sample was characterized by electrochemical measurements. Total scattering data were measured from samples of $Au_{144}(p\text{-MBA})_{60}$ from two different synthesis batches and as shown in Supplementary Fig. 6 the two PDFs are completely reproduced with a small difference residuum of $R_W = 7.8\%$, illustrating reproducibility of the synthesis and reliability of the measurements. We first attempt to use the Form I icosahedral structural model to establish whether this can be made to fit the different PDF by adjusting the refinement parameters. However, the model gives a very poor fit with a large difference between the calculated and measured PDF, and poor fit residuum of $R_W = 36.0\%$ as shown in Fig. 3b. To further confirm that the sample does not simply contain stable clusters of a different size, for example, $Au_{102}(SR)_{44}$ (ref. 3), $Au_{130}(SR)_{50}$ (ref. 42) or $Au_{133}(SR)_{52}$ (ref. 43), we fitted known structural models for these clusters to the *p*-MBA data. In all cases, the models gave very poor agreements with the data (fits shown in Supplementary Figs 7–9), confirming that the samples are not made up of other stable cluster sizes.

Therefore, other models for the $Au_{144}(SR)_{60}$ gold cluster were explored. Initially, we considered only the positions of the 144 gold atoms and ignored the ligands in the model. We based this on the dominating scattering power of gold compared with the thiolates, as discussed further in the Supplementary Note 3. First, a series of close-packed core models were constructed, closely related to bulk fcc gold. These included a 147-atom cuboctahedron, as well as clusters formed by cutting a sphere of ~144 atoms from fcc and hexagonal close-packed (hcp) lattices. The next attempted model was a two-phase fit of the PDFs from cutouts from fcc and hcp, which has been used as a proxy model in PDF modelling for close-packed structures that contain stacking faults[27]. A summary of these simulations is given in Supplementary Table 3 and Supplementary Note 4. None of the fcc- or hcp-based clusters produced completely convincing fits to the observed PDF. However, the fits were significantly better than for the Bahena model, especially for the fcc/hcp

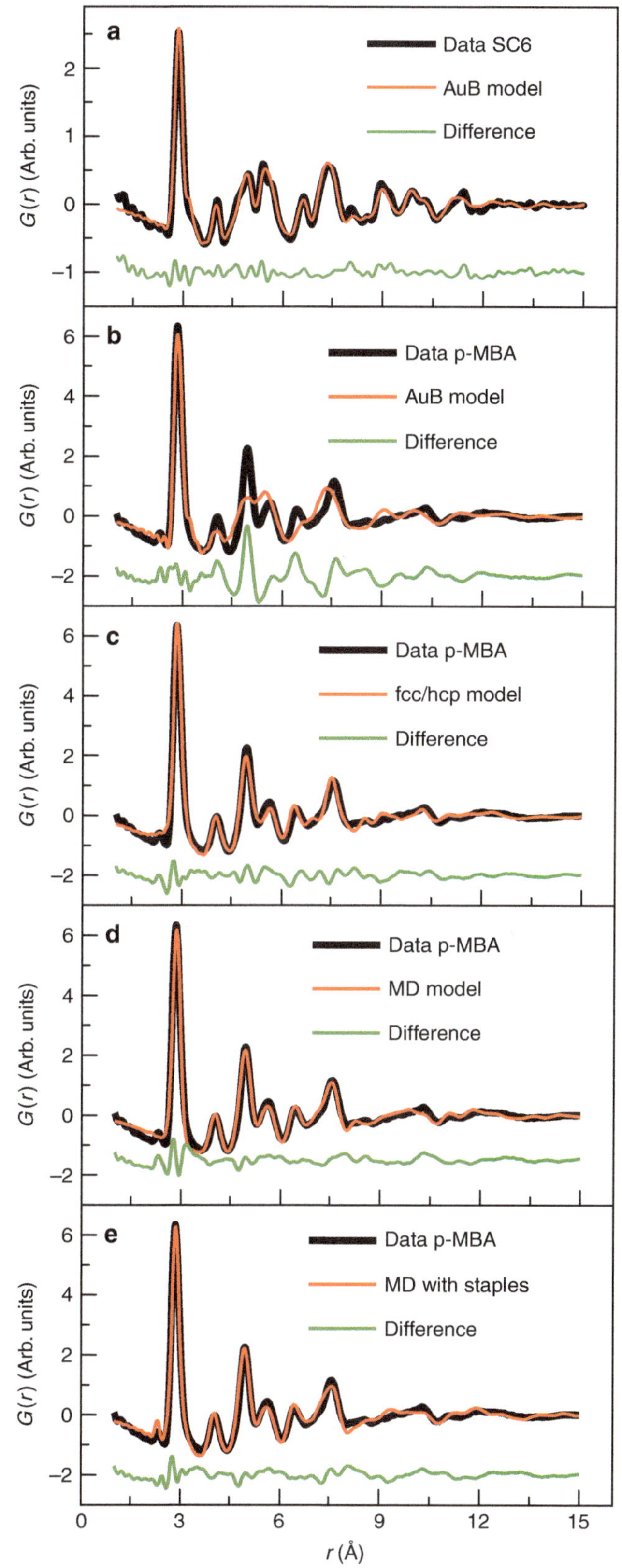

Figure 3 | Fits to experimental PDFs. (**a**) Fit of Bahena model to $Au_{144}(SC6)_{60}$ data. (**b**) Fit of icosahedral model to $Au_{144}(p\text{-MBA})_{60}$ data. (**c**) Fit of fcc/hcp model to $Au_{144}(p\text{-MBA})_{60}$ data. (**d**) Fit of 114 atom decahedral model to $Au_{144}(p\text{-MBA})_{60}$ data. (**e**) Fit of decahedral model with staples to $Au_{144}(p\text{-MBA})_{60}$ data.

mixture as shown in Fig. 3c. The PDF agreement of the fcc/hcp model was remarkably improved after allowing for a separate expansion ratio for the atoms in the outermost shell, giving $R_W = 16.3\%$, a step which was motivated by allowing for a possible surface relaxation. This improved the refinement by fitting the asymmetry in the first Au–Au peak. However, the results indicated that the bond lengths between the atoms in the surface were contracted compared with the bonds in the core and the atomic displacement parameters were excessively large over $0.03\,Å^2$ for the core atoms, suggesting the existence of some atomic relaxations that are not part of these simple models. Furthermore, this model contains 141 atoms in the fcc phase and 147 atoms in the hcp phase. We seek a model that can also explain the high stability of the Au core with 144 atoms, whereas spherical chunks of close-packed bulk material would not have special stability. Nonetheless, the fitting results establish that the structure of $Au_{144}(SR)_{60}$ is much closer to a three-dimensional close-packed structure than the icosahedral, DFT-derived models.

Our search for close-packed structures that have special atom counts led us to explore a series of Marks decahedral structures that are constructed by introducing twin boundaries along the *(110)* planes of the fcc lattice[44,45], as described in more detail in Supplementary Note 5. Closed shell, truncated decahedra can be constructed with a large range of discrete number of atoms, including 144, that is, the exact number of gold atoms in the cluster. This structure and the fit to the experimental PDF are shown in Supplementary Fig. 10, where excellent fits are seen. However, as described above, thiolate ligands are known to create -SR-Au-SR- or -SR-Au-SR-Au-SR- 'staples' on gold surfaces[23,46]. The short staple, that is, -SR-Au-SR- is mainly seen on larger clusters, where the curvature is small, as would be the case in the ~2 nm $Au_{144}(SR)_{60}$ structure, and $Au_{144}(SR)_{60}$ may thus better be represented as $Au_{114}[(SR)\text{-}Au\text{-}(SR)]_{30}$. This pointed us towards a smaller ino decahedral structure as the core of the cluster, as illustrated in Fig. 4a,b. The cluster shown has exactly 114 gold atoms, leaving 30 gold atoms for the staples as required by the putative stoichiometry. The fit of this cluster to the data is shown in Fig. 3d and, as illustrated, the model very well describes the experimental PDF. All distinct sharp peaks up to 8 Å are reproduced and the fit remains very close even at higher *r*-values where the features are broader and less resolved. In studies of smaller clusters, it has been shown that although the core of the cluster is decahedral, a shell of gold atoms may be seen between the core structure and the staple layer[3]. Therefore, we tried stripping down the decahedral structure to a yet smaller core and reattaching the atoms as 'caps' on the remaining structure[3]. However, interestingly, any modification to the 114 atom decahedral core highly deteriorated the PDF fit, making lower symmetry structures unlikely. This makes us confident in a core structure based on the 114 atom decahedron, closely related to the ino decahedron described by Cleveland *et al.*[17]

Various configurations of the staples on the 114-atom decahedral structure cluster were then considered, where one example is presented in Fig. 4c and other selected models are shown in Supplementary Fig. 11. The process of attaching the staples is described in detail in Supplementary Note 6. Staples were placed on the *(111)* surfaces as previously seen[23]; however, to accommodate all ligands to the structure in a physically sensible manner, staples were also attached to the *(100)* surfaces, although this motif has not yet been reported. Several different models were constructed, which all give comparably good fits to the data with R_W values of *ca.* 15–18%, with one example shown in Fig. 3e, where the presence of staples fit to the shoulder of the nearest neighbour Au–Au peak. The PDF refinements were somewhat sensitive to the staple attachment, as subtle differences between the features in the fitted PDF can be observed. However, based on PDF data alone we cannot determine the exact ligand arrangement and further studies combining total scattering with techniques sensitive to the ligand attachment are needed to determine the surface structure with full confidence.

Nevertheless, the PDF analysis clearly shows that the Au_{144} $(p\text{-}MBA)_{60}$ core takes a decahedral structure, unlike the $Au_{144}(SC6)_{60}$ and $Au_{144}(PET)_{60}$ samples described above. We call this second stable structure for $Au_{144}(SR)_{60}$ Form II. The decahedral structure fits well in the thiol stabilized gold cluster structure series. From single-crystal XRD of smaller clusters, a strong effect of ligand on internal structure and allowed nuclearity can be inferred, with close-packed and icosahedral structures both observed. For instance, $Au_{25}(PET)_{18}$ (ref. 47), $Au_{38}(PET)_{24}$ (ref. 48) and $Au_{133}(SPh\text{-}tBu)_{52}$ (ref. 43) have been determined to have icosahedral cores. This is in contrast to $Au_{18}(SC_6H_{11})_{14}$ (ref. 49), $Au_{36}(SPh\text{-}tBu)_{24}$ (ref. 50) and $Au_{102}(p\text{-}MBA)_{44}$ (ref. 3), which have close-packed cores. A cuboctahedron-like structure (which also has closed-packed motifs) was seen for the $Au_{68}(p\text{-}MBA)_{32}$ cluster by advanced single-particle electron microscopy methods[51]. It has furthermore been shown that substituting Au by Ag in ligand-free clusters containing *ca.* 312 metal atoms changes the structure from fcc to icosahedral[52].

Form I and II coexistence. We next attempted to find a trend in ligand type for stabilizing the different structural forms and tested the effects of using linear thiol ligands of different length, namely SC4 and SC12, which we compare with the SC6 and PET samples. The PDF data for SC4, SC12 and PET are shown in Fig. 5a–c along with fits using the Form I model. The refined parameters are given in Supplementary Table 4, where the data from the hexane thiolated sample (SC6) and PET samples show good agreement with the icosahedral Form I model, and the SC4 and SC12 samples give much larger residuum values of 17.9% and 18.6%, respectively. Interestingly, the disagreement between data and model is particularly large around $r = 5$ Å, which is exactly

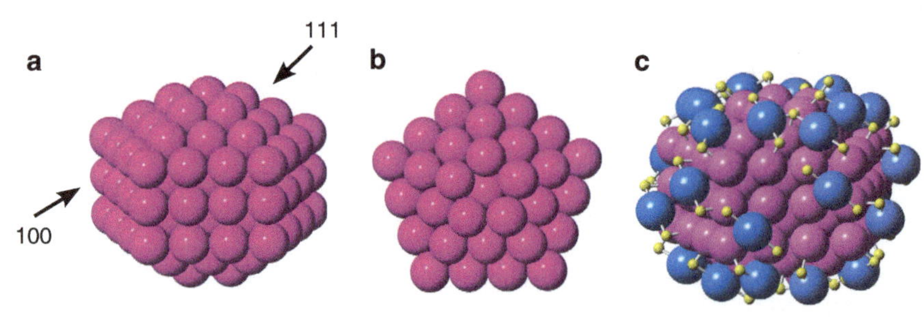

Figure 4 | 114-Atom and 144-atom ino decahedron cluster. (**a**) Side view. (**b**) Top view. (**c**) Decorated with 60 (SR-Au-SR) staples. Pink spheres show gold atoms in the cluster core, whereas blue spheres show gold in the staple structure. Sulfur is shown in yellow. The organic chains have been left out for clarity.

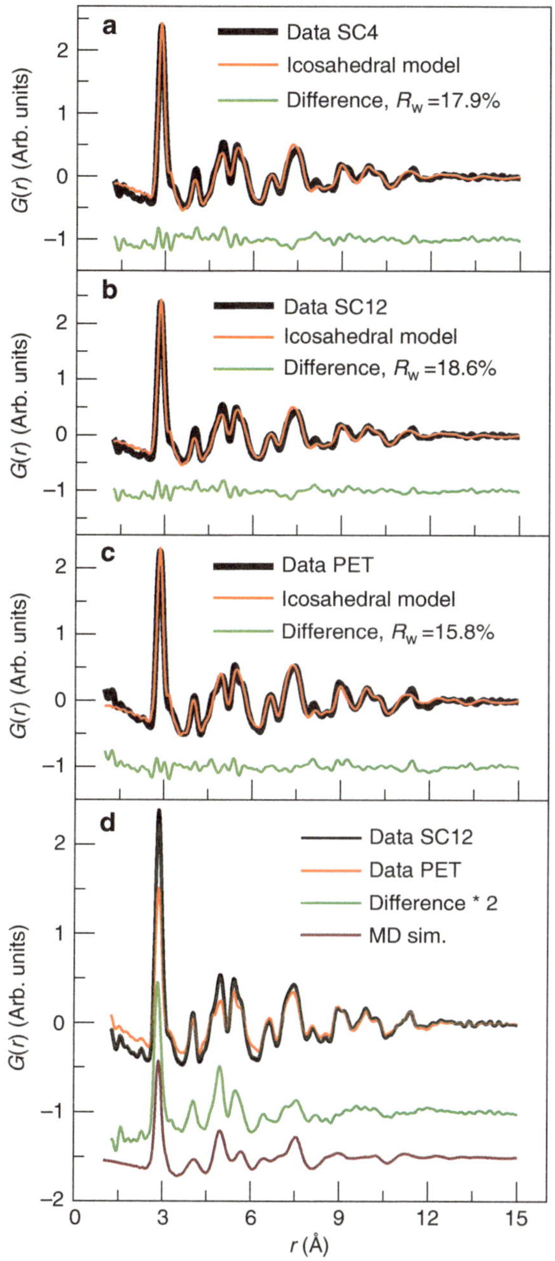

Figure 5 | Fits of the Bahena model to experimental data. (**a**) Fit to $Au_{144}(SC4)_{60}$ data, (**b**) to $Au_{144}(SC12)_{60}$ data and (**c**) to $Au_{144}(PET)_{60}$ data. (**d**) Data for the $Au_{144}(PET)_{60}$ and $Au_{144}(SC12)_{60}$; the difference between them and the calculated PDF from the 114-atom decahedron model. The difference curve has been doubled in scale for clarity.

the position for one of the most dominating peaks in the decahedral PDF. ESI-MS data from both the SC4 and SC12 samples showed $Au_{144}(SR)_{60}$, as well as impurity peaks corresponding to $Au_{137}(SR)_{56}$ in quantities comparable to the PET-protected samples. No other clusters were seen. As the presence of $Au_{137}(SR)_{56}$ in the ESI-MS data did not affect the PDF fits to the PET-protected sample and as no other clusters are identified by ESI-MS, we can rule out that the disagreement is due to the presence of other cluster sizes.

In Fig. 5d, the experimentally derived PET PDF (in Form I) has been subtracted from the SC12 calculated PDF and the difference curve is plotted below. Close inspection indicates that it strongly resembles the PDF of the Form II decahedral core structure, as seen when comparing with the calculated PDF from the 114-atom decahedron plotted along with the data. The difference curve has exactly the same features as seen from the decahedron phase, showing that the sample contains clusters of two distinct structures: Form I and Form II. Similar results are seen for the SC4 sample as illustrated in Supplementary Fig. 12. Two-phase fits showed that the SC4 sample contains 12% decahedral clusters (88% icosahedral clusters), whereas the SC12 sample has 14% decahedral clusters, as listed in Supplementary Table 5 and illustrated in Supplementary Fig. 13. When including the decahedral phase in the fit, the resulting *R*-values are reduced to *ca.* 16%. The results unambiguously show that two polymorphs of the $Au_{144}(SR)_{60}$ cluster are present.

Discussion

The question remains which factors affect the polymorph. Previous studies of gold clusters have indicated that ligand length may influence the structure of the gold core[14]. Our total scattering data cannot confirm this trend, as both the longest (SC12) and shortest (SC4) linear ligand give mainly icosahedral clusters, with a smaller fraction of decahedral clusters present in each sample. The fact that we see both the icosahedral and decahedral clusters in samples made with the same ligands illustrate that the structural diversity is not a simple effect of the ligand chain length, bulkiness or bonding strength. It is a clear indication that the two structures are very close in energy. As discussed above, Wong *et al.*[21] reported ^{1}H-NMR studies of $Au_{144}(p\text{-}MBA)_{60}$ clusters, which showed only one doublet in the aromatic region of the spectrum, suggesting that all ligands are in symmetry equivalent positions. Interestingly, ^{13}C-NMR studies on 29 kDa gold clusters have shown that the NMR signal is highly dependent on the charge state of the nanocluster. The simple NMR signal indicating symmetry equivalent ligands was seen only when the clusters were in charge state $+3$, whereas other signals were seen at lower charge states[53]. In recent work, Tlahuice-Floret *et al.*[54] used DFT to study the effect of charge state on structure of the gold subhalide $Au_{144}Cl_{60}$, which is isoelectronic with $Au_{144}(SR)_{60}$. It was shown that a fully symmetric icosahedral structure is stable at charge states $+2$ and $+4$, but not at neutrality. Furthermore, other studies have experimentally illustrated charge-dependent thermal stability of $Au_{144}(SC_6H_{13})$[55]. All our data have been measured in the uncharged state and, therefore, we cannot comment on a charge-dependent structure in $Au_{144}(p\text{-}MBA)$. However, when considering our new PDF data, applicable for detailed nanostructure analysis, along with the previously published powder x-Ray diffraction (PXRD), STEM and NMR data, this again points to a scenario where the icosahedral core structure and decahedral structure both exist with very similar energies. Small differences in the electronic state of the cluster from, for example, charge or ligand binding, could lead to different structures and, possibly, even switching between the different structural forms.

As noted above, the full staple arrangement on the decahedral clusters cannot be deduced from the PDFs and X-ray scattering data must be combined with techniques sensitive to the organic ligands to establish the total structure. If considering also the experimental PDFs from the icosahedral structures (that is, with SC6 and PET ligands), we note that neither the Bahena *et al.*[22] or Lopez-Acevedo *et al.*[20] models fully capture the details in the PDFs, for example, in the peaks between 4 and 5 Å in Fig. 3a. Some structural details exist, which are not present in the established models. Therefore, further studies of the structure are needed, where scattering is combined with theory[54] and spectroscopy[56], to establish the total structure.

The polymorphism seen in our data suggests many new studies of gold nanoclusters. The close energies between different cluster structures may not be limited to the $Au_{144}(SR)_{60}$ cluster family, but exist in a larger size range and in different materials systems. The presence of polymorphism challenges some of the characterization methods that are used for structure solution. For example, when applying single-crystal XRD, the crystallization process works as a structural sieve that will favour only one cluster polymorph over others that may be present in suspension, resulting in an incomplete picture. As we show, PDF will see the average sample and any structural heterogeneity will be observed in the data. Compared with electron beams used for STEM studies, X-rays are much less perturbing of the system and any structural changes due to beam irradiation are therefore less probable. Furthermore, PDF allows to distinguish between seemingly similar clusters, that is, $Au_{130}(SR)_{50}$, $Au_{133}(SR)_{52}$ and the two forms of $Au_{144}(SR)_{60}$.

In summary, we have shown by means of total scattering PDF analysis on well-characterized samples of $Au_{144}(SR)_{60}$ that the clusters can take two distinct structures: a truncated decahedron structure (Form II) and the previously proposed icosahedral structure (Form I). The two structures have been isolated in samples with *p*-MBA and SC6 ligands, respectively, but in samples with SC4 and SC12, the two structures are seen to coexist, indicating that the energy of the two structures are very close to each other. In recent times, several new metal clusters have been isolated in the size range from 50 to 300 atoms[57,58]. The structures of many of these clusters remain undetermined, owing to difficulties in crystallizing the clusters into a large, single crystal suitable for structure determination. We believe that PDF will be an excellent tool for these studies and, when combined with spectroscopic methods, will be able to provide full structure solutions to many new nanomaterials.

Methods

Synthesis and purification of $Au_{144}(SR)_{60}$. Au_{144}(*p*-MBA)$_{60}$ was synthesized as in Ackerson *et al.*[39] and described in Supplementary Note 7, along with the purification and characterization. $Au_{144}(SR)_{60}$, where SR = PET, S4, S6 and S12, synthesis followed Qian *et al.*[59], also discussed further in Supplementary Note 7. The samples were characterized by ESI–MS (Supplementary Fig. 1 and Supplementary Table 1) and electrochemical methods (Supplementary Fig. 14–16).

X-ray total scattering experiments. Total scattering data were acquired during three different beamtimes at three different facilities. For all samples, the cluster powders were loaded in Kapton tubes with inner diameter of 1 mm. Data for both samples of the Au_{144}(*p*-MBA)$_{60}$ cluster was obtained at ID11 at the European Synchrotron Radiation Facility with an X-ray wavelength of 0.1774 Å at 100 K. For the $Au_{144}(PET)_{60}$, $Au_{144}(SC6)_{60}$, $Au_{144}(SC4)_{60}$ and $Au_{144}(SC12)_{60}$ clusters, data were measured at beamline 11-ID-B at the Advanced Photon Source, at Argonne National Laboratory. Here, data were measured at 100 and 300 K with X-ray wavelength 0.143 Å. Additional data sets for the $Au_{144}(PET)_{60}$ were furthermore measured at the X7B beamline at room temperature with X-ray wavelength of 0.319 Å, as well as at the X17A beamline, X-ray wavelength 0.186 Å at 100 and 300 K, both at the National synchrotron light source facility at Brookhaven National Laboratory.

The experimental powder diffraction patterns were integrated using the programme *Fit2D*[60] and Fourier transformed to obtain the PDF using the programme *PDFgetX3*[61]. Modelling was done using *DiffPy-CMI*. Details on the PDF analysis can be found in Supplementary Note 1.

References

1. Brust, M., Fink, J., Bethell, D., Schiffrin, D. J. & Kiely, C. Synthesis and reactions of functionalized gold nanoparticles. *J. Chem. Soc. Chem. Commun.* 1655–1656 (1995).
2. Brust, M., Walker, M., Bethell, D., Schiffrin, D. J. & Whyman, R. Synthesis of thiol-derivatized gold nanoparticles in a 2-phase liquid-liquid system. *J. Chem. Soc. Chem. Commun.* 801–802 (1994).
3. Jadzinsky, P. D., Calero, G., Ackerson, C. J., Bushnell, D. A. & Kornberg, R. D. Structure of a thiol monolayer-protected gold nanoparticle at 1.1 angstrom resolution. *Science* **318,** 430–433 (2007).
4. Walter, M. *et al.* A unified view of ligand-protected gold clusters as superatom complexes. *Proc. Natl Acad. Sci. USA* **105,** 9157–9162 (2008).
5. Zhu, M., Aikens, C. M., Hollander, F. J., Schatz, G. C. & Jin, R. Correlating the crystal structure of A thiol-protected Au-25 cluster and optical properties. *J. Am. Chem. Soc.* **130,** 5883–5889 (2008).
6. Weissker, H. C. *et al.* Information on quantum states pervades the visible spectrum of the ubiquitous Au-144(SR)(60) gold nanocluster. *Nat. Commun.* **5,** 3785 (2014).
7. Philip, R., Chantharasupawong, P., Qian, H. F., Jin, R. C. & Thomas, J. Evolution of nonlinear optical properties: from gold atomic clusters to plasmonic nanocrystals. *Nano Lett.* **12,** 4661–4667 (2012).
8. Negishi, Y. *et al.* A critical size for emergence of nonbulk electronic and geometric structures in dodecanethiolate-protected Au clusters. *J. Am. Chem. Soc.* **137,** 1206–1212 (2014).
9. Daniel, M. C. & Astruc, D. Gold nanoparticles: assembly, supramolecular chemistry, quantum-size-related properties, and applications toward biology, catalysis, and nanotechnology. *Chem. Rev.* **104,** 293–346 (2004).
10. Jin, R. C. Quantum sized, thiolate-protected gold nanoclusters. *Nanoscale* **2,** 343–362 (2010).
11. Qian, H. F., Zhu, M. Z., Wu, Z. K. & Jin, R. C. Quantum sized gold nanoclusters with atomic precision. *Acc. Chem. Res.* **45,** 1470–1479 (2012).
12. Billinge, S. J. L. & Levin, I. The problem with determining atomic structure at the nanoscale. *Science* **316,** 1698–1698 (2007).
13. Whetten, R. L. *et al.* Nanocrystal gold molecules. *Adv. Mater.* **8,** 428–433 (1996).
14. Schaaff, T. G., Shafigullin, M. N., Khoury, J. T., Vezmar, I. & Whetten, R. L. Properties of a ubiquitous 29 kDa Au: SR cluster compound. *J. Phys. Chem. B* **105,** 8785–8796 (2001).
15. Tian, S. *et al.* Structural isomerism in gold nanoparticles revealed by X-ray crystallography. *Nat. Commun.* **6,** 8667 (2015).
16. Alvarez, M. M. *et al.* Critical sizes in the growth of Au clusters. *Chem. Phys. Lett.* **266,** 91–98 (1997).
17. Cleveland, C. L. *et al.* Structural evolution of smaller gold nanocrystals: the truncated decahedral motif. *Phys. Rev. Lett.* **79,** 1873–1876 (1997).
18. Chaki, N. K., Negishi, Y., Tsunoyama, H., Shichibu, Y. & Tsukuda, T. Ubiquitous 8 and 29 kDa gold: alkanethiolate cluster compounds: mass-spectrometric determination of molecular formulas and structural implications. *J. Am. Chem. Soc.* **130,** 8608–8612 (2008).
19. Qian, H. F. & Jin, R. C. Controlling nanoparticles with atomic precision: the case of Au-144(SCH2CH2Ph)(60). *Nano Lett.* **9,** 4083–4087 (2009).
20. Lopez-Acevedo, O., Akola, J., Whetten, R. L., Gronbeck, H. & Hakkinen, H. Structure and bonding in the ubiquitous icosahedral metallic gold cluster Au-144(SR)(60). *J. Phys. Chem. C* **113,** 5035–5038 (2009).
21. Wong, O. A., Heinecke, C. L., Simone, A. R., Whetten, R. L. & Ackerson, C. J. Ligand symmetry-equivalence on thiolate protected gold nanoclusters determined by NMR spectroscopy. *Nanoscale* **4,** 4099–4102 (2012).
22. Bahena, D. *et al.* STEM electron diffraction and high-resolution images used in the determination of the crystal structure of the Au-144(SR)(60) cluster. *J. Phys. Chem. Lett.* **4,** 975–981 (2013).
23. Hakkinen, H. The gold-sulfur interface at the nanoscale. *Nat. Chem.* **4,** 443–455 (2012).
24. Tran, N. T., Powell, D. R. & Dahl, L. F. Nanosized Pd-145(CO)(x) (PEt3)(30) containing a capped three-shell 145-atom metal-core geometry of pseudo icosahedral symmetry. *Angew. Chem. Int. Ed.* **39,** 4121–4125 (2000).
25. Billinge, S. J. L. & Kanatzidis, M. G. Beyond crystallography: the study of disorder, nanocrystallinity and crystallographically challenged materials with pair distribution functions. *Chem. Commun.* **7,** 749–760 (2004).
26. Young, C. A. & Goodwin, A. L. Applications of pair distribution function methods to contemporary problems in materials chemistry. *J. Mater. Chem.* **21,** 6464–6476 (2011).
27. Yang, X. H. *et al.* Confirmation of disordered structure of ultrasmall CdSe nanoparticles from X-ray atomic pair distribution function analysis. *Phys. Chem. Chem. Phys.* **15,** 8480–8486 (2013).
28. Neder, R. B. & Korsunskiy, V. I. Structure of nanoparticles from powder diffraction data using the pair distribution function. *J. Phys. Condens. Matter* **17,** S125–S134 (2005).
29. Page, K., Hood, T. C., Proffen, T. & Neder, R. B. Building and refining complete nanoparticle structures with total scattering data. *J. Appl. Crystallogr.* **44,** 327–336 (2011).
30. Juhas, P., Cherba, D. M., Duxbury, P. M., Punch, W. F. & Billinge, S. J. L. *Ab initio* determination of solid-state nanostructure. *Nature* **440,** 655–658 (2006).

31. Cliffe, M. J., Dove, M. T., Drabold, D. A. & Goodwin, A. L. Structure determination of disordered materials from diffraction data. *Phys. Rev. Lett.* **104,** 125501 (2010).
32. Kumara, C., Zuo, X. B., Ilavsky, J., Cullen, D. & Dass, A. Atomic structure of Au-329(SR)(84) faradaurate plasmonic nanomolecules. *J. Phys. Chem. C* **119,** 11260–11266 (2015).
33. Kumara, C., Zuo, X. B., Cullen, D. A. & Dass, A. Faradaurate-940: synthesis, mass spectrometry, electron microscopy, high-energy X-ray diffraction, and X-ray scattering study of AU ~ 940 ± 20(SR) ~ 160 ± 4 nanocrystals. *ACS Nano* **8,** 6431–6439 (2014).
34. Kumara, C. *et al.* Super-stable, highly monodisperse plasmonic Faradaurate-500 nanocrystals with 500 gold atoms: Au(~ 500)(SR)(~ 120). *J. Am. Chem. Soc.* **136,** 7410–7417 (2014).
35. Billinge, S. J. L. *et al.* Characterisation of amorphous and nanocrystalline molecular materials by total scattering. *CrystEngComm* **12,** 1366–1368 (2010).
36. Jupally, V. R. *et al.* Au-137(SR)(56) nanomolecules: composition, optical spectroscopy, electrochemistry and electrocatalytic reduction of CO2. *Chem. Commun.* **50,** 9895–9898 (2014).
37. Dharmaratne, A. C. & Dass, A. Au144-xCux(SC6H13)(60) nanomolecules: effect of Cu incorporation on composition and plasmon-like peak emergence in optical spectra. *Chem. Commun.* **50,** 1722–1724 (2014).
38. Salorinne, K. *et al.* Nondestructive size determination of thiol-stabilized gold nanoclusters in solution by diffusion ordered NMR spectroscopy. *Anal. Chem.* **85,** 3489–3492 (2013).
39. Ackerson, C. J., Jadzinsky, P. D., Sexton, J. Z., Bushnell, D. A. & Kornberg, R. D. Synthesis and bioconjugation of 2 and 3 nm-diameter gold nanoparticles. *Bioconjug. Chem.* **21,** 214–218 (2010).
40. Koivisto, J. *et al.* Vibrational perturbations and ligand-layer coupling in a single crystal of Au-144(SC2H4Ph)(60) nanocluster. *J. Phys. Chem. Lett.* **5,** 387–392 (2014).
41. Jeong, I. K., Proffen, T., Mohiuddin-Jacobs, F. & Billinge, S. J. L. Measuring correlated atomic motion using X-ray diffraction. *J. Phys. Chem. A* **103,** 921–924 (1999).
42. Chen, Y. X. *et al.* Crystal structure of barrel-shaped chiral Au-130(p-MBT)(50) nanocluster. *J. Am. Chem. Soc.* **137,** 10076–10079 (2015).
43. Dass, A. *et al.* Au-133(SPh-tBu)(52) nanomolecules: X-ray crystallography, optical, electrochemical, and theoretical analysis. *J. Am. Chem. Soc.* **137,** 4610–4613 (2015).
44. Marks, L. D. Surface-structure and energetics of multiply twinned particles. *Philos. Mag.* **49,** 81–93 (1984).
45. Marks, L. D. Experimental studies of small-particle structures. *Rep. Prog. Phys.* **57,** 603–649 (1994).
46. Jiang, D. E., Tiago, M. L., Luo, W. D. & Dai, S. The 'Staple' motif: a key to stability of thiolate-protected gold nanoclusters. *J. Am. Chem. Soc.* **130,** 2777–2784 (2008).
47. Heaven, M. W., Dass, A., White, P. S., Holt, K. M. & Murray, R. W. Crystal structure of the gold nanoparticle [N(C8H17)(4)][Au-25(SCH2CH2Ph)(18)]. *J. Am. Chem. Soc.* **130,** 3754–3762 (2008).
48. Qian, H. F., Eckenhoff, W. T., Zhu, Y., Pintauer, T. & Jin, R. C. Total structure determination of thiolate-protected Au-38 nanoparticles. *J. Am. Chem. Soc.* **132,** 8280–8284 (2010).
49. Das, A. *et al.* Structure determination of [Au-18(SR)(14)]. *Angew. Chem. Int. Ed.* **54,** 3140–3144 (2015).
50. Das, A. *et al.* Cyclopentanethiolato-protected Au-36(SC5H9)(24) nanocluster: crystal structure and implications for the steric and electronic effects of ligand. *J. Phys. Chem. A* **118,** 8264–8269 (2014).
51. Azubel, M. *et al.* Electron microscopy of gold nanoparticles at atomic resolution. *Science* **345,** 909–912 (2014).
52. Jian, N. & Palmer, R. E. Variation of the core atomic structure of thiolated $(Au_xAg_{1-x})_{312\pm55}$ nanoclusters with composition from aberration-corrected HAADF STEM. *J. Phys. Chem. C* **119,** 11114–11119 (2015).
53. Song, Y., Harper, A. S. & Murray, R. W. Ligand heterogeneity on monolayer-protected gold clusters. *Langmuir* **21,** 5492–5500 (2005).
54. Tlahuice-Flores, A., Black, D. M., Bach, S. B. H., Jose-Yacaman, M. & Whetten, R. L. Structure & bonding of the gold-subhalide cluster I-Au144Cl60[z]. *Phys. Chem. Chem. Phys.* **15,** 19191–19195 (2013).
55. Tofanelli, M. A. & Ackerson, C. J. Superatom electron configuration predicts thermal stability of Au-25(SR)(18) nanoclusters. *J. Am. Chem. Soc.* **134,** 16937–16940 (2012).
56. Yamazoe, S. *et al.* Hierarchy of bond stiffnesses within icosahedral-based gold clusters protected by thiolates. *Nat. Commun.* **7,** 10414 (2016).
57. Li, G., Zeng, C. J. & Jin, R. C. Thermally robust Au-99(SPh)(42) nanoclusters for chemoselective hydrogenation of nitrobenzaldehyde derivatives in water. *J. Am. Chem. Soc.* **136,** 3673–3679 (2014).
58. Nimmala, P. R., Yoon, B., Whetten, R. L., Landman, U. & Dass, A. Au-67(SR)(35) nanomolecules: characteristic size-specific optical, electrochemical, structural properties and first-principles theoretical analysis. *J. Phys. Chem. A* **117,** 504–517 (2013).
59. Qian, H. F. & Jin, R. C. Ambient synthesis of Au-144(SR)(60) nanoclusters in methanol. *Chem. Mater.* **23,** 2209–2217 (2011).
60. Hammersley, A. P., Svensson, S. O., Hanfland, M., Fitch, A. N. & Hausermann, D. Two-dimensional detector software: from real detector to idealised image or two-theta scan. *High Pressure Res.* **14,** 235–248 (1996).
61. Juhas, P., Davis, T., Farrow, C. L. & Billinge, S. J. L. PDFgetX3: a rapid and highly automatable program for processing powder diffraction data into total scattering pair distribution functions. *J. Appl. Crystallogr.* **46,** 560–566 (2013).
62. Jensen, K. M. O. *et al.* Data for polymorphism in magic sized Au144(SR)60 clusters. https://doi.org/10.5281/zenodo.51551 (2016).

Acknowledgements

K.M.Ø.J. acknowledges support from the Villum Foundation. S.J.L.B and P.J. acknowledge funding from Laboratory Directed Research and Development (LDRD) Program 12-007 (Complex Modeling) at Brookhaven National Laboratory, which is funded by the U.S. Department of Energy, Office of Basic Energy Sciences grant DE-SC00112704. We acknowledge the European Synchrotron Radiation Facility for provision of synchrotron radiation facilities at beamline ID11. Use of the Advanced Photon Source at Argonne National Laboratory was supported by the U.S. Department of Energy, Office of Science, Office of Basic Energy Sciences, under contract number DE-AC02-06CH11357. Use of the National Synchrotron Light Source, Brookhaven National Laboratory, was supported by the U.S. Department of Energy, Office of Science, Office of Basic Energy Sciences, under contract number DE-AC02-98CH10886. C.J.A., M.A.T. and C.L.H. acknowledge funding from Colorado State University. C.J.A. was an American Federation for Aging Research New Investigator, while this research was conducted. C.J.A. acknowledges NIH R21 EB014520. We thank Hannu Häkkinen for useful feedback and structure data used at early stages of this work.

Author contributions

M.A.T., C.L.H. and C.H.A. synthesized and characterized the samples. K.M.Ø.J., S.J.L.B. and G.V. collected the total scattering data. K.M.Ø.J. and P.J. analysed and modelled the data. K.M.Ø.J., P.J. and S.J.L.B. wrote the manuscript.

Additional information

Competing financial interests: The authors declare no competing financial interests.

9

Room temperature molecular up conversion in solution

Aline Nonat[1], Chi Fai Chan[2], Tao Liu[1,2], Carlos Platas-Iglesias[3], Zhenyu Liu[4], Wing-Tak Wong[4,5], Wai-Kwok Wong[2], Ka-Leung Wong[2] & Loïc J. Charbonnière[1]

Up conversion is an Anti-Stokes luminescent process by which photons of low energy are piled up to generate light at a higher energy. Here we show that the addition of fluoride anions to a D_2O solution of a macrocyclic erbium complex leads to the formation of a supramolecular $[(ErL)_2F]^+$ assembly in which fluoride is sandwiched between two complexes, held together by the synergistic interactions of the Er-F-Er bridging bond, four intercomplex hydrogen bonds and two aromatic stacking interactions. Room temperature excitation into the Er absorption bands at 980 nm of a solution of the complex in D_2O results in the observation of up converted emission at 525, 550 and 650 nm attributed to Er centred transitions *via* a two-step excitation. The up conversion signal is dramatically increased upon formation of the $[(ErL)_2F]^+$ dimer in the presence of 0.5 equivalents of fluoride anions.

[1] Laboratoire d'Ingénierie Moléculaire Appliquée à l'Analyse, IPHC, UMR 7178 CNRS, Université de Strasbourg, ECPM, Bât R1N0, 25 rue Becquerel, Strasbourg, 67087, France. [2] Department of Chemistry, Hong Kong Baptist University, Hong Kong SAR, Hong Kong. [3] Centro de Investigaciones Científicas Avanzadas, Departamento de Química Fundamental, Universidade da Coruña, Campus da Zapateira, Rúa da Fraga 10, A Coruña 15008, Spain. [4] State Key Laboratory of Chirosciences, Department of Applied Biology and Chemical Technology, Hong Kong Polytechnic University, Kowloon, Hong Kong. [5] Department of Applied Biology and Chemical Technology, Hong Kong Polytechnic University, Hong Kong SAR, Hong Kong. Correspondence and requests for materials should be addressed to W.-T.W. (e-mail: bcwtwong@polyu.edu.hk), or to K.-L.W. (e-mail: klwong@hkbu.edu.hk) or to L.J.C. (e-mail:l.charbonn@unistra.fr).

Conventional luminescence spectroscopy focuses on the emission of photons after absorption of photons of higher energy (that is, down shifting), but processes giving off emission of photons at higher energy than the incident beam and not related to thermal population of the excited states are far less common[1]. These processes gather aspects related to nonlinear optics, such as second and third harmonic generations or two photon absorption[2], and those associated to the cumulative effects of multiple first-order absorption phenomena, also named up conversion (UC)[3], among which are excited state absorption (ESA) and energy transfer up conversion (ETU)[1]. The last two became very popular in recent years, first because they are far more efficient than the others, affording observation of UC with more classical excitation sources, but also because of intensive research on the synthesis and preparation of UC materials, particularly in the field of lanthanide-doped phosphors[3] and more recently with lanthanide-based nanoparticles[4,5]. Lanthanide ions such as Er, Tm or Ho are particularly well-suited for UC processes as they gather two important properties: a ladder-like energy level profile spanning from the Near-infrared (NIR) to the visible region; and long-lived excited states due to Laporte forbidden electric dipole transitions[6].

UC is particularly appealing for biophotonic applications as the emitted signal is devoid of spurious signals such as auto-fluorescence of the sample or excitation bleed through, rendering an almost background-free measurement. Also, the NIR excitation is entirely dedicated to the excitation of the probe, as biological tissues absorb weakly in this region of the spectrum[7], and the penetration of excitation light into the tissues is thus deeper, allowing applications to bioimaging[8,9] or photodynamic therapy[10].

Despite these evident new opportunities, the development of UC devices is still confronted with a scientific challenge: the downscaling of the probes to the molecular scale. In bulk solids and nanoparticles, the phonon energy of the lattice plays a crucial role in luminescence quenching mechanisms. When phonons or their overtones are in resonance with the intermediate excited state levels of the emitting centres, non-radiative phonon-assisted deactivation pathways compete severely with the successive piling up of excited states necessary to reach the higher energy emitting level[11]. At the molecular level, the situation is even worse, where OH, NH and CH oscillators, commonly present in the ligand backbones of the Ln complexes or in their solvation shells, contribute as efficient non-radiative deactivation pathways[12]. As a result, even conventional NIR luminescence of such Ln complexes rarely exceeds quantum efficiencies of a few percent, in the best cases[13,14].

In spite of such apparently insurmountable obstacles, encouraging results were recently published. In 2011, Piguet, Hauser and co-workers reported on the observation of green erbium UC emission upon NIR chromium excitation in a heterotrinuclear triple-stranded helicate composed of two Cr sensitizers and a central Er acceptor in the solid state and solid solutions at room temperature and in frozen acetonitrile up to 150 K (refs 15,16). Unambiguous evidence for the UC process was brought forward by the quadratic dependence of the emitted intensity as a function of excitation power. The study of isostructural compounds of the spectroscopically silent Ga analogue[16] also allowed the authors to get a detailed overview of the kinetic parameters of their system. Within the frame of Förster's theory on energy transfer[17], the energy transfer efficiency is dependent on the inversed sixth power of the donor–acceptor distance and one can anticipate that the rather large distances between the Cr sensitizers and the Er acceptor (8.9 Å on the basis of the analogous Yb compound) may not lead to an efficient energy transfer[18]. In 2012, Faulkner, Beeby and Sorensen achieved multiphoton excitation of different lanthanide salts of Eu, Tb and Sm upon intense excitation of some concentrated ($c > 0.01$ M) triflate salts and complexes in d_6-dimethylsulfoxide solutions[19]. In the case of thulium, they observed a faint excitation band in the multiphoton excitation spectrum that was proposed to be a potential UC process, but lacked details to follow up with the work[20]. Very recently, Hyppänen and co-workers reported on the observation of UC in a system composed of an Er complex and a NIR dye photosensitizer in $CDCl_3$ at room temperature upon strong excitation ($> 3.4\ W\ cm^{-2}$) at 808 nm (ref. 21).

In this contribution, we show that the macrocyclic complex of erbium with ligand L, $[ErL(H_2O)]^+$ can self-assemble in D_2O solution in the presence of fluoride anions to form a supramolecular $[(ErL)_2F]^+$ dimer (Fig. 1). Upon excitation into the NIR absorption band of Er at 980 nm in D_2O, a green emission is observed arising from an UC process, which is largely amplified with the formation of the Er dimer, representing, to the best of our knowledge, the first evidence of molecular UC in heavy water solution at room temperature.

Results

Synthesis and characterization of the complex. The $[ErL(H_2O)](NO_3)$ complex was prepared from ligand L and $Er(NO_3)_3.5H_2O$ (ref. 22) and was fully characterized by elemental analysis, infrared spectroscopy, ^{1}H-NMR and electrospray mass spectrometry (Supplementary Figs 1–4). The presence of the nitrate anions was particularly well evidenced by infrared spectroscopy with new bands at 1,655 and 1,360 cm^{-1}, absent in the chloride complexes of Eu and Tb with L (ref. 22), and characteristic of nitrate anions[23]. In H_2O, the electrospray mass spectrum of the complex displayed a single intense peak with maximum at 714.19 *m/z* units, with an isotopic distribution perfectly fitting that calculated for a $[ErL]^+$ monocharged cation. A very minor species (<2%) could be evidenced as a peak with maximum at 1,488.37 *m/z*, which was related to the presence of two monomers linked by a nitrate anion $[(ErL)_2(NO_3)]^+$ (calculated mass for $C_{56}Er_2H_{68}N_{17}O_{11}^+$: 1,488.39). Finally, the

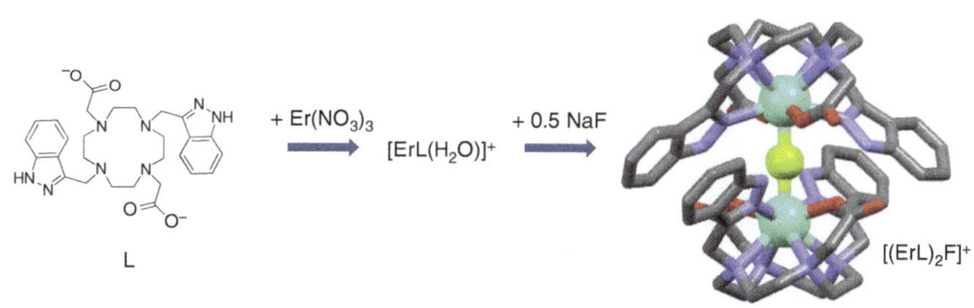

Figure 1 | Formation of the supramolecular erbium dimer. The ligand L react with Er^{3+} salts in water to form an $[ErL(H_2O)]^+$ complex. In the presence of half an equivalent of fluoride anions two complexes form an association in which the F^- anion is sandwiched between the two complexes leading to the $[(ErL)_2F]^+$ dimer. The association is strengthened by two intercomplex stacking interactions and four hydrogen bonds.

complex was also characterized by ^{1}H-NMR spectrometry. Although at the limit of the solubility, and despite the paramagnetic contribution of the Er cation, it was possible to record the spectrum of the complex in D_2O. The spectrum of the paramagnetic complex displayed 15 broad peaks between ca -110 and $+110$ parts per million (p.p.m.), for 16 expected for a C_2 symmetry of the complex (Supplementary Figs 1 and 2). It is surmised that the missing peak is hidden under the peak of the partially deuterated water, HDO. Interestingly, minor peaks corresponding to a second paramagnetic species (<6%) could also be observed in D_2O. Considering that the major species is in a square antiprismatic geometry, as observed for the parent Eu and Yb complex[22], the minor species may be attributed to the presence of some twisted square antiprismatic conformer, as observed in the Ln complexes of 1,4,7,10-tetraazacyclododecane-1,4,7,12-tetraacetic acid (DOTA)[24].

Spectroscopic characterization of the complex. The absorption spectrum of the Er complex in D_2O is composed of a broad absorption band with maximum at 294 nm ($\varepsilon = 9{,}850\ M^{-1}\ cm^{-1}$, Fig. 2), which was attributed to $\pi \rightarrow \pi^*$ transitions centred on the indazolyl moieties[25]. The transmittance spectrum was recorded in the solid state from 400 to 1,800 nm (Supplementary Fig. 5) and displayed numerous Er centred absorption bands attributed to transitions from the ground state to the $^4F_{7/2}$, $^2H_{11/2}$, $^4S_{3/2}$, $^4F_{9/2}$, $^4I_{11/2}$ and $^4I_{13/2}$, respectively at 488, 522, 541, 657, 980 and 1,514 nm.

Upon excitation into the ligand absorption bands (294 nm), the emission spectrum of the complex in D_2O (Fig. 2) displayed a broad emission band, with maximum at 318 nm, which is attributed to some ligand centred fluorescence from the singlet state. A close examination of the visible part of the spectrum also revealed the presence of a structured emission peak between 540 and 550 nm in the low-energy tail of the fluorescence band, which can be ascribed to the $^4S_{3/2} \rightarrow {}^4I_{15/2}$ transition of Er. In the NIR region of the spectrum, a weak and broad band can be found at 980 nm, attributed to the $^4I_{11/2} \rightarrow {}^4I_{15/2}$ transition of Er and a strong and structured emission band could be observed with maximum at 1,535 nm associated to the $^4I_{13/2} \rightarrow {}^4I_{15/2}$ transition of Er. These last emissions arising from Er centred bands point to the presence of a partial ligand to metal energy transfer[26]. Performing the same measurement in H_2O did not reveal any detectable Er emission in the NIR domain, nor in the 540–550 nm region. As previously observed for other lanthanides[22], the replacement of the water molecule in the first coordination sphere of the complex by a D_2O one has a dramatic influence on the luminescence properties, decreasing the non-radiative quenching of OH oscillators[27].

Addition of fluoride to the D_2O solution resulted in a *ca* 60% increase of the Er-centred luminescence intensity in the NIR region (Fig. 2c). At 1,540 nm, the maximum increase was observed at 0.5 equivalent of fluoride, corresponding to the formation of an Er_2F species (Fig. 2d). As it was previously observed for measurements in D_2O instead of H_2O, the coordination of fluoride resulted in a supplementary decrease of the non-radiative quenching processes with a concomitant increase of the luminescence. The addition of larger amounts of fluoride resulted in minor changes in the luminescence intensity, pointing to very similar metal-centred luminescence quantum yields for the species present in solution.

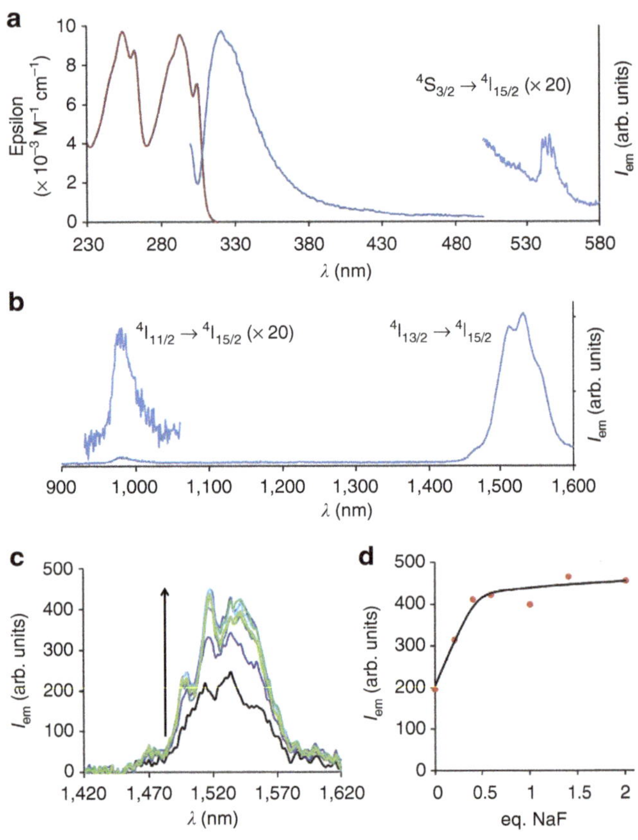

Figure 2 | Spectroscopic properties of the Er complexes. (**a**) The ultraviolet-visible absorption spectrum of the Er complex in D_2O (red, $c = 270\ \mu M$) displayed strong $\pi \rightarrow \pi^*$ absorption bands. When excited into these bands ($\lambda_{exc} = 294$ nm), the emission spectrum of the complex displayed a ligand centred emission band in the visible region (blue), together with a contribution attributed to the $^4S_{3/2} \rightarrow {}^4I_{15/2}$ transition of Er (recorded at 0.5 nm resolution). (**b**) Other Er centred transitions can be observed in the NIR region ($\lambda_{exc} = 294$ nm, emission filter at 850 nm, D_2O). (**c**) The addition of fluoride anions into the D_2O solution of the Er complex resulted in an increase of the Er emission and (**d**) the intensity of the signal at 1,540 nm revealed a maximum increase for 0.5 equivalent (eq.) of fluoride anions, related to the formation of the dimeric $[(ErL)_2F]^+$ complex (the black curve corresponds to fitted data).

UC in D_2O solution. The possibility of an UC process with the Er complex in D_2O was investigated. A 1 mM solution of the complex in D_2O was irradiated at 980 nm into the Er $^4I_{15/2} \rightarrow {}^4I_{11/2}$ transition and the emission spectrum of the solution was recorded from 500 to 680 nm (Fig. 3). In the absence of fluoride, the UC spectrum displayed three weak emission bands at *ca* 525, 550 and 650 nm (Fig. 3a). These transitions were, respectively, assigned to the $^2H_{11/2} \rightarrow {}^4I_{15/2}$, $^4S_{3/2} \rightarrow {}^4I_{15/2}$ and $^4F_{9/2} \rightarrow {}^4I_{15/2}$ emission band from Er. To this solution, increasing amounts of fluoride anions were added in the form of NaF salt. After each addition, the mixture was agitated for 30 min to ensure the equilibration of the solution. As soon as fluoride was added, the intensity of the peaks strongly increased, with a maximum at 0.5 equivalent of added fluoride, and slowly decreased afterwards (up to 4 equivalents). Interestingly, the intensity of the peaks was clearly related to the formation of an Er_2F species, with a 7.7-fold increase at 0.5 equivalent compared with the fluoride-free solution (Fig. 3b).

To further characterize the UC process, the luminescence intensity of the $^2H_{11/2} \rightarrow {}^4I_{15/2}$ and $^4S_{3/2} \rightarrow {}^4I_{15/2}$ transitions from 505 to 580 nm was monitored as a function of the intensity of the incident beam at 980 nm (Fig. 3c) for a mixture containing 0.5 equivalent of fluoride anions (Supplementary Fig. 6)[22]. The calculated slope (2.14 ± 0.12, obtained as the average value of four

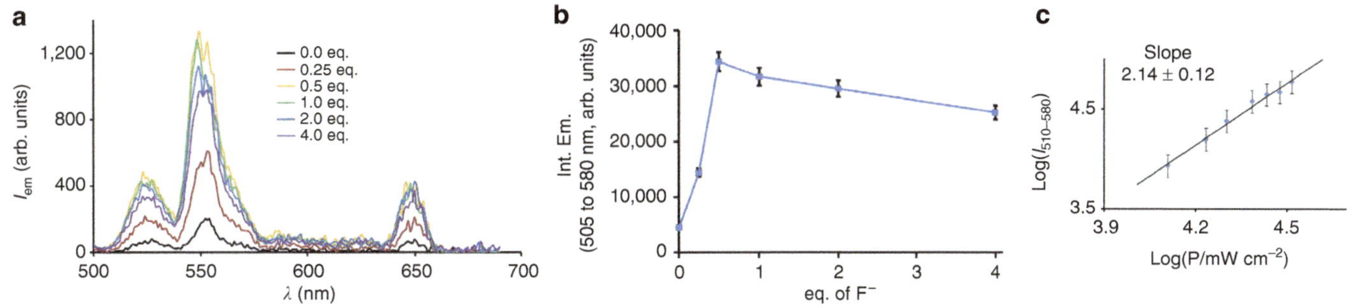

Figure 3 | UC emission spectroscopy. Excitation of a 1 mM solution of the Er complex in D_2O into the Er $^4I_{15/2} \rightarrow {}^4I_{11/2}$ absorption band at 980 nm revealed (**a**) a weak UC emission signal in the visible region associated to Er emission (black curve). Upon titration by fluoride anions, the UC emission strongly increased up to 0.5 equivalent (eq.) of fluoride (orange), as revealed (**b**) by the evolution of the intensity increase (integrated from 505 to 580 nm) as a function of added fluoride anions. (**c**) Plotting the UC emitted intensity (integrated from 505 to 580 nm) as a function of the incident pump intensity at 980 nm in a Log-Log format for the solution containing 0.5 eq. of fluoride anions gave a slope which corresponds to a two photon process (Errors were estimated assuming a Poisson's law for the intensities).

independent measurements) is slightly larger than the expected value of 2, in good agreement with UC processes[28], but such a slope may also be related with a possible ligand-based nonlinear process followed by sensitization of Er. To exclude this last possibility, the spectrum recorded at 0.5 equivalent of added fluoride was enlarged down to the ultraviolet region (from 350 to 700 nm, Supplementary Fig. 7). In these conditions, only the above-mentioned Er centred emission bands could be observed with no traces of ligand centred emission, as would be expected for a ligand-based nonlinear process followed by Er emission, similar to what is observed with one photon excitation of the ligand at 294 nm (Fig. 2). Also, the possibility of an emission at 545 nm resulting from a spurious signal arising from the $^5D_4 \rightarrow {}^7F_5$ transition of Tb impurities can be excluded by comparing the measured UC spectrum with that of the analogous Tb complex (Supplementary Fig. 8). Finally, the intensity of the UC signal at 545 nm was monitored by varying the excitation wavelength from 700 to 1,050 nm (Supplementary Fig. 9), showing a maximum of UC excitation at 980 nm for the $^4I_{15/2} \rightarrow {}^4I_{11/2}$ transition, thereby excluding the possibility of a nonlinear multiphoton process. Considering the relative weakness of the observed signal, an estimation of its quantum efficiency could be made by comparison with Er-doped UC nanoparticles pointing to an upper limit of 10^{-5}.

Additional titration experiments were performed to monitor the one photon emission spectra during Er titration with NaF with excitation into the ligand (294 nm). The visible part of the spectrum showed a weak increase of the ligand centred emission upon fluoride addition, but poor to negligible changes of intensity of the $^4S_{3/2} \rightarrow {}^4I_{15/2}$ transition of Er (at *ca* 550 nm, Supplementary Fig. 10). Following the NIR emission of the $^4I_{13/2} \rightarrow {}^4I_{15/2}$ transition of Er (1,450 to 1,600 nm, Supplementary Fig. 11) upon 980 nm excitation ($^4I_{15/2} \rightarrow {}^4I_{11/2}$) resulted in a maximum increase of ca 70% of the emission intensity, similar to what is obtained upon ligand centred excitation (Supplementary Fig. 12). Altogether, these results showed that the intensity increase of the one photon emission is not appreciably related to an improved ligand to metal energy transfer efficiency, but clearly to a better protection of the Er cation towards non-radiative processes. The weak intensity increase of the $^4S_{3/2} \rightarrow {}^4I_{15/2}$ transition compared with that of $^4I_{13/2} \rightarrow {}^4I_{15/2}$ may be related to important changes in the crystal field splitting of Er (ref. 29) with concomitant changes in the branching ratio.

Speciation of the species formed in solution. The formation of the new species was first monitored qualitatively by electrospray mass spectrometry analysis of the mixtures. Addition of 0.5 equivalent of sodium fluoride to the complex (Supplementary Figs 13 and 14) led to a new peak at 1,449.46 *m/z* units, which was attributed to a $[(ErL)_2F]^+$ singly charged dimer, together with a doubly charged cation with a peak maximum at 736.23 *m/z* units, attributed to the sodium adduct $[(ErL)_2F + Na]^{2+}$. It is interesting to note that each complex is always present with two deuterium atoms, likely as a result of the H–D exchange of the imidazol protons in D_2O. At 1.0 equivalent of NaF (Supplementary Fig. 15), the peak of the singly charged dimer at 1,449.46 became intense in the spectrum, whereas the peak corresponding to the $[ErL]^+$ monomer almost vanished. The other important species were also attributed to the dimer and can be observed with maxima at *m/z* of 725.23 and 736.22, corresponding to the doubly charged dimer with a proton or a sodium cation, respectively. At two equivalents of fluoride (Supplementary Fig. 16), peaks corresponding to the singly charged and doubly charged dimer at 1,449.45 and 736.22 *m/z* units are still present, and the peak at 716.23, assigned to the $[ErL]^+$ complex became important. In these conditions, it is expected that the neutral [ErLF] monomer began to form in solution, and its ionization resulted in the breaking of the Er–F bond, releasing the $[ErL]^+$ species.

The formation of the $[(ErL)_2F]^+$ complex was also monitored by ultraviolet-visible absorption spectroscopy, following the evolution of the absorption spectra of the complex in the presence of increasing amounts of fluoride (Fig. 4) in H_2O solution.

The association of fluoride was fitted to the equilibria:

$$[ErL(H_2O)]^+ + F^- \leftrightarrow [ErLF] + H_2O$$

$$\beta_{1Er} = \frac{[ErLF] \times [H_2O]}{[ErL(H_2O)] \times [F^-]} \qquad (1)$$

$$2[ErL(H_2O)]^+ + F^- \leftrightarrow \left[(ErL)_2F\right]^+ + 2H_2O$$

$$\beta_{2Er} = \frac{\left[(ErL)_2F\right] \times [H_2O]^2}{\left[ErL(H_2O)\right]^2 \times [F^-]} \qquad (2)$$

Taking the water concentration as constant during the titration, analysis of the ultraviolet–visible absorption data by the Specfit software[30] using equations (1) and (2) afforded stability constants of 6.1(3) and 13.0(2) log units, respectively, for β_{1Er} and β_{2Er}. The results for the cumulative stability constant β_{2Er} are in excellent agreement with the one previously reported for other complexes (log $\beta_{2Ln} = 13.0(3)$, 12.5(1.0) and 12.6(1.0), respectively, for Eu, Tb and Yb)[22]. In contrast to what was

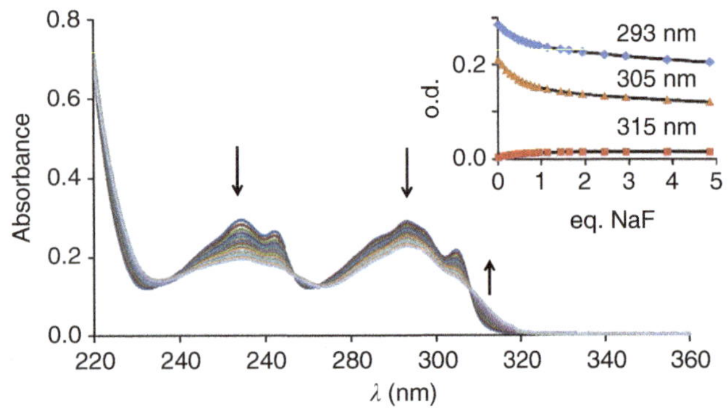

Figure 4 | Ultraviolet-visible titration of [ErL(H_2O)](NO_3). The progress of the dimerization of the erbium complex was followed by ultraviolet-visible absorption spectroscopy by addition of NaF (0–4.8 eq.) to a solution of [ErL(H_2O)](NO_3) ($c = 3.15 \times 10^{-5}$ M) in H_2O. The evolution of the spectra can be fitted with equations (1) and (2) (see the text) and the comparisons between measured and calculated values (black lines) of the absorbances at 293, 305 and 315 nm are presented in the inset.

observed for the other lanthanide cations, the titration of the Er complex allows the observation of the mononuclear [ErLF] intermediate complex. It is worth noting that the ratio of the stepwise association constants, $K_{2Er}/K_{1Er} = 6.3$ is larger than the value of 1.0 expected for a statistical behaviour (see Supplementary Methods for the calculation of the statistical value)[31,32], pointing to a weak positive cooperativity for the formation of the dimer. Also of interest is the value of β_{1Er}, which is larger than other stability constants measured for the coordination of fluoride anions to similar Ln complexes[29,33] (log $\beta_{1Eu} = 4.6$ was measured for pyridine containing cyclen analogue taking water concentration into account)[34]. The details of the calculated concentrations of the species formed during the titration and the recalculated ultraviolet-visible absorption spectra of the species are shown in Supplementary Figs 17 and 18.

Finally, the association was also monitored by ^{1}H-NMR spectroscopy (Supplementary Fig. 19). As previously stated, the ^{1}H-NMR spectrum of the Er complex in D_2O is composed of 16 broad peaks in the +110/−110 p.p.m. region. Addition of fluoride anions up to 0.5 equivalent immediately resulted in the observation of a second set of 16 signals, the two sets coexisting for F/Er ratios smaller than 0.5. This number of peaks is in agreement with the formation of a dimeric complex with D_2 symmetry and the observed variations of the chemical shifts are perfectly corresponding to the observations of Faulkner and co-workers[35] with a change of magnetic anisotropy associated to an anisotropic electronic distribution from prolate to oblate upon fluoride binding. For excess larger than 0.5 equivalent, only minor changes could be observed for the shapes of the signals. In the presence of very large excesses (>5 equivalent), a broadening of the peaks could be noticed. This observation may be related to very minor changes in the coordination of the ligand around Er in the $[(ErL)_2F]^+$ dimer compared with the [ErLF] monomer as expected on the basis of density functional theory (DFT) modelling (see below). Unfortunately, attempts to follow the titration by ^{19}F-NMR proved unsuccessful. The strong paramagnetic contribution of the Er cation with the short Er–F distances in the complexes probably resulted in too broad peaks that we could not observe.

DFT modelling of the dimer. In the absence of available X-ray crystal structure, we turned our attention to DFT calculations for modelling the structure of the [ErLF] monomer and

a

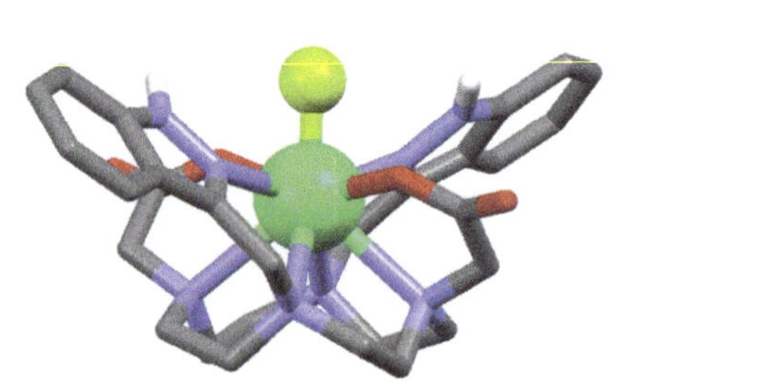

b

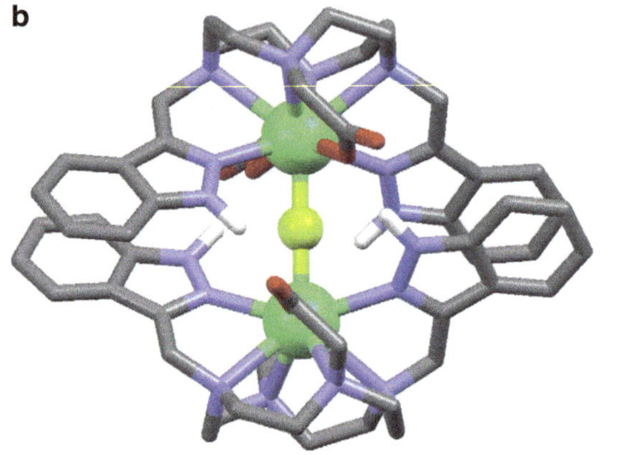

Figure 5 | DFT modelling. The [ErLF] monomer (**a**) and the $[(ErL)_2F]^+$ dimer (**b**) were modelled by DFT calculations and revealed the presence of strong π–π stacking interactions between indazolyl groups in the dimer, together with four hydrogen bonding interactions which, with the formation of Er-F-Er bridge, stabilize the supramolecular dimeric assembly.

$[(ErL)_2F]^+$ dimer (Fig. 5). DFT calculations performed at the M06/LCRECP/6-31G(d,p) level provided an optimized structure for $[(ErL)_2F]^+$ containing a linear Er-F-Er unit with a Er-F distance of 2.242 Å and a Er···Er distance of 4.484 Å. The Er-F distance is slightly shorter in the monomeric [ErLF] entity (2.104 Å). The dimeric structure is supported by π–π stacking interactions involving the indazolyl groups of the two (ErL) entities (distance between centroids 3.891 Å) and hydrogen bonds established between the NH groups of one indazolyl groups of a ErL entity and the oxygen atoms of the carboxylate groups coordinated to the second Er metal ion (N···O 2.751 Å, N-H···O 1.745 Å, N-H···O 162.4°). It is anticipated that the formation of hydrogen bonds and stacking interactions are at the origin of the positive cooperativity observed for the formation of the dimer during the ultraviolet-visible titration experiment (*vide supra*).

Discussion

Based on the stability constants determined above, it was possible to calculate the amounts of the three different species present in D_2O solution, that is, $[ErL(D_2O)]^+$, $[(ErL)_2F]^+$ and [ErLF], as a function of added F^-. Figure 6 represents the calculated concentrations of the different species as a function of the $[F^-]/[Er]$ concentration ratio. In the same graph, the evolution of the integrated intensity of the peak at 550 nm was reported, arbitrarily normalizing the emitted intensity at half an equivalent of added fluoride. The comparison clearly showed that the largest UC intensity increase is directly related to the formation of the dimer content. Interestingly, for larger amounts of fluoride for which the [ErLF] species is formed, the intensity decreased, pointing to a weaker UC process in the [ErLF] monomeric complex. It is expected that the coordination of fluoride anions in the monomeric complex will improve the metal-centred luminescence quantum yield by displacing the non-radiative quenching processes associated to OD oscillators of the coordinated D_2O molecule oscillators of the coordinated heavy water molecule as observed by conventional fluorescence (Fig. 2b,c)[12,36]. This observation emphasizes the improvement of the UC process for the dimer compared with the two monomers, [ErL(D_2O)] and [ErLF].

Considering first-order processes, Fig. 7 summarizes the two possible mechanisms responsible of the UC process in the dimer. For both mechanisms, the first step is a ground-state absorption resulting in the formation of an [Er*FEr] excited state with population of the $^4I_{11/2}$ level of one Er atom. In the case of an ESA (red arrows), the absorption of a second photon by the excited Er atom led to population of the $^4F_{7/2}$ level, [Er**FEr] and relaxation to the $^2H_{11/2}$, $^4S_{3/2}$ and $^4F_{9/2}$ levels results in emission at 525, 550 and 650 nm, respectively. For the second process, the ETU (blue arrows), the ground-state absorption leading to [Er*FEr] is followed by the absorption of a second photon by the second Er atom, so that both Er atoms were in the $^4I_{11/2}$ excited state, forming the [Er*FEr*] excited state. An intramolecular energy transfer from one Er to the other relaxed one Er to the ground state, whereas the second was promoted to a higher energy level such as $^4F_{7/2}$, $^2H_{11/2}$ or $^4S_{3/2}$, depending on whether the transfer process is a phonon-assisted one or not. Relaxation to the ground state also led to emission at 525, 550 and 650 nm.

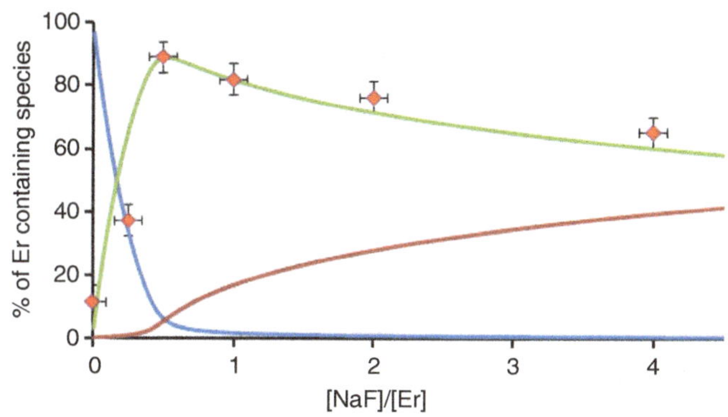

Figure 6 | Speciation in solution. The relative evolution of the percentages of the species present in solution ([ErL(D_2O)]$^+$, blue; [$(ErL)_2F$]$^+$, green; [ErLF], red) can be recalculated from the knowledge of the stability constants and were compared with UC emission intensity (integrated from 505 to 580 nm, red squares) normalized on the concentration at 0.5 equivalent (errors correspond to those of a Poisson's law for the intensity and to ±5% for the relative concentrations). The speciation revealed a marked correlation between the UC intensity and the presence of the Er dimer.

A major difference in the ESA and ETU processes can be found in the risetime of the UC emitted signal. As ESA arise from two absorption phenomena, the risetime is very rapid. In contrast, the ETU process is slowed by the rate of energy transfer from the intermediate excited state and may lead to an observable risetime. Recording the risetime of the emission at 545 nm ($^4S_{3/2} \rightarrow {^4I_{15/2}}$) upon pulsed laser excitation at 980 nm (Supplementary Fig. 20) showed a *ca* 5 μs risetime in the emission, pointing to an ETU mechanism in the Er dimer.

The relationship between the UC efficiency and the concentration of the species in solution points to two observations; the efficiency is directly related to the presence of the Er dimer and decreases as the concentration of [ErLF] monomer increases. Both observations are in favour of an ETU mechanism. If it proceeded via the ESA mechanism, the efficiency of the UC process should not decrease for larger concentrations of [ErLF] monomer as the metal-centred luminescence quantum yield of the monomer was found to be almost similar to that of the dimer (Fig. 2). If one hypothesizes that the crystal field splitting and the transition probabilities of Er are similar for the monomer and the dimer (DFT models point to similar coordination environment around Er in the monomer and the dimer), the UC efficiency should also be the same, which is not the case. Although the UC signals observed for the monomers are probably related to an ESA mechanism, as observed by Faulkner and co-workers[19], the 7.7-fold increase of the UC efficiency associated with the formation of the dimer can hardly be explained only by the ESA mechanism, considering that the one photon luminescence quantum yield only increased by a factor 1.6. These observations,

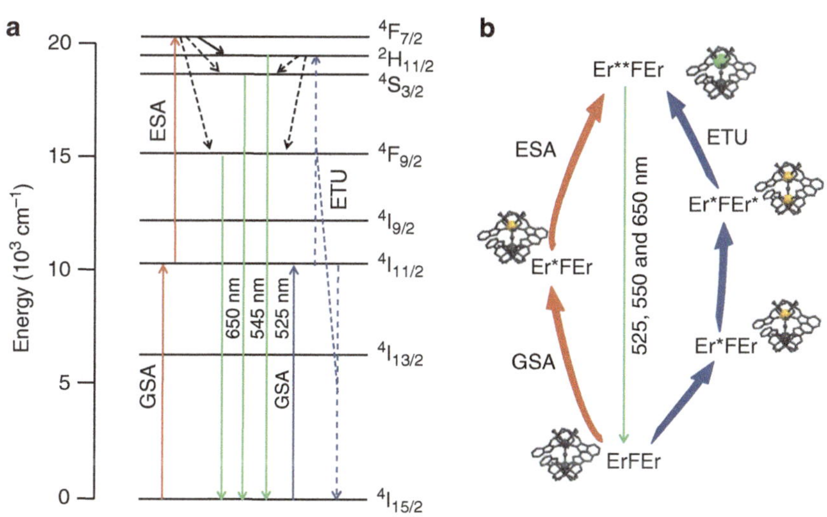

Figure 7 | Energy level diagram and mechanisms of the UC process. (**a**) The different mechanisms (ETU in blue and excited state absorption (ESA) in red) for the UC process are presented in the energy level diagram of the Er ion. In both cases, the first step is the excitation to the $^4I_{11/2}$ level upon excitation at 980 nm. For ESA, the erbium ion in the excited state absorbs a second photon to reach the upper levels and relaxation to the $^2H_{11/2}$, $^4S_{3/2}$ and $^4F_{9/2}$ precedes the emission of the UC signal (green arrows). For ETU, the higher excited state is reached by energy transfer from the second Er atom itself in the $^4I_{11/2}$ excited state. (**b**) Schematic representations of the two mechanisms (the vertical positions are not representative of the energy levels of the different excited states).

together with the evidence of a rising domain in the UC emitted signal, are fully consistent with an ETU mechanism, favoured by the spatial proximity of the two Er centres, as it has been observed in Er doped solids[37].

The coordination of fluoride anions with the $[ErL(H_2O)]^+$ complex resulted in the self-assembly of two complexes around a single fluoride anion. The assembly is characterized by a positive cooperative effect resulting from the formation of hydrogen bonds and aromatic stacking interactions. DFT models of the dimer showed that the intermetallic distance is very short (4.48 Å). In a D_2O solution at room temperature, excitation of the Er complex resulted in the observation of three UC emission bands at 525, 550 and 650 nm, respectively, attributed to the Er centred ${}^2H_{11/2} \rightarrow {}^4I_{15/2}$, ${}^4S_{3/2} \rightarrow {}^4I_{15/2}$ and ${}^4F_{9/2} \rightarrow {}^4I_{15/2}$ transitions, upon excitation in the Er ${}^4I_{15/2} \rightarrow {}^4I_{11/2}$ absorption band at 980 nm. Noticeably, addition of fluoride anions to the solution is accompanied by a very large 7.7-fold increase of the UC emission attributed to the formation of the dimer in a D_2O. These results unambiguously showed that UC can be achieved at the molecular and supramolecular level, in solution under ambient conditions, providing that the system is well-adapted. Although the observed UC may be related to an ESA mechanism in the monomeric complexes, the large increase observed for the dimer and the risetime observed for the UC emitted signal emphasize the importance of the association of the two cations of the dimer in the UC process and strongly indicate an excited state energy transfer mechanism. In this context, the very short intermetallic distance between the two Er cations is believed to be a crucial parameter.

These results are the first evidence of UC at the molecular level in deuterated aqueous solution at room temperature. We can expect that future developments on the structure and compositions of such complexes could lead to similar observations in non-deuterated water, hence opening a brand new window for luminescence tagging in numerous bio-analytical applications.

Methods

Materials and chemical analysis. Solvents and starting materials were purchased from Aldrich, Acros and Alfa Aesar and used without further purification. ^1H and ^{13}C NMR spectra were recorded on Bruker 500, 400 or 300 spectrometers. Chemical shifts are reported in p.p.m., with residual protonated solvent as internal reference[38]. Infrared spectra were recorded on a Perkin Elmer Spectrum One Spectrophotometer as solid samples and only the most significant absorption bands are given in cm^{-1}. Elemental analyses and routine mass spectrometry analysis were carried out by the Service Commun d'Analyses of the University of Strasbourg. High-resolution electrospray ionisation-time of flight (ESI-TOF) mass spectra were recorded using a micro-TOF-Q Applied Biosystems QSTAR Elite spectrometer in the positive mode.

Computational details. All calculations presented in this work were performed employing the Gaussian 09 package (Revision D.01)[39]. Full geometry optimizations of the $[(ErL)_2F]^+$ and [ErLF] systems were performed in aqueous solution employing DFT within the hybrid meta generalized gradient approximation (hybrid meta-GGA), with the M06 exchange-correlation functional[40]. In these calculations, we used the large-core quasirelativistic effective core potential of Dolg and co. and its associated [5s4p3d]-GTO valence basis set for Er[41], whereas the ligand atoms were described by using the standard 6–31G(d,p) basis set. Solvent effects were included by using the CPCM variant of the polarizable continuum model, as implemented in Gaussian 09 (refs 42,43).

Spectroscopic characterizations. Spectroscopic measurements were performed with $10 \times 10\,mm^2$ quartz suprasil certified cells (Helma Analytics). Ultraviolet-visible absorption spectra were recorded on a Perkin Elmer Lambda 950 spectrometer. Steady-state emission spectra were recorded on a FLS920P fluorescence spectrometer (Edimburgh Instrument) equipped with a 450-W continuous wavelength Xe lamp (range from 230 to 900 nm), using Hamamatsu R928 (visible) or R5509-72 (Vis and NIR range) photomultipliers. For steady-state emission in the NIR range, high-pass filters at 850 nm were used to remove second-order artefacts. For UC emission spectra, a 980-nm continuous wavelength laser diode (5 W, Laserwave, Beijing, China) was used as the light source and emission in the visible was detected with a photomultiplier tube (PMT, R928 Hamamatsu). Power dependence of UC emission was measured under a focused excitation light with a spot size (average) of $\approx 0.07\,cm^2$, and laser power increased step by step (0.9–2.3 W). A focus lens (focal length is 10 cm) collimating the laser light (diameter of light cross section is about 5 mm) was placed 10 cm in front of the 980 nm laser, and 5 cm before the aperture. A power metre (FieldMaxII-TO laser power meter and PM30 probe) was placed in a position close to where the cuvette would be to measure the power. Laser power is measured both before and after spectral measurement of solution samples with the power probe placed in the pre-set position. Considering the response time of power metre probe (2 s), laser power is measured after allowing the laser to stabilize for 30 s. The error of the laser power and the power metre is $<1\%$, respectively. The UC excitation spectra and UC kinetic experiments were performed with an ultrafast laser system equipped with R928 Hamamatsu Photomultiplier Tube. Excitation light was generated by an infrared coherent pulsed laser (Ti:sappire compact laser system- 150 fs, with wavelength range 690–1,040 nm), excitation power at every point was calibrated to 750 mW and the UC emission intensity was monitored at 545 nm (${}^4S_{3/2} + {}^2H_{11/2} \rightarrow {}^4I_{15/2}$).

Synthesis of the Er complex. $[ErL(H_2O)](NO_3)$ was synthesized by adaptation of previously published methods[22] using nitrate salts of Er instead of lanthanide chloride salts. Yield: 77%. ^1H NMR (D_2O, 400 MHz, 25 °C, Supplementary Figs 1 and 2): δ 110.16, 80.61, 45.98, 25.64, 21.47, 20.22, 15.50, 14.19, 12.12, 3.09, − 1.52, − 21.61, − 82.97, − 100.84, − 110.54; infrared (attenuated total reflectance (ATR), Supplementary Fig. 3): 3,437 (br, m), 3,216 (br, m), 2,863 (w), 1,655 (m), 1,614 (s), 1,356 (s), 1,324 (s), 1,077 (m), 1,003 (w), 931 (w), 756 (m); ultraviolet-visible (H_2O): $\lambda_{max} = 294$ nm; electrospray mass spectrometry (ES-MS) (ESI^+, Supplementary Fig. 4): $[ErL]^+$ calculated for $C_{28}H_{34}ErN_8O_4$, 714.19; found, 714.20; analysis (calculated, found for $[(ErL)H_2O](NO_3).NaCl.4H_2O$ ($C_{28}H_{44}ClErN_9NaO_{12}$)): C (36.38, 36.49), H (4.80, 4.42); N (13.64, 13.62).

References

1. Auzel, F. Upconversion and Anti-Stokes Processes with f and d Ions in Solids. *Chem. Rev* **104,** 139–173 (2004).
2. Andraud, C. & Maury, O. Lanthanide complexes for nonlinear optics: from fundamental aspects to applications. *Eur. J. Inorg. Chem* 4357–4371 (2009).
3. Gamelin, D. R. & Güdel, H. U. Design of luminescent inorganic materials: new photophysical processes studied by optical spectroscopy. *Acc. Chem. Res* **33,** 235–242 (2000).
4. Wang, F. & Liu, X. Recent advances in the chemistry of lanthanide-doped upconversion nanocrystals. *Chem. Soc. Rev* **38,** 976–989 (2009).
5. Chen, G., Qiu, H., Prasad, P. N. & Chen, X. Upconversion nanoparticles: design, nanochemistry, and applications in theranostics. *Chem. Rev* **114,** 5161–5214 (2014).
6. Bünzli, J.-C.G. Lanthanide luminescence for biomedical analyses and imaging. *Chem. Rev* **110,** 2729–2755 (2010).
7. Weissleder, R. & Ntziachristos, V. Shedding light onto live molecular targets. *Nature Med* **9,** 123–128 (2003).
8. Zako, T. *et al.* Cyclic RGD peptide-labeled upconversion nanophosphors for tumor cell-targeted imaging. *Biochem. Biophys. Res. Comm* **381,** 54–58 (2009).
9. Zhang, T. *et al.* Water-soluble mitochondria-specific ytterbium complex with impressive NIR emission. *J. Am. Chem. Soc* **133,** 20120–20122 (2011).
10. Garaikoetxea Arguinzoniz, A. *et al.* Light harvesting and photoemission by nanoparticles for photodynamic therapy. *Part. Part. Syst. Charact* **31,** 46–75 (2014).
11. Haase, M. & Schäfer, H. Upconverting nanoparticles *Angew. Chem. Int. Ed* **50,** 5808–5829 (2011).
12. Beeby, A. *et al.* Non-radiative deactivation of the excited states of europium, terbium and ytterbium complexes by proximate energy-matched OH, NH and CH oscillators: an improved luminescence method for establishing solution hydration states. *J. Chem. Soc. Perkin Trans* **2,** 493–503 (1999).
13. Comby, S. & Bünzli, J.-C.G. in *Handbook on the Physics and Chemistry of Rare Earths* 37 (eds Gschneidner, Jr K. A., Bünzli, J.-C. G. & Pecharsky, V. K. 217–470 (The Netherlands, 2007).
14. Doffek, C. *et al.* Understanding the quenching effects of aromatic C-H- and C-D oscillators in Near-IR lanthanoid luminescence. *J. Am. Chem. Soc.* **134,** 16413–16423 (2012).
15. Aboshyan-Sorgho, L. *et al.* Near-infrared to visible light upconversion in a molecular trinuclear d-f-d complex *Angew. Chem. Int. Ed* **50,** 4108–4113 (2011).
16. Zare, D. *et al.* Smaller than a nanoparticle with the design of discrete polynuclear molecular complexes displaying near-infrared to visible upconversion. *Dalton Trans* **44,** 2529–2540 (2015).
17. Förster, T. Transfer mechanisms of electronic excitation. *Disc. Faraday Soc* **27,** 7–17 (1959).

18. Andolina, C. M. & Morrow, J. R. Luminescence resonance energy transfer in heterodinuclear LnIII complexes for sensing biologically relevant anions. *Eur. J. Inorg. Chem* 154–164 (2011).
19. Sørensen, T. J., Blackburn, O. A., Tropiano, M. & Faulkner, S. Direct two-photon excitation of Sm^{3+}, Eu^{3+}, Tb^{3+}, Tb.DOTA^{-}, and Tb.propargylDO3A in solution. *Chem. Phys. Lett* **541,** 16–20 (2012).
20. Blackburn, O. A. *et al.* Luminescence and upconversion from thulium(III) species in solution. *Phys. Chem. Chem. Phys* **14,** 13378–13384 (2012).
21. Hyppanen, I. *et al.* Photon upconversion in a molecular lanthanide complex in anhydrous solution at room temperature. *ACS Photonics* **1,** 394–397 (2014).
22. Liu, T. *et al.* Supramolecular luminescent lanthanide dimers for fluoride sequestering and sensing. *Angew. Chem. Int. Ed.* **53,** 7259–7263 (2014).
23. Miller, F. A. & Wilkins, C. H. Infrared spectra and characteristic frequencies of inorganic ions. *Anal. Chem* **24,** 1253–1294 (1952).
24. Aime, S. *et al.* Conformational and coordination equilibria on DOTA complexes of lanthanide metal ions in aqueous solution studied by ^{1}H-NMR spectroscopy. *Inorg. Chem* **36,** 2059–2068 (1997).
25. Starck, M. *et al.* Towards libraries of luminescent lanthanide complexes and labels from generic synthons. *Chem. Eur. J* **17,** 9164–9179 (2011).
26. Nonat, A., Imbert, D., Pecaut, J., Giraud, M. & Mazzanti, M. Structural and photophysical studies of highly stable lanthanide complexes of tripodal 8-hydroxyquinolinate ligands based on 1,4,7-triazacyclononane. *Inorg. Chem* **48,** 4207–4218 (2009).
27. Bünzli, J.-C. G. On the design of highly luminescent lanthanide complexes. *Coord. Chem. Rev* **293-294,** 19–47 (2015).
28. Pollnau, M., Gamelin, D. R., Lüthi, S. R., Güdel, H. U. & Hehlen, M. P. Power dependence of upconversion luminescence in lanthanide and transition-metal-ion systems. *Phys. Rev* **B61,** 3337–3346 (2000).
29. Blackburn, O. *et al.* Spectroscopic and crystal field consequences of fluoride binding by $[Yb.DTMA]^{3+}$ in aqueous solution *Angew. Chem. Int. Ed* **54,** 10783 (2015).
30. Gampp, H., Maeder, M., Meyer, C. J. & Zuberbühler, A. D. Calculation of equilibrium constants from multiwavelenght spectroscopic data. *Talanta* **32,** 95–169 (1985).
31. Ercolani, G., Piguet, C., Borkovec, M. & Hamacek, J. Symmetry numbers and statistical factors in self-assembly and multivalency. *J. Phys. Chem. B* **111,** 12195–12203 (2007).
32. Ercolani, G. & Schiaffino, L. Allosteric, chelate and interannular cooperativity: A mise au point. *Angew. Chem. Int. Ed.* **50,** 1762–1768 (2011).
33. Lima, L. M. P. *et al.* Positively charged lanthanide complexes with cyclen-based ligands: synthesis, solid state and solution structure, and fluoride interaction. *Inorg. Chem* **24,** 12508–12521 (2011).
34. Tripier, R., Platas-Iglésias, C., Boos, A., Morfin, J.-F. & Charbonnière, L. Towards fluoride sensing with positively charged lanthanide complexes. *Eur. J. Inorg. Chem* 2735–2745 (2010).
35. Blackburn, O. A., Kenwright, A. M., Beer, P. D. & Faulkner, S. Axial fluoride binding by lanthanide DTMA complexes alters the local crystal field, resulting in dramatic spectroscopic changes. *Dalton Trans* **44,** 19509–19517 (2015).
36. Butler, S. J. Quantitative determination of fluoride in pure water using luminescent europium complexes. *Chem. Commun* **51,** 10879–10882 (2015).
37. Martín-Rodríguez, R. *et al.* Highly efficient IR to NIR upconversion in Gd_2O_2S: Er^{3+} for photovoltaic applications. *Chem. Mater* **25,** 1912–1921 (2013).
38. Gottlieb, H. E., Kottyar, K. & Nudelman, A. NMR chemical shifts of common laboratory solvents as trace impurities. *J. Org. Chem* **82,** 7512–7515 (1997).
39. Frisch, M. J. *et al.*Gaussian 09, Revision D.01Gaussian, Inc. (2009).
40. Zhao, Y. & Truhlar, D. G. The M06 suite of density functionals for main group thermochemistry, thermochemical kinetics, noncovalent interactions, excited states, and transition elements: two new functionals and systematic testing of four M06-class functionals and 12 other functionals. *Theor. Chem. Acc* **120,** 215–241 (2008).
41. Dolg, M., Stoll, H., Savin, A. & Preuss, H. Energy-adjusted pseudopotentials for the rare earth elements. *Theor. Chim. Acta* **75,** 173–194 (1989).
42. Barone, V. & Cossi, M. Quantum calculation of molecular energies and energy gradients in solution by a conductor solvent model. *J. Phys. Chem. A* **102,** 1995–2001 (1998).
43. Cossi, M., Rega, N., Scalmani, G. & Barone, V. Energies, structures, and electronic properties of molecules in solution with the C-PCM solvation model. *J. Comp. Chem* **24,** 669–681 (2003).

Acknowledgements

L.J.C. thanks the French Ministère des Affaires Etrangères for a grant of the PROCORE program (no. 30707NM and F-HKBU201/13). Jean Marc Strub and Nicolas Busser are gratefully acknowledged for recording ES/MS spectra and for preparing high-quality images, respectively. We gratefully acknowledge the support of the University Research Facility on Chemical and Environmental Analysis (URFCE) of PolyU, HK Polytechnic University Area of Excellent Grants (1-ZVGG) and The Joint Research Program on Bioimaging, Hong Kong Polytechnic University and Hong Kong Baptist University. (RC-ICRS/15-16/02F-WKL02F-WKL).

Author contributions

L.J.C., K.-L.W. and W.-T.W. conceived and supervised the project. A.N. performed the spectroscopic titrations, characterized the complex and measured the spectroscopic properties in solution. T.L. performed the synthesis of the ligand and complex. C.P.-I. performed the DFT modelling. Z.L., C.F.C., W.K.W. and K.L.W. performed the UC experiments.

Additional information

Competing financial interests: The authors declare no competing financial interests.

10

Structural identification of electron transfer dissociation products in mass spectrometry using infrared ion spectroscopy

Jonathan Martens[1], Josipa Grzetic[1], Giel Berden[1] & Jos Oomens[1,2]

Tandem mass spectrometry occupies a principle place among modern analytical methods and drives many developments in the 'omics' sciences. Electron attachment induced dissociation methods, as alternatives for collision-induced dissociation have profoundly influenced the field of proteomics, enabling among others the top-down sequencing of entire proteins and the analysis of post-translational modifications. The technique, however, produces more complex mass spectra and its radical-driven reaction mechanisms remain incompletely understood. Here we demonstrate the facile structural characterization of electron transfer dissociation generated peptide fragments by infrared ion spectroscopy using the tunable free-electron laser FELIX, aiding the elucidation of the underlying dissociation mechanisms. We apply this method to verify and revise previously proposed product ion structures for an often studied model tryptic peptide, $[\text{AlaAlaHisAlaArg} + 2\text{H}]^{2+}$. Comparing experiment with theory reveals that structures that would be assigned using only theoretical thermodynamic considerations often do not correspond to the experimentally sampled species.

[1] Radboud University, Institute for Molecules and Materials, FELIX Laboratory, Toernooiveld 7c, 6525ED Nijmegen, The Netherlands. [2] Van 't Hoff Institute for Molecular Sciences, University of Amsterdam, Science Park 908, 1098XH Amsterdam, The Netherlands. Correspondence and requests for materials should be addressed to J.O. (email: j.oomens@science.ru.nl).

Mass spectrometry-based analysis of peptides and proteins in the bioanalytical and clinical sciences relies on the gas-phase dissociation of their molecular ions to give sequence fragments from which the original primary structure can be inferred. Sequencing by the increasingly popular electron-induced dissociation methods (ExD, such as electron capture and transfer dissociation, ECD and ETD) has recently seen rapid development and widespread application. ECD in Fourier transform ion cyclotron resonance (FT-ICR) mass spectrometry (MS)[1–3] and more recently ETD in a much broader range of mass spectrometers[4–7] constitute the two primary variations of this method. These methods have shown impressive improvements over collision-induced dissociation tandem MS, primarily in the sense that labile post-translational modifications are not detached during activation revealing their position along the backbone, and that sequence coverage is increased, making the sequencing of intact proteins in top-down strategies possible[8–12].

Electron attachment to multiply protonated peptides or proteins leads to extensive backbone fragmentation, which most often takes the form of backbone N–C_α bond cleavages to provide the c- and z-type ion series[1,2,13,14]. The process of electron-induced N–C_α bond cleavage in peptide cation radicals has been the subject of many recent experimental and theoretical studies[13,15–20]. However, the mechanisms involved in electron attachment to peptide ions, the possible role of excited electronic states, and the structural rearrangements and fragmentation reactions that can follow, remain only partially understood. Furthermore, answers to questions regarding the ratios of fragment ions produced and their structures, especially the odd electron fragments, remain, at least partially, elusive. This is undoubtedly related to the uncertainty in the detailed nature of the (open-shell) dissociation products. For example, ExD product ions are known to undergo hydrogen atom migration reactions that increase/decrease the expected masses of the fragment ions, having direct practical implications for ExD-based sequencing applications[21,22].

Several reaction mechanisms have been proposed for ExD of which the Cornell mechanism from the group of McLafferty[2] and the more recent Utah-Washington (UW) mechanism from the groups of Tureček and Simons[19,23–25] are best known. Frison *et al.*[26] have discussed the different structures of c-type fragments that would result from different ECD mechanisms and were able to spectroscopically identify an amide c-type ion. While not being able to exclude the possibility that this structure results from a rearrangement after the ECD process, this assignment appears to support the UW mechanism, rather than mechanisms that would directly produce enol-imine c-type ions.

In ExD dissociation, the radical is typically on the z-type fragment, although H-atom migration reactions in the dissociating molecule may transfer the radical to the c-type fragment, resulting in changes in the *m/z* values of the fragments and complicating the interpretation of ExD MS/MS data. As a result, apparent unit mass shifts of backbone sequence ions in ExD spectra are thus not rare and result in an increased occurrence of ions with overlapping masses. As an illustration, for a peptide with the sequence AlaAlaHisAlaArg, hydrogen atom migration to the open-shell $z_3^{\bullet +}$ fragment would give a closed-shell z_3^+ ion having the same chemical formula as the c_4^+ fragment and makes this mass peak unassignable for sequencing purposes.

Unfortunately, the detailed information about an ion's molecular structure is scarce and often undecipherable from MS/MS data alone. Infrared spectroscopy allows the structural characterization of both the trapped precursor and fragment ions. Infrared ion spectroscopy (IRIS) has been used extensively to determine the gas-phase structures of molecular ions in MS, and specifically in the effort to elucidate peptide fragmentation mechanisms involved in collisional dissociation[27–34]. Tunable infrared free-electron lasers (FELs), such as FELIX at our institute[35], have played an important role in the recent development of IRIS[36], and are especially useful to obtain fingerprint infrared spectra of ionic species on commercial MS platforms[37]. However, to date, only a single example of infrared spectroscopy on ECD-generated fragments has been reported, in which a small closed-shell c-type ion was examined[26].

Structural characterization of peptidic ExD product ions is something that has been highly sought after for a number of years using a variety of methods, including collisional activation[38,39], ion mobility methods[40] and more recently ultraviolet photodissociation studies[41–43]. However, in comparison with infrared spectroscopy these methods provide limited information, regarding the structure and conformation of gas-phase ions. Structural characterization of the dissociation products by infrared spectroscopy and quantum chemistry provides a stringent test to ascertain and confirm uncertain aspects of the fragmentation mechanisms and allows for additional information to be extracted from MS/MS data. Here we present the first direct structural characterization of ETD-generated fragments using IRIS and demonstrate the strengths of this technique for addressing the questions surrounding gas-phase peptide radicals.

Results

ETD MS/MS of [AAHAR + 2H$^+$]$^{2+}$. The ETD MS/MS spectrum of the [AAHAR + 2H]$^{2+}$ (263 *m/z*) ion is presented in red in Fig. 1. We have characterized each of the $z^{\bullet}$-type fragments (depicted in the peptide sequence shown in Fig. 2) from the ETD MS/MS spectrum using infrared spectroscopy, providing a comprehensive identification of their structures and conformations. This detailed characterization forms a basis for analysing the reactions, leading to their formation.

Tryptic peptides of the AAXAR type and their ETD MS/MS behaviour have been extensively studied previously[39,42,44–47]. Here, consistent with previously reported results, we observe that fragment ion intensity is approximately split over the charge reduced ion (ETnoD) and $z^{\bullet}$-type sequence ions (Supplementary Fig. 1). For AAHAR, having arginine in the C-terminal position, it is not surprising that C-terminal $z^{\bullet}$-type fragments primarily retain the proton after dissociation and that they are the principle

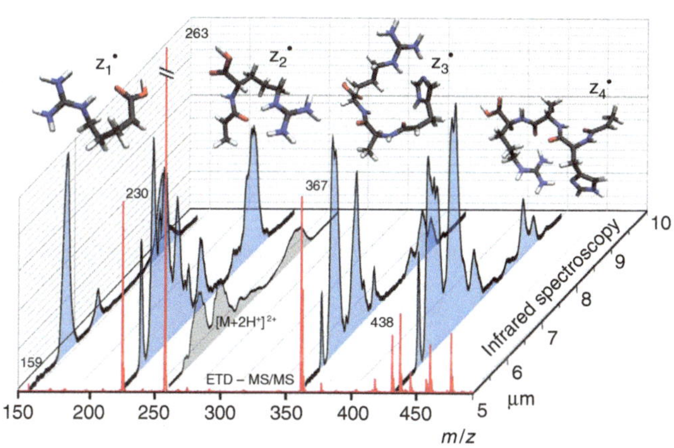

Figure 1 | The ETD MS/MS spectrum of [AAHAR + 2H]$^{2+}$ with the corresponding infrared spectra. The infrared spectra of the ETD-generated fragments are shown in black/blue and that of the precursor peptide in black/grey. Supplementary Fig. 1 contains the comprehensive ETD MS/MS results.

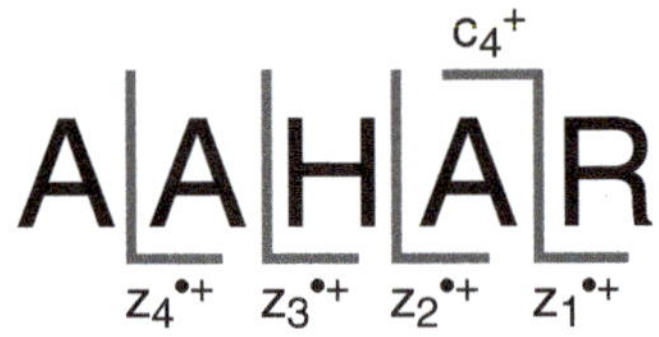

Figure 2 | Dissociation scheme and notation used for product ions. c- and z-type peptide fragments typically result from ETD MS/MS. Here we label only the discussed sequence ions from ETD of $[AAHAR + 2H]^{2+}$. Note that fragments carrying a '•' symbol are open-shell radicals and those without are closed shell.

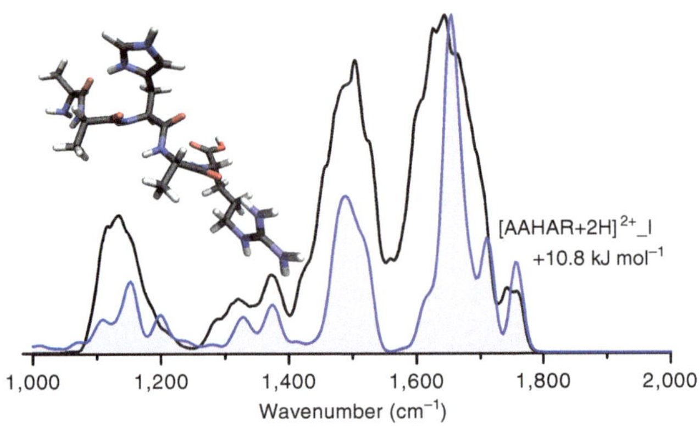

Figure 3 | The infrared spectrum of $[AAHAR + 2H]^{2+}$ 263 *m/z*. The experimental spectrum is presented in black and the spectrum of the assigned calculated structure from this study is shown in blue along with the structure and relative free energy at 298 K.

Table 1 | Summary of structural properties and relative free energies for selected calculated structures.

	m/z	His tautomer	Radical	Rel. ΔG (kJ mol^{-1})
$[AAHAR + 2H^+]^{2+}$	263			
I		—	—	+10.8*
II		—	—	0.0
$z_1^{\bullet +}$	159			
I		—	α	+5.4*
II		—	δ (Arg)	0.0
III		—	β (Arg)	+10.1
IV		—	γ (Arg)	+14.7
$z_2^{\bullet +}$	230			
I		—	α	0.0*
II		—	α	+10.2
III		—	δ (Arg)	+26.5
$z_3^{\bullet +}$	367			
I		N3	α	+22.4*
II		N1	α	0.0
III		N3	β (His)	0.0
$z_4^{\bullet +}$	438			
I		N3	α	0.0*
II		N3	α	+62.5
III		N3	—	+54.4
IV		N3	β (His)	+5.7
z_3^{+}	368			
I		N1	—	+28.5*
c_4_I		—	—	—
II		N3	—	0.0

Rel., relative.
An asterisk (*) indicates the structure assigned spectroscopically.

fragments in the ETD MS/MS spectrum. The electron attachment process is no doubt affected by the protonation sites and conformation of the parent peptide. In its doubly protonated state, this peptide has been shown to protonate on the imidazole group of the histidine side chain and the guanidine group of the arginine side chain[45], and our spectroscopic data and calculations confirm this.

Structure of $[AAHAR + 2H^+]^{2+}$ precursor peptide at *m/z* 263. Figure 3 presents the infrared spectrum obtained for the doubly protonated peptide $[AAHAR + 2H]^{2+}$ at *m/z* 263. Comparison with the calculated spectrum (blue) allows assignment of the protonation sites (His and Arg sidechains) and the conformation to be made. The band just $< 1{,}800\ cm^{-1}$ is consistent with the free (or weakly H-bonded) carboxyl group. Both charged sites are hydrogen bound to backbone carbonyl groups. In terms of hydrogen bonding, involving the charged sites and the C and N termini, this assignment confirms a previously proposed structure[44], although it has an overall more extended conformation, easily distinguished by comparison with the infrared spectra (Supplementary Fig. 2).

For singly charged fragments retaining the His residue, the imidazole ring can tautomerize (having the H on either the N1 or N3 position) and we address this issue for the $z_3^{\bullet +}$ and $z_4^{\bullet +}$ fragments. Note that all fragments carrying a '•' symbol are open-shell radicals and those without are closed shell, where H's and electrons are implicit. Backbone sequence fragments studied here with the structural identification based on our spectroscopic results, as detailed in the following are summarized in Table 1.

$z_1^{\bullet +}$ fragment structure. The $z_1^{\bullet +}$ fragment is the smallest C-terminal fragment obtained upon ETD of the precursor peptide. Calculated structure $z_1^{\bullet}$_I was identified to match most closely with experiment, as demonstrated in the top panel of Fig. 4. This structure is the lowest-energy calculated structure from our selection of ~30 structures from the molecular dynamics (MD) procedure. In structure $z_1^{\bullet}$_I, the radical is located at the α-carbon adjacent to the C terminus and the carboxyl C=O stretch is found at $1{,}635\ cm^{-1}$. On the basis of this structure, radicals at other positions along the carbon chain of this fragment were defined and optimized, most of which were higher in energy. H-atom migration to the δ-carbon of the Arg side chain gave a structure $5.4\ kJ\ mol^{-1}$ more stable; however, a calculated barrier of $115.2\ kJ\ mol^{-1}$ (Supplementary Fig. 3) must be overcome to reach it and it is not consistent with the experimental infrared spectrum. Barriers of $> 100\ kJ\ mol^{-1}$ for H-atom migrations have been reported for different ETD fragments[39,48,49]. With the radical in the α-position, conjugation with the carbonyl occurs giving partial double-bond character to the CC-bond, while lowering the bond order of the carbonyl, causing a significant red shift of its stretching mode relative to a carbonyl stretch $> 1{,}700\ cm^{-1}$ in the β-, γ- and δ-radicals (Supplementary Fig. 4). This is a clear demonstration that the position of the radical can strongly influence the vibrational spectrum, in this case the C=O stretch, highlighting the value of infrared spectroscopy for characterizing open-shell peptide fragments.

$z_2^{\bullet +}$ fragment structure. Figure 5 and Supplementary Fig. 5 present infrared spectra for the $z_2^{\bullet +}$ fragment from ETD of

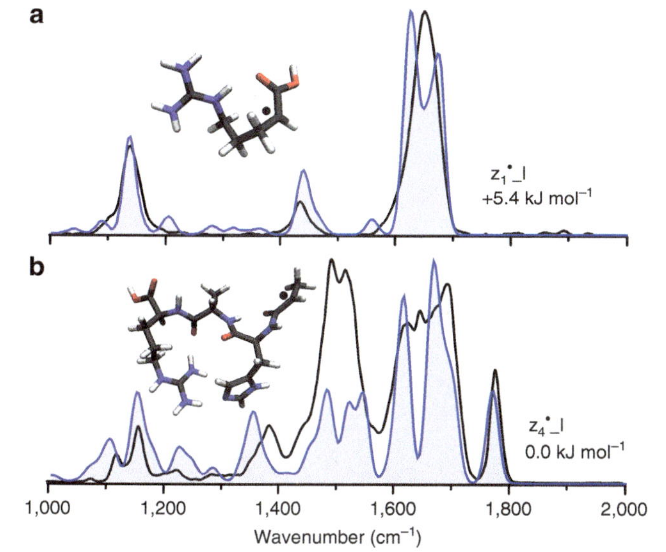

Figure 4 | The infrared spectra of the $z_1^{\bullet+}$ and $z_4^{\bullet+}$ fragments from ETD of [AAHAR + 2H]$^{2+}$. The experimental spectra are presented in black in both cases and assigned calculated structures are shown in blue for (**a**) the $z_1^{\bullet+}$ fragment and (**b**) the $z_4^{\bullet+}$ fragment. The associated relative free energies (298 K) and structures are inlayed with the radical sites labelled by '•'.

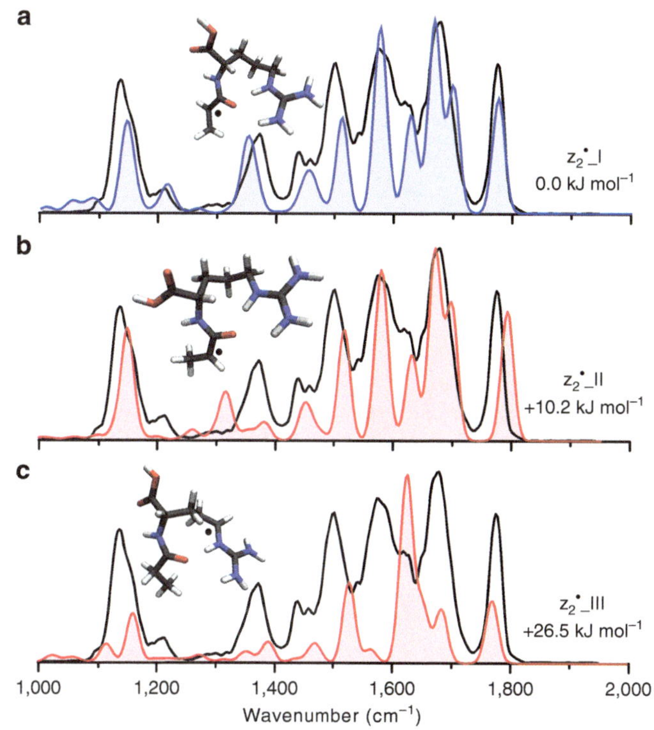

Figure 5 | The infrared spectrum of the $z_2^{\bullet+}$ fragment from ETD of [AAHAR + 2H]$^{2+}$. The experimental spectrum is presented in black and is compared with computed spectra for different low-energy structures. (**a**) The calculated spectrum for the assigned structure is shown in blue. (**b**,**c**) The calculated spectra for structures disregarded on the basis of spectral mismatch are shown in red. Calculated structures and relative free energies (298 K) are inlayed for each plot.

[AAHAR + 2H$^+$]$^{2+}$. The calculated spectrum of $z_2^{\bullet}$_I, presented in the top panel, matches the experimental spectrum well. This structure is the lowest-energy calculation obtained from our computational procedure. A previously proposed structure[39], labelled here as $z_2^{\bullet}$_II and presented in the centre panel, closely resembles $z_2^{\bullet}$_I. However, the alternate hydrogen bonding orientation of the C terminus is distinguished by comparison with experimental and calculated infrared spectra in the 1,300–1,400 cm^{-1} region and the carbonyl stretching region just <1,800 cm^{-1}. This refinement of the conformation reduces the relative energy by 10.2 kJ mol^{-1}. The bottom panel in the figure shows a comparison with the infrared spectrum calculated for the product of hydrogen migration from the α-carbon of the Ala residue to the δ-carbon of the Arg residue, a species 26.5 kJ mol^{-1} higher in energy and readily distinguishable spectroscopically.

$z_3^{\bullet+}$ fragment structure. Figure 6 presents the infrared spectrum for the $z_3^{\bullet+}$ fragment from ETD of [AAHAR + 2H$^+$]$^{2+}$ with the spectrum of the assigned calculated structure $z_3^{\bullet}$_I in blue and spectra of unassigned alternative structures in red. This assignment was made after considering different conformers, imidazole tautomers and products, resulting from hydrogen atom migration interconverting between the α-radical and the β-radical on the His side chain. A low-energy His N1 tautomer ($z_3^{\bullet}$_II) was identified to be 22.4 kJ mol^{-1} lower in energy than $z_3^{\bullet}$_I, a His N3 tautomer; however, the calculated spectrum of this species does not match as well to the experiment, especially in the 1,200–1,400 cm^{-1} region (and the 3400–3600 cm^{-1} region displayed in Supplementary Fig. 6). Furthermore, a β-radical structure ($z_3^{\bullet}$_III) was found to be 22.4 kJ mol^{-1} more stable, but we do not assign this species on the basis of its spectral mismatch (Fig. 6 and Supplementary Fig. 6). The majority of stable structures we identified for each structure/tautomer of the $z_3^{\bullet+}$ ion features a stabilizing hydrogen bonding interaction between the imidazole side chain of His and the guanidinium side chain of Arg, leaving little flexibility for the orientation of the carboxyl group and giving a free C–OH group and a C=O weakly hydrogen bonded to the adjacent amide N–H. Being very sensitive to local environment, the position of the carboxyl C=O stretch just <1,800 cm^{-1} can be used as a diagnostic signature. Structures $z_3^{\bullet}$_II and $z_3^{\bullet}$_III feature hydrogen bonds between the imidazole nitrogen and hydrogens of the two primary nitrogens of the guanidinium group, while in $z_3^{\bullet}$_I the hydrogen bond of the imidazole nitrogen is shared between the hydrogens of the secondary nitrogen and one primary nitrogen.

$z_4^{\bullet+}$ fragment structure. The bottom panel of Fig. 4 presents the infrared spectrum for the $z_4^{\bullet+}$ fragment from ETD of [AAHAR + 2H$^+$]$^{2+}$ and the assigned calculated structure (blue). This is the overall lowest-energy structure obtained after an extensive MD-based search over different conformers for various tautomers and structures obtained by hydrogen migration from the α-radical to the His β-radical (see comparison of $z_4^{\bullet}$_IV in Supplementary Fig. 7). Similar to $z_3^{\bullet}$_I, the assigned structure, $z_4^{\bullet}$_I, has a hydrogen bond between the imidazole side chain and the guanidinium group of the Arg residue. Structure $z_4^{\bullet}$_I offers a refinement over a previously proposed structure[39] (here, re-optimized at the currently applied level of theory), giving both a better spectral match (see $z_4^{\bullet}$_II in Supplementary Fig. 7) and being ~60 kJ mol^{-1} lower in energy.

Fragment ion at *m/z* 368 can be z_3^+ or c_4^+. While inter- and intramolecular (between c and z fragment pairs) hydrogen migration reactions are commonly observed in ETD MS/MS, their behaviour is still relatively weakly understood. For [AAHAR + 2H$^+$]$^{2+}$, only for the $z_3^{\bullet+}$ cation do we observe an

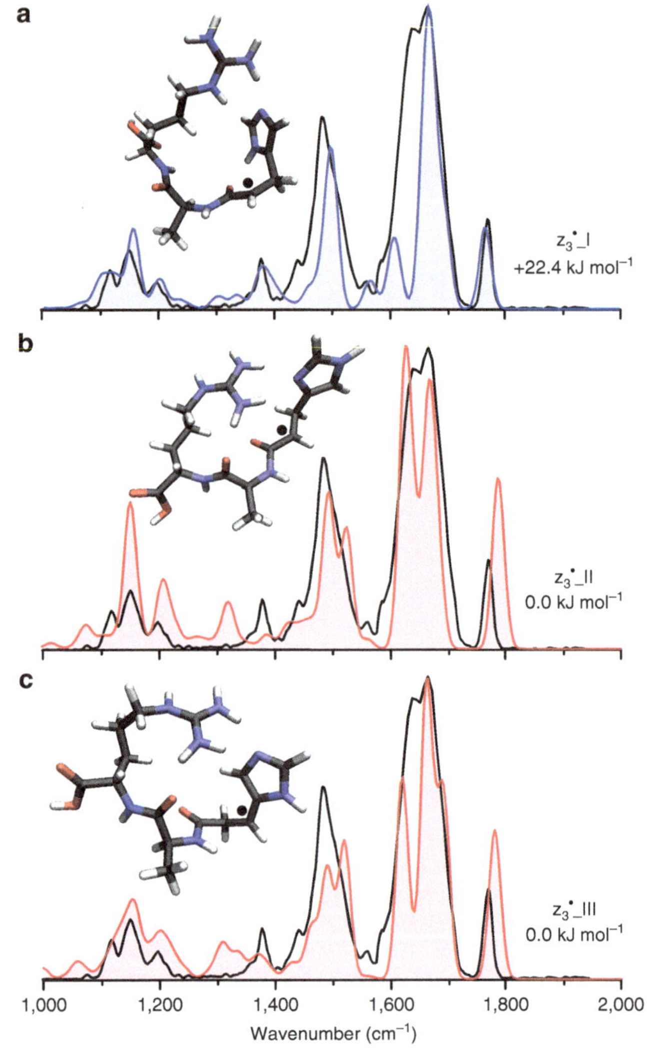

Figure 6 | The infrared spectrum of the $z_3^{\bullet+}$ fragment from ETD of [AAHAR + 2H]$^{2+}$. The experimental spectrum is presented in black and is compared with computed spectra for different low-energy structures. (**a**) The calculated spectrum for the assigned structure is shown in blue. (**b,c**) The calculated spectra for structures disregarded on the basis of spectral mismatch are shown in red. Calculated structures and relative free energies (298 K) are inlayed for each plot.

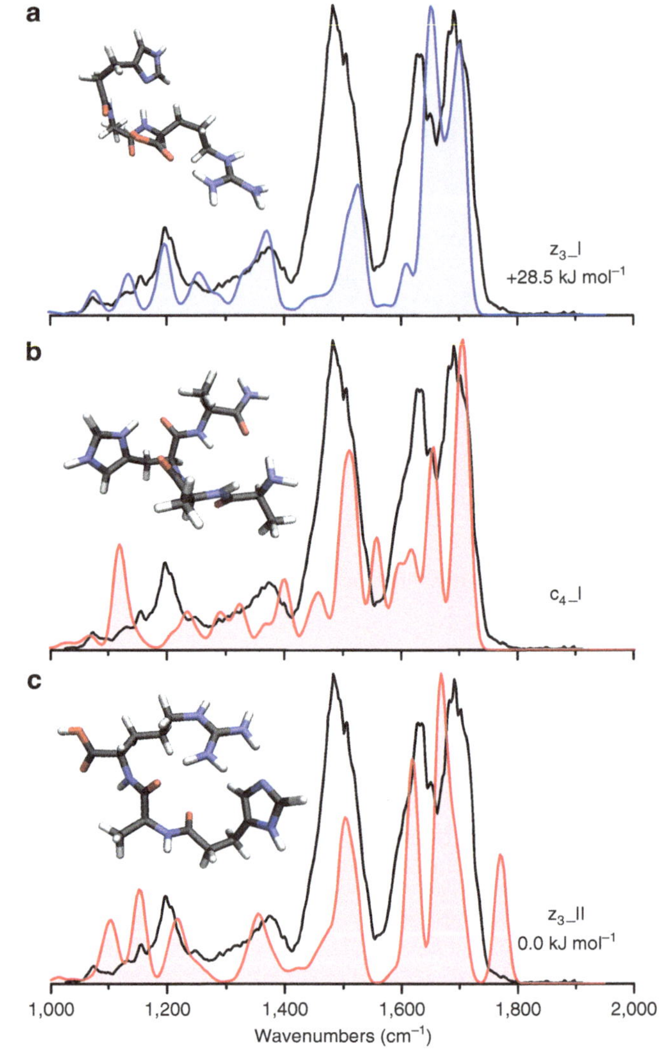

Figure 7 | The infrared spectrum of the *m/z* 368 fragment from ETD of [AAHAR + 2H]$^{2+}$. The experimental spectrum is presented in black and is compared with computed spectra for different low-energy structures. (**a**) The spectrum of the assigned calculated closed-shell z_3^+ structure is shown in blue with the structure and relative free energy (298 K) inlayed. (**b**) c_4_I is a low-energy c_4^+ conformation and its calculated infrared spectrum is presented in red. (**c**) z_3_II is the lowest energy calculated z_3^+ structure identified in this work.

appreciable extent of such a reaction, where we see both the open-shell $z_3^{\bullet+}$ fragment (*m/z* 367) and a closed-shell z_3^+ fragment (*m/z* 368). Highlighting the complications that can arise from hydrogen atom migrations in ETD, the c_4^+ fragment ($C_{15}N_7H_{25}O_4$) has the same chemical formula and overlaps the closed-shell z_3^+ fragment also at *m/z* 368. Identification and consideration of the hydrogen migration products (loss/gain) are important for assigning fragment ions and correct sequencing[38].

In Fig. 7, we identify the *m/z* 368 fragment ion as the z_3^+ species (z_3_I) based on infrared spectral matching. This structure features an alternative hydrogen bonding arrangement in comparison with the open-shell $z_3^{\bullet+}$ and $z_4^{\bullet+}$ fragments described above, most significantly affecting the C terminus (trans configuration) and red shifting the carboxyl C = O stretch for z_3_I away from the position just $<1{,}800\,cm^{-1}$ for the open-shell $z_3^{\bullet+}$ and $z_4^{\bullet+}$ fragments. A closed-shell equivalent of the geometry of the open-shell $z_3^{\bullet}$_I structure is defined as z_3_II and is 28.5 kJ mol^{-1} lower in energy than z_3_I. Calculated spectra for structure z_3_II and c_4_I, a low-energy c_4^+ conformation, are presented in the bottom two panels of Fig. 7 and support the assignment of z_3_I.

Discussion

These results demonstrate the first use of IRIS to characterize the structures of ETD-generated peptide fragments. Using a model tryptic pentapeptide, precursor ion conformation has been related to the observed ETD fragmentation pattern, and the structures and conformations of the various fragment ions. We show that it is possible to distinguish both conformational details and different radical species, when this approach is combined with routine computational modelling.

We conclude that if structural assignments were made only on the basis of theoretical (thermodynamic) considerations, these assignments would in many cases not match the species observed in experiment—highlighting the potential for IRIS to diagnostically identify gas-phase organic radicals and, more specifically, the mechanisms associated with peptide fragmentation in electron attachment methods. Our results suggest that hydrogen atom transfer necessary for radical migration often does not occur after ETD (without additional activation), leading to the frequent observation of non-equilibrium product ions. Understanding intermolecular hydrogen atom migration is also of practical importance, as it causes shifts in ETD fragment masses and makes sequence ion assignments more complicated.

Methods

Ion spectroscopy in a modified ion trap mass spectrometer. The experiment is based on a commercial quadrupole ion trap mass spectrometer (Bruker, AmaZon Speed ETD) coupled to the infrared beam line of the FELIX FEL. $[M+2H]^{2+}$ peptide ions are generated by electrospray ionization. AAHAR (GeneCust (Luxemburg), 95% purity) solutions of 10^{-5}–10^{-6} mol l^{-1} (in 50:50 acetonitrile:water, ~1% formic acid) are introduced at 120 μl h^{-1} flow rates and desolvated by a pressurized nebulizing gas (N_2). The key hardware modifications to the instrument providing optical access to the ion population in the trap were the introduction of a new ring electrode having 3 mm holes at its top and bottom, the installation of mirrors below the trap to direct the beam back out of the instrument and optical windows in the vacuum housing. In ETD experiments, ions were accumulated for 0.1–15 ms in the trap, mass isolated and then reacted with fluoranthene radical anions for ~250 ms. A fragment ion of interest was mass isolated in a subsequent MS/MS stage and irradiated by the tunable infrared beam from the FEL. In the experiments reported here, the FELIX FEL was set to produce infrared radiation in the form of 5–10 μs macropulses at 5 or 10 Hz and of 30–60 mJ (bandwidth ~0.4% of the centre frequency). Resonant absorption of infrared radiation leads to an increase in the internal energy of the molecule aided by intramolecular vibrational redistribution of the absorbed energy. When a sufficient number of photons is absorbed (here, typically in a single macropulse), unimolecular dissociation occurs and produces frequency-dependent fragment ion intensities in the mass spectrometer (Supplementary Note 1). Relating the parent and fragment ion intensities in the observed mass spectral data (yield = ΣI(fragment ions)/ΣI(parent + fragment ions)) generates an infrared vibrational spectrum. The yield at each infrared point is obtained from averaged mass spectra and is linearly corrected for laser power; the frequency is calibrated using a grating spectrometer.

Computational chemistry. We have employed a molecular mechanics (MM)/MD approach using AMBER 12 (refs 50,51). Molecular structures manually defined based on chemical intuition where first optimized for each ion at the B3LYP/6-31 + + G(d,p) level in Gaussian09 (ref. 52). Restrained electrostatic potential (RESP) charges from these initial results were used for parameterization of the nonstandard peptide ions in the antechamber program. After minimization within AMBER, a simulated annealing procedure up to 1,000 K was used with a 1 fs step size. Five hundred structures were obtained as snapshots throughout the procedure and after MM minimization were grouped based on structural similarity using *prtraj* in AMBER. Of these, 30–50 unique structures were then each optimized at the B3LYP/6-31 + + G(d,p) level[51,53,54] and vibrational spectra were calculated within the harmonic oscillator model (vibrational frequencies were scaled by 0.975). This computational approach was applied to all structural isomers considered for each ion, except for the small $z_1^{\bullet+}$ ion, where the MM/MD conformational search was only applied once using the alpha-radical species. Calculated line spectra were broadened using a Gaussian function with a full-width at half-maximum of 25 cm^{-1} to facilitate comparison with experiment. Additional calculations using the LC-BLYP and M06 functionals for a selection of $z_1^{\bullet+}$ and $z_2^{\bullet+}$ structures, and ab initio MP2 calculations for the $z_1^{\bullet+}$ structures were performed to verify the validity of the choice of functional[49,55] and these results are summarized in Supplementary Table 1 and Supplementary Fig. 8. In general, vibrational frequencies were found to be best modelled at the B3LYP/6-31 + + G(d,p) level and calculated free energies are mostly consistent between these levels of theory. Supplementary Data 1-4 contain optimized geometries of the assigned z-type fragments.

References

1. Kruger, N. A., Zubarev, R. A., Horn, D. M. & McLafferty, F. W. Electron capture dissociation of multiply charged peptide cations. *Int. J. Mass. Spectrom.* **185-187,** 787–793 (1999).
2. Zubarev, R. A., Kelleher, N. L. & McLafferty, F. W. Electron capture dissociation of multiply charged protein cations. a nonergodic process. *J. Am. Chem. Soc.* **120,** 3265–3266 (1998).
3. Zubarev, R. A. *et al.* Electron capture dissociation for structural characterization of multiply charged protein cations. *Anal. Chem.* **72,** 563–573 (2000).
4. Syka, J. E. P., Coon, J. J., Schroeder, M. J., Shabanowitz, J. & Hunt, D. F. Peptide and protein sequence analysis by electron transfer dissociation mass spectrometry. *Proc. Natl Acad. Sci. USA* **101,** 9528–9533 (2004).
5. Pitteri, S. J., Chrisman, P. A., Hogan, J. M. & McLuckey, S. A. Electron transfer ion/ion reactions in a three-dimensional quadrupole ion trap: reactions of doubly and triply protonated peptides with SO2•. *Anal. Chem.* **77,** 1831–1839 (2005).
6. Xia, Y. *et al.* Implementation of ion/ion reactions in a quadrupole/time-of-flight tandem mass spectrometer. *Anal. Chem.* **78,** 4146–4154 (2006).
7. Coon, J. J. *et al.* Protein identification using sequential ion/ion reactions and tandem mass spectrometry. *Proc. Natl Acad. Sci. USA* **102,** 9463–9468 (2005).
8. Swaney, D. L., McAlister, G. C. & Coon, J. J. Decision tree-driven tandem mass spectrometry for shotgun proteomics. *Nat. Meth.* **5,** 959–964 (2008).
9. Breuker, K. & McLafferty, F. W. Native electron capture dissociation for the structural characterization of noncovalent interactions in native cytochrome c. *Angew. Chem. Int. Ed.* **42,** 4900–4904 (2003).
10. Breuker, K., Oh, H., Lin, C., Carpenter, B. K. & McLafferty, F. W. Nonergodic and conformational control of the electron capture dissociation of protein cations. *Proc. Natl Acad. Sci. USA* **101,** 14011–14016 (2004).
11. Oh, H. *et al.* Secondary and tertiary structures of gaseous protein ions characterized by electron capture dissociation mass spectrometry and photofragment spectroscopy. *Proc. Natl Acad. Sci. USA* **99,** 15863–15868 (2002).
12. Breuker, K. & McLafferty, F. W. Stepwise evolution of protein native structure with electrospray into the gas phase, 10 − 12 to 102s. *Proc. Natl Acad. Sci. USA* **105,** 18145–18152 (2008).
13. Zubarev, R. A. Reactions of polypeptide ions with electrons in the gas phase. *Mass. Spectrom. Rev.* **22,** 57–77 (2003).
14. Horn, D. M., Zubarev, R. A. & McLafferty, F. W. Automated de novo sequencing of proteins by tandem high-resolution mass spectrometry. *Proc. Natl Acad. Sci. USA* **97,** 10313–10317 (2000).
15. Tureček, F. & Julian, R. R. Peptide radicals and cation radicals in the gas phase. *Chem. Rev.* **113,** 6691–6733 (2013).
16. Simons, J. Mechanisms for S-S and N-Cα bond cleavage in peptide ECD and ETD mass spectrometry. *Chem. Phys. Lett.* **484,** 81–95 (2010).
17. Anusiewicz, I., Skurski, P. & Simons, J. Refinements to the Utah-Washington mechanism of electron capture dissociation. *J. Phys. Chem. B* **118,** 7892–7901 (2014).
18. Li, X., Lin, C., Han, L., Costello, C. E. & O'Connor, P. B. Charge remote fragmentation in electron capture and electron transfer dissociation. *J. Am. Soc. Mass. Spectrom.* **21,** 646–656 (2010).
19. Syrstad, E. A. & Tureček, F. Toward a general mechanism of electron capture dissociation. *J. Am. Soc. Mass. Spectrom.* **16,** 208–224 (2005).
20. Zhurov, K. O., Fornelli, L., Wodrich, M. D., Laskay, U. A. & Tsybin, Y. O. Principles of electron capture and transfer dissociation mass spectrometry applied to peptide and protein structure analysis. *Chem. Soc. Rev.* **42,** 5014–5030 (2013).
21. Savitski, M. M., Kjeldsen, F., Nielsen, M. L. & Zubarev, R. A. Hydrogen rearrangement to and from radical z fragments in electron capture dissociation of peptides. *J. Am. Soc. Mass. Spectrom.* **18,** 113–120 (2007).
22. Liu, J., Liang, X. & McLuckey, S. A. On the value of knowing a z• ion for what it is. *J. Proteome. Res.* **7,** 130–137 (2008).
23. Sobczyk, M. *et al.* Coulomb-assisted dissociative electron attachment: application to a model peptide. *J. Phys. Chem. A* **109,** 250–258 (2004).
24. Anusiewicz, I., Berdys-Kochanska, J. & Simons, J. Electron attachment step in electron capture dissociation (ECD) and electron transfer dissociation (ETD). *J. Phys. Chem. A* **109,** 5801–5813 (2005).
25. Chen, X. & Tureček, F. The arginine anomaly: arginine radicals are poor hydrogen atom donors in electron transfer induced dissociations. *J. Am. Chem. Soc.* **128,** 12520–12530 (2006).
26. Frison, G. *et al.* Structure of electron-capture dissociation fragments from charge-tagged peptides probed by tunable infrared multiple photon dissociation. *J. Am. Chem. Soc.* **130,** 14916–14917 (2008).
27. Lucas, B. *et al.* Investigation of the protonation site in the dialanine peptide by infrared multiphoton dissociation spectroscopy. *Phys. Chem. Chem. Phys.* **6,** 2659–2663 (2004).
28. Erlekam, U. *et al.* Infrared spectroscopy of fragments of protonated peptides: direct evidence for macrocyclic structures of b5 ions. *J. Am. Chem. Soc.* **131,** 11503–11508 (2009).
29. Bythell, B. J., Erlekam, U., Paizs, B. & Maître, P. Infrared spectroscopy of fragments from doubly protonated tryptic peptides. *Chemphyschem* **10,** 883–885 (2009).
30. Polfer, N. C., Oomens, J., Suhai, S. & Paizs, B. Spectroscopic and theoretical evidence for oxazolone ring formation in collision-induced dissociation of peptides. *J. Am. Chem. Soc.* **127,** 17154–17155 (2005).

31. Polfer, N. C., Oomens, J., Suhai, S. & Paizs, B. Infrared spectroscopy and theoretical studies on gas-phase protonated leu-enkephalin and its fragments: direct experimental evidence for the mobile proton. *J. Am. Chem. Soc.* **129,** 5887–5897 (2007).
32. Polfer, N. C. & Oomens, J. Vibrational spectroscopy of bare and solvated ionic complexes of biological relevance. *Mass. Spectrom. Rev.* **28,** 468–494 (2009).
33. Yoon, S. H. *et al.* IRMPD spectroscopy shows that agg forms an oxazolone b2+ ion. *J. Am. Chem. Soc.* **130,** 17644–17645 (2008).
34. Perkins, B. R. *et al.* Evidence of diketopiperazine and oxazolone structures for ha b2+ ion. *J. Am. Chem. Soc.* **131,** 17528–17529 (2009).
35. Oepts, D., van der Meer, A. F. G. & van Amersfoort, P. W. The free-electron-laser user facility FELIX. *Infrared Phys. Technol.* **36,** 297–308 (1995).
36. Oomens, J., Sartakov, B. G., Meijer, G. & Von Helden, G. Gas-phase infrared multiple photon dissociation spectroscopy of mass-selected molecular ions. *Int. J. Mass. Spectrom.* **254,** 1–19 (2006).
37. Bakker, J. M., Besson, T., Lemaire, J., Scuderi, D. & Maître, P. Gas-phase structure of a π-allyl – palladium complex: efficient infrared spectroscopy in a 7T Fourier transform mass spectrometer. *J. Phys. Chem. A* **111,** 13415–13424 (2007).
38. Hamidane, H. B. *et al.* Electron capture and transfer dissociation: peptide structure analysis at different ion internal energy levels. *J. Am. Soc. Mass. Spectrom.* **20,** 567–575 (2009).
39. Ledvina, A., Chung, T., Hui, R., Coon, J. & Tureček, F. Cascade dissociations of peptide cation-radicals. part 2. Infrared multiphoton dissociation and mechanistic studies of z-ions from pentapeptides. *J. Am. Soc. Mass. Spectrom.* **23,** 1351–1363 (2012).
40. Moss, C. L. *et al.* Assigning structures to gas-phase peptide cations and cation-radicals. an infrared multiphoton dissociation, ion mobility, electron transfer, and computational study of a histidine peptide ion. *J. Phys. Chem. B* **116,** 3445–3456 (2012).
41. Shaffer, C. J., Marek, A., Pepin, R., Slovakova, K. & Turecek, F. Combining UV photodissociation with electron transfer for peptide structure analysis. *J. Mass. Spectrom.* **50,** 470–475 (2015).
42. Nguyen, H. T. H., Shaffer, C. J. & Tureček, F. Probing peptide cation–radicals by near-uv photodissociation in the gas phase. structure elucidation of histidine radical chromophores formed by electron transfer reduction. *J. Phys. Chem. B* **119,** 3948–3961 (2015).
43. Nguyen, H. T. H., Shaffer, C. J., Pepin, R. & Tureček, F. U. V. Action spectroscopy of gas-phase peptide radicals. *J. Phys. Chem. Lett.* **6,** 4722–4727 (2015).
44. Tureček, F., Moss, C. L. & Chung, T. W. Correlating ETD fragment ion intensities with peptide ion conformational and electronic structure. *Int. J. Mass. Spectrom.* **330-332,** 207–219 (2012).
45. Tureček, F. *et al.* The histidine effect. electron transfer and capture cause different dissociations and rearrangements of histidine peptide cation-radicals. *J. Am. Chem. Soc.* **132,** 10728–10740 (2010).
46. Zimnicka, M., Moss, C., Chung, T., Hui, R. & Tureček, F. Tunable charge tags for electron-based methods of peptide sequencing: design and applications. *J. Am. Soc. Mass. Spectrom.* **23,** 608–620 (2012).
47. Chung, T., Hui, R., Ledvina, A., Coon, J. & Tureček, F. Cascade dissociations of peptide cation-radicals. part 1. scope and effects of amino acid residues in penta-, nona-, and decapeptides. *J. Am. Soc. Mass. Spectrom.* **23,** 1336–1350 (2012).
48. Chung, T. W. & Tureček, F. Backbone and side-chain specific dissociations of z ions from non-tryptic peptides. *J. Am. Soc. Mass. Spectrom.* **21,** 1279–1295 (2010).
49. Riffet, V., Jacquemin, D. & Frison, G. H-atom loss and migration in hydrogen-rich peptide cation radicals: The role of chemical environment. *Int. J. Mass. Spectrom.* **390,** 28–38 (2015).
50. Case, D.A. *et al.* AMBER 12 (University of California, San Francisco, CA, 2012).
51. Martens, J. K., Grzetic, J., Berden, G. & Oomens, J. Gas-phase conformations of small polyprolines and their fragment ions by IRMPD spectroscopy. *Int. J. Mass. Spectrom.* **377,** 179–187 (2015).
52. Frisch, M. J. *et al.* Gaussian 09 (Gaussian, Inc., Wallingford, CT, 2009).
53. Halls, M. D. & Schlegel, H. B. Comparison of the performance of local, gradient-corrected, and hybrid density functional models in predicting infrared intensities. *J. Chem. Phys.* **109,** 10587–10593 (1998).
54. Kapota, C., Lemaire, J., Maître, P. & Ohanessian, G. Vibrational signature of charge solvation vs salt bridge isomers of sodiated amino acids in the gas phase. *J. Am. Chem. Soc.* **126,** 1836–1842 (2004).
55. Riffet, V., Jacquemin, D., Cauët, E. & Frison, G. Benchmarking DFT and TD-DFT functionals for the ground and excited states of hydrogen-rich peptide radicals. *J. Chem. Theory. Comput.* **10,** 3308–3318 (2014).

Acknowledgements

We gratefully acknowledge the FELIX staff, particularly Dr B. Redlich and Dr A.F.G. van der Meer. In addition, the authors thank Professor F. Tureček and Dr C. Schaffer for discussion. Finally, the authors express their gratitude for the technical assistance from staff at Bruker Daltonics in Bremen, Germany, while implementing this experiment, in particular Dr C. Gebhardt. Financial support for this project was provided by the Chemical Sciences division of the 'Nederlandse organisatie voor Wetenschappelijk Onderzoek' (NWO) under VICI project no. 724.011.002. We also thank NWO Physical Sciences (EW) and the SurfSARA Supercomputer Center for providing the computational resources. This work is part of the research program of FOM, which is financially supported by NWO.

Author contributions

J.M., G.B. and J.O. conceived and designed the experiments. J.M. and J.G. performed the experiments and computations. The manuscript was co-written by all authors.

Additional information

Competing financial interests: The authors declare no competing financial interests.

11

Photoswitchable semiconductor nanocrystals with self-regulating photochromic Förster resonance energy transfer acceptors

Sebastián A. Díaz[1], Florencia Gillanders[1,2], Elizabeth A. Jares-Erijman[2,‡] & Thomas M. Jovin[1]

Photoswitchable molecules and nanoparticles constitute superior biosensors for a wide range of industrial, research and biomedical applications. Rendered reversible by spontaneous or deterministic means, such probes facilitate many of the techniques in fluorescence microscopy that surpass the optical resolution dictated by diffraction. Here we have devised a family of photoswitchable quantum dots (psQDs) in which the semiconductor core functions as a fluorescence donor in Förster resonance energy transfer (FRET), and multiple photochromic diheteroarylethene groups function as acceptors upon activation by ultraviolet light. The QDs were coated with a polymer bearing photochromic groups attached via linkers of different length. Despite the resulting nominal differences in donor-acceptor separation and anticipated FRET efficiencies, the maximum quenching of all psQD preparations was $38 \pm 2\%$. This result was attributable to the large ultraviolet absorption cross-section of the QDs, leading to preferential cyclereversion of photochromic groups situated closer to the nanoparticle surface and/or with a more favourable orientation.

[1] Laboratory of Cellular Dynamics, Max Planck Institute for Biophysical Chemistry, am Fassberg 11, 37077 Göttingen, Germany. [2] Departamento de Química Orgánica, Facultad de Ciencias Exactas y Naturales, Universidad de Buenos Aires, CIHIDECAR, CONICET, 1428 Buenos Aires, Argentina. Correspondence and requests for materials should be addressed to T.M.J. (email: tjovin@gwdg.de).
[‡] Deceased.

Photochromic Förster resonance energy transfer (pcFRET) is a versatile means for creating photomodulatable fluorescent systems. Using selected irradiation wavelengths a photochrome is reversibly converted between a non-acceptor isomer (with a small FRET overlap integral) and an acceptor isomer (with a large FRET overlap integral)[1]. In the latter case, the fluorescence of the donor is quenched by FRET and the conventional FRET formalism can be applied to the donor (D) and acceptor (A) pair. This concept has been widely applied in different areas such as optical data storage[2], sensors[3] and fluorescence microscopy[4]. With the aim of developing new probes for biosensing, our group and others have implemented pcFRET with nanoparticles, particularly quantum dots (QDs)[5–7].

The multiple advantages of QDs as fluorescent probes—broad excitation, narrow emission, extreme photostability and high brightness—account for their widespread application as biosensors[8]. In many instances, the QDs function as energy donors in FRET, exhibiting both specific advantages and disadvantages[9]. QDs can be designed with high quantum yields and long-lived excited-state lifetimes, rendering them efficient as FRET donors[9]. A second characteristic of QDs is their relatively large size (>3 nm in diameter) compared with traditional organic dyes[4]. This feature potentiates the interaction of QD donors with multiple FRET acceptors, albeit at the price of greater complexity and heterogeneity[10]. The size of QDs also imposes a lower limit on the proximity of a potential acceptor to the emission dipole, generally represented as a point centred in the nanoparticle core. The Förster parameter (R_0), the distance at which a given D–A pair exhibits 50% FRET efficiency, is usually in the single-digit nm range, compared with the radii of derivatized QDs that often exceed 10 nm. Nonetheless, due to the large effective Stokes shift resulting from the broad excitation band of QDs, they are optimal donors for measurements based on sensitized emission; that is, a convenient wavelength can generally be found that minimizes the direct excitation of the acceptor[11]. QDs do not constitute perfect point dipoles[12], are often aspherical and surface charges can affect the FRET efficiency[9]. The influence of these variations and other limitations of the conventional FRET formalism[13] are unpredictable yet generally minor, such that most QD FRET systems can and are described adequately by the classical treatments and equations[9,14].

In this report, we feature a family of pcFRET constructs in which the D–A distance (r_{DA}) was varied systematically, with intriguing, unexpected structural and functional consequences. In spite of pronounced differences in individual FRET efficiencies, the end-point ultraviolet-photomodulation of psQD fluorescence was an invariant $38 \pm 2\%$. We attribute this result to FRET-induced cycloreversion, functioning so as to compensate distance–orientation variations by adjustment of the steady-state distribution of acceptor groups.

Results

Design of psQD nanoparticles with photochromic acceptors. The psQDs consisted of a QD FRET donor and a cap of multiple photochromic (PC) diheteroarylethene FRET acceptors attached to an amphiphilic comb-like polymer via linkers of variable length (Fig. 1a). The QDs were commercially available CdSe/CdS/ZnS core–shell–shell nanoparticles[15], and the polymer was based on a commercially available backbone, poly(isobutylene-*alt*-maleic anhydride) (PMA), which allows nucleophilic modification of the anhydride groups. Previous publications from our laboratory have shown that such a system constitutes a photoswitchable nanoparticle, denoted as psQD, that can be utilized in aqueous environments[16,17]. The particular nanoarchitecture and photophysical properties of the new psQDs resulted in a geometric arrangement with 'self-regulating' FRET properties.

The PC moiety exists as two thermally stable isomers interchangeable by selective illumination (Fig. 1b). The open (non-cyclized, non-acceptor) isomer (oPC) has two conformations, parallel and antiparallel, in dynamic equilibrium; only the antiparallel can photocyclize[18]. The closed (acceptor) isomer (cPC) is formed upon ultraviolet irradiation of the oPC form. The conversion is incomplete because the cPC form also absorbs ultraviolet light and undergoes cycloreversion. Thus, sustained exposure to ultraviolet light leads to the attainment of a photostationary state (PS), defined by a maximal fractional conversion to cPC, α_{PS}, which depends on the wavelength of the ultraviolet source (but not its irradiance, which only affects the kinetics of photoconversion) as well as on the local molecular microenvironment. The conversion of the isolated or polymer-associated PC selected for these studies was determined previously for the reference condition of irradiation at 340 nm in a non-polar solvent. The value obtained, $\alpha_{PS,0} = 0.22$, constitutes an absolute, upper limit for the system in the absence of external perturbations[16]. In contrast, cycloreversion induced by exposure to light in the characteristic cPC absorption band at 550 nm is quantitative, that is, leading to 100% oPC, inasmuch as this isomer does not absorb in the visible (Vis) range. The diheteroarylethene photochromes do not undergo spontaneous thermal conversion in either direction. Thus, all combinations of oPC and cPC can be regarded as stationary in the sense of being stable in the absence of illumination.

psQD preparation and characterization. The first-generation psQDs underwent multiple cycles of photoswitching with minimal fatigue (degradation)[16]. The modulation of fluorescence did not exceed 40%; addition of a second non-PC acceptor boosted the value to 50%[17]. In classical FRET systems, the quenching of the donor can be increased by reducing r_{DA}, and we sought to exploit this condition to enhance the QD quenching. A series of photochromic amphiphilic comb-like polymers were synthesized with pendant PC molecules differing only in the length of the linker (Fig. 1; Table 1; Methods; Supplementary Methods). On the basis of our previous studies of such systems, it was anticipated that the longer linkers would place the PCs closer to the QD surface, thereby lowering r_{DA}[16,17]. All the prepared polymers had comparable PC concentrations, with 4.5–8% of the polymer chain monomers modified by the PC and the majority of the remaining monomers coupled to alkyl chains[19].

An experiment was designed to determine *PC*, the total number of PCs per psQD nanoparticle, for the various preparations. The characterized psQD samples were precipitated by acidification and centrifuged, after which the pellet was dissolved in tetrahydrofuran (THF), a solvent with which both the polymer cap and the QD interact preferentially such that the polymer exists as a freely draining dynamic chain[20]. Under these conditions the polymer-bound PC achieved its natural ultraviolet-induced PS ($\alpha_{PS,0} = 0.22$; Supplementary Fig. 1), enabling the calculation of *PC* (Table 1). A decrease in packing efficiency had been anticipated for the polymers with greater backbone modification[20], PMA 8PCaoc 75C12 serving as the extreme example. In fact, polymers containing $>10\%$ of bulky modifications such as PCs are unable to coat nanoparticles.

The rates and extents of photoconversion of oPC to cPC and the corresponding quenching of the QD emissions were monitored by both ultraviolet/Vis absorption and fluorescence spectroscopy following successive pulses of low-intensity ultraviolet (340 ± 10 nm, 1.0 mW cm^{-2}) or Vis (545 ± 10 nm, 6.2 or

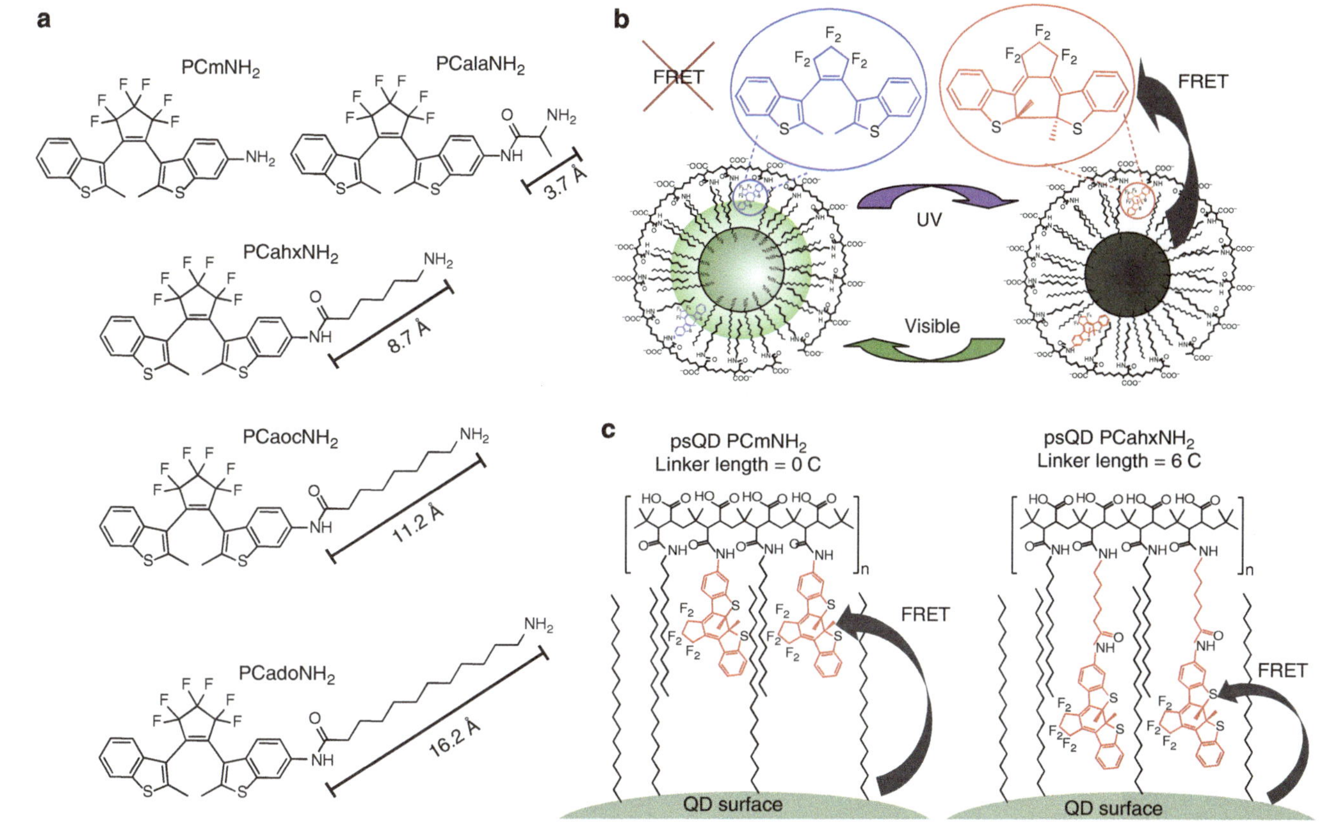

Figure 1 | Photoswitchable QDs with variable separation of the photochromic FRET acceptors. (**a**) Diheteroarylethenes utilized to synthesize photochromic amphiphilic comb-like polymers. (**b**) Schematic of photoswitchable quantum dots and their quenching due to pcFRET. Exposure to ultraviolet light leads to cyclization, manifested by a new absorption band in the visible, which is reversed by irradiation at that wavelength. Adapted, with permission, from ref. 16. Copyright 2011 American Chemical Society. (**c**) Extended linkers place the PC acceptor closer to the QD surface, thereby reducing r_{DA}.

Table 1 | psQD composition.

psQD species	PC utilized*	Linker length (carbons, Å)†	Polymer composition‡	*PC*
1	$PCmNH_2$	0, 0	PMA 6PC 70C12	89 ± 11
2	$PCalaNH_2$	2, 3.7	PMA 6PCala 75C12	77 ± 9
3	$PCahxNH_2$	6, 8.7	PMA 5PCahx 75C12	60 ± 8
4	$PCaocNH_2$	8, 11.2	PMA 8PCaoc 75C12	38 ± 5
5	$PCadoNH_2$	12, 16.2	PMA 4.5PCado 75C12	59 ± 7

PC, photochrome; PMA, poly(isobutylene-*alt*-maleic anhydride; psQD, photoswitchable quantum dot.
*For details see Fig. 1, Methods and Supplementary Methods.
†Estimated linker length based on MM2 minimized Chem3D structures; $PCmNH_2$ is considered as zero length. For distances 1.25 Å was added for each additional sp^3 carbon to the $PCalaNH_2$ value.
‡The adopted nomenclature follows that previously presented in the literature[19]; modifications to the PMA backbone are represented as percentages of anhydride rings coupled with the added molecules.

$3.7\,mW\,cm^{-2}$) irradiation (Fig. 2). The samples were utilized after being subjected to multiple photoconversion cycles to minimize the effect of the inherent, initial photobrightening generally observed with commercial QDs.

The formation of cPC up to the attainment of the PS was monitored by ΔA_{540} (Fig. 2a). This process was accompanied by a pronounced quenching of the QD emission, manifested in both steady-state (Fig. 2b) and time-resolved (fluorescence decay) (Fig. 2c) measurements. However, whereas the dependence of photoconversion (cPC/QD) on the total irradiation time was close to monoexponential (Fig. 2d), the QD quenching was biphasic, with an initial rapid increase (of quenching) and a slower progression to a steady-state plateau value (Fig. 2e). This feature was common to all the psQDs, although the parametric plot (quench[t] versus $cPC[t]$, Fig. 2f) split the five preparations into two main groups, one of which corresponded to the shorter linkers (0, 2 carbons) and the second to the longer linkers (6–12 carbons). A striking finding was that in spite of the kinetic differences, the maximum quenching in the PS, $38 \pm 2\%$, was independent of the linker length. The mean QD fluorescence lifetimes changed accordingly (Supplementary Table 1). The PS also demonstrated a marked decrease in α_{PS} (from 0.19 to 0.07), which correlated with the increase in PC linker length (Table 2), in accordance with the expectation that the D–A separation would decrease for longer linkers, thereby favouring FRET. The decrease in α_{PS} was not due solely to differences in coating efficiency, inasmuch as the values of *PC* did not correlate with the relative changes in α_{PS}.

To rationalize the above observations, we invoked for each psQD preparation: (1) a degree of heterogeneity in the PC groups

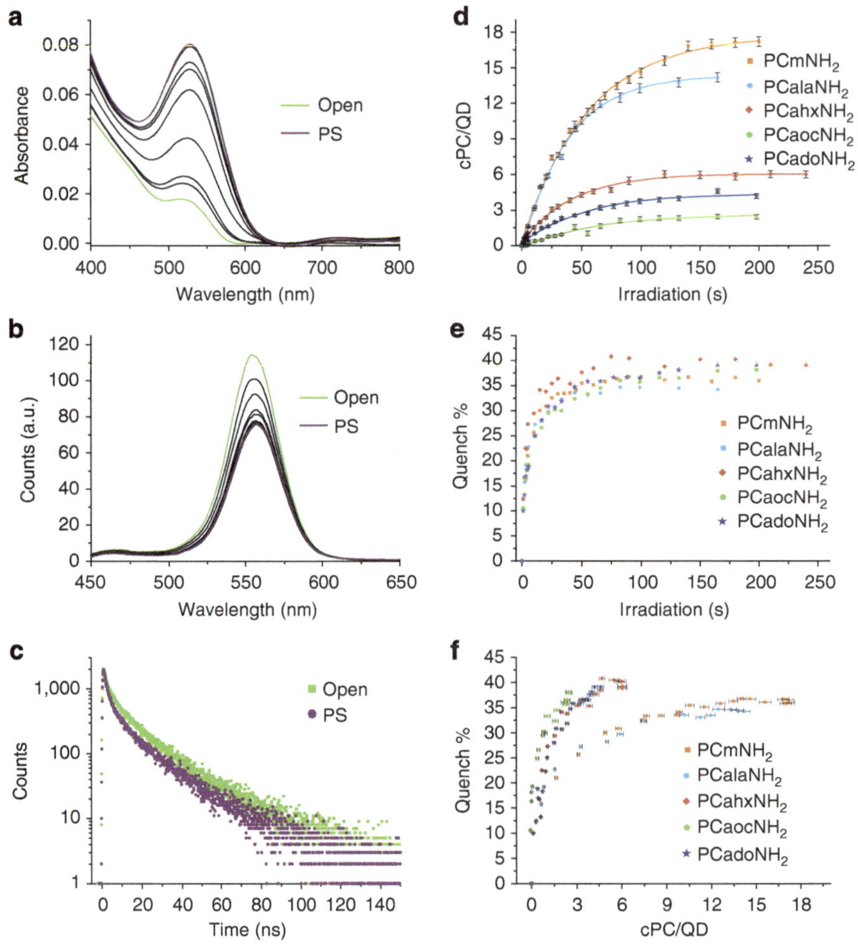

Figure 2 | Photoconversion and quenching of psQDs. (**a**) Absorbance spectra of psQD $PCalaNH_2$. (**b**) Fluorescence spectra of psQD $PCalaNH_2$. (**c**) Fluorescence decay of psQD $PCalaNH_2$, excited at 320 nm; emission collected at 550 nm. (**d**) Evolution of cPC/QD = *cPC* as a function of ultraviolet (1.0 mW cm^{-2}) irradiation time. Solid lines: fits to a monoexponential decay function. (**e**) Quenching of QD fluorescence as a function of ultraviolet (1.0 mW cm^{-2}) irradiation time. (**f**) Quenching of QD fluorescence as a function of cPC/QD. The fits to the data in **d-f** based on equation (2) are given in Supplementary Fig. 6 and Supplementary Table 2. Error bars correspond to measurement uncertainty and error propagation thereof.

Table 2 | Photoconversion kinetics and photostationary end states of psQDs

PC utilized	$10^3 \cdot k'_{oc}$* (s^{-1})	*cPC*† (PS, THF)	*cPC* (PS, SBB)	Quenching of QD (PS, SBB)	α_{PS} (SBB)
$PCmNH_2$	18 ± 1	19.5 ± 2.5	17.2 ± 0.4	0.37	0.19 ± 0.02
$PCalaNH_2$	26 ± 2	17.0 ± 2.0	14.2 ± 0.4	0.36	0.18 ± 0.02
$PCahxNH_2$	24 ± 1	13.2 ± 1.8	6.0 ± 0.3	0.40	0.10 ± 0.01
$PCaocNH_2$	19 ± 1	8.3 ± 1.0	2.5 ± 0.2	0.38	0.06_6 ± 0.01
$PCadoNH_2$	14 ± 2	12.9 ± 1.6	4.2 ± 0.2	0.39	0.07_1 ± 0.01

cPC, closed isomer; PC, photochrome; PS, photostationary state; psQD, photoswitchable quantum dot; QD, quantum dot; SBB, 50 mM Na-borate, pH 9.0; THF, tetrahydrofuran.
*Monoexponential fits of cPC conversion with 340-nm irradiation (1.0 mW cm^{-2}); data of Fig. 2d. The kinetic time courses were only approximately monoexponential due to the time-dependent FRET contribution (see text). In the absence of FRET, the photoconversion rate constant $k'_{oc} = k_1 + k_2$ (see equation (1) and ref. 1).
†Photoconversion of the polymer extracted from the psQD, that is, with individual molecules in a free state, assuming $\alpha_{PS,0} = 0.22$.

and the corresponding cPC populations generated by ultraviolet irradiation, and (2) a mechanism by which FRET-induced cycloreversion compensates and thus buffers distance–orientation variations. In this manner, FRET functions as a virtual photoselection principle establishing the steady-state distribution of cPC. These concepts are developed quantitatively in the following sections. We note that enhanced backconversion of cPC has been achieved with plasmonic enhancement from gold nanoparticles and shown to be dependent on the distance from the nanoparticle surface[21,22].

pcFRET mechanism in psQDs. We have dealt previously[16,17] with the operation of pcFRET in coupled systems of multiple photochrome acceptors assembled about a central QD donor; pcFRET dyads have been investigated by others[23,24]. The interplay between the photochromic acceptor and QD donor manifolds is illustrated in Fig. 3.

The unique feature of pcFRET is the additional mechanism for inducing backconversion (cPC → oPC) via the QD donor. Many other systems incorporate QD donors and an array of centrosymmetric acceptors with presumably invariant transfer

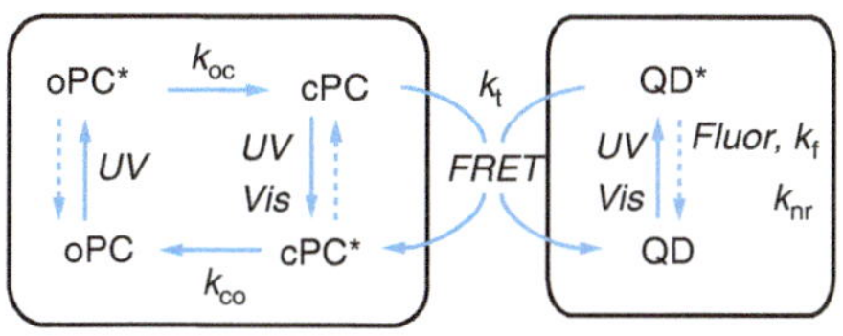

Figure 3 | Photophysical pathways in pcFRET. For simplicity only a single PC moiety is shown. In psQDs, multiple PCs are operative, each characterized by a given resonance energy transfer rate k_t, depending on location and orientation relative to the QD core. Photocylization (k_{oc}) and photoreversal (k_{co}) originate from electronic excited states (oPC* and cPC*, respectively); these quantities differ formally (in definition) from the rate constants denoted with the same symbols elsewhere in the text. QD decay is by emission (k_f), non-radiative pathway(s) (k_{nr}) and FRET (k_t).

cross-sections[25]. In contrast, the transfer efficiency and acceptor distribution in the psQD pcFRET system varies continuously during the photochromic transitions. Thus, α_{PS} depends on the wavelength of the photoconversion light sources in the near-ultraviolet, and the capacity of the QD donor to absorb light in the same region and transfer energy. For a system defined by specific geometric D–A relationships (separation, orientation), the evolution during the kinetic phase of the oPC–cPC interconversion reaction (Fig. 2d) is described by the following differential equation for the time-dependent number of closed form PCs per QD, $cPC[t]$.

$$\begin{aligned} &cPC'[t] = k_1 \cdot oPC[t] - k_2 \cdot cPC[t] - k_3 E[t] \\ &k_1 \equiv k_{ex,oPC}\Phi_{oc}; \quad k_2 \equiv k_{ex,cPC}\Phi_{co}; \quad k_3 \equiv k_{ex,QD}\Phi_{co} \\ &E[t] = \frac{k_t[t]}{k_{d,o} + k_t[t]} = \frac{cPC[t] \cdot \gamma}{1 + cPC[t] \cdot \gamma}; \quad \gamma = \kappa^2 \left(\frac{R^6_{0,-\kappa^2}}{r^6_{DA}} \right) \end{aligned} \tag{1}$$

The first and second terms in equation (1) correspond to the light-dependent cyclization and ring-opening reactions, respectively, and the third term to FRET-induced ring opening; $E[t]$ is the time-dependent transfer efficiency from the QD donor. The three k_{ex} coefficients are the rate constants for direct excitation (by light) of the oPC form, the cPC form and the QD, respectively; they are given numerically by the product of the respective absorption cross-sections and the irradiances, and thus vary with wavelength. Φ_{oc} and Φ_{co} are the quantum yields for ring closure and ring opening, respectively; Φ_{oc} includes the fraction (~ 0.60; Supplementary Methods) of the antiparallel conformation in the oPC population. $k_t[t]$is the time-dependent rate constant for QD→cPC FRET, given by the product of $PC[t]$, the parameter, γ, defined as the 6th power of the ratio of the Förster transfer constant R_0 and the D–A separation, r_{DA}; and the FRET-unperturbed QD excited-state decay rate $k_{d,0}$. R_0^6 is factored explicitly into the term, $R^6_{0,-\kappa^2}$ (lacking κ^2) with κ^2 as a coefficient to accommodate molecular species varying independently in both orientation and D–A separation, while maintaining the same value of γ. Thus, $R^6_{0,-\kappa^2} \propto Jn^{-4}Q$, where J is the D–A overlap integral, n the refractive index of the medium in which transfer occurs (in our case identified with the polymer coat) and Q is the donor (QD) quantum yield (0.2) in the absence of FRET. The invariant $R^6_{0,-\kappa^2}$ computed for FRET occurring within the polymer cap of the psQD systems was 4.1^6 nm^6 (Supplementary Note 1). The issue of distance–orientation coupling is treated in further detail below. Equation (1) incorporates the assumption that in the presence of a single donor (the QD) and i identical acceptors, each with FRET rate k_t, the net transfer rate, $k_{t,tot}$, is given by $i \cdot k_t$ (ref. 26). By definition, the FRET efficiency $E[t]$ is equivalent to the fractional quenching of the QD donor.

The solution to equation (1) yields the kinetic course of the reaction initiated by exposure to ultraviolet or Vis light and the PS established at its conclusion. A number of interesting features revealed by simulations based on equation (1) are featured in Supplementary Note 2; Supplementary Fig. 3. However, the unimodal (single population) model fails to reproduce the very different time courses of quenching and photoconversion exhibited by the psQDs, such that a degree of population heterogeneity with respect to γ (and thus in r_{DA} and/or κ^2) must be invoked. That is, the rapid initial quenching (Fig. 2e) implies that the (sub)populations of PC groups with larger values of γ photoconvert faster, albeit to a reduced extent. Both of these circumstances are attributable to larger values of $k_3E[t]$, which lead to an increase in k'_{oc} and to a corresponding decrease in α_{PS}.

Explicit consideration of the static acceptor condition. Under the assumption of an invariable chemical microenvironment of the PC in the various preparations, the variation in the aliphatic linker length should not affect the inherent photoconversion mechanism. That is, the different psQDs could be considered as FRET systems with common parameters except for the D–A separation r_{DA}. However, during the course of this study it became apparent that another key FRET factor, the orientational distribution of photoconverted cPC acceptors (represented via κ^2, the D–A dipole–dipole orientation factor), must be considered explicitly in quantitative treatments. We have previously utilized the value of 2/3 for κ^2 generally assumed for dynamic, rotationally isotropic systems[17]. However, because QDs and PC groups are deemed to be immobile on the photophysical time scale (ns), the appropriate FRET formalism is that applicable to the 'static orientational regime' in which case it is necessary to consider κ^2 separately from the other parameters constituting the Förster constant R_0^6 (see equation (1)). We have applied concepts and results from seminal publications of van de Meer and Vogel[27–30] to the analysis of the psQDs. It was apparent that the variation in α_{PS}, and thus γ, was incompatible with a constant E, in view of the expression given by Vogel *et al.*[30] for the random, isotropic case: $\langle\gamma\rangle \cong \langle E\rangle/(1-\langle E\rangle)^2$.

QD-based systems differ inherently from the conventional D–A pairs based on organic molecules in that each successive excitation of the semiconductor core of the QD generates an emission dipole that is oriented randomly with respect to any given, individual acceptor. As a consequence, the applicable value of κ^2 requires azimuthal and polar averaging using distinct orientations of the acceptor absorption dipole with respect to the D–A separation vector, **R**, which is colinear with a corresponding particular radius (assuming spherical centrosymmetry) of the QD. Our computations based on these notions indicate that the mean orientation factor, $\langle\kappa^2\rangle$, can vary between $8/(3\pi) = 0.85$ and $2/(3\pi) = 0.21$, respectively, for an acceptor transition moment oriented colinear with or perpendicular to **R** (Supplementary Note 1). One should note that individual κ^2 values can lie between 0 and 4 (see Fig. 4.5 in ref. 28), while r^6_{DA} in the psQD systems reported here can probably vary by a factor of $\sim 10^{1.5-2.5}$. Inclusion of a particular PC group in an acceptor (sub)population defined by a given operational value γ_i requires (only) that it satisfy the condition $\left\langle \left[1 + r^6_{DA}/\left(\kappa^2 R^6_{0,-\kappa^2}\right)\right]^{-1} \right\rangle - 1 = \gamma_i$. We stress that whereas the γ_i are representative of the FRET capability of specific subpopulations of acceptors within a psQD, and are thus invariant once defined by the fits, the mean value $\langle\gamma\rangle$ for the psQD as a whole is dependent on the degree of photoconversion.

Photoconversion heterogeneity. We now return to the observed deviation of the QD-quenching functions from monoexponential behaviour, with large increases at short irradiation times (low *cPC* values) and an asymptotic behaviour as the PS is approached (Fig. 2e). One attributes such behaviour in conventional QD FRET systems to inherent, static heterogeneity with greater relative quenching at low valencies[31]. In the case of the psQDs, it is necessary to invoke not only a static but also a dynamic (time dependent) heterogeneity in population mean FRET efficiencies, such that the final distribution of cPC groups is incongruous with that of the total PC population but biased systematically to lower values of $\langle\gamma\rangle$, that is, greater D–A separations and/or lower $\langle\kappa^2\rangle$.

We have previously considered a simple dual-population acceptor model for psQD constructs[17]. In the present study, we improved data fitting by invoking for each psQD species three PC subpopulations, $i = 1$–3 (this choice is discussed below), defined by the corresponding FRET transfer rates from the QD donor to the cPC acceptor form ($k_{t,i}$), expressed via the respective values of γ_i. The subpopulations generated by the fitting procedure were renumbered in order of increasing values of γ_i. The total numbers of the PC groups (PC_i) of the subpopulations were freely adjustable, although subject to the requirement that they add up to the predetermined experimental *PC* values within their given statistical range. As noted in the previous section, a given value of γ_i can arise from very different combinations of κ^2 and r_{DA}. In the differential equation (2), the underlying principle of equation (1) is extended in that the individual FRET efficiencies are coupled, yielding the following time-dependent relationships for $E_i[t]$, $E[t]$ (the global transfer efficiency), and $\langle\gamma\rangle[t]$, the time-dependent mean global value of γ.

$$E_i[t] = \frac{cPC_i[t]\cdot\gamma_i}{1+\sum_{i=1}^{n} cPC_i[t]\cdot\gamma_i} \qquad i = 1, 2, 3$$

$$E[t] = \frac{\sum_{i=1}^{n} cPC_i[t]\cdot\gamma_i}{1+\sum_{i=1}^{n} cPC_i[t]\cdot\gamma_i} \qquad \langle\gamma\rangle[t] = \frac{\sum_{i=1}^{n} cPC_i[t]\cdot\gamma_i}{cPC[t]} \tag{2}$$

An important assumption underlying equation (2) is that with i distinct subpopulations of acceptors, n_i in number, $k_{t,tot} = \sum n_i \cdot k_{t,i}$. Since there is only a single donor in an individual psQD, the various decay paths mediated by FRET are competitive. That is, the instantaneous quenching produced by the ith acceptor subpopulation, $E_i[t]$, is a function of the other values $E_{k\neq i}[t]$, because the energy transfer probabilities are additive. At equilibrium, the most efficient paths (largest $cPC_{PS,i}\cdot\gamma_i$) lead to the lowest degree of photoconversion, that is, value of $\alpha_{PS,i}$. Furthermore, the net transfer efficiency, E_{PS}, and thus QD quenching, in the PS, is a function of all the *PC* and γ values (equation (3)). Equation (3) is valid for $i = 1$–3 subpopulations (via selection of PC_i) and is of general utility for parametric modelling of psQDs and other related systems based on pcFRET.

$$\text{Solve}\left[1+\frac{c_0+c_1E_{PS}+c_2E_{PS}^2}{(b_1-E_{PS})(b_2-E_{PS})(b_3-E_{PS})} == \frac{1}{1-E_{PS}}, E_{PS}\right], \tag{3}$$

where

$$a_0 = \frac{k_3/k_1}{k_2/k_1-1}; \quad a_i = PC_i\cdot\frac{\gamma_i}{k_2/k_1-1}; \quad b_i = 1 + a_0\cdot\gamma_i;$$

$$c_0 = a_1b_2b_3 + a_2b_1b_3 + a_3b_1b_2;$$

$$c_1 = -[a_1(b_2+b_3) + a_2(b_1+b_3) + a_3(b_1+b_2)];$$

$$c_2 = a_1 + a_2 + a_3.$$

The *cPC*[*t*] and quench[*t*] data for the five psQD classes (Fig. 2d–f) were fit by equations (1) and (2) using the ParametricNDSolve and NonlinearModelFit procedures of Mathematica (Wolfram Research), with PC_i, γ_i, and k_1, k_2 and k_3 as the fitting variables for each psQD. The parameters were to a degree degenerate since *PC* is constrained (the values were allowed to vary within the range given by measured mean ± s.d.) and with $\alpha_{PS,0} = 0.22$, the value of k_2/k_1 was fixed to 3.545. The assumption was that the PCs were identical in photophysical properties, regardless of their position within the polymer cap. The detailed fitting protocols, simulations and results are given in Supplementary Note 3. Selected fit parameters are presented in Table 3 and Fig. 4.

The successful fitting procedures validate the choice of a discrete number of components to represent the heterogeneous population of PC groups in each class of psQDs. Exploration of other potential distribution functions, for example a truncated Gaussian, indicated that the derived global parameters such as $\langle\gamma\rangle$ were not contingent on the adoption of a particular model. Due to the 6th power dependence of γ on distance, three components with parameters freely selectable in the fitting procedure provided sufficient flexibility to account for the overall psQD performance. Fitting with a large number of subpopulations yielded parameters without statistical significance. The instantaneous value of the mean quantity $\langle\gamma\rangle$ is equal to the derivative of quench[*t*], that is, the fractional quenching per unit increment in *cPC*, and thus constitutes a measure of what we denote as 'FRET potency'. Table 3 presents the values of both $\langle\gamma\rangle_0$ and $\langle\gamma\rangle_{PS}$, while the time courses from the initial to the PS state are shown in Fig. 4c.

The decrease in FRET protency during the time course of irradiation underlies the mechanism of energy transfer self-regulation in the psQDs. Qualitatively, it reflects the transition from the maximal achievable rate of photoconversion upon exposure to ultraviolet light, the 'substrate' being the PCs in the oPC form, to the maximal rate of FRET-induced cycloreversion in the PS, in which case the corresponding cPC 'substrate' concentration is maximal. In terms of the photoconversion reaction itself, the system shifts from being determined by k_1 and k_2 to being dominated by k_1 and k_3. An important implication is that one can exploit the very different properties of a given nanoparticle by judicious selection of the operational condition (see Discussion).

The values of $\langle\gamma\rangle_0$, representing the condition under which all PCs are in the oPC form, were quite different for the various psQDs, in accordance with the pcFRET theory. That is, the increase of $\langle\gamma\rangle$ correlated with greater linker length (smaller r_{DA}), reflecting the anticipated closer proximity of the PC groups to the QD surface. For example, the $PCaocNH_2$ linker provides a 10-fold higher FRET potency than $PCmNH_2$, an important distinction for many applications of psQDs. The exemption was the class 5 psQD $PCadoNH_2$ linker, which was longer than the alkyl chains of the QD ligands (TOPO, TOP and HDA). We conclude that the steric nonconformity forced the linker into a foldback configuration, thereby positioning the terminal PC group somewhere between the region corresponding to linker species 3 and 4.

Examination of the subpopulations defined in the fitting procedure leads to additional insights into the functional substructure of the psQD preparations. As in the case of the global parameters of the different psQDs, the values of $\alpha_{PS,i}$ of the corresponding subpopulations were also inversely related to γ_i (Fig. 4a). This feature is also evident in Fig. 4b, featuring the FRET transfer rate, $k_{t,i}/k_{d,0} = cPC_{PS,i}\cdot\gamma_i$, of each subpopulation in the PS. The distribution of FRET transfer activity was biased towards the subpopulations with lower γ_i, thereby creating greater uniformity in the overall system.

Another key parameter that could be derived from the data was the mean donor–acceptor distance for each class of psQDs.

Table 3 | Selected results derived from photoconversion analyses.

PC utilized	PC_i	$cPC_{PS,i}$	γ_i	γ_0	γ_{PS}	$r_{DA,0}$ nm	$r_{DA,PS}$ nm
$PCmNH_2$	20, 61, 19	4, 13, 0.5	0, 0.006, 1.0	0.19 ± 0.01	0.03_4	4.8	6.5
$PCalaNH_2$	70, 8, 9	14, 0.3, 0.07	0.01, 0.7, 3.2	0.39 ± 0.01	0.03_8	4.3	6.4
$PCahxNH_2$	33, 2, 33	5, 0.2, 0.4	0.03, 0.08, 1.1	0.57 ± 0.01	0.11_0	4.1	5.3
$PCaocNH_2$	21, 3, 9	3, 0.05, 0.05	0.10, 2.1, 5.8	1.8 ± 0.2	0.25 ± 0.01	3.4	4.6
$PCadoNH_2$	20, 20, 16	3, 1, 0.1	0.03, 0.3, 2.2	0.72 ± 0.03	0.14 ± 0.01	3.9	5.1

cPC, closed isomer; DA, donor-acceptor; PC, photochrome; PS, photostationary state.
The indicated ranges correspond to s.e. values from the measurements, fits and error propagation thereof. See text for explanation of individual parameters.

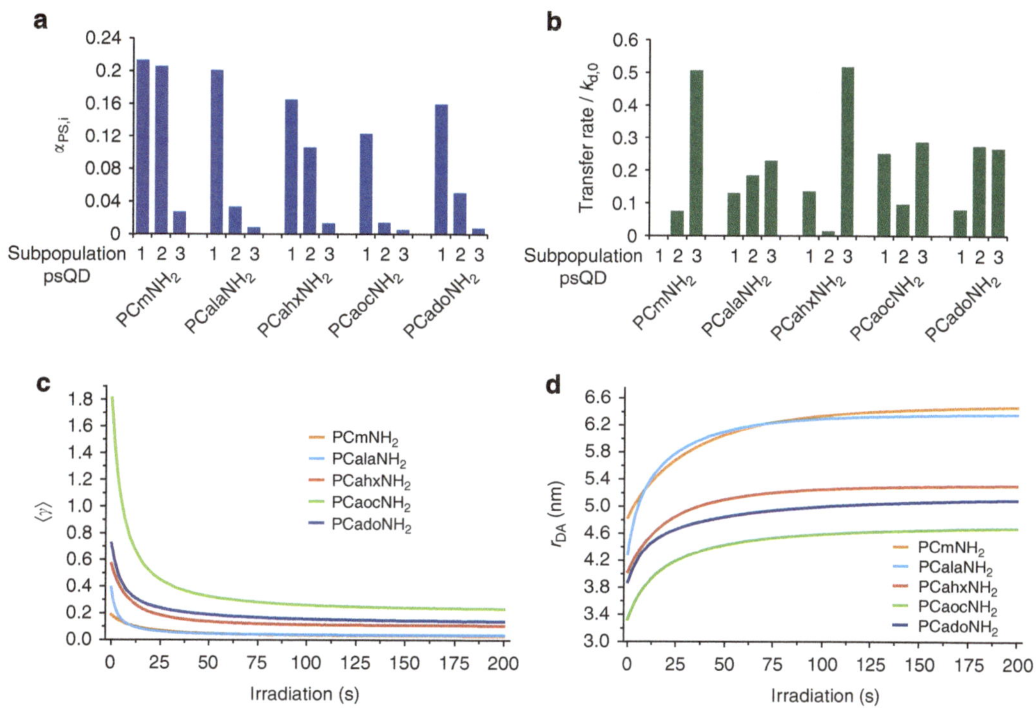

Figure 4 | Parameters for the three subpopulation of the five psQD classes. (**a**) $\alpha_{PS,i}$, subpopulations numbered in order of γ_i. (**b**) Corresponding normalized transfer rates of the acceptor subpopulations. Note that $k_{t,i}/k_{d,0} = cPC_{PS,i} \cdot \gamma_i$. (**c**) Temporal evolution of 'FRET potency', $\langle\gamma\rangle$, as a function of ultraviolet (irradiance $1.0\,mW\,cm^{-2}$) irradiation time. (**d**) Derived time-dependent apparent D-A separation r_{DA} as a function of ultraviolet (irradiance $1.0\,mW\,cm^{-2}$) exposure time.

Equation (4) provides an estimate of r_{DA}based on the mean global FRET parameter $\langle\gamma\rangle[t]$ at any point during the time course of photoconversion.

$$r_{DA}[t] = \sqrt[6]{\langle\kappa^2\rangle R^6_{0,-\kappa^2} \langle\gamma\rangle[t]^{-1}} \qquad (4)$$

A midrange value of $\langle\kappa^2\rangle = 5/(3\pi) = 0.53$ and $R^6_{0,-\kappa^2} = 4.1^6\ nm^6$ were utilized to compute the initial and PS values of r_{DA} (Table 3). The computed distances were compatible with the radius of the QD (3 nm) as well as the assembled psQD in solution, previously determined as 7 nm (ref. 17). According to these estimates, the 'furthermost' PCs, located on the outer edge of the coated psQD, were in $PCmNH_2$ while those of $PCaocNH_2$ were closest to the QD surface. The absolute and relative changes, as well as the rates characterizing the time dependence of $r_{DA}[t]$, differed between the psQD preparations in a manner compatible with considerations presented earlier and the clustering evident in Fig. 2f. However, we note that the shifts in the calculated distances between $r_{DA,0}$ and $r_{DA,PS}$ within the psQD preparations were in the range of 1.2–2.1 nm, values generally larger than those encompassed by the linkers (0.0–1.6 nm). We rationalize this finding by invoking distance-coupled orientational selection according to which $\langle\kappa^2\rangle$ also varied in time.

The quantum yields of photochromism were also determined from the various fit parameters, including those derived from the kinetics of cycloreversion induced by irradiation at 540 nm. Values of 0.3 ± 0.1 (s.d.) were obtained for both cyclization (including the factor for the antiparallel oPC fraction) and cycloreversion. These quantum yield estimates are higher than those we reported previously[16].

Analysis of other photoswitchable semiconductor nanocrystals. We have studied structural and compositional variables of other photoswitchable nanoparticles to test the validity of the model proposed for psQDs.

A photochromic polymer (PMA 8PCaoc 75C12) was coated onto a batch of expired (over a year past expiration date) QDs, which had lost >80% of their initial luminescence (Supplementary Fig. 2). It is unclear whether the effect was homogenous or heterogeneous in the population of nanoparticles; however, this consideration does not affect the interpretation of the results. It was assumed that the QD parameters other than the decreased quantum yield were unchanged. For example, coating the expired QDs occurred readily, as with the original QDs, implying that they had neither aggregated nor shed excess surface ligands. The expired sample had a faster cyclization rate k'_{oc} ($0.037 \pm 0.001\ s^{-1}$) and a higher α_{PS} (0.17 compared with 0.07; Table 2) than the psQDs with higher quantum yields (Supplementary Fig. 2), a result supporting the attribution of a dominant influence of QD-mediated pcFRET (k_3, E) on the cycloreversion of cPC.

Another fluorescent semiconductor nanocrystal (Series A plus 560 nm, CAN GmbH, Hamburg), elongated CdSe/CdS core–shell nanorods (24 nm length, 4.6 nm diameter) with higher extinction coefficients (ε_{340} 5,500; ε_{400} 4,000; ε_{540} 227 $mM^{-1} cm^{-1}$), provided an additional test for the proposed psQD model. The quantum rods (QRs) had an emission spectrum similar to that of the original QDs (Figs 2b and 5b). They were coated and purified in a similar manner as the psQDs, eluting earlier from the size-exclusion chromatography column due to their larger size. The resultant photoswitchable nanorods (psQR) differed physically from the psQDs in having a rod shape and a sixfold increase in surface area. The pcFRET properties were also different: $PC = 125 \pm 20$, $R_0 = 6.0$ nm (utilizing a 2/3 estimation for $\langle \kappa^2 \rangle$ and a higher quantum yield), and an emission source no longer representable as a single point dipole at the crystal centre. The polymer coating of the psQR necessarily lacks radial symmetry such that detailed evaluations of κ^2 are difficult. However, the predictions of equations (1)–(3) were fulfilled in that the psQR showed lower values of α_{PS} (PCalaNH$_2$, 0.10 to PCaocNH$_2$, 0.02) and corresponding E_{PS} (0.25 to 0.15). As an example, the psQR prepared with PCadoNH$_2$ polymer had $\alpha_{PS,0} = 0.04$ and $E_{PS} = 0.15$, implying a 19-fold increase in k_3/k_1, a value close to the 16-fold larger extinction coefficient of the QR.

Since backconversion due to FRET limited the quenching of the QD, a system was designed in which the nanoparticle was coated with a pre-irradiated photochromic polymer. This strategy shifted α_{PS} (and thus E_{PS}) to larger values than were achievable in the ultraviolet-induced PS state of the psQDs. The QRs were selected for this experiment since they exhibited increased stability, less photobrightening and higher brightness than the QDs. Two psQR samples were prepared with the PMA 6PCala 75C12 polymer at the same time and under the same conditions, including protection from light (except red), to avoid photoconversion during purification. One sample was irradiated before coating so as to achieve $\alpha_{PS,0}$ (0.22) while the other had all PC groups in the open form, thus corresponding to $\alpha = 0$. After purification the fluorescence and absorbance of the samples were measured, and they were then exposed to pulses of either ultraviolet or Vis light (Fig. 5).

The psQR with $\alpha = 0$ exhibited the expected characteristics: ultraviolet irradiation led to a quenching of the psQR by $25 \pm 3\%$ and Vis irradiation restored the fluorescence to the initial state. The sample exhibited photobrightening and served as a control,

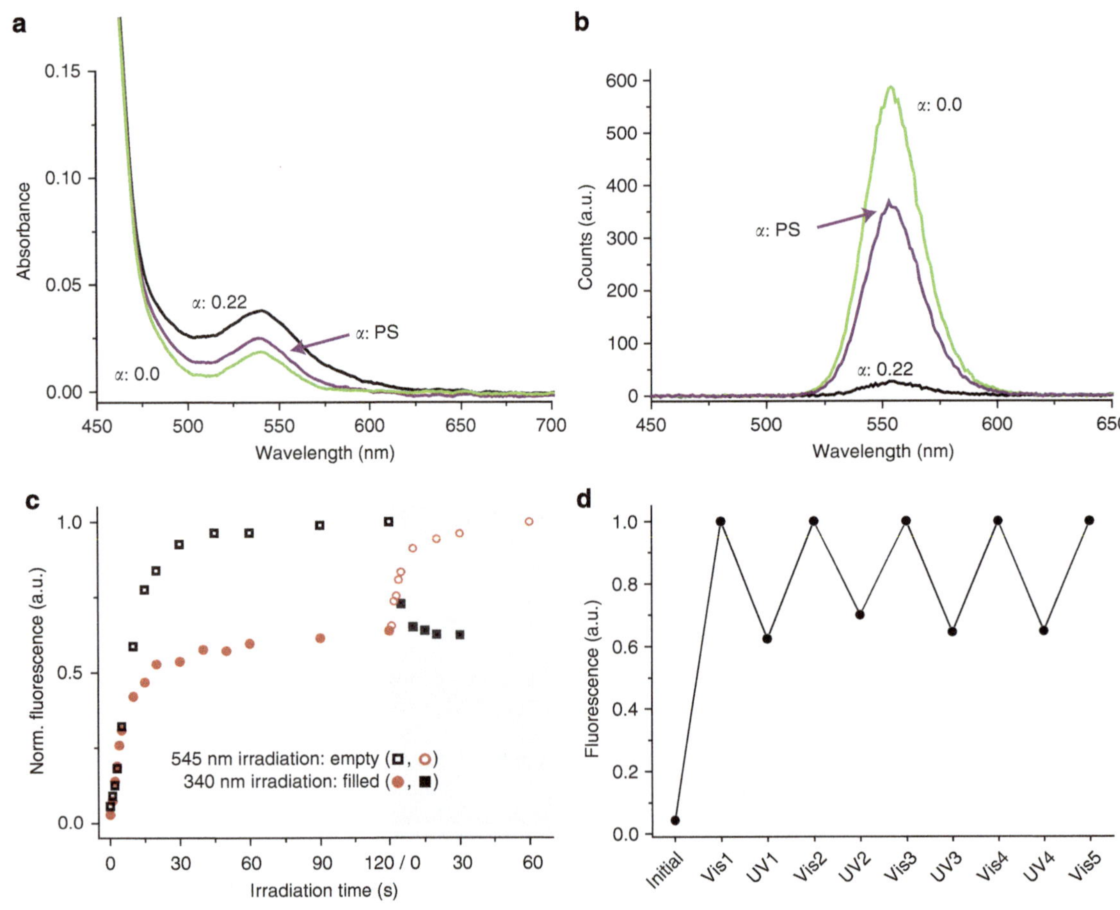

Figure 5 | Spectral characterization of psQR in different photoconversion states. Pre-irradiated polymer (ultraviolet, $\alpha = 0.22$; visible, $\alpha = 0$). Photostationary (ultraviolet) psQR, $\alpha_{PS} = 0.10$. (**a**) Absorbance spectra. (**b**) Fluorescence spectra, excitation at 400 nm. (**c**) Normalized fluorescence of two psQR samples that had been previously irradiated into the PS state ($\alpha_{PS} = 0.22$) and then exposed to pulses of ultraviolet (filled symbols) or visible (empty symbols) light. The shaded region indicates the time after the change in irradiation wavelength. (**d**) Fluorescence of a psQR sample undergoing cycles of irradiation demonstrating that the pre-irradiated state ($\alpha = 0.22$) with almost complete quenching could not be restored. Irradiation times, 180 s.

allowing for the normalization in Fig. 5b–d. The pre-irradiated, ($\alpha_{PS} = 0.22$) psQR had an initial fluorescence of only 2–5% (95–98% quenching) and increased in intensity upon irradiation with either Vis or ultraviolet light. After restoring the open form with green-light irradiation, the 100% reference fluorescence of the QR was observed, but it was not possible to achieve complete quenching during subsequent cycling (Fig. 5d). Ultraviolet irradiation also increased the fluorescence intensity of the pre-irradiated sample, but only until the α_{PS} value was obtained, at which point Vis light was required to regenerate 100% fluorescence (Fig. 5c). Absorbance measurements and polymer-stripping experiments indicated that the maximum number of *cPC* was 17, and that α decreased from an intial value of 0.21 to 0.10 after exposure to light.

By pre-irradiating the PC acceptors and then coating the QR, we succeeded in bypassing the self-regulation of the system imposed by FRET, thereby achieving a much higher fractional conversion to the cPC form, as well as exploiting the higher FRET potency of γ_0 compared with $\langle\gamma\rangle_{PS}$. Upon cycling, this circumstance could no longer be achieved, such that quenching was finite. The pre-irradiation strategy allows the development of a completely photoactivatable QD probe inasmuch as the diheteroarylethenes lack thermal backconversion. This circumstance is similar to other caged fluorescence QDs presented in the literature, with which the destruction[32] or release[33,34] of the quencher decages the QD. Our system presents the additional advantages that the initial fluorescence increase can be realized with a broad range of excitation wavelengths <550 nm, and that upon dequenching the sample can still undergo photomodulation. The samples in Fig. 5 were measured 3 days after preparation, but others stored up to a month showed a similar response.

Discussion

A key finding of this study is that in the ultraviolet-induced PS the photomodulation of psQD by pcFRET is operationally distance independent. The ultraviolet light used to photocyclize the PC to the cPC form excites the QD, which transfers energy to a pre-existing cPC acceptor and drives it back to the oPC conformation. Photocyclization is unaffected by FRET, while cycloreversion is highly dependent on the FRET efficiency and cPC content. In the other PS achieved by exposure to Vis light, the PCs exist exclusively in the open, acceptor-incompetent form because photocyclization is excluded. Under this condition, the 'FRET potency', expressed as fractional QD quenching per extent of acceptor photoconversion, can be substantially accentuated by increasing linker length, that is, by 'engineering (reducing)' r_{DA}. Thus, the psQD family members of this study differ under one condition but not the other.

The relative contribution of cycloreversion to the reaction under ultraviolet irradiation increases as α progresses from 0 (all PCs in the oPC form) to α_{PS} in the PS state, due in part to the large discrepancy in extinction coefficients of the various molecular species at 340 nm (ε, $mM^{-1} cm^{-1}$: QD, 342; oPC, 2.0; cPC, 9.1). Thus, in psQDs the single QD donor undergoes 2–4 more cycles of excitation per unit time than the entire ensemble of PC groups. FRET 'photoselection' can also be regarded as a means for using light to selectively nanoposition distinct chemical groups (in this case oPC and cPC) on the surface of a nanoparticle. That is, while the distribution of the PCs as a whole is determined by ground-state chemistry, that of the cPC subpopulation depends on photonic principles and photochemistry.

As FRET increases, the QD emits fewer photons (is more quenched), yet the content of cPC groups, that is, the operational acceptors, is also reduced. Quenching is independent of the D–A distance as long as the QD absorbance is considerably higher than that of the total oPC content at the irradiation wavelength. It follows that the development of novel FRET-based photomodulatable probes requires consideration of more than the D–A spectral overlap and r_{DA}, inasmuch as the rate constants for the forward and reverse reactions at the conversion wavelength(s) affect the kinetics and extent of quenching in a complex manner.

Expanding the dynamic range (achieving greater quenching) in the psQD system requires larger values of $\alpha_{PS,0}$, and thus a reduction in k_2/k_1 [$\alpha_{PS,0} = (1 + k_2/k_1)^{-1}$] and/or in k_3/k_1. These aims can be achieved by optimizing (i) the choice of PC (greater oPC absorption cross-section, lower quantum yield of cycloreversion); (ii) nanoparticle composition, size and shape; and/or (iii) the ultraviolet photoconversion wavelength. Increasing *PC*, the ostensibly most obvious way to promote greater quenching, is limited by the finite amount of polymer that can bind irreversibly to the QD as well as by the necessity of maintaining PC modification of the polymer to <10% to ensure successful transfer of the the QD to an aqueous medium. We have demonstrated that capping with a pre-irradiated polymer can create virtually completely quenched QDs, which can serve as binary photoactivatable cellular probes. Numerous approaches for achieving spatial and as well as temporal super-reresolution can be considered on the basis of the principles applied in the existing psQD constructs.

Polymer modification and capping is extremely versatile in other ways. One can envision numerous extensions of the system featured in this report, for example, the use of other PC groups and modalities: PCs with a fluorescent cPC form[35]; chemically gated PCs[36]; invisible (blue-shifted cPC) PCs[37]; multicolour PCs[38]; and enantioselective photochromism[39]. The capabilities offered by other photonic strategies directed at the distinct 'writing, reading, erasing' functions of psQDs (or more generally photoswitchable nanoparticles) are extensive: simultaneous or alternating[40] wavelengths for photoconversion and monitoring, which permit biasing the system at any desired PS-operating point; switching between low and high irradiance regimes to achieve saturation for enhancing/suppressing FRET[41] and/or resolution[42]; multiphoton cycloreversion[43]; and lock-in detection for enhancing contrast and selectivity in microscopy[44], thus exploiting the distinctive 'FRET potencies' achievable by engineering the polymer-linker cap. The latter feature constitutes a means for implementing a single-spectrum barcode.

Methods

Synthesis. The PC and photochromic comb-like polymers were synthesized as previously described in the literature[20,45] (for details see Supplementary Methods).

psQD preparation. The psQD preparation was slightly modified from that reported in the literature[19,46]. The QDs (Series A CSS 540 nm QDs, CAN GmbH) were precipitated from the solvent by a procedure supplied by the manufacturer and resuspended in anhydrous $CHCl_3$. A solution of the photochromic polymer was prepared in $CHCl_3$ (~10 mg ml^{-1}) and added to a QD or QR solution in a glass flask in a previously determined optimum proportion (1 mg polymer per ~900 pmol of the nanoparticle). The solution was mixed at 65 °C for 2 h, after which the solvent was slowly evaporated. The dry sample was resuspended in an excess of 50 mM SBB and stirred slowly overnight.

psQD purification. Sample purification was realized by filtration with 0.2-μm inorganic filters, after which the solutions were concentrated to ~1 μM using Amicon 100-kDa cutoff filters (Millipore) and a 50 mM Na-borate, pH 9.0 buffer. Injections of 20 or 50 μl (depending on concentration) were applied to a Superdex 200 analytical column (GE Healthcare Life Sciences) in an HPLC system with ultraviolet–Vis detection. Elution was with 50 mM Na-borate, pH 9, starting with a flow of 50 μl min^{-1} for 3 min and then 20 μl min^{-1}. The average run time was 70 min. The QD fraction was collected in the 45-min range, while the QR fraction was collected at 42 min. In some cases a second injection was required to obtain the desired purity.

The relevant molar extinction coefficients (ε, $mM^{-1} cm^{-1}$ at 340 nm) are: QD, 342; oPC, 2.0; cPC, 9.1. At 540 nm, ε ($mM^{-1} cm^{-1}$) of cPC = 17.5.

Polymer stripping. An aliquot of each psQD (150 µl) was acidified by addition of 1 N HCl (25 µl, final pH 2) and left to aggregate overnight at room temperature. The samples were then centrifuged at 12,000 r.p.m. for 4 min, obtaining a pellet. The supernatant was separated and the pellet was then dissolved in an equivalent volume of THF. The sample was allowed to settle for 2 h to eliminate scattering due to micelle formation, after which photoconversion of the polymer was measured as stated above.

Measurements. Measurements were carried out as reported previously[16]. A custom-built absorption/fluorescence spectrometer with built-in photoconversion capabilities was utilized for the expired QD and some of the pre-irradiated measurements[35].

References

1. Giordano, L., Jovin, T. M., Irie, M. & Jares-Erijman, E. A. Diheteroarylethenes as thermally stable photoswitchable acceptors in photochromic fluorescence resonance energy transfer (pcFRET). *J. Am. Chem. Soc.* **124,** 7481–7489 (2002).
2. Bianco, A., Perissinotto, S., Garbugli, M., Lanzani, G. & Bertarelli, C. Control of optical properties through photochromism: a promising approach to photonics. *Laser Photon. Rev.* **5,** 711–736 (2011).
3. Zhang, J., Zou, Q. & Tian, H. Photochromic materials: more than meets the eye. *Adv. Mater.* **25,** 378–399 (2013).
4. Jares-Erijman, E. A. & Jovin, T. M. FRET imaging. *Nat. Biotechnol.* **21,** 1387–1395 (2003).
5. Jares-Erijman, E., Giordano, L., Spagnuolo, C., Lidke, K. & Jovin, T. M. Imaging quantum dots switched on and off by photochromic fluorescence resonance energy transfer (pcFRET). *Mol. Cryst. Liquid Cryst.* **430,** 257–265 (2005).
6. Wu, T., Boyer, J.-C., Barker, M., Wilson, D. & Branda, N. R. A "plug-and-play" method to prepare water-soluble photoresponsive encapsulated upconverting nanoparticles containing hydrophobic molecular switches. *Chem. Mater.* **25,** 2495–2502 (2013).
7. Jung, H.-y., You, S., Lee, C., You, S. & Kim, Y. One-pot synthesis of monodispersed silica nanoparticles for diarylethene-based reversible fluorescence photoswitching in living cells. *Chem. Commun.* **49,** 7528–7530 (2013).
8. Petryayeva, E., Algar, W. R. & Medintz, I. L. Quantum dots in bioanalysis: a review of applications across various platforms for fluorescence spectroscopy and imaging. *Appl. Spectrosc.* **67,** 215–252 (2013).
9. Medintz, I. L. & Mattoussi, H. Quantum dot-based resonance energy transfer and its growing application in biology. *Phys. Chem. Chem. Phys.* **11,** 17–45 (2009).
10. Algar, W. R., Kim, H., Medintz, I. L. & Hildebrandt, N. Emerging non-traditional Förster resonance energy transfer configurations with semiconductor quantum dots: Investigations and applications. *Coordin. Chem. Rev.* **263-264,** 65–85 (2014).
11. Ishikawa-Ankerhold, H. C., Ankerhold, R. & Drummen, G. P. C. Advanced fluorescence microscopy techniques—FRAP, FLIP, FLAP, FRET and FLIM. *Molecules* **17,** 4047–4132 (2012).
12. Chung, I. H., Shimizu, K. T. & Bawendi, M. G. Room temperature measurements of the 3D orientation of single CdSe quantum dots using polarization microscopy. *Proc. Natl Acad. Sci. USA* **100,** 405–408 (2003).
13. Beljonne, D., Curutchet, C., Scholes, G. D. & Silbey, R. J. Beyond Förster resonance energy transfer in biological and nanoscale systems. *J. Phys. Chem. B* **113,** 6583–6599 (2009).
14. Hua, Z. *et al.* Energy transfer from a single semiconductor nanocrystal to dye molecules. *ACS Nano* **8,** 7060–7066 (2014).
15. Talapin, D. V. *et al.* CdSe/CdS/ZnS and CdSe/ZnSe/ZnS core – shell – shell nanocrystals. *J. Phys. Chem. B* **108,** 18826–18831 (2004).
16. Díaz, S. A. *et al.* Photoswitchable water-soluble quantum dots: pcFRET based on amphiphilic photochromic polymer coating. *ACS Nano* **5,** 2795–2805 (2011).
17. Díaz, S. A., Giordano, L., Jovin, T. M. & Jares-Erijman, E. A. Modulation of a photoswitchable dual-color quantum dot containing a photochromic FRET acceptor and an internal standard. *Nano Lett.* **12,** 3537–3544 (2012).
18. Irie, M. Photochromism of diarylethene single molecules and single crystals. *Photochem. Photobiol. Sci.* **9,** 1535–1542 (2010).
19. Fernández-Argüelles, M. T. *et al.* Synthesis and characterization of polymer-coated quantum dots with integrated acceptor dyes as FRET-based nanoprobes. *Nano Lett.* **7,** 2613–2617 (2007).
20. Diaz, S. A., Giordano, L., Azcarate, J. C., Jovin, T. M. & Jares-Erijman, E. A. Quantum dots as templates for self-assembly of photoswitchable polymers: small, dual-color nanoparticles capable of facile photomodulation. *J. Am. Chem. Soc.* **135,** 3208–3217 (2013).
21. Nishi, H., Asahi, T. & Kobatake, S. Plasmonic enhancement of gold nanoparticles on photocycloreversion reaction of diarylethene derivatives depending on particle size, distance from the particle surface, and irradiation wavelength. *Phys. Chem. Chem. Phys.* **14,** 4898–4905 (2012).
22. Ouhenia-Ouadahi, K. *et al.* Photochromic-fluorescent-plasmonic nanomaterials: Towards integrated three-component photoactive hybrid nanosystems. *Chem. Commun.* **50,** 7299–7302 (2014).
23. Spangenberg, A. *et al.* Photoswitchable interactions between photochromic organic diarylethene and surface plasmon resonance of gold nanoparticles in hybrid thin films. *Phys. Chem. Chem. Phys.* **15,** 9670–9678 (2013).
24. Ouhenia-Ouadahi, K. *et al.* Fluorescence photoswitching and photoreversible two-way energy transfer in a photochrome-fluorophore dyad. *Photochem. Photobiol. Sci.* **11,** 1705–1714 (2012).
25. Kim, H., Ng, C. Y. W. & Algar, W. R. Quantum dot-based multidonor concentric FRET system and its application to biosensing using an excitation ratio. *Langmuir* **30,** 5676–5685 (2014).
26. Fabian, A. I., Rente, T., Szollosi, J., Matyus, L. & Jenei, A. Strength in numbers: effects of acceptor abundance on FRET efficiency. *ChemPhysChem* **11,** 3713–3721 (2010).
27. van der Meer, B. W. Kappa-squared: from nuisance to new sense. *Rev. Mol. Bioctechnol.* **82,** 181–196 (2002).
28. van der Meer, BW, van der Meer, DM & Vogel, SS. in: *FRET—Förster Resonance Energy Transfer.* (eds Medintz, I & Hildebrandt, N) 63–104 (Wiley-VCH, 2013).
29. Vogel, S. S., Nguyen, T. A., van der Meer, B. W. & Blank, P. S. The impact of heterogeneity and dark acceptor states on FRET: implications for using fluorescent protein donors and acceptors. *PLoS ONE* **7,** e49593 (2012).
30. Vogel, S. S., van der Meer, B. W. & Blank, P. S. Estimating the distance separating fluorescent protein FRET pairs. *Methods* **66,** 131–138 (2014).
31. Pons, T., Medintz, I. L., Wang, X., English, D. S. & Mattoussi, H. Solution-phase single quantum dot fluorescence resonance energy transfer. *J. Am. Chem. Soc.* **128,** 15324–15331 (2006).
32. Tagliazucchi, M., Amin, V. A., Schneebeli, S. T., Stoddart, J. F. & Weiss, E. A. High-contrast photopatterning of photoluminescence within quantum dot films through degradation of a charge-transfer quencher. *Adv. Mater.* **24,** 3617–3621 (2012).
33. Han, G., Mokari, T., Ajo-Franklin, C. & Cohen, B. E. Caged quantum dots. *J. Am. Chem. Soc.* **130,** 15811–15813 (2008).
34. Impellizzeri, S., McCaughan, B., Callan, J. F. & Raymo, F. M. Photoinduced enhancement in the luminescence of hydrophilic quantum dots coated with photocleavable ligands. *J. Am. Chem. Soc.* **134,** 2276–2283 (2012).
35. Gillanders, F., Giordano, L., Diaz, S. A., Jovin, T. M. & Jares-Erijman, E. Photoswitchable fluorescent diheteroarylethenes: substituent effects on photochromic and solvatochromic properties. *Photochem. Photobiol. Sci.* **13,** 603–612 (2014).
36. Ohsumi, M., Fukaminato, T. & Irie, M. Chemical control of the photochromic reactivity of diarylethene derivatives. *Chem. Commun.* 3921–3923 (2005).
37. Fukaminato, T., Tanaka, M., Kuroki, L. & Irie, M. Invisible photochromism of diarylethene derivatives. *Chem. Commun.* 3924–3926 (2008).
38. Morimoto, M., Kobatake, S. & Irie, M. Multicolor photochromism of two- and three-component diarylethene crystals. *J. Am. Chem. Soc.* **125,** 11080–11087 (2003).
39. Yamamoto, S., Matsuda, K. & Irie, M. Absolute asymmetric photocyclization of a photochromic diarylethene derivative in single crystals. *Angew. Chem. Int. Ed.* **42,** 1636–1639 (2003).
40. Lee, N.K. *et al.* Three-color alternating-laser excitation of single molecules: monitoring multiple interactions and distances. *Biophys. J.* **92,** 303–312 (2007).
41. Beutler, M. *et al.* satFRET: estimation of Förster resonance energy transfer by acceptor saturation. *Eur. Biophys. J.* **35,** 115–129 (2008).
42. Harke, B. *et al.* Resolution scaling in STED microscopy. *Opt. Express* **16,** 4154–4162 (2008).
43. Ishibashi, Y. *et al.* Multiphoton-gated cycloreversion reactions of photochromic diarylethene derivatives with low reaction yields upon one-photon visible excitation. *Photochem. Photobiol. Sci.* **9,** 172–180 (2010).
44. Marriott, G. *et al.* Optical lock-in detection imaging microscopy for contrast-enhanced imaging in living cells. *Proc. Natl Acad. Sci. USA* **105,** 17789–17794 (2008).
45. Jańczewski, D., Tomczak, N., Han, M.-Y. & Vancso, G.J. Synthesis of functionalized amphiphilic polymers for coating quantum dots. *Nat. Protoc.* **6,** 1546–1553 (2011).
46. Pellegrino, T. *et al.* Hydrophobic nanocrystals coated with an amphiphilic polymer shell: a general route to water soluble nanocrystals. *Nano Lett.* **4,** 703–707 (2004).

Acknowledgements

This article is dedicated to the memory of E.A.J.-E., who died on 29 September 2011. This study was supported by the Max Planck Society. S.A.D. and F.G. were supported by the DAAD (German Academic Exchange Service) and the Ministerio de Educación, Argentina. We thank the Department of Physical Biochemistry, MPIbpc, for the use of the TCSPC fluorimeter.

Author contributions

S.A.D. planned and carried out the experiments, analysed the data and wrote the paper. F.G. built the spectrophotometer and helped with the synthesis and measurements. E.A.J.-E. supervised the initial steps of the project. T.M.J. supervised the project, analysed the data, modelled the system and wrote the paper.

Additional information

Competing financial interests: The authors declare no competing financial interests.

12

Superconcentrated electrolytes for a high-voltage lithium-ion battery

Jianhui Wang[1,*], Yuki Yamada[1,2,*], Keitaro Sodeyama[2,3,4], Ching Hua Chiang[1], Yoshitaka Tateyama[2,4] & Atsuo Yamada[1,2]

Finding a viable electrolyte for next-generation 5 V-class lithium-ion batteries is of primary importance. A long-standing obstacle has been metal-ion dissolution at high voltages. The $LiPF_6$ salt in conventional electrolytes is chemically unstable, which accelerates transition metal dissolution of the electrode material, yet beneficially suppresses oxidative dissolution of the aluminium current collector; replacing $LiPF_6$ with more stable lithium salts may diminish transition metal dissolution but unfortunately encounters severe aluminium oxidation. Here we report an electrolyte design that can solve this dilemma. By mixing a stable lithium salt $LiN(SO_2F)_2$ with dimethyl carbonate solvent at extremely high concentrations, we obtain an unusual liquid showing a three-dimensional network of anions and solvent molecules that coordinate strongly to Li^+ ions. This simple formulation of superconcentrated $LiN(SO_2F)_2$/dimethyl carbonate electrolyte inhibits the dissolution of both aluminium and transition metal at around 5 V, and realizes a high-voltage $LiNi_{0.5}Mn_{1.5}O_4$/graphite battery that exhibits excellent cycling durability, high rate capability and enhanced safety.

[1] Department of Chemical System Engineering, University of Tokyo, 7-3-1, Hongo, Tokyo 113-8656, Japan. [2] Elements Strategy Initiative for Catalysts and Batteries, Kyoto University, 1-30, Goryo-Ohara, Kyoto 615-8245, Japan. [3] JST PRESTO, 4-1-8 Honcho Kawaguchi, Saitama 332-0012, Japan. [4] International Center for Materials Nanoarchitectonics, National Institute for Materials Science, 1-1, Namiki, Tsukuba 305-0044, Japan. * These authors contributed equally to this work. Correspondence and requests for materials should be addressed to A.Y. (email: yamada@chemsys.t.u-tokyo.ac.jp).

Lithium-ion batteries, having received great commercial success in the portable power source market, are being aimed for large-scale energy-storage application in electric vehicles[1–3]. To approach the high energy-density requirements for automobiles, a pragmatic approach is to elevate the operating voltage of batteries, from the present 4 V to around 5 V (refs 4,5). This allows the direct application of the mature fabrication technology of 4 V-class lithium-ion batteries, the well-developed negative electrodes (for example, graphite and graphite/silicon), and high-voltage positive electrodes (for example, spinel $LiNi_{0.5}Mn_{1.5}O_4$ and some layered oxides). However, new challenges—which mainly arise from the electrolyte—hinder the practical application of the next-generation 5 V-class battery.

One major problem is metal dissolution from the positive electrode at high voltages, which poses a serious dilemma in designing an electrolyte. In state-of-the-art lithium-ion electrolyte, chemically unstable $LiPF_6$ is an essential component to suppress anodic (oxidative) dissolution of an aluminium current collector because its hydrolysis product of hydrofluoric acid (HF) contributes to an insoluble AlF_3 passivation film[6,7]. However, the generated HF accelerates the dissolution of transition metals from the active electrode materials, which causes severe capacity decay upon cycling, especially at high voltages and elevated temperatures[8,9]. Using $LiNi_{0.5}Mn_{1.5}O_4$ as an example, the dissolved Mn^{2+} and Ni^{2+} ions, albeit <1% of the total amount, deposit on the surface of the graphite negative electrode, which thicken the solid electrolyte interphase (SEI) by catalysing the reductive decomposition of the electrolyte, and consume the limited lithium reserve in the battery to result in a >50% capacity loss in 100 charge/discharge cycles[10,11]. Diversified functional additives and/or alternative solvents have been explored[12–17] but improvements are still unsatisfactory. Efforts have tried more stable salts (less tendency to generate HF) to replace $LiPF_6$, such as lithium perfluorosulfonylamide (shortened to 'amide')[18]. However, the chemically stable amide does not participate in the reaction with Al to form a stable passivation film, thus causing severe anodic dissolution of the Al current collector at >4 V (refs 19–22). As a result, it remains a dilemma for electrolyte design to suppress both the Al dissolution (requiring an unstable salt) and transition metal dissolution (avoiding an unstable salt). Recently, increasing the concentration of amide salts was reported to alleviate anodic Al dissolution[23–25], but the operating voltage of a half-cell is still limited below 4.3 V, presumably owing to some or all of the following reasons: insufficient salt concentration[23], too low ionic conductivity[24] and too low oxidative stability of the solvent[26].

In this work, we report an electrolyte system to resolve the dilemma. We select stable yet dissociative lithium bis(fluorosulfonyl)amide (LiFSA) as the salt and oxidation-stable carbonate esters as the solvent. We demonstrate an unusual liquid with a peculiar three-dimensional structural network obtained at extremely high salt concentrations. The superconcentrated electrolyte not only effectively suppresses the anodic Al dissolution but also remarkably inhibits the transition metal dissolution and, thus, realizes a safe, stable and fast-rate high-voltage $LiNi_{0.5}Mn_{1.5}O_4$|graphite battery.

Results

Physicochemical properties. LiFSA salt was dissolved at various concentrations into three different carbonate ester solvents: dimethyl carbonate (DMC), ethylene carbonate (EC) and mixed EC:DMC. All the mixtures are transparent liquids at room temperature (see Fig. 1a as an example). Their basic physicochemical properties are presented in Supplementary Table 1. Figure 1b shows their viscosity as a function of salt concentration. Independent of the solvents used, the viscosity increases exponentially with increasing the LiFSA mole fraction (X_{LiFSA}). Among the three groups of solutions, the group with DMC as the solvent shows the lowest viscosity because pure DMC has a lower viscosity than pure EC or mixed EC:DMC. For electrolytes with similar solvation radiuses of mobile ions, the ionic conductivity is proportional to the number of mobile ions and inversely proportional to the viscosity of the medium[18]. As shown in Fig. 1c, at dilute concentrations of $X_{LiFSA}<0.14$ (below 1.5 mol dm^{-3}), the use of the EC:DMC mixture shows the highest ionic conductivity owing to a synergistic effect: the high-dielectric-constant EC increases the mobile ion number by promoting salt dissociation; the low-viscosity DMC increases the ion mobility by decreasing the solution viscosity. This is why the mixed solvents of EC and linear carbonates are generally adopted in conventional electrolytes of the lithium-ion battery[18]. However, when X_{LiFSA} is above 0.14, the solution with DMC as the sole solvent shows an even higher ionic conductivity than that with EC:DMC, which should result from the much lower viscosity of the former at high concentrations. This result suggests that the viscosity becomes the decisive factor on the ionic conductivity for a concentrated solution, wherein intensive ionic association exists independent of the solvents used, showing a distinct departure from the conventional electrolyte design strategy on the basis of dilute concentrations. For the LiFSA/DMC solution, a commercially acceptable ionic conductivity of 1.12 mS cm^{-1} (30 °C) is obtained even at a 'super-high' concentration with salt-to-solvent molar ratio of 1:1.1 despite a high viscosity of 238.9 mPa s. Although the ionic conductivity is lower than that of the commercial dilute electrolyte, it does not compromise the rate capability of the battery (shown later).

On the other hand, the drawbacks of the high volatility and high flammability of linear carbonate solvents can be overcome to a large degree owing to the much lower content of organic solvents in the concentrated solutions. Thermogravimetry measurements (Supplementary Fig. 1) show that the weight loss of the superconcentrated 1:1.1 LiFSA/DMC solution is only 1.5 wt% after elevating the temperature to 100 °C, which is considerably lower than those of a dilute 1:10.8 LiFSA/DMC solution (65.5 wt%, corresponding to 1.0 mol dm^{-3}) and a commercial electrolyte (28.7 wt%). As demonstrated in the flame tests (Fig. 1d,e), the 1:1.1 LiFSA/DMC solution does not burn as fiercely as the commercial dilute electrolyte. The superior thermal stability and flame retardant ability of the concentrated electrolytes contribute to the remarkably improved safety properties as compared with the dilute electrolytes.

Reversible reaction of a 5 V-class electrode. Anodic dissolution of the Al current collector and/or oxidative decomposition of solvent may be encountered in the high-voltage application of amide-based electrolytes. To exclude the possible influence of the anodic Al dissolution, we initially used platinum foil as the current collector for the $LiNi_{0.5}Mn_{1.5}O_4$ electrode in a three-electrode cell (Supplementary Fig. 2). The results showed that both dilute (1:10.8) and superconcentrated (1:1.1) LiFSA/DMC electrolytes enabled a reversible Li^+ de-intercalation/intercalation on the $LiNi_{0.5}Mn_{1.5}O_4$|Pt electrode, indicating a reasonably good compatibility between the present electrolyte system and $LiNi_{0.5}Mn_{1.5}O_4$ material at ~5 V.

However, when applied in a coin cell using the conventional Al current collector, low concentrations of LiFSA/DMC electrolytes encountered problems, confirming the critical drawback of anodic Al dissolution for the amide-based electrolytes. As shown in Fig. 2a, the first charge on the $LiNi_{0.5}Mn_{1.5}O_4$|Al electrode is impossible in the dilute 1:10.8 LiFSA/DMC electrolyte owing to

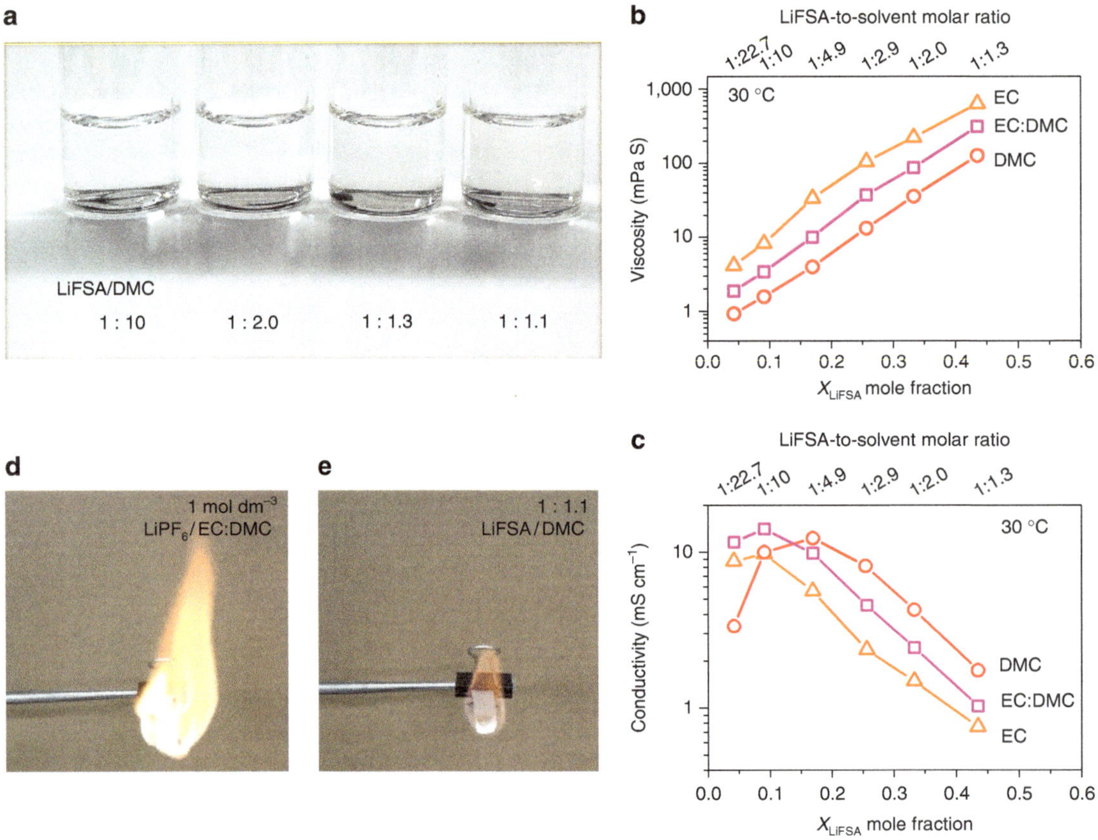

Figure 1 | Physicochemical properties dependent on solution concentration. (**a**) Images of various salt-to-solvent molar ratios of LiFSA/DMC solutions. Viscosity (**b**) and ionic conductivity (**c**) for solutions of LiFSA in DMC, EC and EC:DMC (1:1 by mol.) at 30 °C. The X_{LiFSA} mole fraction is the molar amount of LiFSA salt divided by the total molar amount of the salt and solvents. The LiFSA-to-solvent molar ratios of the solutions are shown on the upper axis. (**d**) Flame tests of a commercial dilute electrolyte of 1.0 mol dm^{-3} $LiPF_6$/EC:DMC (1:1 by vol.) and (**e**) the lab-made superconcentrated electrolyte of 1:1.1 LiFSA/DMC.

the continuous Al dissolution at 4.3 V. In the concentrated 1:1.9 LiFSA/DMC electrolyte (Fig. 2b), the charge/discharge cycling becomes possible up to the cutoff voltage of 4.9 V, but the large irreversible capacity indicates the parasitic Al dissolution remains. The Al dissolution subsequently deteriorates the electrical contacts between the $LiNi_{0.5}Mn_{1.5}O_4$ material and the Al current collector, and results in a fast capacity decay (Fig. 2d). Actually, the poor cycling performance on the $LiNi_{0.5}Mn_{1.5}O_4$ electrode was generally observed in other concentrated 1:2 LiFSA/carbonate ester electrolytes, such as LiFSA in EC, propylene carbonate (PC), ethyl methyl carbonate (EMC) and diethyl carbonate (DEC; Supplementary Fig. 3), indicating this concentration is not sufficient to fully inhibit Al dissolution at 5 V.

In contrast, the superconcentrated 1:1.1 LiFSA/DMC electrolyte enables a reversible Li^+ de-intercalation/intercalation reaction on the $LiNi_{0.5}Mn_{1.5}O_4$ electrode even at a high voltage of 5.2 V (Fig. 2c). In a charge/discharge cycling test, the capacity retention after 100 cycles was over 95% (Fig. 2d), and the coulombic efficiency was close to 100% (Supplementary Fig. 4), evidencing an effective inhibition of anodic Al dissolution. Similarly, using the super-high concentration of 1:1.3, all the LiFSA/carbonate electrolytes enabled stable charging/discharging cycling of the $LiNi_{0.5}Mn_{1.5}O_4$ electrode (see Fig. 2e for example). Especially, the electrolytes using low-dielectric-constant and low-viscosity linear carbonate solvents (for example, DMC, EMC and DEC) showed a faster rate capability as compared with those using high-dielectric-constant and high-viscosity cyclic carbonate solvents (for example, EC, PC and their corresponding mixtures), which is at least partly owing to the much higher ionic conductivity of the former. These results demonstrate that the salt-superconcentrated strategy is a simple, effective and fruitful approach to various safe and stable high-voltage electrolytes. To the best of our knowledge, this is the first time that stable and fast charge/discharge cycling of a 5 V-class electrode using amide salt-based organic electrolytes has been achieved.

The progressive inhibition of anodic Al dissolution with increasing salt concentration is further proved by linear sweep voltammetry (LSV) of an Al electrode and the subsequent scanning electron microscopy observation on the polarized Al surface (see Fig. 3 for details). This notable concentration effect was recently reported but with a debate on whether a stable surface film on Al (ref. 23) or the elimination of uncoordinated (free) solvents of electrolyte[24,25] plays the key role. We conducted a surface analysis of the Al electrodes polarized in various concentrations of LiFSA/DMC electrolytes by X-ray photoemission spectroscopy (XPS) as well as a comparative LSV study between fresh and polarized Al electrodes (see Supplementary Figs 5 and 6 for details). We were unable to obtain any essential evidence to support the existence of a stronger surface film generated in the concentrated electrolyte. Instead, we found that the LiFSA salt readily decomposes and produces LiF upon Ar^+ etching (Supplementary Fig. 7). The previous observation in ref. 23—a much thicker surface film of LiF produced in a higher concentration of electrolyte—is likely to arise from the decomposition of un-rinsed amide salt induced by Ar^+ etching in the XPS measurement.

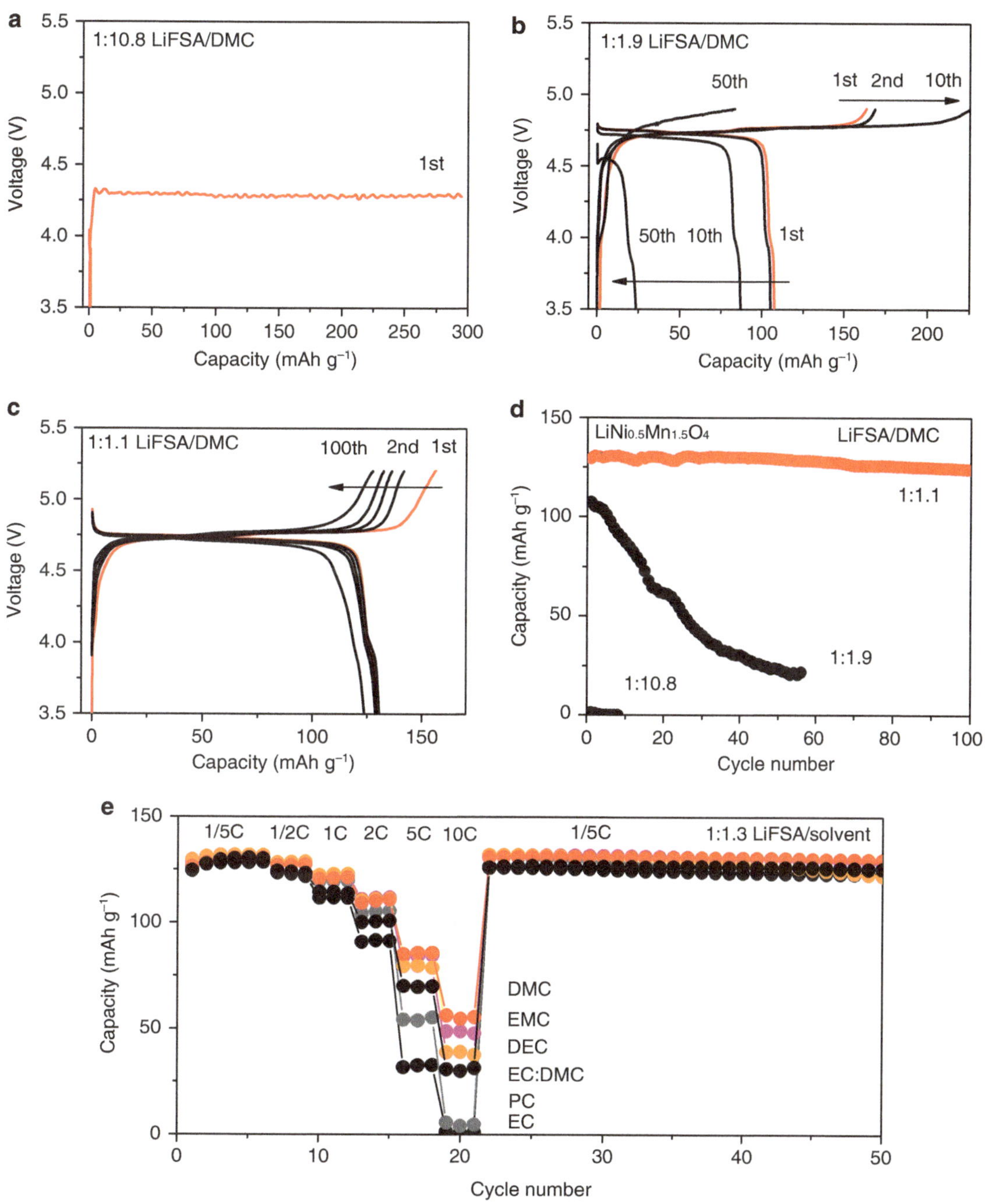

Figure 2 | Performance of 5 V-class $LiNi_{0.5}Mn_{1.5}O_4$ electrode in a half-cell. Charge-discharge voltage curves of $LiNi_{0.5}Mn_{1.5}O_4$|lithium metal half-cells using (**a**) dilute 1:10.8, (**b**) moderately concentrated 1:1.9 and (**c**) superconcentrated 1:1.1 LiFSA/DMC electrolytes at a C/5 rate. Some/all curves of 1st, 2nd, 10th, 50th and 100th cycles are shown. (**d**) Discharge (Li^+ intercalation) capacity retention of the half-cells using different concentrations of LiFSA/DMC electrolytes at a C/5 rate. (**e**) Rate capacity and subsequent cycling retention of the half-cells using 1:1.3 LiFSA-based electrolytes with different carbonate solvents. Charge-discharge tests were conducted at 25 °C with a cutoff voltage of 3.5–4.9 V and a maximum-time restriction of 10 h except for that using the 1:1.1 LiFSA/DMC electrolyte whose cutoff voltage was 3.5–5.2 V. The 1C-rate corresponds to 147 mA g^{-1} on the weight basis of the $LiNi_{0.5}Mn_{1.5}O_4$ electrode.

We now study the solution structure of the electrolytes using Raman spectroscopy observation and density functional theory molecular dynamics simulation (DFT-MD). As shown in the Raman spectra (Fig. 4b left), a free DMC molecule exhibits an O-CH_3 stretching vibration band at 910 cm^{-1} (ref. 27). This band shifts up to 930–935 cm^{-1} when DMC participates in Li^+ solvation. In dilute 1:10.8 LiFSA/DMC, the majority of DMC molecules exist in a free state because the solvent-to-salt molar ratio (10.8) is much larger than a typical four- or fivefold coordination of Li^+ in aprotic solvents. As the LiFSA concentration increases, the population of free DMC decreases and that of Li^+-coordinated DMC increases; the Li^+-FSA^- association simultaneously intensifies through the formation of contact ion pairs (CIPs, FSA^- coordinating to one Li^+) and aggregate clusters (AGGs, FSA^- coordinating to two or more Li^+). The latter is evidenced from a remarkable upshift of the FSA^- band (700–780 cm^{-1}, Fig. 4b right), which is typically observed in the amide-based concentrated solutions[24,25,28–31]. For the moderately concentrated 1:2 LiFSA/DMC solution, the Raman band corresponding to free DMC is remarkably

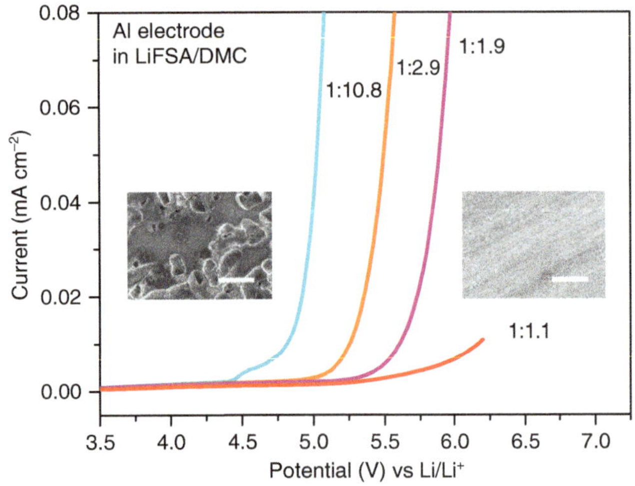

Figure 3 | Oxidation stability of an aluminium electrode. LSV of an aluminium electrode in various concentrations of LiFSA/DMC electrolytes in a three-electrode cell. The scan rate was 1.0 mV s^{-1}. The insets are scanning electron microscopy images of the Al surface polarized in the dilute 1:10.8 (left of panel) and superconcentrated 1:1.1 (right of panel) electrolytes. Many corroding pits cover the surface of the Al electrode polarized in the dilute electrolyte, showing a severe anodic Al dissolution. In contrast, no corroding pits appear on the surface of the Al electrode polarized in the superconcentrated electrolyte, indicating a good inhibition of anodic Al dissolution. The white scale bar represents 20 μm.

weakened, suggesting that the majority of DMC are solvating to Li^+. This is consistent with the DFT-MD simulation, which shows ca. 90% DMC are coordinating to Li^+ with the rest as free solvent (marked as light blue in Fig. 4d). Moreover, the simulation illustrates that all FSA^- anions are coordinating to Li^+ with ca. 20% as CIPs and ca. 80% as AGGs (marked as orange and dark blue in Fig. 4d, respectively). The coordination environment is shown in Supplementary Fig. 8. For the superconcentrated 1:1.1 LiFSA/DMC solution, both DMC and FSA^- bands further upshift substantially, indicating both Li^+-DMC and Li^+-FSA^- interactions enhanced compared with those in 1:2 LiFSA/DMC. The DFT-MD simulation reveals that all DMC molecules, together with all FSA^- anions, are coordinating to Li^+ (no free solvent or anion). Interestingly, besides oxygen, significant amount of nitrogen on FSA^- anions also participate in the coordination with Li^+, which is hardly observed at the lower concentrations. The contribution of nitrogen coordinating to Li^+ is shown in Supplementary Fig. 8. More importantly, almost all FSA^- anions remain in AGG states during the whole DFT-MD simulation time (0.1 fs $\times$ 100,000 steps), and a CIP state is rarely observed, demonstrating the unusual solution structural feature with AGG clusters as the predominant components in the superconcentrated LiFSA/DMC solution. It is noteworthy that each FSA^- anion coordinates to 2–3 Li^+ and each Li^+ is coordinated by 2–3 FSA^- in 1:1.1 LiFSA/DMC. Hence, FSA^- anions in the superconcentrated solution connect with each other through the intensive association with Li^+, leading to a reinforced three-dimensional network (shown in Fig. 4e). This feature is different from the less concentrated solutions, wherein significant amount of CIPs and free solvents divide the solution structure into relatively small-size parts.

Generally, the anodic metal dissolution requires three steps: first, oxidation of the metal to a metal cation; second, coordination of the metal cation by solvents or anions; and finally, the diffusion of the solvated metal cation to the bulk electrolyte[32]. At high voltages of ~5 V, the first step proceeds more rapidly and extensively than at the conventional operating voltage of 4 V. Thereby, the subsequent coordination and diffusion must be strongly inhibited by the nature of electrolyte solutions to suppress the metal ion dissolution. In the moderately concentrated 1:2 LiFSA/carbonate electrolytes, the presence of significant free solvents and CIPs (with two or more coordination sites remaining vacant) could coordinate to Al cations and fail to inhibit Al dissolution completely at 5 V. In contrast, the superconcentrated 1:1.1 LiFSA/DMC electrolyte effectively inhibits Al dissolution even over 5 V, which can be ascribed to its peculiar AGGs-predominant solution structure: (i) all DMC solvents and all FSA^- anions strongly coordinate to Li^+ cations and thus have a much lower probability of coordinating to other metal cations; (ii) the resulting reinforced three-dimensional network further retards the diffusion rate of the metal cations, particularly, those with multiple charge.

Stable cycling of a 4.6 V $LiNi_{0.5}Mn_{1.5}O_4$|graphite battery. In addition to the excellent performance achieved on the $LiNi_{0.5}Mn_{1.5}O_4$-positive electrode, the superconcentrated 1:1.1 LiFSA/DMC electrolyte also realized ultra-stable charge/discharge cycling on the natural graphite-negative electrode despite the absence of conventional SEI-forming agent of EC (Supplementary Fig. 9): the application in a graphite|Li half-cell exhibits a capacity retention of 99.6% after 100 cycles with coulombic efficiency of ~99.8%, and with rate capability comparable with that using a commercial dilute electrolyte. Accordingly, the superconcentrated electrolyte was further applied in the high-voltage $LiNi_{0.5}Mn_{1.5}O_4$|graphite full cell, a much harsher condition than in the half-cell, because the active lithium resource is limited and a new underlying problem arises from the transition metal dissolution from the $LiNi_{0.5}Mn_{1.5}O_4$ especially at high voltages and elevated temperatures. Figure 5a,b shows charge/discharge voltage profiles of $LiNi_{0.5}Mn_{1.5}O_4$|natural graphite full cells at 40 °C using a state-of-the-art commercial electrolyte and the lab-made superconcentrated electrolyte, respectively. The cell with the commercial electrolyte suffers from a severe capacity decay during cycling, that is, only 18% of the initial capacity left after 100 cycles (Fig. 5a,c), which is consistent with previous reports[11,33]. In contrast, the capacity retention using the 1:1.1 LiFSA/DMC electrolyte is over 90% after 100 cycles, exhibiting a remarkably improved cycling durability (Fig. 5b,c and Supplementary Figs 10 and 11). Notably, the superiority of the superconcentrated electrolyte becomes even more marked at 55 °C (Supplementary Fig. 12). It is generally accepted that the poor cycling performance of the $LiNi_{0.5}Mn_{1.5}O_4$|graphite battery originates from the dissolution of transition metals from $LiNi_{0.5}Mn_{1.5}O_4$ into the electrolyte, as introduced at the beginning of this article. Indeed, energy dispersive X-ray spectroscopy (EDS) observation shows a much lower content of Mn and Ni on the graphite electrode of the full cell cycled in the superconcentrated electrolyte than that in the commercial electrolyte, which provides evidence for the effective inhibition of transition metal dissolution in the former. There are two main reasons for the improved performance: (i) LiFSA is less reactive to produce HF as compared with $LiPF_6$, which alleviates the corrosion of $LiNi_{0.5}Mn_{1.5}O_4$ and, thus, reduces the formation of soluble Mn^{2+} and Ni^{2+}; (ii) even if some Mn^{2+} and Ni^{2+} are formed on the surface of $LiNi_{0.5}Mn_{1.5}O_4$, they can hardly dissolve in and transport through the AGGs-predominant superconcentrated electrolyte owing to the same functional manner for the inhibition of Al dissolution. Moreover, it is worth noting that the rate performance of the full cell using the

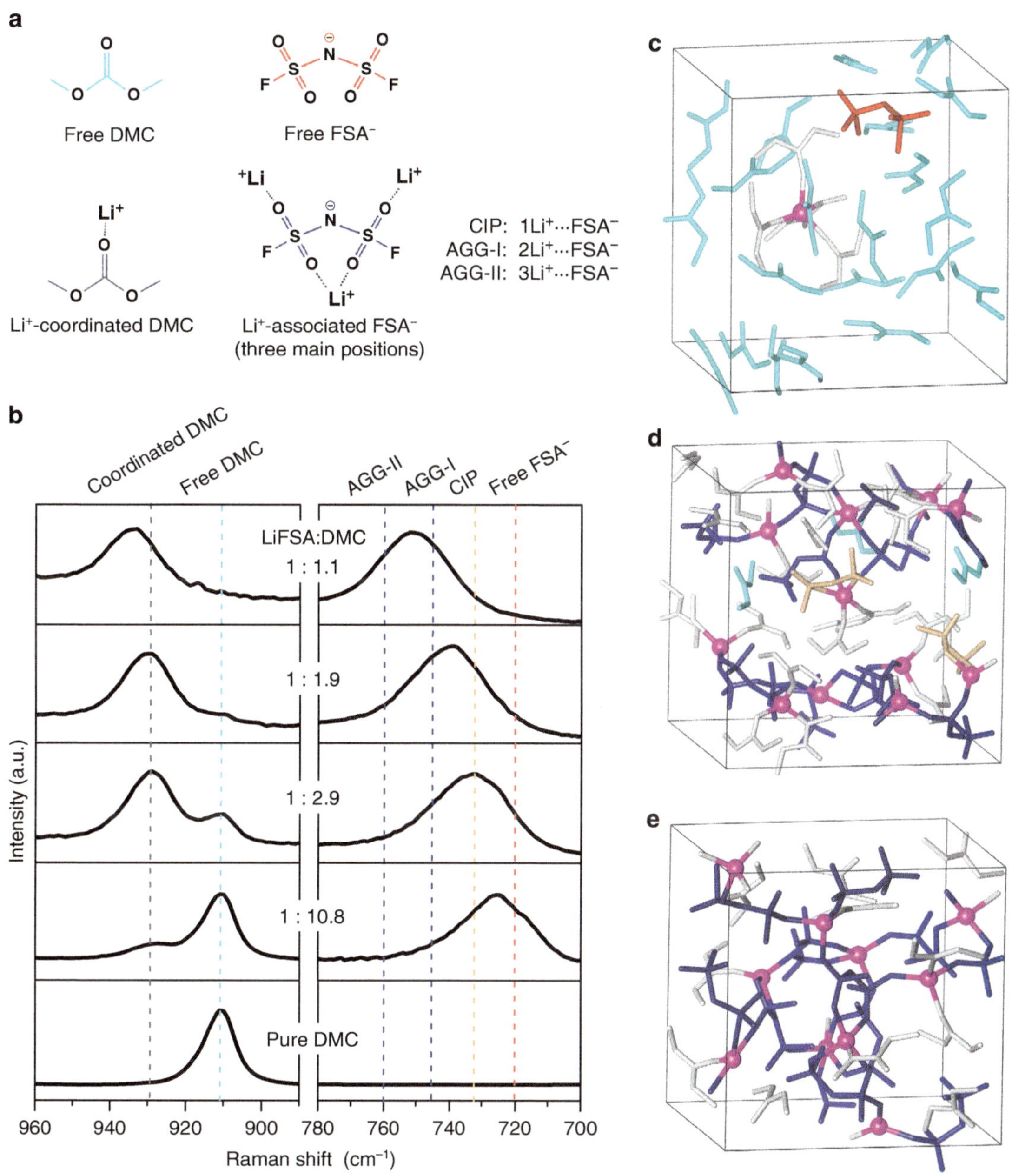

Figure 4 | Li salt – solvent coordination structure dependent on salt concentration. (**a**) The several main species in the LiFSA/DMC solutions. (**b**) Raman spectra of LiFSA/DMC solutions with various salt-to-solvent molar ratios in the range of 890-900 cm^{-1} (O-CH_3 stretching mode of the DMC solvent) and 700-780 cm^{-1} (S-N stretching mode of the FSA^- anion). Snapshots of typical equilibrium trajectories obtained by DFT-MD simulations: (**c**) dilute solution (1 LiFSA/25 DMC, <1 mol dm^{-3}), (**d**) moderately concentrated solution (12 LiFSA/24 DMC, ca. 4 mol dm^{-3}), and (**e**) superconcentrated solution (10 LiFSA/11 DMC, ca. 5.5 mol dm^{-3}). The coordination of Li^+ – DMC and Li^+ – FSA^- is supposed to build up when the involved atoms locate within 2.5 Å from Li^+. The coordination numbers of solvents and anions to Li^+ are shown in Supplementary Fig. 8. Li cations are marked in purple. Free and coordinated DMC molecules are marked in light blue and grey, respectively. Free, CIP and AGG states of FSA^- anions are marked in red, orange and dark blue, respectively. Hydrogen atoms are not shown.

superconcentrated 1:1.1 LiFSA/DMC is comparable with that using the commercial electrolyte (Fig. 5d), although the former shows an ionic conductivity one-order lower than the latter. The mechanistic understanding on the high-rate capability of the superconcentrated electrolyte is underway in our laboratory. To the best of our knowledge, this is the first case that, an electrolyte with such an ultra-simple formulation—a single salt and a single solvent without any additive—realizes stable cycling of a high-voltage lithium-ion battery.

Discussion

The conventional dilute $LiPF_6$/EC-based electrolytes have dominated the electrolyte market of 4 V-class lithium-ion batteries over the past 25 years; however, they show difficulties in satisfying the requirements of next-generation 5 V-class batteries in terms of both safety and stability. Our work demonstrates a number of electrolytes with a reinforced three-dimensional network that are obtained by simple mixing of a stable salt with a conventional carbonate solvent at 'super-high' concentrations. Owing to its

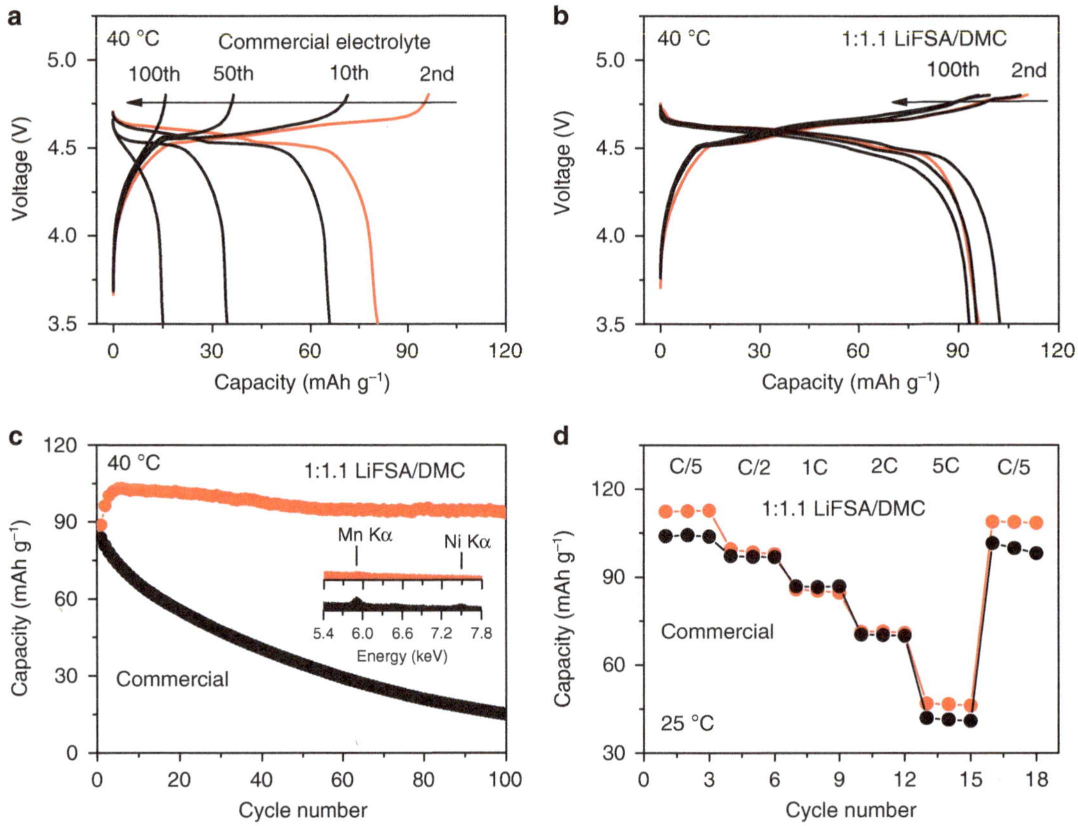

Figure 5 | Performance of a high-voltage $LiNi_{0.5}Mn_{1.5}O_4$|natural graphite battery. Charge-discharge voltage curves of $LiNi_{0.5}Mn_{1.5}O_4$|graphite full cells using (**a**) a commercial 1.0 mol dm^{-3} $LiPF_6$/EC:DMC (1:1 by vol.) electrolyte and (**b**) lab-made superconcentrated 1:1.1 LiFSA/DMC electrolyte at a C/5 rate and 40 °C. The curves of 2nd, 10th, 50th and 100th cycle are shown. (**c**) Discharge capacity retention of the full cells at a C/5 rate. The inset shows EDS spectra on the graphite electrode surface (200 × 200 μm^2 area) after 8-day cycling tests, which is equivalent to the operating time of 100 and 20 cycles for the battery using the commercial and superconcentrated electrolytes, respectively. (**d**) Discharge capacity of the full cell at various C-rates and 25 °C. All charge-discharge cycling tests were conducted with a cutoff voltage of 3.5–4.8 V. 1C-rate corresponds to 147 mA g^{-1} on the weight basis of the $LiNi_{0.5}Mn_{1.5}O_4$ electrode.

peculiar structural characteristics, the superconcentrated electrolytes overcomes the longstanding challenge faced by the unstable $LiPF_6$-based electrolytes at high voltages (passivating the Al current collector versus accelerating the transition metal dissolution of the active material), thus, enables a stable operation of a 5 V-class battery. Emphasis is on the fact that the peculiar solution structure and functionalities are unique to such super-high concentrations (solvent/salt ≈ 1.1), and cannot be achieved in moderately high concentrations (solvent/salt > 1.8) as in previous reports[24,30,31,34,35]. Besides the positive electrode side, the superconcentrated electrolytes also show a good compatibility with the natural graphite-negative electrode even in the absence of EC. It breaks through the limitation of a general requirement of EC for a SEI formation for a lithium-ion electrolyte, and diversifies the electrolyte formulation towards various EC-free electrolytes. Different from the conventional electrolyte design that requires a high-dielectric-constant (usually high-viscosity) solvent, the superconcentrated electrolyte prefers a low-viscosity solvent. Although the ionic conductivity of the superconcentrated electrolyte is lower than that of the conventional dilute electrolyte, it does not necessarily compromise the rate capability of the battery. Clarification of the corresponding mechanism would be enlightening for developing novel high-power batteries. Furthermore, the superconcentrated electrolytes show superior thermal stability and flame retardant ability, alleviating the safety risk for a high-voltage battery using conventional dilute electrolytes. Finally, it is noteworthy that our reported superconcentrated electrolytes do not contain any additives, signifying the potential to further enhance the performance. These desirable features above outperform the conventional dilute electrolytes; meanwhile, the wide-temperature window of the liquid state (ensuring a good contact with the electrode materials), as well as the convenience of the approach, surpass the solid-state electrolytes. Therefore, the superconcentrated electrolytes might offer opportunities to build safe and stable high-voltage batteries that are not limited to the lithium-ion.

Methods

Preparation of electrolytes and electrodes. LiFSA (Nippon shokubai) and all solvents (DMC, DEC, EMC, EC and PC, Kishida Chemical Co. Ltd) were lithium battery grade and used without purification. Electrolyte solutions were prepared by mixing a given amount of LiFSA with solvents in an Ar-filled glove box. The commercial electrolyte of 1.0 mol dm^{-3} $LiPF_6$/EC:DMC (1:1 by vol) was purchased from Kishida Chemical Co. Ltd and used as the reference. Both the lab-made LiFSA-based electrolytes and as-received commercial $LiPF_6$-based electrolyte were dried by molecular sieve before tests. The water content was less than 2 p.p.m., as detected by a coulometric Karl Fischer Titrator.

The electrodes were fabricated by first well mixing the active materials of $LiNi_{0.5}Mn_{1.5}O_4$ (Hosen Corp., mean particle size $\hat{R} = 5$ μm, no surface treatment) and natural graphite (SEC Carbon Ltd., $\hat{R} = 10$ μm), polyvinylidene difluoride (PVdF) and/or Denka black (AB, HS-100) in *N*-methylpyrrolidone with weight ratios of 80:10:10 ($LiNi_{0.5}Mn_{1.5}O_4$:PVdF:AB) and 90:10 (graphite:PVdF). The resultant slurry was cast on the Al or Pt foil (20 μm thickness) for the $LiNi_{0.5}Mn_{1.5}O_4$ electrode and on the Cu foil (10 μm thickness) for the graphite electrode using a 50 μm doctor blade. All those electrodes were dried at 120 °C under vacuum for 12 h. The active material mass loading was 0.7–2 mg cm^{-2} with a thickness of ~15–20 μm, unless otherwise mentioned. The use of relatively low mass loading was to spotlight the critical issue of anodic Al dissolution as the content ratio of metallic Al components (Al current collector and Al positive cap) to the active electrode material becomes much higher in a coin cell. Nevertheless, thick electrodes with a high mass loading of $\sim 10\,mg\,cm^{-2}$ were also tested. The results are shown in Supplementary Figs 11 and 12.

Electrochemical measurements. LSV was performed by VMP-3 (BioLogic) in a beaker cell with an Al belt ($1 \times 4\,cm^2$, 0.6 cm soaked in the electrolyte) as a working electrode and lithium metal as the reference and counter electrodes (shown in Supplementary Fig. 5 inset). $LiNi_{0.5}Mn_{1.5}O_4$|Li half-cells and $LiNi_{0.5}Mn_{1.5}O_4$|-graphite full cells were assembled in the standard 2032-type coin cell hardware in an Ar-filled glove box. A combined separator, composed of cellulose separator (Nippon Kodoshi, placed on the positive electrode side) and glass fibre separator (Advantec GB-50, placed on the negative electrode side), was used. The amount of electrolyte in a coin cell was ca 160 µl to fully wet the separators and electrodes. In the full cells, the weight ratio of $LiNi_{0.5}Mn_{1.5}O_4$:graphite was ~2:1, which corresponds to ~1:1.3 of their theoretical capacity ratio. Galvanostatic charge/discharge cycling and rate capability tests were conducted on a charge/discharge unit (TOSCAT). Charge and discharge were conducted at the same C-rate without using a constant-voltage mode at both ends of the charge and discharge.

Characterization. The density and viscosity of solution samples were evaluated with a DMA 35 density meter and a Lovis 2000 M viscometer, respectively. The ionic conductivity was measured by AC impedance spectroscopy at 1 kHz (Solartron 147055BEC) in a symmetric cell (Pt|electrolyte|Pt). The flammability was tested on an electrolyte-soaked glass fibre filter (Advantec GB-100).

The solution structure was studied by a Raman spectroscopy with an exciting laser of 514 nm (NRS-5100). The samples were sealed in a quartz cell in the glove box to avoid any contamination from the air.

The morphology of Al electrodes after LSV tests were observed by a field-emission scanning electron microscopy at 2.0 kV. The transition metals deposited on the graphite electrode in the $LiNi_{0.5}Mn_{1.5}O_4$|graphite full cells after charge/discharge cycling were estimated by an EDS. The cells were disassembled in the glove box. The obtained electrodes were rinsed in DMC and dried in the glove box. The sample was exposed in air for <1 min at sample loading.

The experimental details for thermogravimetric analysis and XPS measurements are shown in Supplementary Fig. 1 legend and Supplementary Methods, respectively.

Simulations. Car-Parrinello type DFT-MD simulations were carried out using CPMD code[36]. LiFSA/DMC solutions with salt-to-solvent molar ratios of 1:25, 1:2 and 1:1.1 were calculated in cubic supercells with 15.05, 17.03 and 14.34 Å linear dimensions, respectively. A fictitious electric mass of 500 a.u. and a time step of 4 a.u. (0.10 fs) were chosen. The temperature was controlled using a Nosé thermostat with a target temperature of 30 °C. After 5 ps equilibration steps, statistical averages were computed from trajectories of at least 10 ps in length. The electronic wave function was quenched to the Born-Oppenheimer surface approximately every 1 ps to maintain adiabaticity. The energy cutoff of the plane wave basis is set to 90 Ry. Goedecker–Teter–Hutter type norm-conserving pseudopotentials for C, H, O, N, S, F and Li were used[37].

References

1. Goodenough, J. B. & Kim, Y. Challenges for rechargeable Li batteries. *Chem. Mater.* **22,** 587–603 (2010).
2. Etacheri, V., Marom, R., Elazari, R., Salitra, G. & Aurbach, D. Challenges in the development of advanced Li-ion batteries: a review. *Energy Environ. Sci.* **4,** 3243–3262 (2011).
3. Amine, K., Kanno, R. & Tzeng, Y. Rechargeable lithium batteries and beyond: Progress, challenges, and future directions. *MRS Bull.* **39,** 395–401 (2014).
4. Patoux, S. *et al.* High voltage spinel oxides for Li-ion batteries: From the material research to the application. *J. Power Sources* **189,** 344–352 (2009).
5. Croy, J. R., Abouimrane, A. & Zhang, Z. Next-generation lithium-ion batteries: the promise of near-term advancements. *MRS Bull.* **39,** 407–415 (2014).
6. Myung, S.-T., Sasaki, Y., Sakurada, S., Sun, Y.-K. & Yashiro, H. Electrochemical behavior of current collectors for lithium batteries in non-aqueous alkyl carbonate solution and surface analysis by ToF-SIMS. *Electrochim. Acta* **55,** 288–297 (2009).
7. Zhang, X. & Devine, T. M. Identity of passive film formed on aluminum in Li-ion battery electrolytes with LiPF6. *J. Electrochem. Soc.* **153,** B344–B351 (2006).
8. Aurbach, D. *et al.* Review on electrode–electrolyte solution interactions, related to cathode materials for Li-ion batteries. *J. Power Sources* **165,** 491–499 (2007).
9. Zhan, C. *et al.* Mn(II) deposition on anodes and its effects on capacity fade in spinel lithium manganate-carbon systems. *Nat. Commun.* **4,** 2437 (2013).
10. Kim, J.-H. *et al.* Understanding the capacity fading mechanism in $LiNi_{0.5}Mn_{1.5}O_4$/graphite Li-ion batteries. *Electrochim. Acta* **90,** 556–562 (2013).
11. Pieczonka, N. P. W. *et al.* Understanding transition-metal dissolution behavior in $LiNi_{0.5}Mn_{1.5}O_4$ high-voltage spinel for lithium ion batteries. *J. Phys. Chem. C* **117,** 15947–15957 (2013).
12. Zhang, Z. *et al.* Fluorinated electrolytes for 5 V lithium-ion battery chemistry. *Energy Environ. Sci.* **6,** 1806–1810 (2013).
13. Hu, L., Zhang, Z. & Amine, K. Fluorinated electrolytes for Li-ion battery: An FEC-based electrolyte for high voltage $LiNi_{0.5}Mn_{1.5}O_4$/graphite couple. *Electrochem. Commun.* **35,** 76–79 (2013).
14. Pieczonka, N. P. W. *et al.* Impact of lithium bis(oxalate)borate electrolyte additive on the performance of high-voltage spinel/graphite Li-ion batteries. *J. Phys. Chem. C* **117,** 22603–22612 (2013).
15. Song, Y.-M., Han, J.-G., Park, S., Lee, K. T. & Choi, N.-S. A multifunctional phosphite-containing electrolyte for 5 V-class $LiNi_{0.5}Mn_{1.5}O_4$ cathodes with superior electrochemical performance. *J. Mater. Chem. A* **2,** 9506–9513 (2014).
16. Xu, M. *et al.* Improved performance of high voltage graphite/$LiNi_{0.5}Mn_{1.5}O_4$ batteries with added lithium tetramethyl borate. *ECS Electrochem. Lett.* **4,** A83–A86 (2015).
17. Xu, K. Electrolytes and Interphases in Li-Ion Batteries and Beyond. *Chem. Rev.* **114,** 11503–11618 (2014).
18. Xu, K. Nonaqueous liquid electrolytes for lithium-based rechargeable batteries. *Chem. Rev.* **104,** 4303–4417 (2004).
19. Krause, L. J. *et al.* Corrosion of aluminum at high voltages in non-aqueous electrolytes containing perfluoroalkylsulfonyl imides; new lithium salts for lithium-ion cells. *J. Power Sources* **68,** 320–325 (1997).
20. Wang, X., Yasukawa, E. & Mori, S. Inhibition of anodic corrosion of aluminum cathode current collector on recharging in lithium imide electrolytes. *Electrochim. Acta* **45,** 2677–2684 (2000).
21. Li, L. *et al.* Transport and electrochemical properties and spectral features of non-aqueous electrolytes containing LiFSI in linear carbonate solvents. *J. Electrochem. Soc.* **158,** A74–A82 (2011).
22. Kramer, E. *et al.* Mechanism of anodic dissolution of the aluminum current collector in 1 M LiTFSI EC:DEC 3:7 in rechargeable lithium batteries. *J. Electrochem. Soc.* **160,** A356–A360 (2013).
23. Matsumoto, K. *et al.* Suppression of aluminum corrosion by using high concentration LiTFSI electrolyte. *J. Power Sources* **231,** 234–238 (2013).
24. McOwen, D. W. *et al.* Concentrated electrolytes: decrypting electrolyte properties and reassessing Al corrosion mechanisms. *Energy Environ. Sci.* **7,** 416–426 (2014).
25. Moon, H. *et al.* Solvent activity in electrolyte solutions controls electrochemical reactions in Li-ion and Li-sulfur batteries. *J. Phys. Chem. C* **119,** 3957–3970 (2015).
26. Yoshida, K. *et al.* Oxidative-stability enhancement and charge transport mechanism in glyme-lithium salt equimolar complexes. *J. Am. Chem. Soc.* **133,** 13121–13129 (2011).
27. Katon, J. E. & Cohen, M. D. The vibrational spectra and structure of dimethyl carbonate and its conformational behavior. *Can. J. Chem.* **53,** 1378–1386 (1975).
28. Seo, D. M., Borodin, O., Han, S.-D., Boyle, P. D. & Henderson, W.A. Electrolyte solvation and ionic association II. Acetonitrile-lithium salt mixtures: highly dissociated salts. *J. Electrochem. Soc.* **159,** A1489–A1500 (2012).
29. Yamada, Y., Yaegashi, M., Abe, T. & Yamada, A. A superconcentrated ether electrolyte for fast-charging Li-ion batteries. *Chem. Commun.* **49,** 11194–11196 (2013).
30. Yamada, Y. *et al.* Unusual stability of acetonitrile-based superconcentrated electrolytes for fast-charging lithium-ion batteries. *J. Am. Chem. Soc.* **136,** 5039–5046 (2014).
31. Yamada, Y. & Yamada, A. Review-Superconcentrated electrolytes for lithium batteries. *J. Electrochem. Soc.* **162,** A2406–A2423 (2015).
32. Yamada, Y. *et al.* Corrosion prevention mechanism of aluminum metal in superconcentrated electrolytes. *ChemElectroChem* **2,** 1687–1694 (2015).
33. Glatthaar, S., Dorit, N., Binder, J. R. & Andres, H. Evidence of loss of active lithium in titanium-doped $LiNi_{0.5}Mn_{1.5}O_4$/graphite cells. *J. Power Sources* **274,** 1267–1275 (2015).
34. Jeong, S.-K., Inaba, M., Iriyama, Y., Abe, T. & Ogumi, Z. Electrochemical intercalation of lithium ion within graphite from propylene carbonate solutions. *Electrochem. Solid State Lett.* **6,** A13–A15 (2003).
35. Suo, L., Hu, Y.-S., Li, H., Armand, M. & Chen, L. A new class of solvent-in-salt electrolyte for high-energy rechargeable metallic lithium batteries. *Nat. Commun.* **4,** 1481 (2013).
36. CPMD. http://www.cpmd.org:81/manual/node4.html, Copyright IBM Corp (1990-2015), Copyright MPI für Feskörperforschung Stuttgart (1997-2001).
37. Goedecker, S., Teter, M. & Hutter, J. Separable dual-space Gaussian pseudopotentials. *Phys. Rev. B* **54,** 1703–1710 (1996).

Acknowledgements

This work was partially supported by JSPS Grant-in-Aid for Young Scientists (A) (No. 26708030) and JSPS Specially Promoted Research (No. 15H05701). The calculations were carried out at the super-computer centres of National Institute for Materials Science, the University of Tokyo, and the K computer at the RIKEN through the HPCI

System Research Projects (hp150209). We thank Keisuke Kikuchi, Reiko Kawakami and Dr Kouhei Okitsu for their assistance in the experiments, and specially thank Dr Sai-Cheong Chung for his valuable suggestions on the manuscript.

Author contributions

J.W. and Y.Y. contributed equally to this work. Y.Y. and A.Y. proposed the concept. J.W. and Y.Y. designed the experiments. J.W. and C.H.C. carried out the experiments and analysed the data. K.S. and Y.T. designed and conducted the theoretical calculations. J.W., Y.Y. and A.Y. wrote the manuscript. A.Y. supervised the whole project.

Additional information

Competing financial interests: The authors declare no competing financial interests.

13

Interface strain in vertically stacked two-dimensional heterostructured carbon-MoS_2 nanosheets controls electrochemical reactivity

Landon Oakes[1,2], Rachel Carter[1], Trevor Hanken[1], Adam P. Cohn[1], Keith Share[1,2], Benjamin Schmidt[3] & Cary L. Pint[1,2,3]

Two-dimensional (2D) materials offer numerous advantages for electrochemical energy storage and conversion due to fast charge transfer kinetics, highly accessible surface area, and tunable electronic and optical properties. Stacking of 2D materials generates heterogeneous interfaces that can modify native chemical and physical material properties. Here, we demonstrate that local strain at a carbon-MoS_2 interface in a vertically stacked 2D material directs the pathway for chemical storage in MoS_2 on lithium metal insertion. With average measured MoS_2 strain of $\sim$0.1% due to lattice mismatch between the carbon and MoS_2 layers, lithium insertion is facilitated by an energy-efficient cation-exchange transformation. This is compared with low-voltage lithium intercalation for unstrained MoS_2. This observation implies that mechanical properties of interfaces in heterogeneous 2D materials can be leveraged to direct energetics of chemical processes relevant to a wide range of applications such as electrochemical energy storage and conversion, catalysis and sensing.

[1] Department of Mechanical Engineering, Vanderbilt University, Nashville, Tennessee 37235, USA. [2] Interdisciplinary Materials Science Program, Vanderbilt University, Nashville, Tennessee 37235, USA. [3] Vanderbilt Institute of Nanoscale Science and Engineering, Vanderbilt University, Nashville, Tennessee 37235, USA. Correspondence and requests for materials should be addressed to C.L.P. (email: cary.l.pint@vanderbilt.edu).

2D graphene and transition metal dichalcogenide (TMDC) materials have captivated researchers in the past decade owing to a set of unique physical and chemical properties that deviate from their bulk analogues and the exploitation of these properties in broad applications[1]. Specifically for electronics, vertical integration of 2D materials enables logic component design with high on-off ratio and promise for three-dimensional (3D) electronics that progress beyond silicon[2–8]. In semiconductor electronics, engineering 2D TMDC materials using strain applied at an interface has been shown to strongly modulate the bandgap and band structure, which results in modified electrical and optical properties for strained materials[9–14]. For monolayer MoS_2, the bandgap is observed to shift by up to 15 meV under tensile strain of up to 4.8% (ref. 15), and for multilayered WSe_2 the bandgap transitions from indirect to direct under strain of up to 2% (ref. 16). Theoretical efforts have further emphasized strain-enabled broadband absorption and photodetection in MoS_2, even though this has not yet been experimentally realized[17]. The intersection of strain-engineered properties of 2D materials and heterostructured vertically integrated interfaces of 2D materials presents an engaging research area for next-generation electronics. Nonetheless, such materials and research directions remain elusive for many other applications of 2D materials.

On this front, 2D materials and specifically 2D TMDCs have demonstrated excellent performance in a range of electrochemical applications including lithium- and sodium-ion batteries[18–20], photocatalytic conversion[21–23] and biosensing applications[24–26]. Unlike electronic devices, these applications require bulk-like quantities of 2D materials—a challenge aided by recent developments in the liquid exfoliation and assembly of layered TMDCs[27–29]. However, addressing how properties in heterostructured or complex 2D materials can impact chemical processes responsible for electrochemical applications is hampered by uniform material fabrication routes that can be used on scales required for electrochemical measurements. As a result, the impact of strained interfaces in 2D materials on chemical and electrochemical processes remains virtually unstudied. Only recently has an observation emerged that compressive strain on Pt catalysts can improve the oxygen reduction reaction capability of Pt relevant to fuel cells[30]. In this manner, 2D materials provide an ideal test bed for the understanding of how interfaces and strain can impact electrochemical processes, motivated by pioneering efforts in the field of semiconductor electronics.

This is the focus of this report, which documents that interface strain measured over statistical quantities of stacked 2D carbon-MoS_2 heterostructured nanosheets can be directly correlated with distinct differences in chemical processes occurring in 2D materials. In particular, combining optical and electrochemical techniques we demonstrate that strain engineering of interfaces can enable control of the energetic pathway for the chemical conversion of MoS_2 into electrochemically active Mo and Li_2S_n species during reaction with lithium. This highlights the important role that mechanical strain can have in engineering energy storage processes in materials.

Results

Fabrication of interface-strained materials. MoS_2 nanosheets were produced through liquid exfoliation of bulk MoS_2 powders in *n*-methyl-2-pyrrolidone (NMP) solvents and subsequent centrifugation. Transmission electron microscopy (TEM) of a representative exfoliated MoS_2 nanosheet is shown in Fig. 1a,b. Interlayer spacing of ~0.61 nm is observed for MoS_2 nanosheets, with thicknesses ranging from 2–15 atomic layers (Supplementary Fig. 1, Supplementary Note 1). Ultrathin carbon layers are grown directly on the MoS_2 surface through MoS_2 catalysed decomposition of C_2H_2 precursors using a temperature ramp chemical vapour deposition process (Supplementary Fig. 2) that is capable of gram-scale batch processing. This generates vertically stacked architectures where ultrathin carbon layers are formed on both sides of the MoS_2 nanosheets, with a representative TEM image of this architecture shown in Fig. 1c, and corresponding elemental analysis map in Fig. 1d. To produce an appreciable mass of material for electrochemical tests, electrophoretic deposition was used to assemble the vertically stacked carbon-interfaced MoS_2 (C-MoS_2) nanosheets into conformal films on metal substrates from NMP dispersions following chemical vapour deposition (Fig. 1e; Supplementary Fig. 3, Supplementary Note 2). All aspects of synthesis and processing of C-MoS_2 materials are chosen to be compatible with scalable processing, and ongoing research efforts to improve the simplicity and scalability of liquid exfoliation can further improve this[31]. Raman spectroscopy analysis (Fig. 1f) of vertically stacked C-MoS_2 nanosheets indicates that vertical stacks maintain an identical signature of crystalline MoS_2 following the carbon synthesis based on E_{2g} and A_{1g} modes (300–500 cm^{-1}). The carbon layers exhibit a significant amount of sp^3 hybridized carbon atoms (~1,320 cm^{-1}, or D-band) relative to sp^2 carbon species (~1,580 cm^{-1}, or G-band). Whereas this indicates the presence of carbon, X-ray photoelectron spectroscopy (XPS) indicates the emergence of both a peak in the sulfur $2p_{3/2}$ and $2p_{5/2}$ spectra at low-binding energies, as well as shoulders in both Mo $3d_{5/2}$ and $3d_{3/2}$ peaks that represent a modulation to the local bonding environment in the crystal (Fig. 1g). Notably, the existence of the low-energy shoulder in the sulfur *2p* and Mo *3d* peaks is consistent with the formation of the metallic 1T-phase of MoS_2 produced from the strain-induced distortion of sulfur atoms observed previously (Supplementary Fig. 4, Supplementary Note 3)[32]. This analysis collectively supports the formation of a distinct carbon-MoS_2 stacked interface, where mismatch of the in-plane lattice spacing between MoS_2 and carbon will induce significant interface strain that propagates into the MoS_2 nanosheet from the C-MoS_2 interface.

To assess strain in the vertically stacked 2D material, statistical Raman spectroscopy mapping comprising over 200 individual scans in separate areas was performed on the E_{2g} and A_{1g} modes of MoS_2, which are highly sensitive to tensile or compressive strain[33–35]. (Fig. 2a,b) On the basis of the peak-to-peak analysis of Raman mode distributions, blue-shifts of ~0.66 cm^{-1} and ~0.59 cm^{-1} were observed for the E_{2g} and A_{1g} modes, respectively. Asymmetry in these modes is expected and attributed to stronger electronic coupling to the A_{1g} mode[36]. This yields an ~0.1–0.2% compressive strain based on relative A_{1g} mode shifts in accordance with previous studies[9,34], supporting the presence of interface-induced compressive strain on the MoS_2 material. This is further confirmed using X-ray diffraction which demonstrates a similar ~0.1% compressive strain due to vertical stacking based on analysis of the (100) and (110) low-index planes of MoS_2 (Fig. 2c). We anticipate this measured strain to be present at both the interface and on the interior of the nanosheets as they possess a thickness well below the critical layer thickness at which strain relaxation occurs[37].

Controlling electrochemical processes using strain. To assess how the vertically stacked architecture and interface strain influences electrochemical processes, we combined electrophoretically assembled vertically stacked 2D C-MoS_2 nanosheets with Li metal electrodes, and a 1.0 M lithium hexafluorophosphate solution in ethylene carbonate and diethyl carbonate electrolyte, and compared the electrochemical

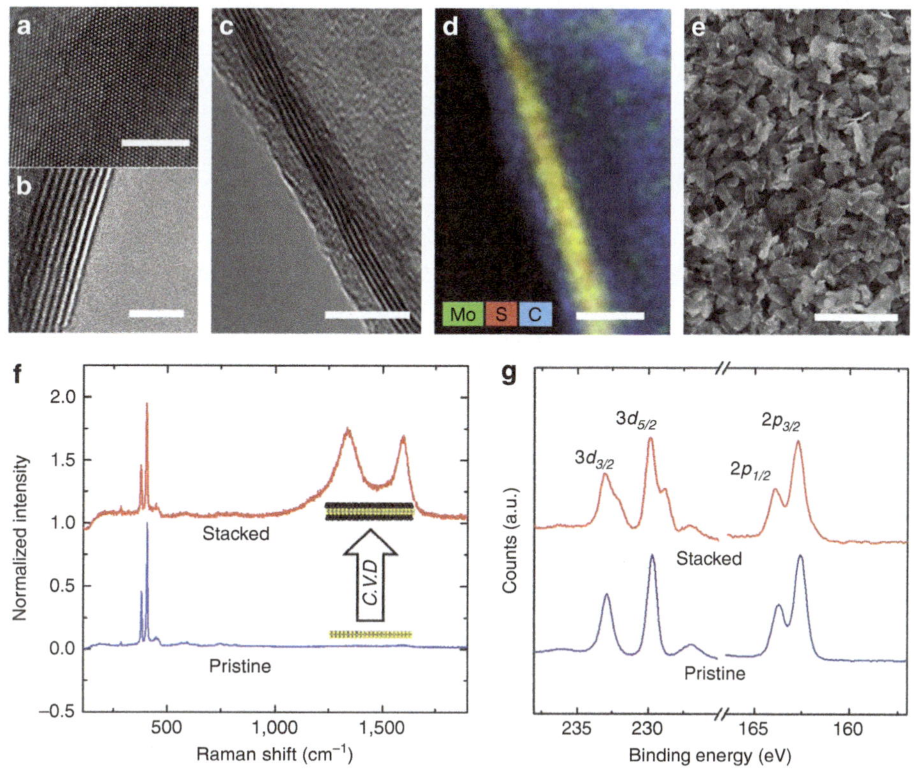

Figure 1 | Synthesis of vertically stacked carbon-MoS_2 heterostructured nanosheets. (**a**,**b**) TEM images of pristine exfoliated MoS_2 nanosheets. (**c**) Edge-view TEM image of a vertically stacked carbon-MoS_2 (C-MoS_2) nanosheet. (**d**) Energy-dispersive X-ray spectroscopy analysis of a stacked C-MoS_2 nanosheet. The coloured inset describes the colour assignment of each mapped element, overlapping Mo and S signals appear yellow in the image. (**e**) Scanning electron microscopy image of an electrophoretically assembled electrode material formed with stacked C-MoS_2 nanosheets. Scale bars, 4 nm (**a**,**b**) 10 nm (**c**,**d**) and 1 µm (**e**). (**f**) Raman spectroscopy and (**g**) X-ray photoelectron spectroscopy of pristine MoS_2 nanosheets and stacked C-MoS_2 nanosheets.

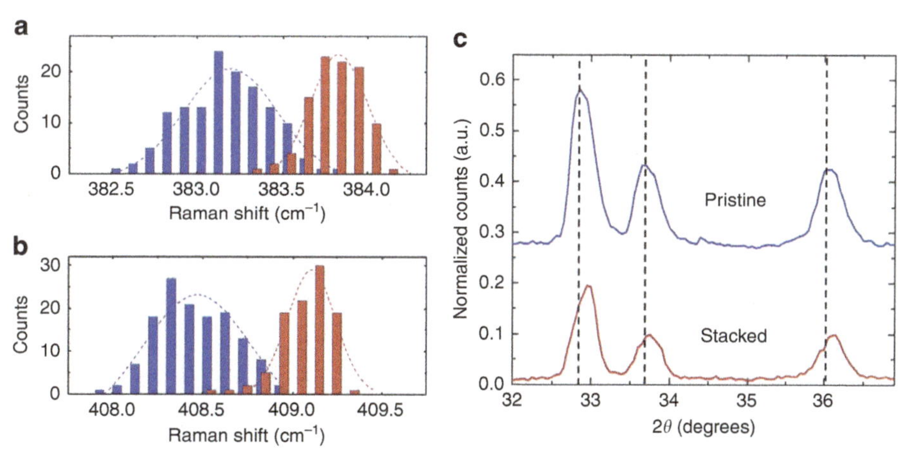

Figure 2 | Interface strain in vertically stacked 2D C-MoS_2 nanosheets. Distributions from Raman spectroscopy maps comprising > 100 individual scans showing average shifts in the MoS_2 (**a**) E_{2g} and (**b**) A_{1g} modes due to strain induced by a lattice mismatched carbon-MoS_2 interface. The dotted lines overlaying each distribution represent Gaussian fits to the distribution data. (**c**) X-ray diffraction analysis of vertically stacked C-MoS_2 nanosheets indicating stacking-induced strain in low-index planes. The following planes can be assigned to the X-ray diffraction spectra: $2\theta = 32.6°$ is (100), $2\theta = 33.5°$ is (101) and $2\theta = 35.8°$ is (102). The vertical dotted lines emphasize the difference in peak positions between the two spectra.

properties against similar electrode materials produced with pristine MoS_2 nanosheets (Supplementary Figs 5,6, Supplementary Notes 4,5). Cyclic voltammetry scans of these electrodes and corresponding differential capacity plots based on galvanostatic measurements are shown in Fig. 3. As these two electrode materials differ only by the presence of a vertically stacked C-MoS_2 interface, electrochemical data indicates significant changes to the chemical processes occurring on lithium insertion into the MoS_2 material. For pristine MoS_2 nanosheets, two subsequent reactions are observed, with the first one at ~1.1 V and the second at ~0.55 V versus Li/Li^+. (Fig. 3a,b) This is consistent with the known pathways for insertion of lithium into pristine MoS_2, which occurs first through an intercalation reaction (1.1 V) that follows

$$MoS_2 + xLi^+ + xe^- \rightarrow Li_xMoS_2 \quad (1)$$

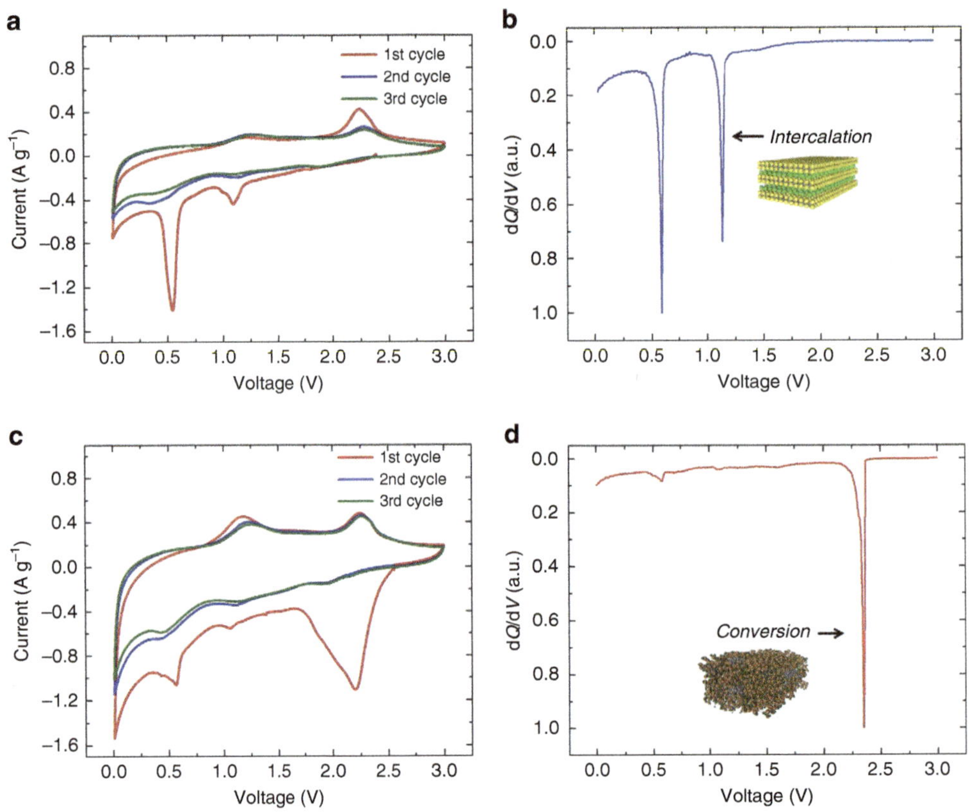

Figure 3 | Electrochemical assessment of strained carbon-MoS_2 heterostructured nanosheets. (**a**) CV and (**b**) normalized differential capacity measurements for MoS_2 under lithium insertion and extraction, with an arrow indicating the potential of the intercalation reaction that occurs in pristine MoS_2 materials. (**c**) CV and (**d**) normalized differential capacity plots for vertically stacked C-MoS_2 nanosheets, with an arrow indicating the electrochemical signature of direct chemical conversion.

where MoS_2 undergoes a transition from a semiconducting 2H phase to the metallic 1 T phase due to strain-induced deformation of the crystal from lithium ion insertion. Following this transition, the MoS_2 can undergo a subsequent conversion reaction at lower voltages (0.55 V) that follows

$$Li_xMoS_2 + 4Li^+ + 4e^- \rightarrow 2Li_2S + Mo/Li_y \qquad (2)$$

In contrast to this, insertion of lithium into the vertically stacked C-MoS_2 nanosheets yields a chemical reaction evident at $\sim$2.3 V versus Li/Li^+ based on both CV and diferential capacity curves (Fig. 3c,d) which is close to the open-circuit voltage (OCV) of the device. This reaction proceeds in the absence of Li intercalation into MoS_2 and no significant signature of lithium insertion at lower voltages occurs such as in pristine MoS_2 nanosheet electrodes. This highlights the presence of a chemical storage process occurring in vertically stacked C-MoS_2 materials that is not observed in pristine MoS_2 materials (Fig. 4a) and that interface-induced strain can directly trigger chemical conversion at practical voltages. It is important to note that these results distinguish the stacked C-MoS_2 configuration from simple mixtures of carbon nanomaterials and MoS_2 nanosheets which have been reported previously[38–42], in two key ways, namely that the grown carbon layer induces interface strain not possible in weakly adhered mechanically mixed materials that modulates the chemical reactivity of the MoS_2 nanosheet toward chemical conversion, and that the stacked configuration provides a physical barrier to inhibit polysulfide dissolution into the electrolyte in a manner consistent with a yolk-shell confinement strategy[43]. However, the benefits reported in other studies by combining MoS_2 and carbons through mixing are also maintained in our study as evidenced by galvanostatic studies indicating better rate capability and improved storage capacity at high rates for the C-MoS_2 electrodes (Supplementary Fig. 7). In addition, whereas the correlation between X-ray diffraction, Raman spectroscopy and XPS indicate the presence of strain due to the mismatch of in-plane lattice parameters of the stacked materials below the critical thickness, charge transfer effects between the C and MoS_2 materials can play a role in the origin of strain measured using these techniques. We also observe the lack of any electrochemical signature of lithium intercalation into the ultrathin carbon coating, implying that stacked C-MoS_2 nanosheets exhibit differences from pristine nanosheets that can be explained by compressive strain imposed on the 2D MoS_2 material.

To address the effect of the C-MoS_2 interface on the electrochemical properties during lithium insertion, Raman spectroscopy of the electrophoretically assembled MoS_2-based electrodes at different cathodic potentials was carried out (Fig. 4b). This analysis is possible due to the preparation of electrodes in a manner that does not require binder materials often used in conventional battery electrodes that can overwhelm the desired Raman spectroscopic features. At OCV conditions, both electrode materials exhibit only the native Raman modes of MoS_2. Cathodic scans from OCV conditions to 1.75 V versus Li/Li^+—an energy below the Li^+ insertion reaction observed for vertically stacked C-MoS_2 nanosheets, yields no change for the pristine MoS_2 nanosheets, but the emergence of a distinct Raman peak at 746 cm^{-1} is observed for the stacked C-MoS_2 nanosheets. This mode is due to the formation of lithium polysulfides (Li_2S_n for $4 \leq n \leq 8$) that can be attributed to chemical conversion of the MoS_2 into reaction products in a manner consistent with equation (2) (ref. 44). Correspondingly, a decrease of the MoS_2

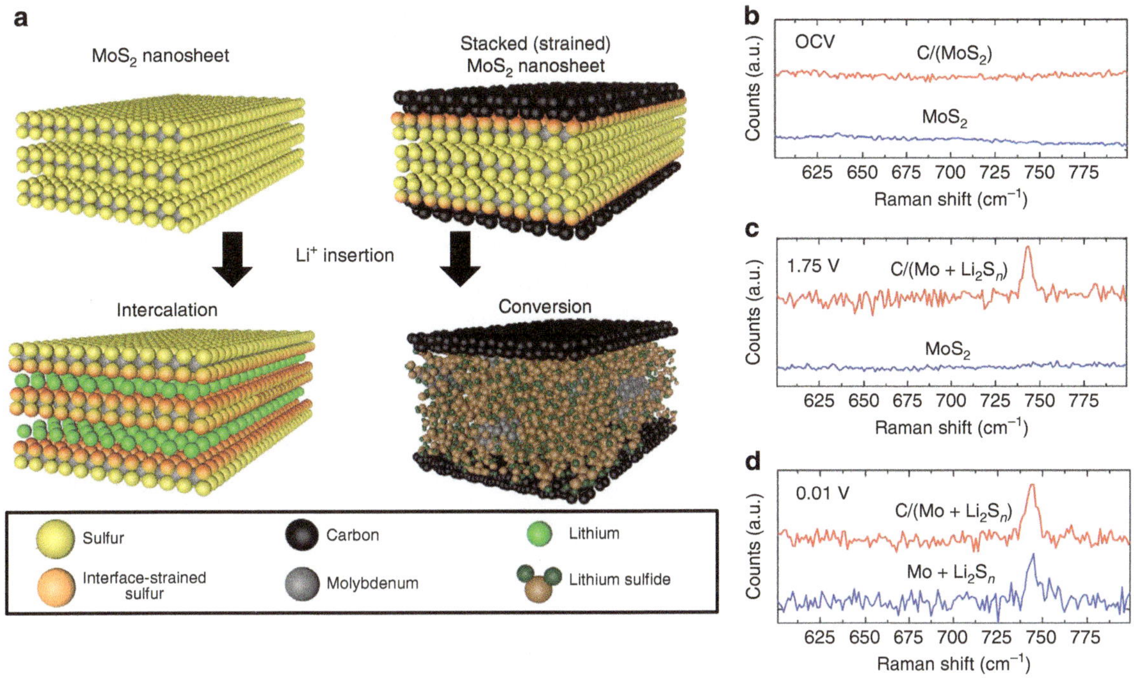

Figure 4 | Strain-induced modulation of chemical conversion in vertically stacked C-MoS_2 nanosheets. (**a**) Scheme showing the different chemical states between interface-strained MoS_2 nanosheets and pristine MoS_2 nanosheets after Li insertion, showing chemical conversion for stacked C-MoS_2 and intercalation for pristine MoS_2. *Ex situ* Raman spectroscopy confirming the electrochemical signature of chemical conversion based on the Raman mode of polysulfides at 746 cm^{-1} is presented for both materials at (**b**). Open-circuit voltage (OCV), (**c**) 1.75 V and (**d**) 0.01 V versus Li/Li^+. Notably, at 1.75 V versus Li/Li^+, the vertically stacked C-MoS_2 nanosheets have undergone chemical conversion, whereas the pristine MoS_2 nanosheets remain unconverted.

A_{1g} and E_{2g} modes are observed as expected from conversion (Supplementary Fig. 8, Supplementary Note 6). This confirms that the high voltage ~2.3 V signature in the CV and differential capacity curves for vertically stacked C-MoS_2 nanosheets (Fig. 3c,d) is a chemical conversion process similar to that which is known to occur at low voltages (0.55 V) in pristine MoS_2. To further support this, cathodic scans were continued down to 0.01 V versus Li/Li^+ for the pristine MoS_2 electrode, where the signature of conversion is evident due to the presence of the Raman mode at 746 cm^{-1}. Despite the difference in voltage of the conversion reactions between MoS_2 and C-MoS_2, similar Raman spectroscopic signatures of the polysulfide conversion product and similar post-conversion redox energetics in both cases implies that the resulting chemical state of the material in both conversion reactions is the same. To further support this, we performed *in situ* electrochemical impedance spectroscopy (EIS, Supplementary Fig. 9, Supplementary Table 1), where EIS analysis was performed at the same voltages as those assessed using Raman spectroscopy. EIS studies elucidate a picture consistent with Raman spectroscopy results that indicates the electrochemical signature of conversion in different voltage regimes for the pristine MoS_2 and the stacked C-MoS_2, respectively (Supplementary Note 7).

Discussion

On the basis of the combined spectroscopic and electrochemical analysis, a picture emerges emphasizing the role of interface strain on lithium insertion in vertically stacked C-MoS_2 materials. In the pristine semiconducting 2H phase of MoS_2, conversion of MoS_2 nanosheets cannot occur at voltages above where intercalation occurs (1.1 V versus Li/Li^+) and can only proceed in a two-step chemical process following an intercalation reaction. As many researchers have discussed the promise of MoS_2 nanosheets for battery cathodes, the requirement of a conversion reaction that is near the reduction potential of lithium is highly impractical in full-cell architectures. In a vertically stacked C-MoS_2 nanosheet, the interface strain due to lattice mismatch in a carbon-MoS_2 solid–solid interface leads to an average ~0.1% compressive strain that propagates into the MoS_2 nanosheet lattice and enables control of the energetics of chemical conversion. In this regard, interface strain provides the appropriate energetic landscape to sustain direct conversion at voltages that enable pairing with conventional anode materials such as silicon or graphite which give promise to practical incorporation of this device in a full-cell battery configuration. This distinctive difference in the chemical pathways achieved during lithium insertion as a result of interface strain is illustrated in Fig. 4a, where the chemical state of the 2D materials is drastically different following a cathodic scan to 1 V.

The broader use of controlled interface mechanics to modify chemical pathways in low-dimensional materials presents an unexplored horizon for engineered low-dimensional nanostructures. This utilizes 2D materials as a framework to merge concepts of strain engineering with mechanochemistry to control the energetics of chemical processes relevant to energy storage, energy conversion, catalysis, and sensing applications. Building from our results showing that strained interfaces in vertically stacked C-MoS_2 nanosheets can modulate electrochemical reactivity, we envision a new paradigm for material design that exploits interface strain and mechanics as a versatile toolbox to modulate performance of designer 2D materials in diverse applications.

Methods

MoS_2 nanosheet synthesis. In all, 500 mg of bulk MoS_2 powder (Aldrich, particle size < 2 μm) was added to 50 ml of 1-methyl-2-pyrrolidinone (Aldrich, 99.5% anhydrous). The solution was sonicated in a bath sonicator for 12 h using 8 × 90 min intervals. After sonication, the solution was centrifuged at 2,000 r.p.m. for 40 min. The upper 2/3 of the supernatant was removed and subsequently centrifuged for 1 h at 5,000 r.p.m. before the excess supernatant was removed and the accumulated nanosheets were placed under vacuum overnight to completely evaporate the solvent.

Vertically stacked carbon-MoS_2 nanosheet synthesis. MoS_2 nanosheets contained within an alumina boat were placed into a 1-inch tube furnace and the tube was evacuated to 2 mTorr. A gas mixture of 100 standard cubic centimeters per minute (sccm) Ar and 20 sccm H_2 maintained at atmospheric pressure were introduced during ramping to a temperature 750 °C. At 750 °C, 2 sccm of acetylene (C_2H_2) was introduced for 10 min followed by a temperature ramp to 850 °C and an additional 10 min soak, and a final ramp to 950 °C for 10 min. After this process, the acetylene was turned off and the furnace cooled back down under a flow of argon and hydrogen, where materials were removed.

Electrophoretic deposition. A total of 20 mg of vertically stacked C-MoS_2 or pristine MoS_2 nanosheets were suspended in 20 ml NMP and sonicated for 30 min. Following this, a stainless steel spacer was placed at 0.5 cm from a 1 × 1 cm stainless steel counter electrode. A voltage of 30 V was applied for 30 min before carefully removing the coated steel electrode from solution and placing this into a vacuum chamber overnight to dry the sample.

Electrochemical device fabrication and testing. Electrochemical half-cell devices were assembled in an argon glovebox using CR 2,032 stainless steel coin cells purchased from MTI. The coated steel discs were separated from a lithium metal anode by a 2,500 Celgard separator saturated with a 1 M $LiPF_6$ in 1 g per 1 ml solution of ethylene carbonate and diethyl carbonate. Cyclic voltammetry scans and galvanostatic measurements were performed using a Metrohm autolab multichannel testing system.

Material characterization. Raman measurements were performed using a Renishaw inVia confocal Raman spectrometer. Micro-Raman maps were collected using a 532 nm laser. Before characterizing the electrodes after electrochemical testing, coin cells were disassembled in an Ar filled glovebox and the electrodes washed with an EC/DEC solution. TEM analysis was performed by drop casting a solution of pristine or stacked MoS_2 materials onto amorphous carbon TEM grids and then imaged using an FEI Osiris TEM at a beam voltage of 200 kV. X-ray diffraction was performed on powder samples using a Scintag XGEN 4,000 system with a CuKα (wavelength = 0.154 nm) radiation source. XPS was collected using a PHI 5,000 Versaprobe. Measurements were collected at a 45° takeoff angle with a 100 μm spot size with charge neutralization.

References

1. Liu, J. H. & Liu, X. W. Two-dimensional nanoarchitectures for lithium storage. *Adv. Mater.* **24,** 4097–4111 (2012).
2. Gong, Y. J. *et al.* Vertical and in-plane heterostructures from WS_2/MoS_2 monolayers. *Nat. Mater.* **13,** 1135–1142 (2014).
3. Wang, X. & Xia, F. Van der Waals heterostructures: stacked 2D materials shed light. *Nat. Mater.* **14,** 264–265 (2015).
4. Yu, W. J. *et al.* Vertically stacked multi-heterostructures of layered materials for logic transistors and complementary inverters. *Nat. Mater.* **12,** 246–252 (2013).
5. Jariwala, D., Sangwan, V. K., Lauhon, L. J., Marks, T. J. & Hersam, M. C. Emerging device applications for semiconducting two-dimensional transition metal dichalcogenides. *ACS Nano* **8,** 1102–1120 (2014).
6. Britnell, L. *et al.* Strong light-matter interactions in heterostructures of atomically thin films. *Science* **340,** 1311–1314 (2013).
7. Hong, X. *et al.* Ultrafast charge transfer in atomically thin MoS_2/WS_2 heterostructures. *Nat. Nanotechnol.* **9,** 682–686 (2014).
8. Lee, G.-H. *et al.* Flexible and transparent MoS_2 field-effect transistors on hexagonal boron nitride-graphene heterostructures. *ACS Nano* **7,** 7931–7936 (2013).
9. Li, M.-Y. *et al.* Epitaxial growth of a monolayer WSe_2-MoS_2 lateral pn junction with an atomically sharp interface. *Science* **349,** 524–528 (2015).
10. Georgiou, T. *et al.* Vertical field-effect transistor based on graphene-WS_2 heterostructures for flexible and transparent electronics. *Nat. Nanotechnol.* **8,** 100–103 (2013).
11. He, J. *et al.* Electron transfer and coupling in graphene–tungsten disulfide van der Waals heterostructures. *Nat. Commun.* **5,** 5622 (2014).
12. Roy, K. *et al.* Graphene-MoS2 hybrid structures for multifunctional photoresponsive memory devices. *Nat. Nanotechnol.* **8,** 826–830 (2013).
13. Castellanos-Gomez, A. *et al.* Local strain engineering in atomically thin MoS_2. *Nano Lett.* **13,** 5361–5366 (2013).
14. He, K., Poole, C., Mak, K. F. & Shan, J. Experimental demonstration of continuous electronic structure tuning via strain in atomically thin MoS_2. *Nano Lett.* **13,** 2931–2936 (2013).
15. Liu, Z. *et al.* Strain and structure heterogeneity in MoS_2 atomic layers grown by chemical vapour deposition. *Nat. Commun.* **5,** 5246 (2014).
16. Desai, S. B. *et al.* Strain-induced indirect to direct bandgap transition in multi layer WSe_2. *Nano Lett.* **14,** 4592–4597 (2014).
17. Feng, J., Qian, X., Huang, C.-W. & Li, J. Strain-engineered artificial atom as a broad-spectrum solar energy funnel. *Nat. Photon.* **6,** 866–872 (2012).
18. Hwang, H., Kim, H. & Cho, J. MoS_2 nanoplates consisting of disordered graphene-like layers for high rate lithium battery anode materials. *Nano Lett.* **11,** 4826–4830 (2011).
19. Lacey, S. D. *et al.* Atomic force microscopy studies on molybdenum disulfide flakes as sodium-ion anodes. *Nano Lett.* **15,** 1018–1024 (2015).
20. Hu, L. R., Ren, Y. M., Yang, H. X. & Xu, Q. Fabrication of 3D hierarchical MoS_2/polyaniline and MoS_2/C architectures for lithium-ion battery applications. *ACS Appl. Mater. Inter.* **6,** 14644–14652 (2014).
21. Kibsgaard, J., Chen, Z., Reinecke, B. N. & Jaramillo, T. F. Engineering the surface structure of MoS_2 to preferentially expose active edge sites for electrocatalysis. *Nat. Mater.* **11,** 963–969 (2012).
22. Chou, S. S. *et al.* Understanding catalysis in a multiphasic two-dimensional transition metal dichalcogenide. *Nat. Commun.* **6,** 8311 (2015).
23. Wang, Y. *et al.* Monolayer $PtSe_2$, a new transition-metal-dichalcogenide, grown by direct selenization, shows promise for photocatalysis and valleytronics. *Nano Lett.* **15,** 4013–4018 (2015).
24. Sarkar, D. *et al.* MoS_2 field-effect transistor for next-generation label-free biosensors. *ACS Nano.* **8,** 3992–4003 (2014).
25. Perkins, F. K. *et al.* Chemical vapor sensing with monolayer MoS_2. *Nano Lett.* **13,** 668–673 (2013).
26. Lee, J. *et al.* Two-dimensional layered MoS_2 biosensors enable highly sensitive detection of biomolecules. *Sci. Rep.* **4,** 7352 (2014).
27. Shen, J. *et al.* Liquid phase exfoliation of two-dimensional materials by directly probing and matching surface tension components. *Nano Lett.* **15,** 5449–5454 (2015).
28. Nicolosi, V., Chhowalla, M., Kanatzidis, M. G., Strano, M. S. & Coleman, J. N. Liquid exfoliation of layered materials. *Science* **340,** 1226419 (2013).
29. Haar, S. *et al.* Enhancing the liquid-phase exfoliation of graphene in organic solvents upon addition of *n*-octylbenzene. *Sci. Rep.* **5,** 16684 (2015).
30. Du, M., Cui, L., Cao, Y. & Bard, A. J. Mechanoelectrochemical catalysis of the effect of elastic strain on a platinum nanofilm for the ORR exerted by a shape memory alloy substrate. *J. Amer. Chem. Soc.* **23,** 7397–7403 (2015).
31. Paton, K. R. *et al.* Scalable production of large quantities of defect-free few-layer graphene by shear exfoliation in liquids. *Nat. Mater.* **13,** 624–630 (2014).
32. Eda, G. *et al.* Photoluminescence from chemically exfoliated MoS_2. *Nano Lett.* **11,** 5111–5116 (2011).
33. Conley, H. J. *et al.* Bandgap engineering of strained monolayer and bilayer MoS_2. *Nano Lett.* **13,** 3626–3630 (2013).
34. Hui, Y. Y. *et al.* Exceptional tunability of band energy in a compressively strained trilayer MoS_2 sheet. *ACS Nano* **7,** 7126–7131 (2013).
35. Rice, C. *et al.* Raman-scattering measurements and first-principles calculations of strain-induced phonon shifts in monolayer MoS_2. *Phys. Rev. B.* **87,** 081307 (2013).
36. Lanzillo, N. A. *et al.* Temperature-dependent phonon shifts in monolayer MoS_2. *Appl. Phys. Lett.* **103,** 093102 (2013).
37. MacManus-Driscoll, J. L. Self-assembled heteroepitaxial oxide nanocomposite thin film structures: designing interface-induced functionality in electronic materials. *Adv. Funct. Mater.* **20,** 2035–2045 (2010).
38. Stephenson, T., Li, Z., Olsen, B. & Mitlin, D. Lithium ion battery applications of molybdenum disulfide (MoS_2) nanocomposites. *Energy Environ. Sci.* **7,** 209–231 (2014).
39. Chang, K. & Chen, W. L-cysteine-assisted synthesis of layered MoS_2/graphene composites with excellent electrochemical performances for lithium ion batteries. *ACS Nano* **5,** 4720–4728 (2011).
40. Jiang, H. *et al.* 2D monolayer MoS_2–carbon interoverlapped superstructure: engineering ideal atomic interface for lithium ion storage. *Adv. Mater.* **27,** 3687–3695 (2015).
41. Zhu, C., Mu, X., van Aken, P. A., Maier, J. & Yu, Y. Fast Li storage in MoS_2-graphene-carbon nanotube nanocomposites: advantageous functional integration of 0D, 1D, and 2D nanostructures. *Adv. Energy Mater.* **5,** 4 (2015).
42. Zhu, C., Mu, X., van Aken, P. A., Yu, Y. & Maier, J. Single-layered ultrasmall nanoplates of MoS_2 embedded in carbon nanofibers with excellent electrochemical performance for lithium and sodium storage. *Angew. Chem. Int. Ed.* **53,** 2152–2156 (2014).
43. Li, W. *et al.* High-performance hollow sulfur nanostructured battery cathode through a scalable, room temperature, one-step, bottom-up approach. *Proc. Natl Acad. Sci. USA* **110,** 7148–7153 (2013).

44. Yeon, J.-T. *et al.* Raman spectroscopic and X-ray diffraction studies of sulfur composite electrodes during discharge and charge. *J. Electrochem. Soc.* **159,** A1308–A1314 (2012).

Acknowledgements
We thank Nitin Muralidharan and Professor Rizia Bardhan for insightful discussions. This work was supported by the National Science Foundation under CMMI grant #1334269 and Vanderbilt start-up funds. TEM images were obtained from an instrument made possible by NSF grant EPS 1004083. A.P.C. and K.S. are supported by an NSF graduate fellowship under grant # 1445197.

Author contributions
L.O. and C.L.P jointly conceived the project and designed the experiments. L.O. and T.H. performed material fabrication with insights from K.S., and L.O. carried out device testing and performance analysis. R.C. performed TEM imaging and analysis, and A.P.C. performed X-ray diffraction measurements and analysis. B.S. performed XPS measurements and analysis. L.O. and C.L.P. wrote the manuscript and all authors participated in discussion and reviewed the manuscript before submission.

Additional information

Competing financial interests: The authors declare no competing financial interests.

14

Three-dimensional porous carbon composites containing high sulfur nanoparticle content for high-performance lithium-sulfur batteries

Guoxing Li[1], Jinhua Sun[1], Wenpeng Hou[1], Shidong Jiang[1], Yong Huang[1] & Jianxin Geng[1]

Sulfur is a promising cathode material for lithium-sulfur batteries because of its high theoretical capacity (1,675 mA h g^{-1}); however, its low electrical conductivity and the instability of sulfur-based electrodes limit its practical application. Here we report a facile *in situ* method for preparing three-dimensional porous graphitic carbon composites containing sulfur nanoparticles (3D S@PGC). With this strategy, the sulfur content of the composites can be tuned to a high level (up to 90 wt%). Because of the high sulfur content, the nanoscale distribution of the sulfur particles, and the covalent bonding between the sulfur and the PGC, the developed 3D S@PGC cathodes exhibit excellent performance, with a high sulfur utilization, high specific capacity (1,382, 1,242 and 1,115 mA h g^{-1} at 0.5, 1 and 2 C, respectively), long cycling life (small capacity decay of 0.039% per cycle over 1,000 cycles at 2 C) and excellent rate capability at a high charge/discharge current.

[1] Technical Institute of Physics and Chemistry, Chinese Academy of Sciences, 29 Zhongguancun East Road, Haidian District, Beijing 100190, China. Correspondence and requests for materials should be addressed to J.G. (email: jianxingeng@mail.ipc.ac.cn).

Lithium–sulfur (Li–S) batteries have recently attracted great interest as promising electrochemical devices for energy conversion and storage applications because of the abundance, low cost, environmental friendliness and high theoretical capacity (1,675 mA h g^{-1}) of sulfur[1-4]. Despite these advantages, the practical application of Li–S batteries is still handicapped by the following problems: (1) the low electrical conductivities of sulfur (5×10^{-30} S cm^{-1} at 25 °C), intermediate polysulphides and Li_2S; (2) the dissolution of lithium polysulphides, which results in a shuttling effect and in the deposition of insoluble lithium sulfide on the anode in each of the charge/discharge cycles and eventually the complete loss of capacity of the sulfur cathode; and (3) severe volume changes in the active electrode materials during the lithiation/delithiation processes[1,4-7], resulting in the pulverization of the electrode materials.

To overcome these problems, various carbon materials, including graphene[8-15], carbon nanotubes[16,17], porous carbon[18-26] and carbon nanofibres[27-29], have been tested in recent years as supporting materials for sulfur cathodes to improve the electrochemical performance of Li–S batteries. Carbon frameworks improve the electrical conductivity of sulfur cathodes and trap soluble polysulphides during cycling. In addition, yolk–shell structures such as a sulfur–TiO_2 yolk–shell[30] and a sulfur–polyaniline yolk–shell[31] have been developed to address the large volume changes of sulfur during the lithiation/delithiation processes. Recently, Choi *et al.* fabricated a polydopamine-coated S/C composite cathode with a high sulfur loading, which exhibited a high areal capacity (9 mA h cm^{-2}) (ref. 32). In a pioneering work, Pyun and co-workers prepared sulfur-containing polymers that exhibited high electrochemical activity and suitability as cathode materials for Li–S batteries[33,34]. However, despite these research efforts, no strategy has satisfactorily solved the aforementioned problems. Most seriously, the long-cycle stability under high charge/discharge rates remains a major challenge for sulfur-based cathodes, especially for composites with relatively high sulfur content. To date, conventional methods for the preparation of carbon–sulfur composites[8,16,19-21,35], in which carbon materials are impregnated with sulfur by diffusion after the carbon structures are prepared, still face challenges. These problems include the complexity of multistep operations, the low sulfur content of the composites and the out-diffusion of lithium polysulphide into the electrolyte during the charge/discharge cycles owing to the diffusion process used for the incorporation of sulfur into the carbon materials. Furthermore, certain unresolvable trade-offs have been found in previous studies. For example, a relatively high sulfur content in the sulfur/carbon hybrid structures is always accompanied by larger sulfur particles[11,30,36], which severely reduces the rate of sulfur utilization because of the long diffusion path for electrons and lithium ions[3]. Although a very high specific capacity (>1,000 mA h g^{-1}) can be obtained with an electrode that has a low sulfur content[37-40], the low sulfur content greatly reduces the overall volumetric capacity and energy density of the cathode. Therefore, it is crucial to design high-sulfur-content composites for use as cathode materials in Li–S batteries, in which the composite cathodes maintain a high sulfur utilization rate, a high specific capacitance, a long cycling life and good rate capability. This may be achievable by controlling the existing state and distribution of sulfur in the hybrid structures.

Herein, we report a facile and scalable strategy for the *in situ* synthesis of sulfur nanoparticles in three-dimensional (3D) porous graphitic carbon (PGC) (designated 3D S@PGC) with a tuneable sulfur content and demonstrate the utility of the 3D S@PGC as a cathode material for Li–S batteries. Compared with the conventional methods for the preparation of carbon–sulfur composites[8,16,19-21,35], our strategy facilitates access to composites that have the advantages of a high sulfur content (up to 90 wt%) and nanoscale distribution of the sulfur particles, as well as covalent bonding between the sulfur nanoparticles and the PGC network, which ensures the efficient utilization of the loaded sulfur. The sulfur content of the composite can be readily tuned by changing the Na_2S/glucose ratio. Because of the C–S bonds, unique interconnected hierarchical porous structures, high sulfur content and nanoscale sulfur particles, our 3D S@PGC (90% S) composite exhibits significantly improved electrochemical performance as a cathode material for Li–S batteries. In particular, this material has a high specific capacity (1,382, 1,242 and 1,115 mA h g^{-1} at 0.5, 1 and 2 C, respectively), long cycling life (small capacity decay of 0.039% per cycle over 1,000 cycles at 2 C), and excellent rate capability at a high charge/discharge current.

Results

Synthesis of the 3D S@PGC composites. Figure 1 displays the scheme for the preparation of the 3D S@PGC composites. The self-stacking of water-soluble NaCl and Na_2S crystals was utilized to form a hard template for the 3D PGC network. Na_2S, which reacts with $Fe(NO_3)_3$ to form sulfur, was used as the sulfur precursor. The reaction is described by the equation $Na_2S + 2Fe(NO_3)_3 \rightarrow 2Fe(NO_3)_2 + 2NaNO_3 + S$. NaCl, Na_2S and glucose were first dissolved in deionized (DI) water to obtain a homogeneous solution, which was subsequently subjected to freeze-drying. During the freezing process, the sizes of the NaCl and Na_2S crystals were restricted, and the crystals were uniformly coated with an ultrathin glucose film. The NaCl crystals surrounded the Na_2S crystals because of the high NaCl/Na_2S molar ratio (10:1). The hybrid structure is hereafter referred to as 3D NaCl-Na_2S@glucose. Upon heating at a high temperature (750 °C) in an argon atmosphere, the glucose underwent carbonization to form graphitic carbon (GC) with micro- and

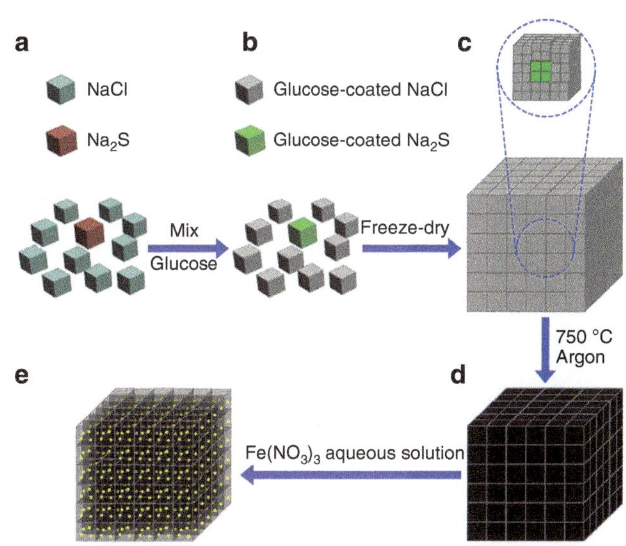

Figure 1 | Schematic illustration of an *in situ* strategy for the preparation of 3D S@PGC composites. (**a**) NaCl and Na_2S crystals. (**b**) Glucose-coated NaCl and Na_2S crystals. (**c**) Self-stacking of the glucose-coated NaCl and Na_2S crystals (3D NaCl-Na_2S@glucose), with Na_2S crystals surrounded by NaCl crystals. (**d**) Self-stacking of GC-coated NaCl and Na_2S crystals (3D NaCl-Na_2S@GC). (**e**) 3D S@PGC composite formed through the simultaneous dissolution of the NaCl crystals and oxidation of the Na_2S with $Fe(NO_3)_3$.

mesopores[41,42], thus leading to 3D NaCl–Na_2S@GC. Finally, the immersion of 3D NaCl–Na_2S@GC in an aqueous $Fe(NO_3)_3$ solution dissolved the NaCl template, leaving macropores in the composite. Concurrently, sulfur nanoparticles formed *in situ* and deposited on the walls of the macropores through the oxidation of Na_2S by $Fe(NO_3)_3$. During this process, the Na_2S that was dissolved in the aqueous $Fe(NO_3)_3$ solution in the 3D PGC macropores diffused into the micro- and mesopores because of capillarity, allowing the micro- and mesopores to fill with sulfur. Hence, the sulfur nanoparticles were uniformly distributed in the 3D PGC through an *in situ* chemical deposition process. According to previous reports[18,19], the PGC content, 3D porous architecture and sulfur content are crucial parameters that determine the properties of batteries such as the electrical conductivity of the electrode, electrolyte transport in the electrode and specific capacity. In our research, NaCl and Na_2S crystals were used as a template to fabricate the 3D porous architecture, and the sulfur content of the composites could be readily tuned by changing the Na_2S/glucose ratio. A 3D S@PGC composite with a sulfur content of up to 90 wt% [3D S@PGC (90% S)] was obtained by optimizing the Na_2S/glucose ratio (1:0.4). Therefore, our method has advantages over conventional methods[8,16,19–21,35] because the composition and structure of the composites can be easily controlled.

Structure and morphology of the 3D S@PGC composites. The crystallographic structure of the 3D S@PGC (90% S) composite was first analysed using X-ray diffraction (Fig. 2a). The X-ray diffraction pattern of elemental sulfur has three prominent Bragg reflections at 23.1, 25.9 and 27.8° (bottom trace), which can be indexed as the (222), (026) and (040) planes of the *fddd* orthorhombic structure[43]. As expected, the 3D S@PGC (90% S) composite produced a similar pattern, confirming the crystalline feature of the sulfur nanoparticles in the 3D PGC. The broad reflection peak, which can be clearly identified in the range of 15–35° (inset of Fig. 2a and Supplementary Fig. 1), is attributed to PGC. The peak for PGC in the 3D S@PGC (90% S) is weak because of the relatively low amount of PGC in the composite. A thermogravimetric analysis (TGA) indicated that the sulfur content of the 3D S@PGC (90% S) composite is ~90 wt% (Fig. 2b).

The morphology of the 3D S@PGC (90% S) composite was investigated via electron microscopy. Figure 3a displays an overview of the composite, with its honeycomb-like porous surface, obtained via scanning electron microscopy (SEM). At higher magnification (Fig. 3b), the unique network of the composite, which is composed of interconnected submicron-sized macropores, can be seen. With a further increase in magnification (Fig. 3c), sulfur nanoparticles with sizes ranging from 18 to 54 nm, homogeneously and densely anchored onto the walls of the PGC network, can be observed (Supplementary Fig. 2). A comparison of these morphologies with those of the 3D NaCl–Na_2S@GC and the pure 3D PGC framework (Supplementary Figs 3 and 4) led us to conclude that the 3D self-stacking of the NaCl and Na_2S crystals had been well preserved after carbonization. Therefore, the SEM data support the formation mechanism of the 3D S@PGC composites (Fig. 1). An image obtained via transmission electron microscopy (TEM) reveals that the pores overlap each other and form a continuous 3D network with ultrathin GC walls (Fig. 3d). In good agreement with the SEM data, the ultrathin carbon walls are uniformly covered with sulfur nanoparticles (Fig. 3e). The composites that were subjected to an elongated period of sonication resulted in similar TEM images (Supplementary Fig. 5). The TEM results indicate that the attachment of the sulfur nanoparticles to the walls of the PGC network was robust because the materials used for TEM observation were subjected to long periods of sonication during the sample preparation. Moreover, the 3D porous structure remained intact, demonstrating the high mechanical flexibility of the 3D S@PGC composites. A high-resolution TEM image reveals the crystalline feature of the sulfur nanoparticles (Fig. 3f). Clear lattice fringes with an interlayer spacing of 0.38 nm corresponding to the (222) planes are readily observable, as the (222) Bragg reflection has the greatest intensity in the X-ray diffraction pattern (Fig. 2a). Energy-dispersive X-ray spectroscopy (EDS) elemental mapping confirmed the presence of carbon and sulfur in the 3D S@PGC (90% S) composite (Fig. 3g–i), as well as the homogenous distribution of sulfur nanoparticles in the PGC framework (Fig. 3i, sulfur mapping). A nanosized distribution is extremely important for the application of the sulfur particles as a cathode material for Li–S batteries, as the utilization rate of sulfur is higher for smaller sulfur particles because of the short diffusion path of the electrons and lithium ions[3]. Compared with previously reported hybrid structures that have a comparably high sulfur content (~70 wt%) (refs 10,11,30), the sulfur particles in the 3D S@PGC (90% S) were much smaller and had a much more uniform distribution. These features can be attributed to the simultaneous formation of the porous structures and the *in situ* deposition of sulfur nanoparticles through the oxidation of Na_2S.

To further investigate the interactions between sulfur and 3D PGC, X-ray photoelectron spectroscopy (XPS) was performed. For comparison, XPS data were also collected for pure 3D PGC,

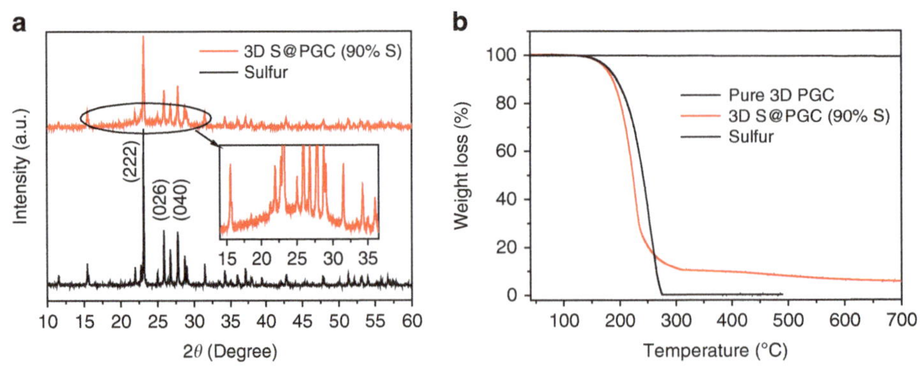

Figure 2 | Structural analysis of the 3D S@PGC (90% S) composite. (**a**) X-ray diffraction patterns of sulfur (bottom trace) and 3D S@PGC (90% S) composite (top trace), which confirm the crystalline feature of the sulfur nanoparticles in the 3D PGC. (**b**) TGA curves of sulfur, pure 3D PGC and the 3D S@PGC (90% S) composite. The 3D PGC apparently did not undergo weight loss up to 700 °C; the calculated sulfur content of the composite is thus ~90 wt%.

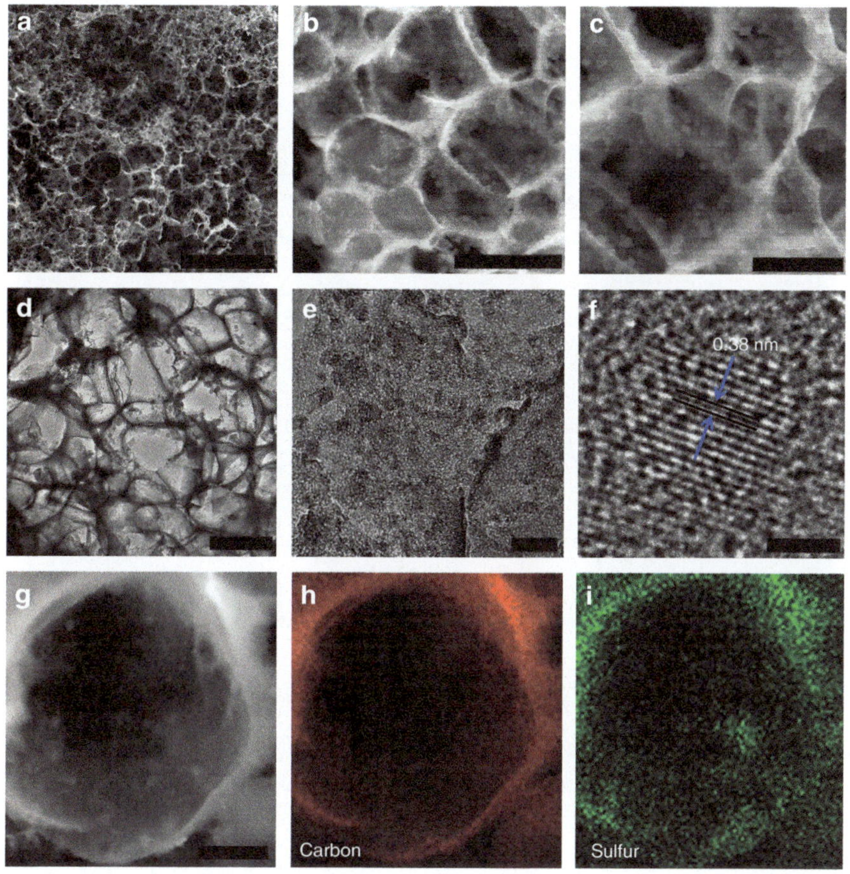

Figure 3 | Morphology of the 3D S@PGC (90% S) composite. (**a–c**) SEM images and (**d,e**) TEM images of the 3D S@PGC (90% S) composite at different magnifications. (**f**) HRTEM image of a sulfur nanoparticle in the composite. (**g**) TEM image of the 3D S@PGC (90% S) composite. (**h,i**) EDS elemental maps of (**h**) carbon and (**i**) sulfur, which were collected from the entire area shown in **g**. Scale bars in **a,b** and **c**: 20; 1; and 0.5 μm. Scale bars in **d,e,f** and **g**: 500; 50; 2; and 200 nm. The SEM and TEM images indicate that the composite possesses a 3D network consisting of interconnected submicron-sized macropores. From the SEM images, the sulfur nanoparticles anchored to the walls of the PGC network were calculated to have a size distribution of 18–54 nm. The EDS results indicate that the sulfur is uniformly distributed in the composite. HRTEM, high-resolution TEM.

which was prepared by immersing NaCl–Na_2S@GC in water to remove NaCl and Na_2S (Fig. 4a). The C 1*s* XPS spectrum of pure 3D PGC has a major peak at 284.7 eV, corresponding to sp^2 hybridized carbon, as well as three weak peaks at 286.4, 287.2 and 288.9 eV, which can be ascribed to C–O, C=O and O–C=O species, respectively[14]. The survey XPS spectrum of the 3D S@PGC (90% S) composite confirms the presence of sulfur in 3D PGC (Fig. 4b). In contrast to the C 1*s* XPS spectrum of pure 3D PGC, that of the 3D S@PGC (90% S) composite has an additional peak at 285.5 eV, which is ascribed to C–S bonds (Fig. 4c)[14]. This finding reveals the presence of covalent bonding between sulfur and PGC. The S 2*p* XPS peaks, that are characterized by an S $2p_{3/2}$ and $2p_{1/2}$ doublet with an energy separation of 1.2 eV, reconfirm the presence of C–S bonds (Fig. 4d), as the binding energy of the S $2p_{3/2}$ peak (163.5 eV) is lower than that of elemental sulfur (164.0 eV)[14,44]. The weak peak at 168.6 eV is due to sulfate species formed by the oxidation of sulfur in air[14]. The presence of C–S bonds is also supported by Fourier transform infrared (FTIR) spectroscopy because the vibration characteristic of C–S bonds was detected at 671 cm^{-1} (Supplementary Fig. 6)[45]. The C–S bonds could be formed through the addition of various reactive intermediates, including free radicals (for example, $HS^{\bullet}$) and radical anions (for example, $S^{\bullet -}$ and $S_X^{\bullet -}$)[46], to the unsaturated carbon–carbon double bonds of the PGC as well as through the nucleophilic attack of transient negatively charged polysulphides (for example, $S_X{}^{2-}$) with residual oxygen-containing functional groups present in the PGC (see Supplementary Note 1). Therefore, the C−S bonds were formed during the oxidation of Na_2S by $Fe(NO_3)_3$ because free radicals, radical anions and negatively charged polysulphides were the intermediate products of the oxidation reaction[46].

To reconfirm the presence of C−S bonds, the 3D S@PGC (90% S) composite was subjected to Soxhlet extraction using CS_2. The TGA curve of the extracted sample revealed a continual weight loss up to 700 °C (Supplementary Fig. 7); such a weight loss could be assigned to the removal of bonded sulfur[47], and the percentage of bonded sulfur was estimated to be ~48 wt%. By comparing the sulfur content of the as-prepared 3D S@PGC (90% S) composite to that of the Soxhlet-extracted sample, the bonded and unbonded sulfur content were calculated to be ~9 and 81 wt% in the 3D S@PGC (90% S) composite (Supplementary Table 1). The X-ray diffraction pattern of the extracted sample did not have any sulfur peaks (Supplementary Fig. 8), which is in line with the formation of C−S bonds[48]. Compared with previously reported physical strategies for confining sulfur[21,23,25,49,50], covalent bonding between sulfur nanoparticles and the PGC framework should effectively prevent the loss of the active materials and stabilize the cycling life of the corresponding Li–S batteries.

The sulfur content of our 3D S@PGC composites is readily tuneable. To tune the amount of sulfur, we used different Na_2S/glucose ratios (that is, 1:0.5, 1:0.45 and 1:0.4) during the

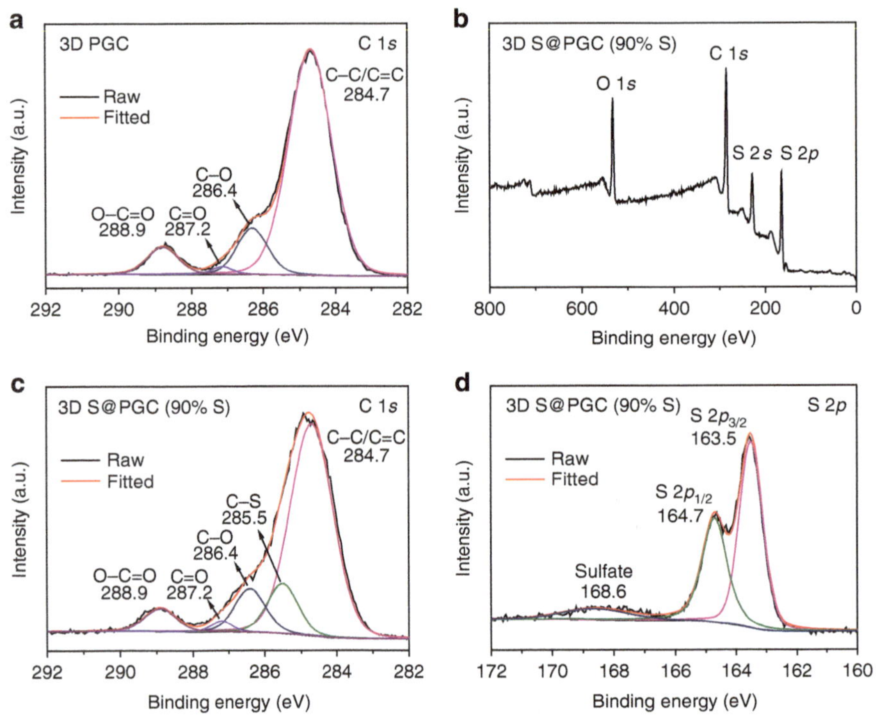

Figure 4 | XPS results for pure 3D PGC and the 3D S@PGC (90% S) composite. (**a**) C 1s XPS spectrum of pure PGC. (**b**) XPS survey spectrum of the 3D S@PGC (90% S) composite. (**c**) C 1s and (**d**) S 2*p* XPS spectra of the 3D S@PGC (90% S) composite. The data indicate the presence of C-S bonds in the 3D S@PGC (90% S) composite.

preparation while maintaining a constant Na_2S content (2.0 g). Figure 5 displays the TGA, X-ray diffraction and Raman spectroscopy data of 3D S@PGC composites prepared with different Na_2S/glucose ratios. The sulfur content calculated from the TGA curves for the 3D S@PGC composites prepared with the aforementioned ratios was ~64, 70 and 90 wt%, respectively (Fig. 5a). The 3D S@PGC composites exhibited weight losses at different stages. Compared with pure sulfur, the 3D S@PGC composites exhibited slightly lower starting evaporation temperatures. The evaporation of sulfur in this temperature range was ascribed to the sulfur nanoparticles attached to the walls of the PGC without covalent bonding. This lowering of the evaporation temperature was mainly ascribed to the nanosized distribution of the sulfur particles, which possessed excess surface free energy and facilitated heat transfer because of the large contact area between the sulfur nanoparticles and the 3D PGC (ref. 51). The 3D S@PGC composites also exhibited evaporation temperature ranges that were higher than that of pure sulfur. This phenomenon was caused by the sulfur that filled the micro- and mesopores[52] and covalently bonded to the wall surfaces of the PGC[47]. The 3D S@PGC (64% S) composite exhibited the most obvious weight loss at a relatively higher temperature because the proportion of sulfur contained within the micro- and mesopores and bonded to the PGC relative to the total amount of sulfur was the highest among the three composites. The X-ray diffraction patterns of the composites contain reflection peaks for both sulfur (23.1°, 25.9° and 27.8°) and GC (broad peak from 15 to 35°) (Fig. 5b and Supplementary Fig. 1), indicating the presence of sulfur in the 3D PGC. With the decrease in the PGC content of the composites, the intensity of the PGC peak became weaker. Collectively, the sulfur content of the 3D S@PGC composites increased with the amount of Na_2S used in the synthetic procedure (Supplementary Table 1).

The structural features of sulfur and the carbon matrix of the 3D S@PGC composites were further investigated by Raman spectroscopy (Fig. 5c). Elemental sulfur produced characteristic peaks at 471, 216 and 151 cm^{-1}. In agreement with the XRD result, the Raman spectra of the 3D S@PGC composites contain characteristic peaks of sulfur. Furthermore, the Raman spectra of the composites present peaks at ~1,350 and 1,588 cm^{-1}, which correspond to the D and G bands of carbon materials[53]. The presence of such bands in the Raman spectra suggests the conversion of glucose into GC. With the decrease in the PGC content of the composites, the D and G bands also became weaker.

The SEM and TEM images (Supplementary Figs 9 and 10) reveal that 3D S@PGC composites with different amounts of sulfur display similar morphologies, indicating the retention of the unique honeycomb-like network in the composites. Although the sizes of the sulfur nanoparticles calculated from the SEM images increased with the sulfur content (that is, 6–24 nm for 3D S@PGC (64% S), 12–35 nm for 3D S@PGC (70% S) and 18–54 nm for 3D S@PGC (90% S); Supplementary Fig. 2 and Supplementary Table 1), the sulfur particle size found in the 3D S@PGC (90% S) composite (18–54 nm) still resulted in high sulfur utilization, as discussed below. The sizes of the sulfur nanoparticles calculated from the Scherrer formula using the (222) Bragg reflection at 23.1° (~28, 33 and 44 nm for the 3D S@PGC (64% S), 3D S@PGC (70% S) and 3D S@PGC (90% S) composites, respectively) are in agreement with the values obtained from the SEM images. The FTIR spectra and XPS data confirm that C-S bonds also exist in the 3D S@PGC (70% S) and 3D S@PGC (64%) composites (Supplementary Figs 6 and 11). The pore structures of the 3D S@PGC composites were also characterized using N_2 physisorption measurements, providing insight into the micro- and mesopore structures of the composites (Supplementary Fig. 12 and Supplementary Note 2). All the 3D S@PGC composites had smaller surface areas and pore volumes than the corresponding PGC frameworks obtained by immersing NaCl-Na_2S@GC in water. The marked decreases in the specific

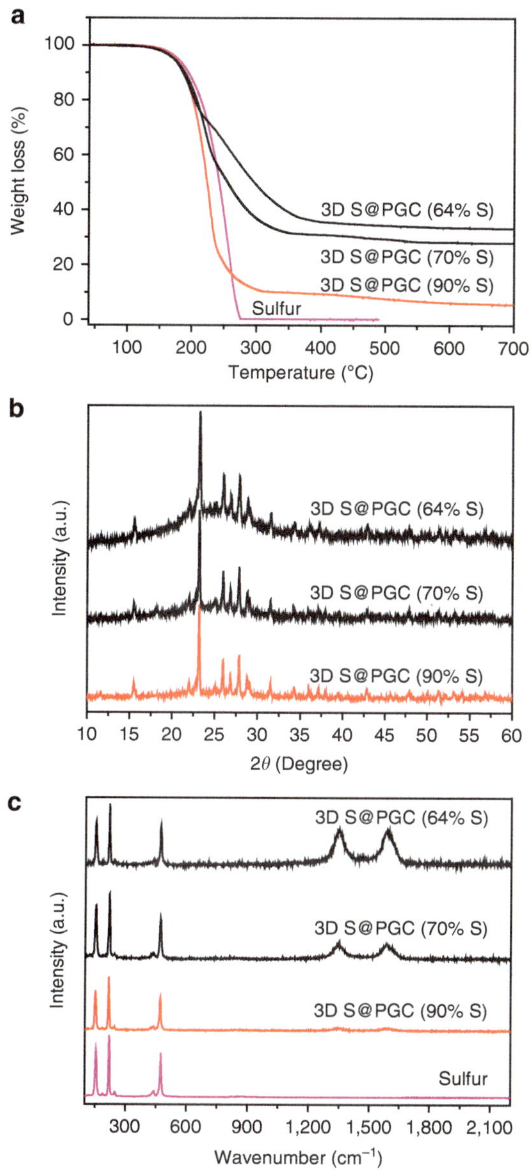

Figure 5 | Characterization of 3D S@PGC composites with various sulfur contents. (**a**) TGA curves, (**b**) X-ray diffraction patterns and (**c**) Raman spectra reveal that the sulfur content in the 3D S@PGC composites can be readily tuned by changing the Na_2S/glucose ratio and that the sulfur nanoparticles in the composites have the same crystalline structures.

surface area and total pore volume upon the deposition of sulfur nanoparticles indicated that the sulfur occupied the volumes of the pores in the 3D PGC frameworks. A comparison of the pore-size distributions of the 3D S@PGC composites and the corresponding 3D PGC frameworks indicated that the latter contained numerous micro- and mesopores, with sizes ranging from 1.5 to 25 nm, which were filled during formation of the composites. Combining the data from the SEM, TEM and N_2 physisorption analyses, we can conclude that the sulfur situates both in the micro-/mesopores and on the walls of the macropores of the 3D PGC.

The 3D S@PGC composites for Li-S battery cathode materials. Coin cells with Li foil as an anode were fabricated to evaluate the electrochemical performance of the 3D S@PGC (90% S) composite as a cathode material. Profiles obtained by cyclic voltammetry conducted at a scan rate of $0.1\,mV\,s^{-1}$ are presented in Fig. 6a. Well-defined reduction peaks were observed at 2.37 and 2.04 V, indicating that the sulfur was reduced in two stages. According to a reported mechanism[15], the peak at 2.37 V corresponds to the reduction of elemental sulfur to lithium polysulphides (Li_2S_n, $4 \leq n \leq 8$), and the peak at 2.04 V corresponds to the further reduction of the lithium polysulphides to Li_2S_2 and, eventually, to Li_2S. The weak peak at ~1.7 V is assigned to the reduction of $LiNO_3$. Likewise, the oxidation of the cathode also proceeded through two stages: the conversion of Li_2S_2/Li_2S to Li_2S_n ($n>2$), associated with the oxidation peak at 2.43 V, and the final formation of elemental sulfur, corresponding to the oxidation peak at 2.47 V. The galvanostatic charge/discharge behaviour of the 3D S@PGC (90% S) cathode was studied at charge/discharge rates of 0.5, 1, 2 and 4 C within a potential window of 1.5–3.0 V versus Li^+/Li (Fig. 6b–e). Consistent with the two reduction peaks at 2.37 and 2.04 V in the cathodic sweep, two plateaus were observed at ~2.3 and 2.0 V in the discharge process at 0.5 C, corresponding to the two-stage reduction of elemental sulfur to lithium polysulphides (Li_2S_{4-8}) and then to Li_2S_2/Li_2S, respectively. Discharge curves obtained at higher currents (1, 2 and 4 C) also clearly contain the two plateaus, indicating that the electrochemical reactions at higher charge/discharge rates follow processes similar to those occurring at lower rates. However, the non-conductive nature of Li_2S_2 and Li_2S subjects the conversion of Li_2S_{4-8} to Li_2S_2/Li_2S to higher polarization at higher charge/discharge rates, that is, decreased voltage at the second plateau[54]. As previously noted[16,47,55,56], the charge/discharge profiles can be sloppy at the ends. This phenomenon could be ascribed to the presence of C–S bonds[16,47] and the strong absorption of sulfur in the micro- and mesopores of the 3D PGC[55,56], as well as the irreversible reduction of $LiNO_3$ (refs 57,58).

The cycling performance of the 3D S@PGC (90% S) composite cathode was first studied at 0.5 C (Supplementary Fig. 13). The material achieved an initial discharge capacity as high as $1,382\,mA\,h\,g^{-1}$, which corresponds to 82.5% sulfur utilization based on the theoretical value of sulfur ($1,675\,mA\,h\,g^{-1}$). After 200 cycles, the calculated capacity retention was 62%. The high utilization rate of sulfur is because of the hierarchical porous structures and the nanosized sulfur particles distributed on the GC walls. The macropores highly favour the rapid access to the electrode interior by the electrolyte, while the nanosized sulfur particles and ultrathin GC walls with micro- and mesopores facilitate the efficient transport of ions into the deeper portions of the sulfur nanoparticles because of the short pathway[18,23,41]. Figure 6f displays the cycling performance of the 3D S@PGC (90% S) composite cathodes at charge/discharge rates of 1, 2 and 4 C. The initial discharge capacity was $1,242\,mA\,h\,g^{-1}$ at 1 C. After 200 cycles, a capacity of $917\,mA\,h\,g^{-1}$ remained, corresponding to a capacity retention of 74%. When the charge/discharge rate was raised to 2 C, the initial capacity obtained ($1,115\,mA\,h\,g^{-1}$) became lower than the corresponding value at 1 C, but the cycling stability improved. A capacity of $920\,mA\,h\,g^{-1}$ corresponding to a capacity retention of 83% was obtained after 200 cycles at 2 C. When the charge/discharge rate was further elevated to 4 C, the measured initial capacity was $638\,mA\,h\,g^{-1}$. After 200 cycles, the capacity remained at $548\,mA\,h\,g^{-1}$, which corresponds to a capacity retention of 86%. The capacities of the composite obtained at current densities of 0.5, 1 and 2 C (that is, 1,382, 1,242 and $1,115\,mA\,h\,g^{-1}$ at 0.5, 1 and 2 C, respectively) are much higher than those of previously reported sulfur–carbon composites ($430–1,100\,mA\,h\,g^{-1}$ at 0.5 C, $400–980\,mA\,h\,g^{-1}$ at 1 C and $450–900\,mA\,h\,g^{-1}$ at 2 C)[11,21,36,37,52,59–62]. Although some capacities reported at

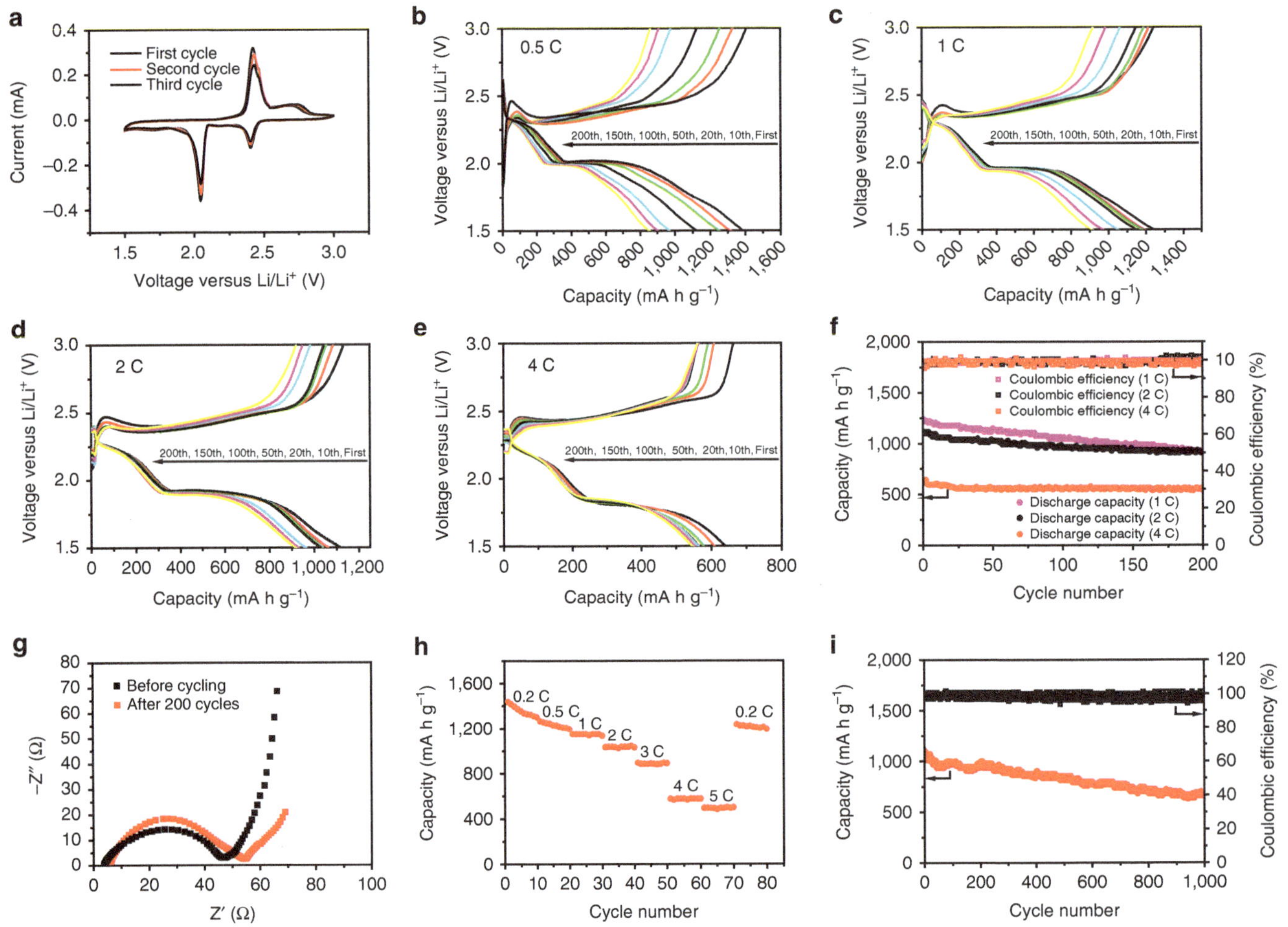

Figure 6 | Electrochemical performance of the 3D S@PGC (90% S) composite as a cathode material for Li-S batteries. (**a**) CV profiles of the 3D S@PGC (90% S) composite cathode. (**b-e**) Charge/discharge curves of the cathode at charge/discharge rates of (**b**) 0.5 C, (**c**) 1 C, (**d**) 2 C and (**e**) 4 C. (**f**) Cycling performance of the cathode at charge/discharge rates of 1, 2 and 4 C. (**g**) EIS curves of the cathode before and after 200 cycles at 2 C. (**h**) Rate performance of the cathode. (**i**) Cycling performance of the cathode over 1,000 cycles at a charge/discharge rate of 2 C. The data indicate that the 3D S@PGC (90% S) composite displays excellent electrochemical performance, in particular, a high sulfur utilization, high specific capacity (1,115 $mA\,h\,g^{-1}$ at 2 C), long cycling life (small capacity decay of 0.039% per cycle over 1,000 cycles at 2 C), and excellent rate capability at a high charge/discharge current. CV, cyclic voltammetry. EIS, electrochemical impedance spectroscopy.

higher current densities (for example, 4 and 5 C) were relatively high (650–750 $mA\,h\,g^{-1}$), the sulfur content of their composites was relatively low (40–65 wt%)[21,60]. The S@PGC (90% S) composite cathode exhibited high coulombic efficiencies (>98%) at all the current rates tested, confirming that the shuttling effect of the polysulphides was efficiently suppressed because of the C–S bonding between the sulfur nanoparticles and 3D PGC as well as the unique porous structure of the 3D S@PGC composite. The use of $LiNO_3$ in the electrolyte also helped suppress polysulphide shuttling, improving the coulombic efficiency and enhancing the rate capability[63,64], as demonstrated by experiments without $LiNO_3$ in the electrolyte (Supplementary Fig. 14 and Supplementary Table 2). The relatively lower capacity retention at a lower charge/discharge rate was likely due to the incomplete oxidation of the insulating Li_2S_2 and Li_2S (refs 55,65) because the shuttling effect was efficiently suppressed and the coulombic efficiency was found to be higher than 98%.

The stability of the 3D S@PGC (90% S) composite cathode was evidenced by electrochemical impedance spectroscopy (EIS) before and after 200 cycles at 2 C (Fig. 6g). The corresponding Nyquist profiles were fitted to a widely used equivalent circuit (Supplementary Fig. 15)[66]. The electrolyte resistance (R_e) and the charge-transfer resistance at the electrode/electrolyte interface (R_{ct}) were determined to be 3.6 and 43.2 Ω before cycling and 4.6 and 47.9 Ω after cycling, respectively. The slight changes in R_e and R_{ct} after the charge/discharge cycles, which could be ascribed to the formation of a passive layer on the carbon frameworks[67], suggest the high conductivity and good stability of the 3D S@PGC (90% S) cathode due to the stable 3D PGC framework and the C–S bonds between the sulfur nanoparticles and the 3D PGC framework. TEM observations also indicated the stability of the 3D S@PGC (90% S) composite electrode, as the porous structures of the 3D PGC remained intact and the sulfur nanoparticles were still firmly and homogeneously anchored to the PGC walls after 200 charge/discharge cycles at 2 C (Supplementary Fig. 16). In addition, the FTIR data of the electrode material indicated the retention of the C − S bonds after cycling (Supplementary Fig. 17), consistent with a previous study that reported the retention of C − S bonds after cycling[68].

The rate capability of the 3D S@PGC (90% S) composite is depicted in Fig. 6h. The discharge capacity gradually decreased as the current rate increased from 0.2 to 5 C. At the maximum charge/discharge rate tested (5 C), the specific capacity remained high (500 mA h g^{-1}) and the cycling remained stable; when the current rate was restored to 0.2 C, the composite recovered most of its capacity. Most importantly, the composite exhibited stable cycling performance over 1,000 charge/discharge cycles at 2 C (Fig. 6i). A high specific capacity (670 mA h g^{-1}) was retained after 1,000 cycles. The calculated capacity retention is 61%, which corresponds to a very small capacity decay (0.039% per cycle). This finding demonstrates the excellent cycling stability of our Li–S batteries. Although other composites have been reported to exhibit a long cycling life when used as cathodes in Li–S batteries, the sulfur content employed in such materials was relatively low (ranging from 30 to 80%)[30,37,63,69]. Therefore, 3D S@PGC is a high-sulfur-content (up to 90%) cathode material that exhibits excellent cycling stability at a high current density. The pure sulfur cathode (the control) used under the same conditions exhibited a much lower specific capacity and worse cycling stability than those of the 3D S@PGC (90% S) composite (Supplementary Fig. 18).

Discussion

The electrochemical performance of the 3D S@PGC (64% S) and 3D S@PGC (70% S) composites as cathodes in Li–S batteries were also evaluated. As summarized in Supplementary Fig. 19 and Supplementary Tables 3 and 4, 3D S@PGC composites with a lower sulfur content exhibited better performance: specifically, higher specific capacities, higher capacity retention and higher rate performance (Supplementary Note 3). These results are attributable to the smaller sizes of the sulfur nanoparticles in the 3D S@PGC composites with a relatively low sulfur content. Indeed, smaller sulfur nanoparticles have larger specific contact areas with the 3D PGC framework, which helps alleviate the shuttling effect and improves the cycle stability. Smaller particle sizes also facilitate electron and Li^+ diffusion and lead to better sulfur utilization and a higher specific capacity. The larger specific surface areas effectively reduce the discharging current densities and the Li^+ flux, thereby limiting the formation of a Li_2S blocking layer at high charge/discharge rates[51]. Although the 3D S@PGC composites with a lower sulfur content exhibited higher specific capacities calculated on the basis of sulfur, the relatively low sulfur content reduced the overall volumetric capacity and energy density of the corresponding cathodes. Therefore, 3D S@PGC composites with relatively high sulfur content may be promising candidates for use in practical applications.

The excellent overall electrochemical performance of the 3D S@PGC composites can be attributed to the following factors that stem from the design of the materials. First, the *in situ* chemical deposition method allows access to composites with high sulfur content (up to 90 wt%) and affords the nanoscale distribution of the sulfur particles in the resultant 3D PGC network. As described above, nanosized sulfur particles facilitate a high sulfur utilization rate (82.5% for 3D S@PGC (90% S), 84.5% for 3D S@PGC (70% S) and 86% for 3D S@PGC (64% S) at 0.5 C). Second, the C–S bonds formed between the sulfur nanoparticles and 3D PGC can effectively prevent agglomeration of the sulfur nanoparticles, minimize the loss of lithium polysulphides to the electrolyte and suppress the shuttling effect during the charge/discharge cycles. Third, the 3D PGC networks that display high electrical conductivities, large surface areas and high mechanical flexibility confer high electrical conductivity and structural integrity to the electrodes. The numerous walls between the interconnected macropores may function as multilayered barriers to further mitigate the dissolution of polysulphides into the electrolyte. Finally, the unique interconnected hierarchical pores in the 3D PGC network facilitate access to the sulfur nanoparticles by the electrolyte and preserve the rapid transport of Li^+ to the active material.

In conclusion, we report a new methodology that is facile and scalable and allows the *in situ* preparation of 3D S@PGC composites with a high sulfur content. The strategy utilizes Na_2S as a sulfur precursor and NaCl and Na_2S as a template for the porous structure of the resultant composite. The sulfur nanoparticles were homogenously distributed and covalently bonded to 3D PGC, as confirmed by various spectroscopic and microscopic techniques. Li–S batteries prepared using the composites as cathodes exhibited excellent performance; specifically, high sulfur utilization, high specific capacities, good cycling stabilities and high rate capabilities were observed. Notably, Li–S batteries prepared using 3D S@PGC (90% S) as a cathode displayed a long cycling stability, with a capacity decay of only 0.039% per cycle over 1,000 cycles at a high charge/discharge current (2 C). Overall, the methodology described herein offers a new avenue for the fabrication of cathode materials based on carbon–sulfur hybridized nanostructures for use in high-performance Li–S batteries. We believe that the strategy may also inspire the preparation of other 3D porous structures for use in other areas, including applications in catalysis, selective adsorption, separations and sensing.

Methods

Materials. All reagents were purchased from commercial sources and used without further purification. All solvents used were purified using standard procedures.

Representative synthesis of 3D S@PGC composites. In a typical synthesis, $Na_2S \cdot 9H_2O$ (2.0 g), NaCl (5.0 g) and glucose (0.8 g) were dissolved in DI water (15 ml). The resultant solution was frozen in liquid nitrogen, and the water in the mixture was removed via freeze-drying. The resultant gel was ground into a fine powder and then heated at 750 °C for 2 h under an atmosphere of argon. A black powder was obtained and subsequently stirred in an aqueous solution of $Fe(NO_3)_3$ (20 g $Fe(NO_3)_3 \cdot 9H_2O$ in 150 ml DI water) for 40 h to dissolve the residual NaCl crystals and to deposit the sulfur. Afterwards, the black powder product was washed several times with DI water and centrifuged to afford the desired composite. Composites with various sulfur contents were synthesized by using different amounts of glucose (0.9 and 1.0 g) in the aforementioned procedure.

Characterization. X-ray diffraction data were collected on a Bruker D8 Focus diffractometer using an incident wavelength of 0.154 nm (Cu Kα radiation) and a Lynx-Eye detector. Raman spectra were recorded on a Renishaw inVia-Reflex confocal Raman microscope at an excitation wavelength of 532 nm. TGA measurements were carried out using a TGA Q50 at a scanning rate of 10 °C min^{-1}. SEM observations were performed on a field-emission SEM (Hitachi S-4800) equipped with EDS. TEM images were obtained using a JEOL-2100F microscope operated under an accelerating voltage of 200 kV. EDS analysis was also performed on Tecnai F20 scanning transmission electron microscope operated at 200 keV using an Oxford detector with a beam current of ~1 nA. N_2 adsorption–desorption isotherms and pore-size distribution were obtained at 77 K using a QuadraSorb SI MP apparatus. The total specific surface areas of the samples were calculated via the Brunauer–Emmett–Teller method. The pore-size distribution was calculated via the density functional theory model. XPS spectra were recorded on a PHI Quantera Scanning X-ray Microprobe using monochromated Al Kα radiation (1486.7 eV). FTIR spectra were recorded on an Excalibur 3100 spectrometer with a resolution of 0.2 cm^{-1} using KBr pellets.

Electrochemistry. The 3D S@PGC composites were combined with conductive carbon and poly(vinylidene fluoride) as a binder with a mass ratio of 80:10:10 and milled into a slurry with N-methylpyrrolidone. The slurry was then blade cast onto a carbon-coated Al foil and dried at 50 °C for 10 h in a vacuum oven. The loading density of sulfur was ca. 2.36 mg cm^{-2}. CR2032 coin cells were assembled in an argon-filled glove box employing the 3D S@PGC-coated Al foil as the cathode, a porous membrane (Celgard 3501) as the separator, and lithium foil as the reference/counter electrode. The electrolyte used was lithium bis(trifluoro methane)sulphonimide (0.38 M) and lithium nitrate (0.31 M) in a solvent mixture of 1,3-dioxolane and 1,2-dimethoxy ethane (1:1 v/v). Pristine sulfur electrodes were fabricated under similar conditions. Cyclic voltammetry curves were collected using a CHI 660E electrochemical workstation at a scan rate of 0.1 mV s^{-1} from 3.0 to 1.5 V. Cycling tests of the batteries were galvanostatically performed at various charge/discharge rates within a potential window of 1.5–3.0 V versus

Li^+/Li. The electrochemical impedance spectroscopy data were recorded using a Zennium 40088 electrochemical workstation by applying a sine wave with an amplitude of 10 mV over a frequency range of 100 kHz to 10 mHz.

References

1. Bruce, P. G., Freunberger, S. A., Hardwick, L. J. & Tarascon, J. M. Li-O_2 and Li-S batteries with high energy storage. *Nat. Mater.* **11,** 19–29 (2012).
2. Choi, N. S. *et al.* Challenges facing lithium batteries and electrical double-layer capacitors. *Angew. Chem. Int. Ed. Engl.* **51,** 9994–10024 (2012).
3. Evers, S. & Nazar, L. F. New approaches for high energy density lithium-sulfur battery cathodes. *Acc. Chem. Res.* **46,** 1135–1143 (2013).
4. Manthiram, A., Fu, Y. Z. & Su, Y. S. Challenges and prospects of lithium-sulfur batteries. *Acc. Chem. Res.* **46,** 1125–1134 (2013).
5. Yamin, H., Gorenshtein, A., Penciner, J., Sternberg, Y. & Peled, E. Lithium sulfur battery—oxidation reduction-mechanisms of polysulfides in THF solutions. *J. Electrochem. Soc.* **135,** 1045–1048 (1988).
6. Peled, E., Sternberg, Y., Gorenshtein, A. & Lavi, Y. Lithium-sulfur battery—evaluation of dioxolane-based electrolytes. *J. Electrochem. Soc.* **136,** 1621–1625 (1989).
7. Jin, B., Kim, J. U. & Gu, H. B. Electrochemical properties of lithium-sulfur batteries. *J. Power Sources* **117,** 148–152 (2003).
8. Chen, R. J. *et al.* Graphene-based three-dimensional hierarchical sandwich-type architecture for high-performance Li/S batteries. *Nano Lett.* **13,** 4642–4649 (2013).
9. Huang, J. Q. *et al.* Entrapment of sulfur in hierarchical porous graphene for lithium-sulfur batteries with high rate performance from − 40 to 60 °C. *Nano Energy* **2,** 314–321 (2013).
10. Wang, H. L. *et al.* Graphene-wrapped sulfur particles as a rechargeable lithium-sulfur battery cathode material with high capacity and cycling stability. *Nano Lett.* **11,** 2644–2647 (2011).
11. Zhou, W. D. *et al.* Amylopectin wrapped graphene oxide/sulfur for improved cyclability of lithium-sulfur battery. *ACS Nano* **7,** 8801–8808 (2013).
12. Wang, C. *et al.* Macroporous free-standing nano-sulfur/reduced graphene oxide paper as stable cathode for lithium-sulfur battery. *Nano Energy* **11,** 678–686 (2015).
13. Wang, C. *et al.* Slurryless Li_2S/reduced graphene oxide cathode paper for high-performance lithium sulfur battery. *Nano Lett.* **15,** 1796–1802 (2015).
14. Wang, Z. Y. *et al.* Enhancing lithium-sulfur battery performance by strongly binding the discharge products on amino-functionalized reduced graphene oxide. *Nat. Commun.* **5,** 5002 (2014).
15. Zhao, M. Q. *et al.* Unstacked double-layer templated graphene for high-rate lithium-sulfur batteries. *Nat. Commun.* **5,** 3410 (2014).
16. Guo, J. C., Xu, Y. H. & Wang, C. S. Sulfur-impregnated disordered carbon nanotubes cathode for lithium-sulfur batteries. *Nano Lett.* **11,** 4288–4294 (2011).
17. Han, S. C. *et al.* Effect of multiwalled carbon nanotubes on electrochemical properties of lithium sulfur rechargeable batteries. *J. Electrochem. Soc.* **150,** A889–A893 (2003).
18. He, G., Ji, X. L. & Nazar, L. High 'C' rate Li-S cathodes: sulfur imbibed bimodal porous carbons. *Energy Environ. Sci.* **4,** 2878–2883 (2011).
19. Ji, X. L., Lee, K. T. & Nazar, L. F. A highly ordered nanostructured carbon-sulfur cathode for lithium-sulfur batteries. *Nat. Mater.* **8,** 500–506 (2009).
20. Zhang, B., Qin, X., Li, G. R. & Gao, X. P. Enhancement of long stability of sulfur cathode by encapsulating sulfur into micropores of carbon spheres. *Energy Environ. Sci.* **3,** 1531–1537 (2010).
21. Xin, S. *et al.* Smaller sulfur molecules promise better lithium-sulfur batteries. *J. Am. Chem. Soc.* **134,** 18510–18513 (2012).
22. Xin, S., Yin, Y. X., Guo, Y. G. & Wan, L. J. A high-energy room-temperature sodium-sulfur battery. *Adv. Mater.* **26,** 1261–1265 (2014).
23. Schuster, J. *et al.* Spherical ordered mesoporous carbon nanoparticles with high porosity for lithium-sulfur batteries. *Angew. Chem. Int. Ed. Engl.* **51,** 3591–3595 (2012).
24. Song, J. X. *et al.* Strong lithium polysulfide chemisorption on electroactive sites of nitrogen-doped carbon composites for high-performance lithium-sulfur battery cathodes. *Angew. Chem. Int. Ed. Engl.* **54,** 4325–4329 (2015).
25. Zhang, C. F., Wu, H. B., Yuan, C. Z., Guo, Z. P. & Lou, X. W. Confining sulfur in double-shelled hollow carbon spheres for lithium-sulfur batteries. *Angew. Chem. Int. Ed. Engl.* **51,** 9592–9595 (2012).
26. Su, Y. S. & Manthiram, A. Lithium-sulfur batteries with a microporous carbon paper as a bifunctional interlayer. *Nat. Commun.* **3,** 1166 (2012).
27. Choi, Y. J., Kim, K. W., Ahn, H. J. & Ahn, J. H. Improvement of cycle property of sulfur electrode for lithium/sulfur battery. *J. Alloy. Compd.* **449,** 313–316 (2008).
28. Ji, L. W. *et al.* Porous carbon nanofiber-sulfur composite electrodes for lithium/sulfur cells. *Energy Environ. Sci.* **4,** 5053–5059 (2011).
29. Zheng, G. Y., Yang, Y., Cha, J. J., Hong, S. S. & Cui, Y. Hollow carbon nanofiber-encapsulated sulfur cathodes for high specific capacity rechargeable lithium batteries. *Nano Lett.* **11,** 4462–4467 (2011).
30. Seh, Z. W. *et al.* Sulfur-TiO_2 yolk-shell nanoarchitecture with internal void space for long-cycle lithium-sulfur batteries. *Nat. Commun.* **4,** 1331 (2013).
31. Zhou, W. D., Yu, Y. C., Chen, H., DiSalvo, F. J. & Abruna, H. D. Yolk-shell structure of polyaniline-coated sulfur for lithium-sulfur batteries. *J. Am. Chem. Soc.* **135,** 16736–16743 (2013).
32. Kim, J. S., Hwang, T. H., Kim, B. G., Min, J. & Choi, J. W. A lithium-sulfur battery with a high areal energy density. *Adv. Funct. Mater.* **24,** 5359–5367 (2014).
33. Chung, W. J. *et al.* The use of elemental sulfur as an alternative feedstock for polymeric materials. *Nat. Chem* **5,** 518–524 (2013).
34. Griebel, J. J., Li, G. X., Glass, R. S., Char, K. & Pyun, J. Kilogram scale inverse vulcanization of elemental sulfur to prepare high capacity polymer electrodes for Li-S batteries. *J. Polym. Sci. A Polym. Chem* **53,** 173–177 (2015).
35. Chen, S. R. *et al.* Ordered mesoporous carbon/sulfur nanocomposite of high performances as cathode for lithium-sulfur battery. *Electrochim. Acta* **56,** 9549–9555 (2011).
36. Ji, L. W. *et al.* Graphene oxide as a sulfur immobilizer in high performance lithium/sulfur cells. *J. Am. Chem. Soc.* **133,** 18522–18525 (2011).
37. Lu, S. T., Cheng, Y. W., Wu, X. H. & Liu, J. Significantly improved long-cycle stability in high-rate Li-S batteries enabled by coaxial graphene wrapping over sulfur-coated carbon nanofibers. *Nano Lett.* **13,** 2485–2489 (2013).
38. Wang, D. W. *et al.* A microporous-mesoporous carbon with graphitic structure for a high-rate stable sulfur cathode in carbonate solvent-based Li-S batteries. *Phys. Chem. Chem. Phys.* **14,** 8703–8710 (2012).
39. Xin, S., Yin, Y. X., Wan, L. J. & Guo, Y. G. Encapsulation of sulfur in a hollow porous carbon substrate for superior Li-S batteries with long lifespan. *Part. Part. Syst. Char.* **30,** 321–325 (2013).
40. Zhang, Z. W. *et al.* 3D interconnected porous carbon aerogels as sulfur immobilizers for sulfur impregnation for lithium-sulfur batteries with high rate capability and cycling stability. *Adv. Funct. Mater.* **24,** 2500–2509 (2014).
41. Qin, J. *et al.* Graphene networks anchored with Sn@graphene as lithium ion battery anode. *ACS Nano* **8,** 1728–1738 (2014).
42. Zhu, T., Chen, J. S. & Lou, X. W. Glucose-assisted one-pot synthesis of FeOOH nanorods and their transformation to Fe_3O_4@carbon nanorods for application in lithium ion batteries. *J. Phys. Chem. C* **115,** 9814–9820 (2011).
43. Wang, Q. Q. *et al.* Improve rate capability of the sulfur cathode using a gelatin binder. *J. Electrochem. Soc.* **158,** A775–A779 (2011).
44. Zheng, S. *et al.* High performance C/S composite cathodes with conventional carbonate-based electrolytes in Li-S battery. *Sci. Rep.* **4,** 4842 (2014).
45. Yu, X. U. *et al.* Stable-cycle and high-capacity conductive sulfur-containing cathode materials for rechargeable lithium batteries. *J. Power Sources* **146,** 335–339 (2005).
46. Steudel, R. Mechanism for the formation of elemental sulfur from aqueous sulfide in chemical and microbiological desulfurization processes. *Ind. Eng. Chem. Res.* **35,** 1417–1423 (1996).
47. Zhang, Y. Z., Liu, S., Li, G. C., Li, G. R. & Gao, X. P. Sulfur/polyacrylonitrile/carbon multi-composites as cathode materials for lithium/sulfur battery in the concentrated electrolyte. *J. Mater. Chem. A* **2,** 4652–4659 (2014).
48. Zu, C. X. & Manthiram, A. Hydroxylated graphene-sulfur nanocomposites for high-rate lithium-sulfur batteries. *Adv. Energy Mater* **3,** 1008–1012 (2013).
49. Liang, C. D., Dudney, N. J. & Howe, J. Y. Hierarchically structured sulfur/carbon nanocomposite material for high-energy lithium battery. *Chem. Mater.* **21,** 4724–4730 (2009).
50. Ye, H., Yin, Y. X., Xin, S. & Guo, Y. G. Tuning the porous structure of carbon hosts for loading sulfur toward long lifespan cathode materials for Li-S batteries. *J. Mater. Chem. A* **1,** 6602–6608 (2013).
51. Chen, H. W. *et al.* Monodispersed sulfur nanoparticles for lithium sulfur batteries with theoretical performance. *Nano Lett.* **15,** 798–802 (2015).
52. Li, Z. *et al.* A highly ordered meso@microporous carbon-supported sulfur@smaller sulfur core-shell structured cathode for Li-S batteries. *ACS Nano* **8,** 9295–9303 (2014).
53. Sun, J. *et al.* Fluorine-doped SnO_2@graphene porous composite for high capacity lithium-ion batteries. *Chem. Mater.* **27,** 4594–4603 (2015).
54. Zhang, S. S. New insight into liquid electrolyte of rechargeable lithium/sulfur battery. *Electrochim. Acta* **97,** 226–230 (2013).
55. Li, G. C., Li, G. R., Ye, S. H. & Gao, X. P. A polyaniline-coated sulfur/carbon composite with an enhanced high-rate capability as a cathode material for lithium/sulfur batteries. *Adv. Energy Mater.* **2,** 1238–1245 (2012).
56. Li, G. C., Hu, J. J., Li, G. R., Ye, S. H. & Gao, X. P. Sulfur/activated-conductive carbon black composites as cathode materials for lithium/sulfur battery. *J. Power Sources* **240,** 598–605 (2013).
57. Zhang, S. S. Role of $LiNO_3$ in rechargeable lithium/sulfur battery. *Electrochim. Acta* **70,** 344–348 (2012).
58. Zhang, S. S. Effect of discharge cutoff voltage on reversibility of lithium/sulfur batteries with $LiNO_3$-contained electrolyte. *J. Electrochem. Soc.* **159,** A920–A923 (2012).

59. Li, B., Li, S. M., Liu, J. H., Wang, B. & Yang, S. B. Vertically aligned sulfur-graphene nanowalls on substrates for ultrafast lithium-sulfur batteries. *Nano Lett.* **15,** 3073–3079 (2015).
60. Zhou, G. M. *et al.* A graphene-pure-sulfur sandwich structure for ultrafast, long-life lithium-sulfur batteries. *Adv. Mater.* **26,** 625–631 (2014).
61. Jayaprakash, N., Shen, J., Moganty, S. S., Corona, A. & Archer, L. A. Porous hollow carbon@sulfur composites for high-power lithium-sulfur batteries. *Angew. Chem., Int. Ed. Engl.* **50,** 5904–5908 (2011).
62. Zhao, Y., Wu, W. L., Li, J. X., Xu, Z. C. & Guan, L. H. Encapsulating MWNTs into hollow porous carbon nanotubes: a tube-in-tube carbon nanostructure for high-performance lithium-sulfur batteries. *Adv. Mater.* **26,** 5113–5118 (2014).
63. Song, M. K., Zhang, Y. G. & Cairns, E. J. A long-life, high-rate lithium/sulfur cell: a multifaceted approach to enhancing cell performance. *Nano Lett.* **13,** 5891–5899 (2013).
64. Choudhury, S. *et al.* Nanoporous cathodes for high-energy Li-S batteries from gyroid block copolymer templates. *ACS Nano* **9,** 6147–6157 (2015).
65. Evers, S. & Nazar, L. F. Graphene-enveloped sulfur in a one pot reaction: a cathode with good coulombic efficiency and high practical sulfur content. *Chem. Commun.* **48,** 1233–1235 (2012).
66. Narayanan, S. R., Shen, D. H., Surampudi, S., Attia, A. I. & Halpert, G. Electrochemical impedance spectroscopy of lithium-titanium disulfide rechargeable cells. *J. Electrochem. Soc.* **140,** 1854–1861 (1993).
67. Zhao, M. Q. *et al.* Hierarchical vine-tree-like carbon nanotube architectures: *in situ* CVD self-assembly and their use as robust scaffolds for lithium-sulfur batteries. *Adv. Mater.* **26,** 7051–7058 (2014).
68. Fanous, J., Wegner, M., Grimminger, J., Andresen, A. & Buchmeiser, M. R. Structure-related electrochemistry of sulfur-poly(acrylonitrile) composite cathode materials for rechargeable lithium batteries. *Chem. Mater.* **23,** 5024–5028 (2011).
69. Liang, X. *et al.* A highly efficient polysulfide mediator for lithium-sulfur batteries. *Nat. Commun.* **6,** 5682 (2015).

Acknowledgements

This work was supported by the 'Hundred Talents Program' of Chinese Academy of Sciences and the National Natural Science Foundation of China (21404108, U1362106).

Author contributions

G.L. and J.G. conceived and designed the experiments. G.L. performed the laboratory experiments, characterization of materials and analysis of the results. S.J. conducted part of the TEM characterization. J.S. and W.H. participated in the discussion of electrochemical results. G.L. and J.G. co-wrote the paper and all authors discussed the results and commented on the manuscript. J.G. and Y.H. supervised the project.

Additional information

Competing financial interests: The authors declare no competing financial interests.

15

Failure mechanisms of single-crystal silicon electrodes in lithium-ion batteries

Feifei Shi[1,2], Zhichao Song[1], Philip N. Ross[2], Gabor A. Somorjai[2,3], Robert O. Ritchie[1,2,4] & Kyriakos Komvopoulos[1]

Long-term durability is a major obstacle limiting the widespread use of lithium-ion batteries in heavy-duty applications and others demanding extended lifetime. As one of the root causes of the degradation of battery performance, the electrode failure mechanisms are still unknown. In this paper, we reveal the fundamental fracture mechanisms of single-crystal silicon electrodes over extended lithiation/delithiation cycles, using electrochemical testing, microstructure characterization, fracture mechanics and finite element analysis. Anisotropic lithium invasion causes crack initiation perpendicular to the electrode surface, followed by growth through the electrode thickness. The low fracture energy of the lithiated/unlithiated silicon interface provides a weak microstructural path for crack deflection, accounting for the crack patterns and delamination observed after repeated cycling. On the basis of this physical understanding, we demonstrate how electrolyte additives can heal electrode cracks and provide strategies to enhance the fracture resistance in future lithium-ion batteries from surface chemical, electrochemical and material science perspectives.

[1] Department of Mechanical Engineering, University of California, Berkeley, California 94720, USA. [2] Materials Sciences Division, Lawrence Berkeley National Laboratory, Berkeley, California 94720, USA. [3] Department of Chemistry, University of California, Berkeley, California 94720, USA. [4] Department of Materials Science and Engineering, University of California, Berkeley, California 94720, USA. Correspondence and requests for materials should be addressed to R.O.R. (email: roritchie@lbl.gov) or to K.K. (email: kyriakos@me.berkeley.edu).

Despite significant research devoted to the exploration of new types of batteries[1–3], lithium-ion batteries (LIBs) remain the most extensively used power source for various applications, such as portable electronics, electric vehicles and long-term energy storage. In common with traditional batteries, LIBs have two electrodes that reversibly host lithium-ion insertion and extraction[4]. Novel electrode materials, such as silicon, have been proposed as a promising upgrade for the current graphite carbon-based electrodes because of their improved gravimetric energy density[5]. However, the larger capacity density implies that the silicon electrode must host more insertion and extraction of lithium ions during lithiation/delithiation cycling. As a result, the silicon electrode experiences excessive volume expansion and contraction cyclically, which induces irreversible electrode deformation and fracture. Consequently, the mechanical degradation of the silicon electrode results in severe capacity and power fade, thereby greatly limiting the battery's long-term durability for critical applications, such as power systems of electric vehicles. Accordingly, here we seek to provide a fundamental understanding of the failure mechanisms of silicon electrodes in LIBs over extended cycles, to provide guidance for new and improved electrode design with minimal capacity decay.

It has been suggested that decreasing electrode material size to nanostructures in the form of nanoparticles, nanowires or nanotubes is a promising strategy for preventing electrode failure[5–7]. In addition, critical sizes of 'fracture-free' nanoparticle and nanowire electrodes have actually been proposed[8,9]. These delicate nanostructured silicon electrodes, however, are still not ready for commercialization due to several reasons. First, the coulombic efficiency of these nanostructures in the first cycle is naturally very poor because their high surface-to-volume ratio causes more lithium ions to become trapped in the solid electrolyte interphase (SEI). Second, the much higher manufacturing cost of nanostructured electrodes may offset the competitive advantages associated with their presumed improved capacity. Third, the size of silicon particles that commercial vendors can manufacture with reasonable control is still larger than the estimated 'fracture-free' size. Consequently, the salient mechanisms controlling the fracture and hence the lifetime of silicon electrodes of LIBs remain an unavoidable issue that must be understood. To this end, various theoretical models have been developed and experimental studies performed to elucidate the fracture mechanisms of both crystalline and amorphous silicon electrodes[10–13]. However, there have been few attempts to investigate the physical mechanisms underlying fracture behaviour of these electrodes at more realistic larger scales (that is, from micrometres to centimetres), especially during long-term cycling[14].

In this study, we combine chemical and electrochemical experiments with fracture mechanics and the finite element method (FEM) to investigate the electrochemical and mechanical response of single-crystal silicon electrodes subjected to long-term cycling. Single-crystal silicon is chosen because it provides an ideal model surface and bulk material; moreover, as standard electrochemistry measurements can be readily made, it is possible to track the development of a crack in the electrode and, most importantly, identify its trajectory over extended cycles. The simulated stress/strain contours and predicted progress of the crack paths are shown to be consistent with the experimental observations. Specifically, with increasing number of cycles, perpendicular cracks initiate at the electrode surface, propagate in the vertical (thickness) direction and eventually deflect along the lithiation boundary causing delamination. We believe that these results have significant implications to the understanding of the progressive failure of silicon electrodes and provide guidance for the development of design strategies that can mitigate the degradation and failure of silicon electrodes in LIBs.

Results

Voltammetry and evolution of electrode surface morphology. Figure 1a–c shows cyclic voltammetry curves of a p-type boron-doped Si(100) electrode subjected to 30 cycles of voltage between 2.0 and 0.01 V at a scan rate of $0.1\,mV\,s^{-1}$. (All the electrochemical tests were performed in a custom-made reaction cell shown in Supplementary Fig. 1.) The onset of the reduction current is observed in the first cycle at ~0.1 V, corresponding to the initial alloying of the crystalline silicon with lithium, while in the second cycle the lithiation reduction peak shifts to ~0.3 V (Fig. 1a). The two anodic peaks at ~0.3 and ~0.5 V are associated with Li dissolution[15–17]. The magnitude of the oxidation and reduction peaks increases with the number of cycles (Fig. 1b), which can be attributed to the continuous increase of the amount of electrochemically active silicon in each cycle. The kinetics of lithiation in silicon is analogous to the model of SiO_2 layer formation on silicon[18]. During the initial cycles (Fig. 1a,b), the increase of active silicon is controlled by the lithiation reaction rate; thus the current density increases with the number of cycles (that is, lithiation time). Both the lithiation depth and the time needed for lithium to transport from the surface to the reaction front increase with cycling. After 30 cycles, the time for lithium transport approaches the cycle time and the cyclic voltammetry curve stabilizes (Fig. 1c). With continued cycling of the silicon electrode, the redox current area begins to decrease, indicating the loss of active silicon material.

Here we characterize the evolution of the surface morphology of the silicon electrode using top-view scanning electron microscope (SEM) images obtained after 3, 8 and 50 cycles (Fig. 1d–f) and magnified SEM images of the electrode surface obtained after 30 cycles (Fig. 1g–i). The dominant feature is the development of surface cracks in two orthogonal directions, resulting in isolated small squares that form a regular repetitive pattern on the electrode surface. These crack patterns further account for the formation of square cavities (Fig. 1f). Apparently, the damage to the electrode surface is not a rapid and catastrophic process, but occurs in a gradual and cumulative fashion. While damage of the electrode surface is insignificant during the first few cycles, shallow surface cracks initiate after a critical number of cycles and propagate with further cycling, eventually causing delamination at the electrode surface.

Formation of orthogonal surface cracks. To explain the origin of these orthogonal surface cracks, we need to consider that lithiation in single-crystal silicon is a strongly orientation-dependent process. Although the entry of lithium ions into the silicon electrode occurs perpendicular to the electrode surface (that is, the (100) direction), further lithium flux in the electrode bulk occurs in all directions. In particular, lithiation invasion preferentially occurs in the <110> direction and is significantly less in the <100> and <111> directions[19]. Figure 2a shows a top-view SEM image of $3 \times 3 \times 8\,\mu m^3$ micropillars fabricated by photolithography on a Si(100) substrate with exposed {110} lateral surfaces and round corners in the <100> direction. (Top-view and three-dimensional micropillar images are also shown in Supplementary Fig. 2.) With increasing lithiation, the straight edges of the micropillars preferentially expand in the <110> directions (Fig. 2b–d), while the round corners expand significantly less. This anisotropic expansion makes the 'rotation' of the micropillars in the <100> direction to vanish (Fig. 2d,e) and the neighbouring micropillars to merge (Fig. 2f). If a micropillar is presumed to represent a 'unit cell' of the solid

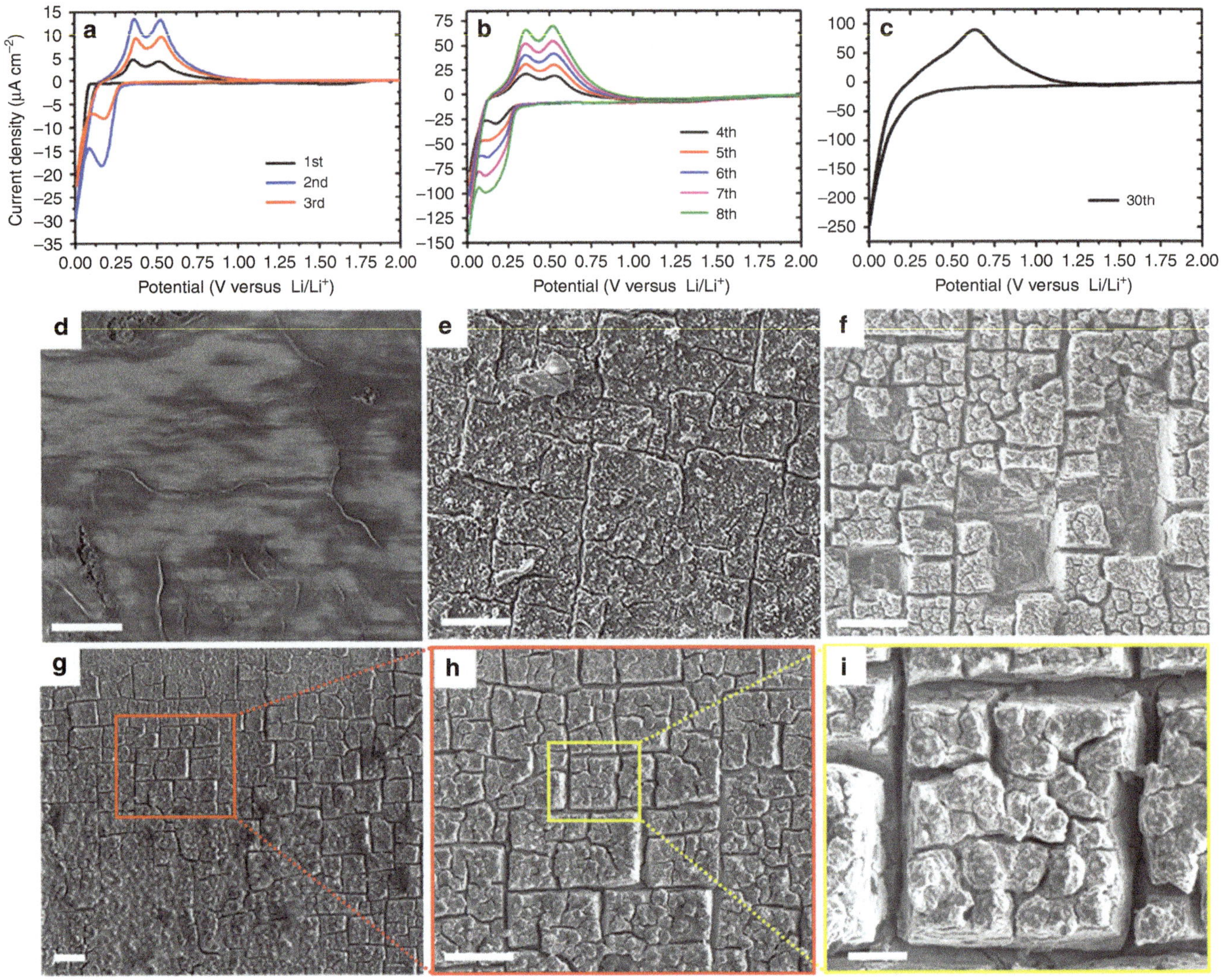

Figure 1 | Electrode cyclic voltammetry response and surface morphology. Current-voltage curves of a Si(100) electrode cycled between 2.0 and 0.01 V at a scan rate of $0.1\,mV\,s^{-1}$ for (**a**) 1-3, (**b**) 4-8 and (**c**) 30 successive lithiation/delithiation cycles. Top-view scanning electron microscope images of electrode surface morphology obtained after (**d**) 3, (**e**) 8, (**f**) 50 and (**g-i**) 30 cycles. Scale bar, 20 μm (**d-h**); 3 μm (**i**).

Si(100) electrode, the volumetric expansion during cycling will be confined by the surrounding material, resulting in the development of a compressive force at the micropillar edges. Because of the significantly higher lithiation rate in the <110> directions than the <100> directions, the 'unit cell' is subjected to a higher compressive force on all the {110} planes, which, in turn, induces a high shear stress and plasticity along the ±45°directions, that is, along the {100} planes.

To validate the above hypothesis, a planar octahedral model of the 'unit cell' of the silicon electrode was developed and analysed with the FEM (Fig. 2g). (More details about the 'unit cell' model and associated boundary conditions can be found in Supplementary Methods and related Supplementary Fig. 3.) Lithiation-induced volumetric expansion was simulated as a thermal expansion process. A moving temperature field was applied to model lithium invasion and migration into the silicon electrode. The unlithiated (crystalline) silicon (c-Si) was modelled as an isotropic elastic material of elastic modulus $E = 185\,GPa$ and Poisson's ratio $v = 0.22$, whereas the lithiated (amorphous) silicon (a-Si) was modelled as an elastic–plastic material with material properties depending on lithium-ion concentration (that is, $E = 50–102\,GPa$, $v = 0.22$ and yield strength $Y = 0.5–3.0\,GPa$)[13]. Because the pillar height is much larger than the in-plane pillar dimensions and the expansion in the (100) direction is less than that along the in-plane <110> directions, plane-strain conditions were assumed in all simulations. Although the axial elongation is not exactly zero, cross-sectional SEM images of the lithiated crystalline pillars revealed a negligible change in pillar height compared with in-plane expansion, providing experimental evidence of the validity of the plane-strain assumption. Figure 2g shows the deformed configuration of a partially lithiated micropillar, including contours of equivalent plastic strain. Due to the anisotropic lithiation rate (lithium invasion in the (110) direction is 6.4 times faster than that in the (100) direction[11]), the expansion is anisotropic, in agreement with our experimental observations (Fig. 2b,c) and those of others[10,18]. This anisotropic expansion leads, in turn, to the development of a high shear stress, which is responsible for crack initiation. Figure 2h shows the equivalent plastic strain in a 'unit cell' of the electrode after full lithiation. Because volumetric expansion of the 'unit cell' is fully constrained by the surrounding material, large plastic strains develop along the (100) edges, contributing to the formation of perpendicular stress bands along the <100> directions.

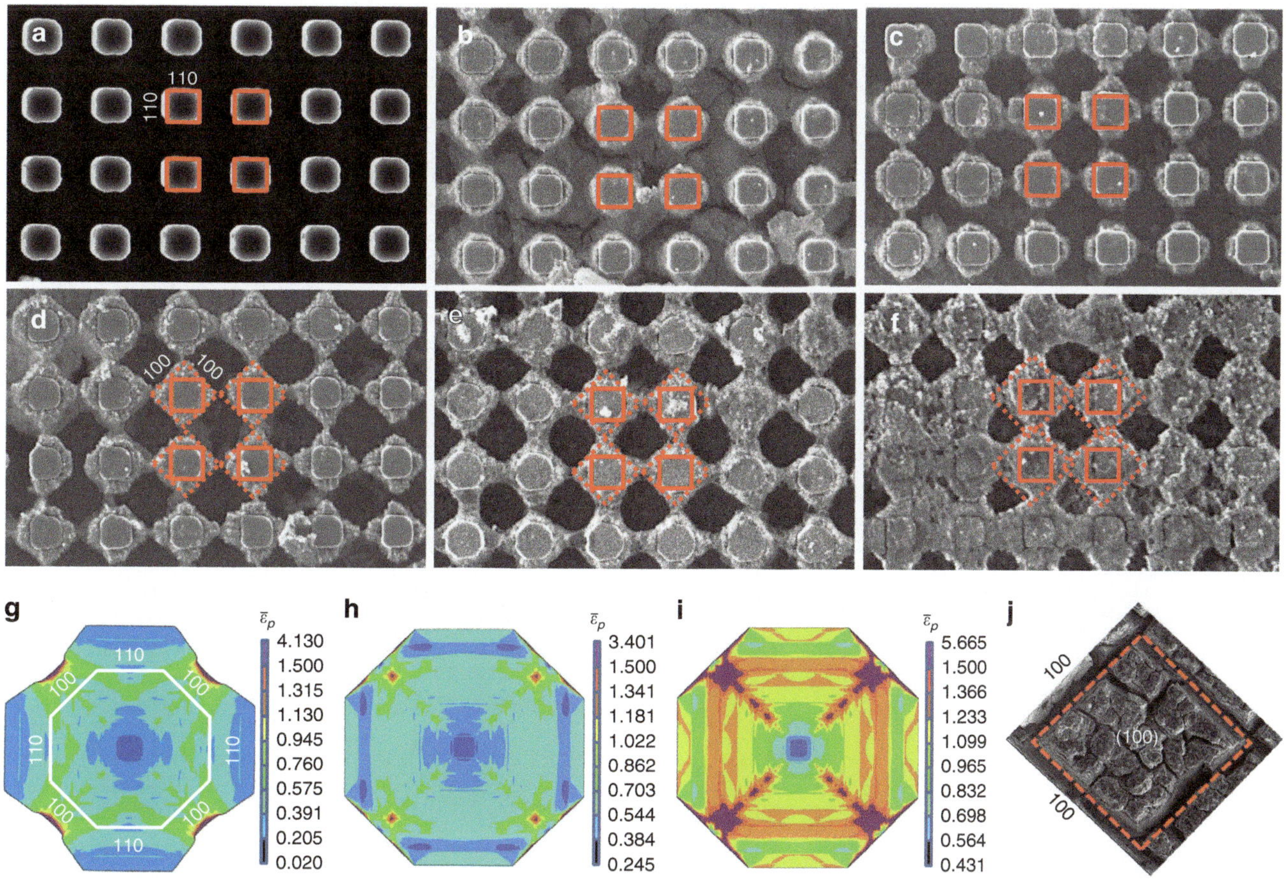

Figure 2 | Anisotropic in-plane deformation and fracture of lithiated Si(100) micropillars and electrodes. (a–f) Scanning electron microscope (SEM) images of square micropillars fabricated on a p-type Si(100) wafer obtained at various lithiation stages (the size of the red solid squares is $3 \times 3\,\mu m^2$). Finite element results of in-plane equivalent plastic strain $\bar{\varepsilon}_p$ distribution in **(g)** lithiated (unconstrained) micropillar and **(h)** lithiated and **(i)** delithiated (constrained) octahedral 'unit cell' of the silicon electrode. **(j)** SEM image showing the crack pattern on the surface of a lithiated silicon electrode revealing crack formation in the <100> directions.

Figure 2i shows equivalent plastic strain contours after full delithiation, indicating that during delithiation the locations of large plastic strain continue to accumulate plasticity as a pair of perpendicular deformation bands. The field of the equivalent plastic strain in the electrode bulk can be assembled by packing the equivalent plastic strain contours in the 'unit cell' to construct the high-plastic-strain bands at larger scales, in agreement with the experimental observation of repeated perpendicular surface cracks at different length scales.

Surface crack initiation, propagation and deflection. After the formation of cracks perpendicular to the electrode surface (Fig. 1e,g–i), small squares of electrode material defined by these surface cracks begin to delaminate (Fig. 1f). To understand the origin of this process, it is necessary to consider how these surface cracks form and progressively propagate with cycling before causing electrode material to delaminate. This can be accomplished by analysing the experimental evidence of surface crack growth during the first 50 lithiation/delithiation cycles. Figure 3a–d (left) shows cross-sectional SEM images across a typical surface crack after 3, 8, 30 and 50 cycles, respectively. The crack initiates at the electrode surface (Fig. 3b), consistent with the fracture of single-crystal silicon nanopillars[19], propagates in the depth direction with further cycling (Fig. 3c) and, after reaching a depth of ~ 5–$8\,\mu m$, is deflected laterally (Fig. 3d). The reason for the formation of square delamination fragments of electrode material is the lateral deflection of neighbouring cracks after prolonged cycling (Fig. 1f).

Despite the clear evidence from cross-sectional SEM images of the observed crack trajectories, the question that remains is the physics and mechanics underlying such behaviour; specifically, why do cracks initiate at the electrode surface and why do such initially vertical cracks deflect laterally? To seek answers to these questions, we analysed the lithiation/delithiation-induced stress and strain fields in the cross-section of the silicon electrode over multiple cycles. A cohesive zone model (Supplementary Fig. 4; Supplementary Methods; Supplementary Note 1) was used to represent multiple interfaces, which were allowed to separate at a critical stress and thus simulate crack initiation and growth. These interfaces are characterized by the cohesive strength σ_c and the interface work of adhesion G_c (fracture energy). An approximately linear lithium concentration profile (Supplementary Fig. 5; Supplementary Note 2) was simulated by the gradual advancement of the lithiation boundary with increasing lithiation/delithiation cycles.

As shown in Fig. 3a (middle column), after three cycles both the lithiation depth d ($\sim 1\,\mu m$) and plastic deformation are mainly confined at the electrode surface. With continued cycling, the lithiation depth increases with more plasticity accumulating below the surface owing to further lithium insertion. During the subsequent delithiation, the tensile residual stresses generated in the large plastic region induce face separation along the cohesive interface, resulting in the initiation and propagation of vertical

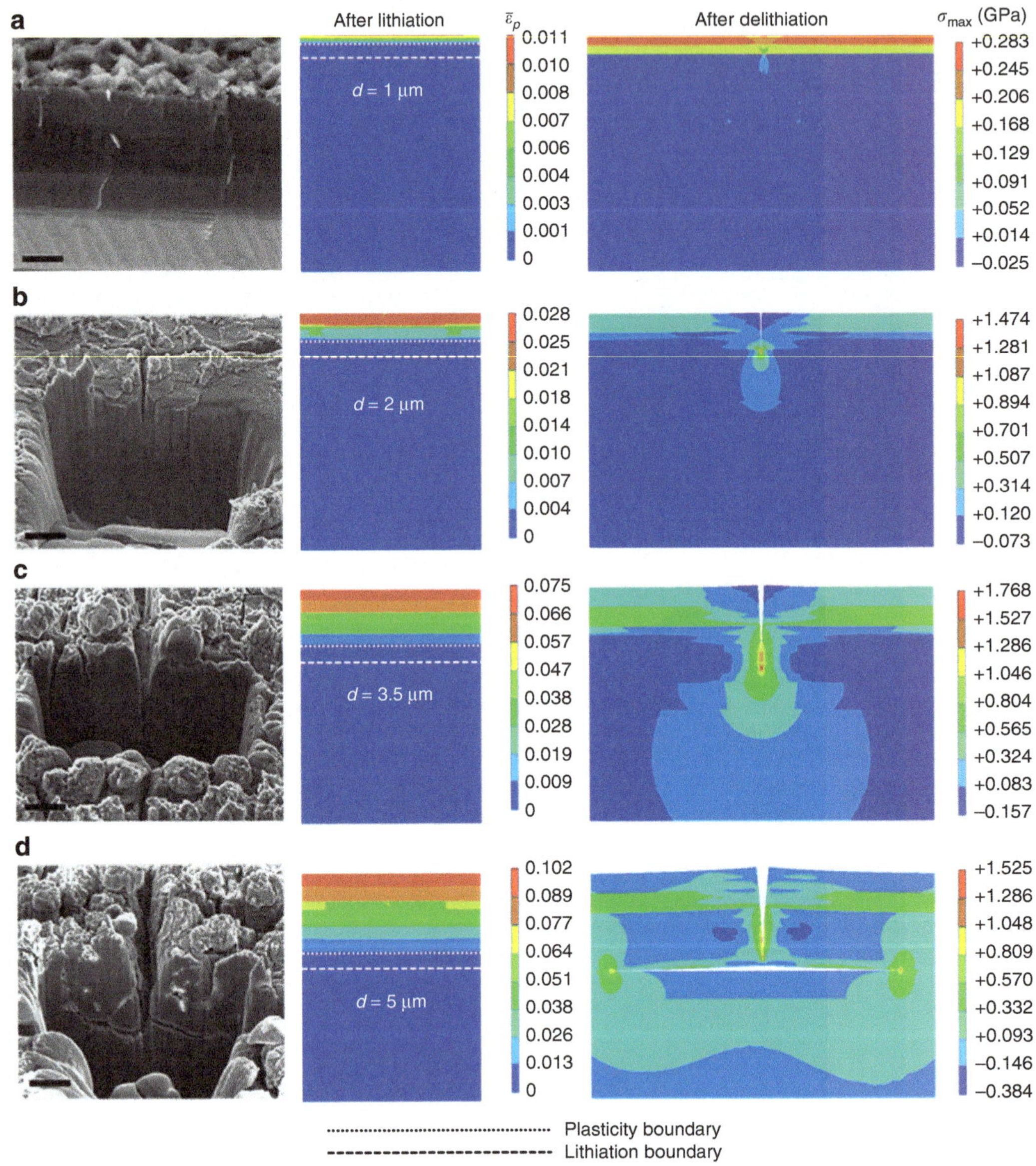

Figure 3 | Experimental and numerical results of electrode fracture. Cross-sectional focused ion beam (FIB)-scanning electron microscope (SEM) images (left column) and corresponding finite element method (FEM) results (middle and right columns) of a Si(100) electrode obtained after (**a**) 3, (**b**) 8, (**c**) 30 and (**d**) 50 lithiation/delithiation cycles. The FIB-SEM images show crack growth through the electrode thickness followed by crack deflection and propagation along the lithiation boundary (lithiated/unlithiated interface). The FEM results show the equivalent plastic strain $\bar{\varepsilon}_p$ after lithiation and the maximum principal stress σ_{max} after delithiation, illustrating crack initiation, propagation through the electrode thickness and lateral deflection along the lithiation boundary for different lithiation/delithiation cycles. Scale bar, 0.5 μm (**a**); 3 μm (**b**-**d**).

cracks (Fig. 3b,c, right). When the crack tip approaches the lithiation boundary, the crack is abruptly deflected laterally and continues to propagate along the a-Si/c-Si interface (Fig. 3d, right), which is consistent with the experimental evidence (Fig. 3d, left). Such a marked deflection in crack trajectory can be interpreted in terms of the mutual competition between the direction of maximum mechanical driving force and the weakest microstructural path[20]. Specifically, from a mechanics perspective, cracks in nominally brittle (elastic) materials follow the path of maximum strain energy release rate. This is essentially consistent with a $K_{II} = 0$ crack trajectory, where K_{II} is the stress intensity factor for in-plane shear crack displacement, although this is mitigated in real materials by the nature of the microstructure encountered by the crack. In the current system pertaining to silicon electrodes of LIBs, although the elastic modulus of c-Si (185 GPa) is higher than that of a-Si (50–102 GPa), its fracture toughness $G_c \approx 9\,J\,m^{-2}$ ($K_{IC} \approx 1.3\,MPa\,m^{1/2}$) is an order of magnitude higher than that of the a-Si/c-Si interface in the (100) direction, whose estimated fracture toughness is $G_c \approx 1\,J\,m^{-2}$ ($K_{IC} \approx 0.3\,MPa\,m^{1/2}$)[21]; accordingly, the crack is deflected along the a-Si/c-Si interface, because this is both energetically more favourable and the weakest microstructural path.

Fracture mechanics-based interpretation. Such deliberations of the behaviour of cracks impinging on dissimilar (elastic) material interfaces can be explained in mechanics terms by the analysis of He and Hutchinson[22], which provides a quantitative criterion to predict whether a singular crack will penetrate through or deflect along a linear-elastic bimaterial interface. As depicted in Fig. 4, the crack path depends on two principal factors, namely, the elastic

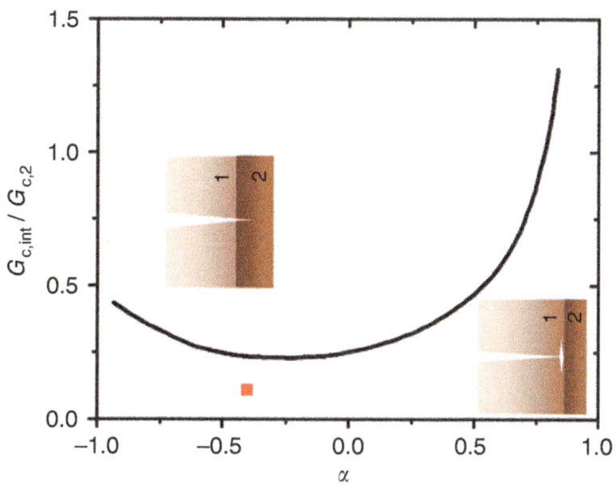

Figure 4 | Crack deflection at a bimaterial interface. Linear-elastic solution of interface to material beyond the interface toughness $G_{c,int}/G_{c,2}$ ratio versus Dundurs' parameter $\alpha = (E_1 - E_2)/(E_1 + E_2)$, where E_1 and E_2 are the elastic modulus of material 1 and 2, respectively[22]. Whether the crack penetrates through the interface or is deflected along the interface depends on the elastic modulus mismatch (represented by the Dundurs' parameter) and interface-to-material 2 (c-Si) toughness ratio. For the a-Si/c-Si interface, $\alpha \approx -0.4$ and $G_{c,int}/G_{c,2} \approx 0.11$ (red square point). The theory predicts that a crack growing in the a-Si will not propagate into the c-Si, but will be deflected along the a-Si/c-Si interface, in agreement with the experimental observations (Fig. 3, left column).

modulus mismatch across the interface, described by the Dundurs' parameter $\alpha = (E_1 - E_2)/(E_1 + E_2)$, where E_1 and E_2 are the respective elastic moduli of material 1 and 2, and the ratio of the toughness (critical strain energy release rate) of the interface to that of the material beyond the interface $G_{c,int}/G_{c,2}$. By substituting the material and interfacial properties of the present system (given in the preceding paragraph) into this analysis, we can calculate that the modulus mismatch α for the lithiated/unlithiated a-Si/c-Si interface is approximately equal to -0.4, with a toughness ratio $G_{c,int}/G_{c,2} \approx 0.11$ (red square point in Fig. 4). Accordingly, it is clear from Fig. 4 that, from a mechanical driving force perspective, the crack will definitely deflect along the a-Si/c-Si interface.

Inspired by the mechanistic understanding of the silicon electrode failure, we proceeded to investigate if electrolyte additive, which is typically used to modify the chemical properties of the SEI[23,24], can affect crack growth and, in turn, electrode degradation. A Fourier transform infrared (FTIR) spectroscopy investigation has shown that the reduction product of vinylene carbonate (VC) additive is poly-VC, whereas that of the fluoroethylene carbonate (FEC) additive is alkyl carbonate $ROCO_2Li$ salt[25–27]. Both of these reaction products can chemically passivate the electrode surface before the decomposition of the electrolyte solvent. The poly-VC aggregates fill the vertical surface cracks (Fig. 5a,b) and promote crack-face bridging (Fig. 5c). As a result, the fracture toughness is increased and further crack propagation is inhibited. The crack depth after 30 cycles is ~100 nm, which is significantly smaller than the 5–8 μm crack depths observed with additive-free electrolyte consisting of 1 M lithium hexafluorophosphate ($LiPF_6$) in ethylene carbonate (EC)/diethyl carbonate (DEC) (EC:DEC = 1:2 v/v). The $ROCO_2Li$ salt uniformly precipitates on the electrode surface (Fig. 5d,e), as evidenced by the roughening of the electrode surface. Although surface cracks are still visible, they do not perfectly align in perpendicular directions as for the additive-free electrolyte (Fig. 1g), because $ROCO_2Li$ continues to form on the exposed edges and faces (Fig. 5f). These $ROCO_2Li$ salts serve to fill and passivate the surface cracks, thereby significantly decelerating crack growth. We believe that these findings will stimulate research to formulate new electrolyte additives for enhancing the electrode fracture resistance.

Discussion

In addition to the electrolyte additive strategy described above, the fracture mechanics model introduced in this study can also

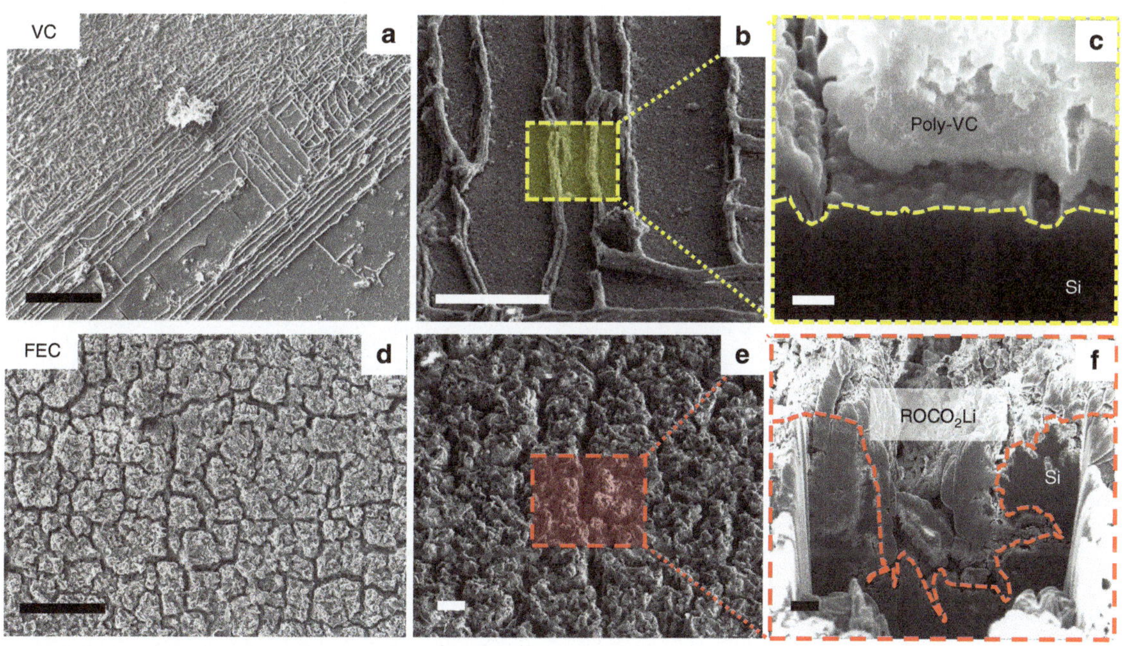

Figure 5 | Electrode crack arrest by the reduction products of electrolyte additives. Top-view and cross-sectional scanning electron microscope images of a Si(100) electrode cycled between 2.0 and 0.01 V for 30 cycles at a scan rate of 0.01 mV s^{-1} in EC/DEC/lithium hexafluorophosphate ($LiPF_6$) electrolyte with (**a–c**) 5% vinylene carbonate (VC) additive and (**d–f**) 5% fluoroethylene carbonate (FEC) additive. The dashed rectangles shown in **b** and **e** indicate the focused-ion-beam milling area. The reduction products of the VC and FEI additives (poly-VC and $ROCO_2Li$, respectively) modify the mechanical properties of the solid electrolyte interphase (SEI) film and inhibit crack growth. Scale bar, 100 μm (**a,d**); 20 μm (**b,e**); 300 nm (**c**); 3 μm (**f**).

lead to other approaches for improving the electrode robustness and durability, including: (i) Surface modification (artificial SEI design). Recent FTIR spectroscopy studies[28–30] have shown that the main components of the SEI forming on silicon electrodes have high solubility in the electrolyte and very weak mechanical strength. A thin protective film of enhanced chemical stability, high fracture toughness and strong adhesion to the electrode surface can minimize the surface crack initiation by reducing the surface plasticity[7]. (ii) Electrode material selection. Despite its high capacity, the silicon electrode material is naturally brittle and prone to fracture. Emerging ductile electrode materials (for example, silicon-metal alloy) and structurally optimized electrode materials (for example, porous/hollow structure composite with tuned binder) may significantly reduce the risk of electrode fracture[6,31,32]. (iii) Charging profile optimization. Secondary ion mass spectrometry (SIMS) shows approximately linear decrease in lithium concentration through the electrode thickness (Supplementary Fig. 5). By tuning the charging profile, we can obtain the same capacity with a uniform lithium distribution through the electrode thickness. This would allow a lower surface lithium concentration, thereby reducing surface plasticity, which is a key factor for suppressing crack initiation and propagation.

The present work provides a mechanistic explanation of the root cause of premature silicon electrode failure in LIBs following prolonged lithiation/delithiation cycling. On the basis of this insight, we demonstrate how electrolyte additives change the SEI, tune the mechanical properties of the cohesive layer and improve the electrode fracture resistance. In addition, the present failure analysis may also guide the optimization of other strategies, including surface modification, electrode material development and tuning of the charge profile to increase the material ductility and suppress the high surface stress. We consider the knowledge of the failure mechanisms of silicon electrodes gained from the present study as a foundation for future design improvements and the aforementioned strategies as main paths towards the development of more robust and durable electrodes for next-generation LIBs.

Methods

Electrode preparation. The silicon electrodes used in this investigation were p-type boron-doped Si(100) wafers of 500 μm thickness and 0.001 Ω cm electrical resistivity (MTI Co.). The native oxide film was removed by first treating with diluted 5% hydrofluoric acid and then rinsing with ultrapure water of 18.2 MΩ cm electrical resistivity for 2 min. The $3 \times 3 \times 8\ \mu m^3$ micropillar electrodes used in this study (Supplementary Fig. 2) were fabricated by standard photolithography. Reactive ion etching was performed with the Bosch process.

Electrochemical measurements. All of the electrochemical experiments were performed in a single-compartment teflon cell (Supplementary Fig. 1). The electrolyte consisted of 1 M $LiPF_6$/EC/DEC (EC:DEC = 1:2 v/v) (Novolyte Technologies, BASF) with or without 5% VC or FEC additive. Dimethyl carbonate (DMC) (>99.9%, high-performance liquid chromatography grade, Sigma-Aldrich) was used as the rinsing solvent. The counter and reference electrodes were made of Li, whereas the working electrodes were cut from Si(100) wafers with or without micropillars. All potentials reported in this work are referred to the Li/Li^+ redox couple. All electrochemical tests were performed inside a glove box filled with Ar gas (H_2O and O_2 contents <10 p.p.m.). A multi-channel potentiostat (Multistat 1480, Salartron Analytical) was used for potential/current control.

Surface morphology characterization. After electrochemical treatment, the silicon samples were first rinsed with DMC to remove the residual electrolyte and then transferred to an SEM (JSM-6700F, JEOL) for imaging. The electrode microstructure was observed with a focused ion beam (Nova 600i Dual Beam, FEI). The cycled electrodes were cross-sectioned with a Ga^+ ion beam and observed with the SEM.

Secondary ion mass spectrometry. SIMS measurements of the lithiated silicon electrodes were obtained by Evans Analytical Group. The Cameca IMS 4f magnetic sector of the SIMS instrument was operated at a pressure of 2.67×10^{-7} Pa (2×10^{-9} torr). Elemental concentration depth profiles were obtained by analysing positively charged CsLi and CsSi secondary ions. The Cs^+ current was typically equal to 10 nA. The sputtering depth was determined from the depth of the sputtering craters measured with a profilometer. Data acquisition and post-processing analysis were performed with the SIMSview software of the SIMS instrument. The lithiated samples were sealed in a glove box and transferred to the SIMS spectrometer within 1 min.

Finite element analysis. The general purpose FEM code ABAQUS/Standard (version 6.14) was used to determine the in-plane stress and strain fields resulting from anisotropic lithiation/delithiation and to analyse crack initiation, propagation, deflection and delamination over multiple lithiation/delithiation cycles. The unlithiated silicon (c-Si) was modelled as an isotropic elastic material with $E = 185$ GPa and $\nu = 0.23$, whereas the lithiated silicon (a-Si) was modelled as an isotropic elastic-perfectly plastic material with elastic modulus and yield strength varying with the lithium-ion concentration.

References

1. Tarascon, J.-M. & Armand, M. Issues and challenges facing rechargeable lithium batteries. *Nature* **414**, 359–367 (2001).
2. Armand, M. & Tarascon, J.-M. Building better batteries. *Nature* **451**, 652–657 (2008).
3. van Noorden, R. A better battery. *Nature* **507**, 26–28 (2014).
4. Goodenough, J. B. & Park, K.-S. The Li-ion rechargeable battery: a perspective. *J. Am. Chem. Soc.* **135**, 1167–1176 (2013).
5. Chan, C. K. *et al.* High-performance lithium battery anodes using silicon nanowires. *Nat. Nanotechnol.* **3**, 31–35 (2008).
6. Magasinski, A. *et al.* High-performance lithium-ion anodes using a hierarchical bottom-up approach. *Nat. Mater.* **9**, 353–358 (2010).
7. Wu, H. *et al.* Stable cycling of double-walled silicon nanotube battery anodes through solid-electrolyte interphase control. *Nat. Nanotechnol.* **7**, 310–315 (2012).
8. Liu, X. H. *et al.* Size-dependent fracture of silicon nanoparticles during lithiation. *ACS Nano* **6**, 1522–1531 (2012).
9. Ryu, I., Choi, J. W., Cui, Y. & Nix, W. D. Size-dependent fracture of Si nanowire battery anodes. *J. Mech. Phys. Solids* **59**, 1717–1730 (2011).
10. Pharr, M., Zhao, K., Wang, X., Suo, Z. & Vlassak, J. J. Kinetics of initial lithiation of crystalline silicon electrodes of lithium-ion batteries. *Nano Lett.* **12**, 5039–5047 (2012).
11. Pharr, M., Suo, Z. & Vlassak, J. J. Measurements of the fracture energy of lithiated silicon electrodes of Li-ion batteries. *Nano Lett.* **13**, 5570–5577 (2013).
12. Chon, M. J., Sethuraman, V. A., McCormick, A., Srinivasan, V. & Guduru, P. R. Real-time measurement of stress and damage evolution during initial lithiation of crystalline silicon. *Phys. Rev. Lett.* **107**, 045503 (2011).
13. Berla, L. A., Lee, S. W., Cui, Y. & Nix, W. D. Mechanical behavior of electrochemically lithiated silicon. *J. Power Sources* **273**, 41–51 (2015).
14. McDowell, M. T., Lee, S. W., Nix, W. D. & Cui, Y. Understanding the lithiation of silicon and other alloying anodes for lithium-ion batteries. *Adv. Mater.* **25**, 4966–4985 (2013).
15. Green, M., Fielder, E., Scrosati, B., Wachtler, M. & Moreno, J. S. Structured silicon anodes for lithium battery applications. *Electrochem. Solid-State Lett.* **6**, A75–A79 (2003).
16. Long, B. R., Chan, M. K. Y., Greeley, J. P. & Gewirth, A. A. Dopant modulated Li insertion in Si for battery anodes: theory and experiment. *J. Phys. Chem. C* **115**, 18916–18921 (2011).
17. Schroder, K. W., Celio, H., Webb, L. J. & Stevenson, K. J. Examining solid electrolyte interphase formation on crystalline silicon electrodes: influence of electrochemical preparation and ambient exposure conditions. *J. Phys. Chem. C* **116**, 19737–19747 (2012).
18. Deal, B. E. & Grove, A. S. General relationship for the thermal oxidation of silicon. *J. Appl. Phys.* **36**, 3770–3778 (1965).
19. Lee, S. W., McDowell, M. T., Berla, L. A., Nix, W. D. & Cui, Y. Fracture of crystalline silicon nanopillars during electrochemical lithium insertion. *Proc. Natl. Acad. Sci. USA* **109**, 4080–4085 (2012).
20. Ritchie, R. O., Cannon, R. M., Dalgleish, B. J., Dauskardt, R. H. & McNaney, J. M. Mechanics and mechanisms of crack growth at or near ceramic-metal interfaces: interface engineering strategies for promoting toughness. *Mater. Sci. Eng. A* **166**, 221–235 (1993).
21. Jung, S. C., Choi, J. W. & Han, Y.-K. Anisotropic volume expansion of crystalline silicon during electrochemical lithium insertion: an atomic level rationale. *Nano Lett.* **12**, 5342–5347 (2012).
22. He, M.-Y. & Hutchinson, J. W. Crack deflection at an interface between dissimilar elastic materials. *Int. J. Solids Struct.* **25**, 1053–1067 (1989).
23. Chen, L., Wang, K., Xie, X. & Xie, J. Enhancing electrochemical performance of silicon film anode by vinylene carbonate electrolyte additive. *Electrochem. Solid-State Lett.* **9**, A512–A515 (2006).

24. Choi, N.-S. *et al.* Effect of fluoroethylene carbonate additive on interfacial properties of silicon thin-film electrode. *J. Power Sources* **161,** 1254–1259 (2006).
25. Shi, F., Zhao, H., Komvopoulos, K., Somorjai, G. & Ross, P. N. in *227th Electrochemical Society Meeting (Abstract no 49912)* 24-28 May, (Chicago, IL, 2015).
26. Xu, C. *et al.* Improved performance of the silicon anode of Li-ion batteries: understanding the surface modification mechanism of fluoroethylene carbonate as an effective electrolyte additive. *Chem. Mater.* **27,** 2591–2599 (2015).
27. Dalavi, S., Guduru, P. & Lucht, B. L. Performance enhancing electrolyte additives for lithium ion batteries with silicon anodes. *J. Electrochem. Soc.* **159,** A642–A646 (2012).
28. Shi, F. *et al.* A catalytic path for electrolyte reduction in lithium-ion cells revealed by *in situ* attenuated total reflection-Fourier transform infrared spectroscopy. *J. Am. Chem. Soc.* **137,** 3181–3184 (2015).
29. Shi, F. *et al.* Identification of diethyl 2,5-dioxahexane dicarboxylate and polyethylene carbonate as decomposition products of ethylene carbonate based electrolytes by Fourier transform infrared spectroscopy. *J. Phys. Chem. C* **118,** 14732–14738 (2014).
30. Shi, F., Komvopoulos, K., Ross, P. N. & Somorjai, G. A. In-situ ATR-FTIR investigation of the solid electrolyte interface (SEI) on single-crystal Si anodes for Li ion batteries. *Prepr. Pap. - Am. Chem. Soc. Div. Energy Fuels* **59,** 486 (2014).
31. Mazouzi, D. *et al.* Critical roles of binders and formulation at multiscales of silicon-based composite electrodes. *J. Power Sources* **280,** 533–549 (2015).
32. Han, Z.-J. *et al.* High-capacity Si-graphite composite electrodes with a self-formed porous structure by a partially neutralized polyacrylate for Li-ion batteries. *Energy Environ. Sci.* **5,** 9014–9020 (2012).

Acknowledgements

This work was supported by the Assistant Secretary for Energy Efficiency and Renewable Energy, Office of Freedom CAR and Vehicle Technologies, US Department of Energy under contract no. DE-AC02 05CH1123. K.K. also acknowledges the funding provided for this work by the UCB – KAUST Academic Excellence Alliance Program. The potentiostat instrumentation was purchased with funding from the Director, Office of Science, Office of Basic Energy Sciences, Materials Science and Engineering Division, US Department of Energy, which also provided support to R.O.R. We thank C. Shen for assistance in micropillar sample preparation.

Authors contributions

F.S. designed and performed the experiments, Z.S. performed the FEM simulations, F.S., Z.S, P.N.R., G.A.S., R.O.R and K.K. analysed the results, F.S., Z.S, R.O.R and K.K. prepared and revised the manuscript, and K.K. supervised and directed the work of F.S. and Z.S. throughout this project.

Additional information

Competing financial interests: The authors declare no competing financial interests.

Visualizing non-equilibrium lithiation of spinel oxide via *in situ* transmission electron microscopy

Kai He[1,*], Sen Zhang[2,*], Jing Li[1], Xiqian Yu[1], Qingping Meng[1], Yizhou Zhu[3], Enyuan Hu[1], Ke Sun[1], Hongseok Yun[2], Xiao-Qing Yang[1], Yimei Zhu[1], Hong Gan[1], Yifei Mo[3], Eric A. Stach[1], Christopher B. Murray[2] & Dong Su[1]

Spinel transition metal oxides are important electrode materials for lithium-ion batteries, whose lithiation undergoes a two-step reaction, whereby intercalation and conversion occur in a sequential manner. These two reactions are known to have distinct reaction dynamics, but it is unclear how their kinetics affects the overall electrochemical response. Here we explore the lithiation of nanosized magnetite by employing a strain-sensitive, bright-field scanning transmission electron microscopy approach. This method allows direct, real-time, high-resolution visualization of how lithiation proceeds along specific reaction pathways. We find that the initial intercalation process follows a two-phase reaction sequence, whereas further lithiation leads to the coexistence of three distinct phases within single nanoparticles, which has not been previously reported to the best of our knowledge. We use phase-field theory to model and describe these non-equilibrium reaction pathways, and to directly correlate the observed phase evolution with the battery's discharge performance.

[1] Brookhaven National Laboratory, Upton, New York 11973, USA. [2] Department of Chemistry, University of Pennsylvania, Philadelphia, Pennsylvania 19104, USA. [3] Department of Materials Science and Engineering, University of Maryland, College Park, Maryland 20742, USA. * These authors contributed equally to this work. Correspondence and requests for materials should be addressed to C.B.M. (email: cbmurray@sas.upenn.edu) or to D.S. (email: dsu@bnl.gov).

The spinel transition metal oxide family—which includes $LiMn_2O_4$, $Li_4Ti_5O_{12}$, Fe_3O_4, Co_3O_4, Mn_3O_4, among others—is an important group of compounds, and one that sees considerable use as electrode materials in lithium-ion batteries[1–7]. Specifically, magnetite (Fe_3O_4) is a promising candidate as an anode material, as it is both inexpensive and non-toxic. Importantly, it also has a high electronic conductivity and can store up to eight Li ions per formula unit, which leads to a high theoretical specific capacity of 926 mAh g^{-1} (refs 8–20). The structure of Fe_3O_4 is inverse spinel ($(Fe^{3+})_{8a}[Fe^{2+}Fe^{3+}]_{16d}O_4$, denoted by Wyckoff notation), where there are 64 tetrahedral sites in a unit cell (one-eighth of them are 8a sites occupied by 8 Fe^{3+} cations) and 32 octahedral sites (half of them are 16d sites occupied by 8 Fe^{2+} and 8 Fe^{3+} cations), leaving behind 56 empty tetrahedral (8b and 48f) sites and 16 empty octahedral (16c) sites in the interstitial space to allow for accommodation of guest Li^+ ions[17–20]. The lithiation of spinel iron oxide was investigated by Thackeray *et al.*[16,17] using a combination of open-circuit voltage (OCV) measurements and X-ray diffraction, and they posited that the overall lithiation of Fe_3O_4 follows the reaction equation shown in equation (1):

$$8Li + Fe_3O_4 \rightarrow 4Li_2O + 3Fe \tag{1}$$

They suggested that during the initial stage of lithium insertion, the Fe^{3+} cations at tetragonal 8a sites are distorted into the nearby octahedral 16c sites by inserting Li^+ ions into the octahedral 16c sites, which leads to a lattice reconstruction of the spinel structure to form the rocksalt structure, that is, $(Li^+{}_xFe^{3+})_{16c}[Fe^{2+}Fe^{3+}]_{16d}O_4$ (refs 16–21). This is an intercalation process, and thus it does not lead to a significant volume change. They proposed that further lithiation triggers destruction of the rocksalt structure along with an extrusion of metallic Fe from the material, which is commonly known as a conversion reaction. During the conversion reaction, the rocksalt structure transforms into a composite which is composed of both metallic Fe and Li_2O. Critically, this conversion leads to a drastic volume expansion[9–15].

Fe_3O_4 shows dramatic differences in electrochemical performance as a function of discharge rate ('C rate'), with the intercalation process not readily observed in the discharge voltage profiles[9–11]. These differences cannot be understood from simple consideration of the reaction equation, which describes how these processes occur in an equilibrium condition. In one related example, the positive electrode material $LiFePO_4$ (LFP) exhibits a solid-solution transformation at high rates when undergoing an intercalation reaction, specifically when the two-phase transition process is suppressed above a certain overpotential[22,23]. Thus, it is suggested that for Fe_3O_4 the overall charge-discharge performance depends strongly on the kinetics of specific discharge mechanisms. This in turn determines the realistic redox reaction characteristics that the battery sees in working conditions.

Finally, it is well-known that the rate capability of electrodes can be improved by reducing the material size from the micrometre to the nanometre regime. This improvement is due to both the increase in surface area and the reduction of electron/ion diffusion length[24–26]. However, in the case of nanoscale electrodes, the overall kinetics of lithiation is further complicated by local inhomogeneities that can occur in the electrochemical conditions. Besides the presence of both intercalation and conversion reactions, all of these facts underscore the importance of understanding the phase transitions of Fe_3O_4 that occur in non-equilibrium conditions, as it is these parameters that must be optimized when considering spinel-based electrode materials in electrochemical applications[22].

In this work, we investigate how non-equilibrium intercalation and conversion reactions proceed by studying the lithiation process in real time. X-ray scattering or Mössbauer spectroscopy is most often used to identify the phase transformations that occur in nanosized electrode materials[16,19], but these techniques have the disadvantage of providing only average information, and generally they lack the ability to probe reaction kinetics in real time. Critically, it is important to undertake lithiation studies *in situ*, as *ex situ* studies may provide inaccurate information due to changes that may occur during removal of the electrode materials from an actively lithiated condition[22,27]. In contrast, transmission electron microscopy (TEM) allows direct, real-time information to be obtained from structures at the atomic scale. The need for *in situ* observations of this type has been increasingly recognized, and thus there have been many recent reports of the use of *in situ* TEM techniques to study alloying and conversion reactions[28–38]. However, directly visualizing phase transitions that occur during intercalation reactions is challenging because the volume change that occurs during this type of reaction is negligible[39–42]. This prevents the use of standard diffraction contrast imaging techniques. In this work, we utilize a strain-sensitive, bright-field scanning TEM (BF-STEM) imaging technique to overcome this challenge. This approach allows us to observe the phase changes that occur in monodisperse Fe_3O_4 nanoparticles *in situ* during the intercalation reaction, and follow the subsequent conversion reaction directly. By integrating *ex situ* synchrotron X-ray and *in situ* TEM approaches, we capture in detail the mechanisms by which lithiation occurs and relate this directly to the reaction kinetics. Furthermore, we support these observations with both density functional theory (DFT) calculations and non-equilibrium phase-field theory to describe the atomistic processes from first principles and phenomenological perspective, respectively. This work highlights the importance of exploiting advanced TEM techniques to reveal reaction inhomogeneities at the nanoscale.

Results

Material structures and electrochemical properties. The as-synthesized Fe_3O_4 nanoparticles are *ca.* 80 nm truncated octahedrons, as shown in Fig. 1a and Supplementary Fig. 1. Atomic-resolution high-angle annular dark-field (HAADF) STEM images (as in Fig. 1b) indicate that the pristine Fe_3O_4 nanoparticles are single-crystalline with the inverse spinel structure ($Fd\bar{3}m$ space group) and terminated with {111} crystal planes. To characterize the electrochemical performance, voltage profiles at different C-rates (1C, C/10 and C/200, or 926 mA g^{-1}, 92.6 mA g^{-1} and 4.63 mA g^{-1}, respectively) are shown in Fig. 1e. The first discharge capacity is 1,080 mAh g^{-1} at C/200, and it decreases to 916 mAh g^{-1} and 914 mAh g^{-1} for C/10 and 1C, respectively, close to the theoretical capacity of 926 mAh g^{-1}. The capacity excess in the initial discharge is generally believed due to side reactions involving electrolyte decomposition and formation of solid electrolyte interphases[43–45], while the following cyclic tests show a rate performance (Supplementary Fig. 2) that is comparable to previous reports[10–12]. After full discharge (lithiation), the electrode materials have changed their structure and morphology dramatically to form a nanocomposite consisting of ultrafine (<5 nm) Fe nanoparticles and an amorphous Li_2O matrix phase (Fig. 1c), as confirmed by electron diffraction (Fig. 1d). It is noted that two distinct plateaus appear in the C/200 discharge curve, corresponding to the intercalation reaction and conversion reaction, respectively[13–15,17], whereas the intercalation plateau is absent from the relatively faster discharge curves at rates of 1C and C/10. This can be due to enhancements in the reaction

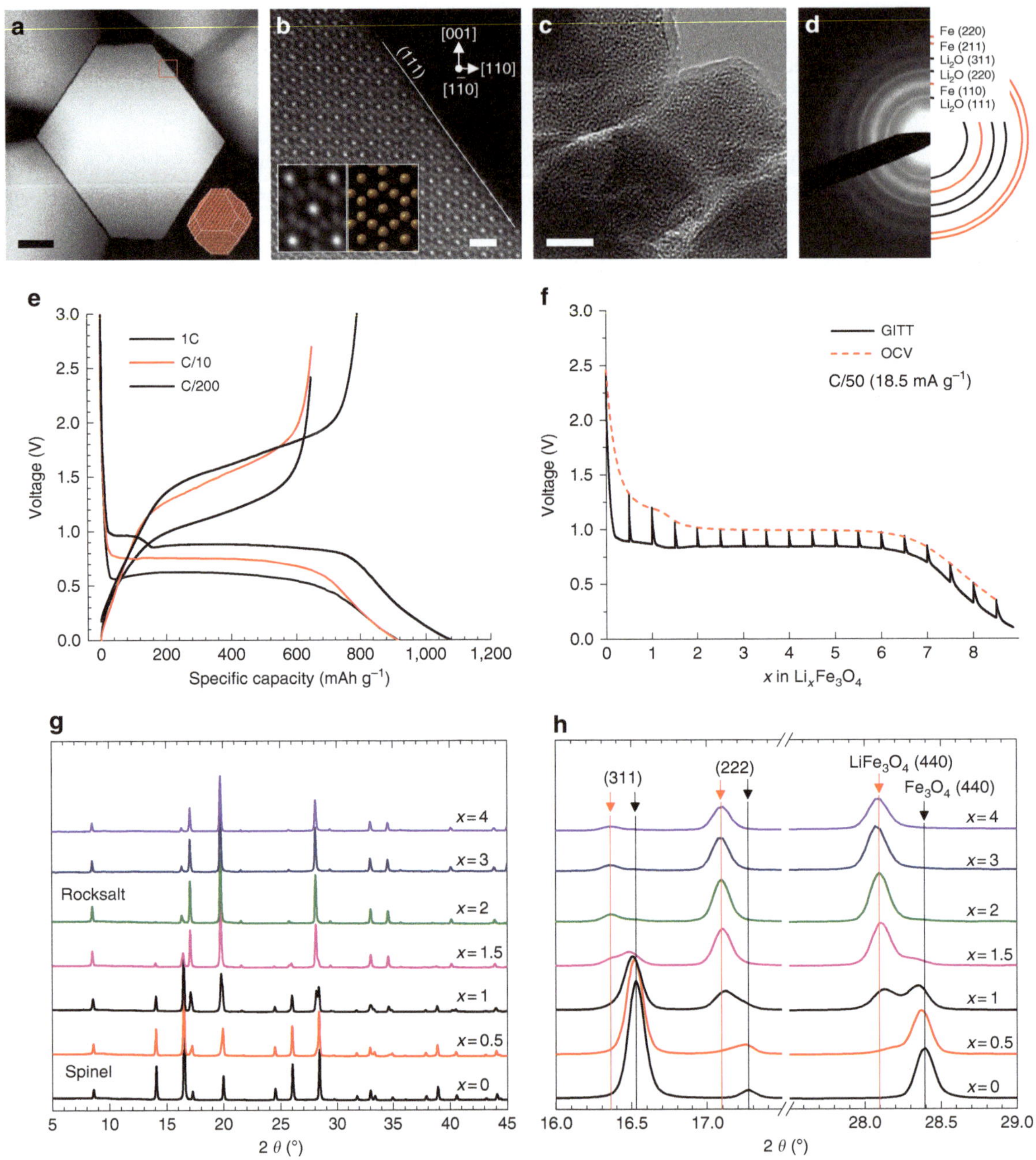

Figure 1 | Electrode materials and electrochemical performance. (**a**) STEM image of the pristine Fe_3O_4 single crystals. Inset showing the crystals in the truncated octahedral shape. (**b**) High-resolution HAADF-STEM image showing spinel structure along [1-10] zone axis. Inset showing enlarged STEM image compared with the atom model. (**c**) TEM image and (**d**) SAED pattern of the fully lithiated electrode, confirming the nanocomposite of Fe nanocrystals in amorphous Li_2O matrix. Scale bars, (**a**,**c**) 20 nm; (**b**) 1 nm. (**e**) First-cycle charge/discharge profiles of coin-cell batteries at rates of 1C, C/10 and C/200 (that is, 926 mA g^{-1}, 92.6 mA g^{-1} and 4.63 mA g^{-1}). (**f**) GITT and OCV profiles in the first discharge cycle measured at C/50 (18.5 mA g^{-1}). GITT measurements were performed by applying an intermittent current of C/50 for 3.125 h followed by a 24 h relaxation period. (**g**) SXRD patterns of $Li_xFe_3O_4$ at various SOCs (x = 0, 0.5, 1, 1.5, 2, 3, 4). (**h**) Enlarged SXRD patterns showing detailed peak shifts of (311), (222) and (440) reflections. Fe_3O_4 and $LiFe_3O_4$ phases are marked by black and red arrows. X-ray wavelength is 0.72768 Å.

kinetics that can occur at high discharge current densities (such as voltage polarization). To justify this hypothesis, we employed the galvanostatic intermittent titration technique (GITT)[46] for a discharge cycle at C/10 to measure the OCV profile: this reflects the equilibrium redox potentials at different states of charge (SOCs). Figure 1f clearly indicates the voltage polarization between the OCV and regular discharge profile, and confirms the existence of an intercalation plateau which is related to the charge rate. The intercalation process can generally be expressed as equation (2):

$$x\text{Li} + \text{Fe}_3\text{O}_4(\text{spinel}) \rightarrow \text{Li}_x\text{Fe}_3\text{O}_4(\text{rocksalt}) \qquad (2)$$

where x is the lithium content per formula. For an intercalation reaction, x is supposedly between 0 and 2 (refs 16–18), and its upper limit depends on the total amount of lithium ions accommodated at the tetragonal 8a sites[16,17]. Further lithiation would trigger the conversion reaction. This is also confirmed by high-resolution TEM (HRTEM) imaging of *ex situ* lithiated

$Li_xFe_3O_4$ with Li content $x=2, 3, 4, 5, 6, 7, 8$ (Supplementary Fig. 3). These HRTEM images and selected area electron diffraction (SAED) patterns show the transition from rocksalt $Li_xFe_3O_4$ to $Fe+Li_2O$ composite in a gradually proceeding manner. Since the intercalation reaction only involves Li content $x<2$, it is crucial to focus our study on the first half of the lithiation. To study the intercalation process further, we utilized synchrotron X-ray diffraction (SXRD) to identify the phase changes at various SOCs of $x=0$, 0.5, 1, 1.5, 2, 3 and 4, as shown in Fig. 1g,h. The series of SXRD patterns show that the electrode materials transfer from spinel structure to rocksalt structure with increasing Li composition (detailed indexing is presented in Supplementary Fig. 4). From the enlarged patterns in Fig. 1h, we indeed identified the two stable Fe_3O_4 and $LiFe_3O_4$ phases. The nanosized Fe phase that resulted from the conversion reaction cannot be captured by SXRD even up to $x=4$; however, its presence was confirmed using TEM (Supplementary Fig. 5). These observations clearly indicate that SXRD—due to its lack of sensitivity in detecting nanoscale particles, and its averaging of the overall phase information—is not able to precisely determine either the initiation of the conversion reaction or to follow its evolution. In addition, the absence of an intercalation plateau in the cycling profiles of 1C and C/10 implies that the intercalation process is bypassed at high rates. Although the SXRD results clearly show the existence of a rocksalt intermediate phase, which arises from the intercalation reaction, the presence of this phase might also be due to a possible structure relaxation and phase separation that occurs post-mortem. Effects of this type have been reported in LFP system, where *ex situ* characterization is shown to inaccurately describe the reaction processes that are occurring[21]. In addition, the intercalation phase transition can either follow a two-phase model or solid-solution process depending on the details of the reaction kinetics, a process which cannot be resolved from the *ex situ* study. To solve these issues, we have utilized *in situ* electron microscopy approaches to determine which reaction pathways occur in spinel oxide.

Phase evolution by *in situ* electron diffraction. We utilized a dry cell set-up for our *in situ* TEM investigation[36–38]. This approach can precisely track the phase evolution at very high spatial resolution. Since there are no liquid organic electrolyte involved in the reactions, to verify the consistency in the phase transitions between the *ex situ* and *in situ* experiments, we tracked the dynamical phase evolution using *in situ* electron diffraction throughout the entire lithiation process, as shown in Fig. 2a and Supplementary Movie 1 The radially integrated intensity profiles from a series of time-sequenced SAED patterns are plotted as a function of lithiation time, with the intensity represented by false colours (Fig. 2a). It is obvious that both the position and intensity of the Bragg reflections change as the lithiation proceeds, verifying the overall phase transformation from pristine Fe_3O_4 to the eventual Fe and Li_2O composite (and in the same manner as the *ex situ* results). Using the intensity profiles of Fe_3O_4 (311) and Fe (110) SAED peaks as a measure, the gradual evolution of the phase transformation becomes clear, as depicted in Fig. 2b. It is also worth noting that the Bragg peaks of the Fe_3O_4 phase display a negative shift in reciprocal space (as indicated by arrows in Fig. 2a) during the initial stage of lithiation (up to ~1,000 s), which corresponds to an increase in the lattice parameter as Li^+ ions are inserted into the Fe_3O_4 lattice (Fig. 2c). This lattice parameter increase is quantitatively consistent with the phase transition from Fe_3O_4 to $LiFe_3O_4$, as extracted from our SXRD measurements. After ~1,000 s, the lattice expansion

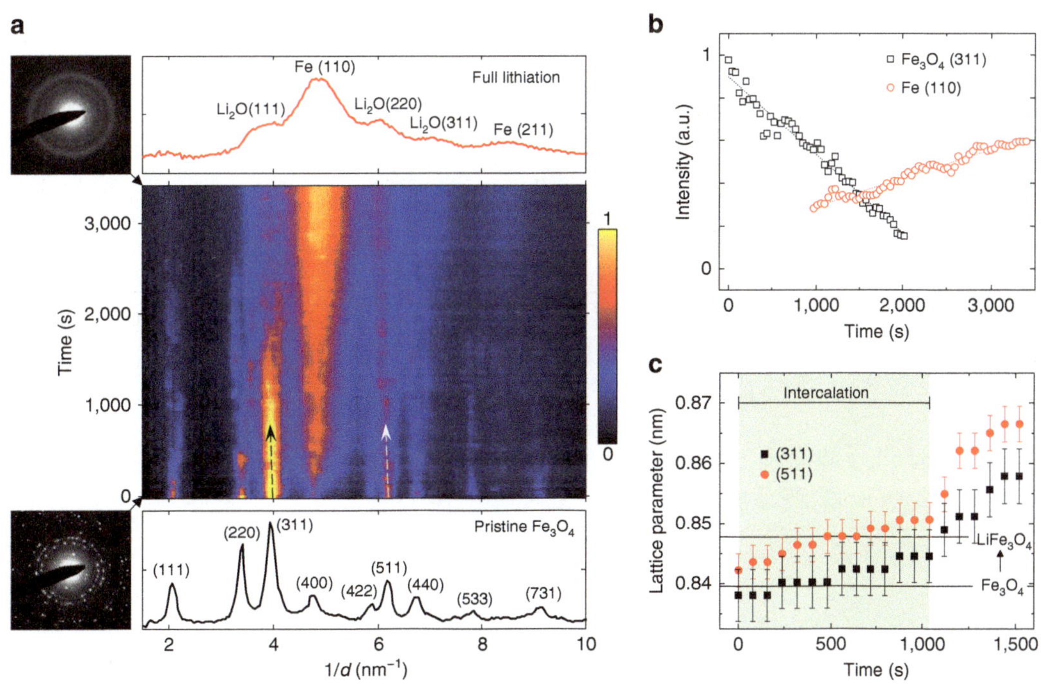

Figure 2 | Phase evolution probed by *in situ* electron diffraction. (**a**) Electron diffraction intensity profile (the colour map in the middle) as a function of reaction time during an *in situ* lithiation of Fe_3O_4 nanoparticles (see Supplementary Movie 1). The SAED patterns and corresponding radially integrated intensity profiles obtained at pristine (0 s) and fully lithiated (3400 s) states are shown below and above the colour map, respectively. Arrows indicate diffraction peak shift towards a lower value in the reciprocal space. The sample was lithiated at a rate of ~980 mA g^{-1}. (**b**) Intensity profiles of Fe_3O_4 (311) and Fe (110) Bragg reflections during the lithiation, indicating the phase evolution of Fe_3O_4 and Fe. (**c**) Lattice parameter measured from Fe_3O_4 (311) and (511) reflections showing an incremental shift as a function of lithiation time, which quantitatively agrees with the SXRD results. Error bars are given by 1 pixel precision of SAED measurements. Blue lines indicate lattice parameters of Fe_3O_4 and $LiFe_3O_4$ measured by SXRD in Fig. 1g.

becomes more severe, indicating that the subsequent lithiation was dominated by the conversion reaction, as per equation (3).

$$(8-x)\mathrm{Li}+\mathrm{Li}_x\mathrm{Fe}_3\mathrm{O}_4 \rightarrow 4\mathrm{Li}_2\mathrm{O}+3\mathrm{Fe} \qquad (3)$$

Visualization of two-step lithiation by *in situ* BF-STEM. After verifying that the phase transformation occurs using *in situ* SAED, we performed an *in situ* BF-STEM observation to follow how the structure evolves during the entire lithiation process in real space, as shown in Fig. 3a and Supplementary Movie 2. The time-sequenced BF-STEM images record the intensity of the transmitted electrons scattered to lower angles, as well as the direct beam (involving the most coherent electron scattering). This signal is strain sensitive and thus enables direct visualization of the small lattice changes that occur during the intercalation process (Supplementary Fig. 6)[47]. We show that this is essential observing intercalation reactions in real time. This is in contrast to other, more conventional imaging techniques (such as HAADF-STEM or BF/DF TEM) which do not have the requisite image sensitivity to directly observe this process (sensitivity to strain is not satisfied due to incoherent contrast or bending contour). The first intercalation step follows the reaction described in equation (2) to generate the lithium-inserted $Li_xFe_3O_4$ phase (shown with a lighter contrast). These regions will be further lithiated in the subsequent conversion reaction, which is also accompanied by the extrusion of ultrafine metallic Fe nanoparticles to form a composite with the amorphous Li_2O. For better visualization, we use false colours to distinguish the pristine Fe_3O_4 (red), Li-inserted Fe_3O_4 (blue) and completely conversion composite (green) phases and their evolution as a function of time. This was also verified using electron energy-loss spectroscopy (EELS) in STEM (Supplementary Fig. 7). We note that, under realistic electrochemical environment, the lithiation process shows a non-equilibrium reaction pathway, that is, the conversion reaction starts to take place before the intercalation completely finishes. This may be likely due to the fact that the diffusion of lithium on the Fe_3O_4 particle surface is faster than that through the lithiated $Li_xFe_3O_4$ phase, which can cause a sufficiently large lithium concentration at the particle surface and thereby trigger the conversion reaction to happen at an early stage (but still after the intercalation reaction). It is obvious that both intercalation and conversion reactions follow a 'shrinking-core' mode, proceeding from the outer surface to the inner region[37]. Specifically, the initial intercalation trajectory does not exhibit any preferential directions, in accord with the zigzag Li^+ diffusion paths in three-dimensional (3D) tunnels[16], whereas the subsequent conversion tends to propagate through the outer facets, that is, {111} planes. As an approximate measure of reaction kinetics, we quantified the projected areas of the three phases versus the lithiation time, displayed in Fig. 3b. This figure shows that the propagation speed of the intercalation process is about 1 order of magnitude faster than the following conversion reaction (Fig. 3c). We note that these measurements are from one single particle, where the electrochemical conditions are supposed to be identical. This is consistent with the fact that full conversion needs multiple Li-ion transfers, when compared with the intercalation process. We do not believe this to be an effect resulting from local electrochemical inhomogeneity.

HRTEM characterization. For a fundamental understanding of the intercalated phase, we conducted HRTEM imaging to track the atomic structure changes for both *in situ* and *ex situ* scenarios, as shown in Fig. 4. A new phase with the rocksalt structure was observed *in situ* as Li^+ ions were inserted into the spinel Fe_3O_4 phase (Fig. 4a,b). Using two sets of diffraction spots from the fast Fourier transform (FFT, Fig. 4a insets), we can separately map the distribution of the spinel (red) and rocksalt (green) phases and thus visualize the propagation of the intercalation front in the

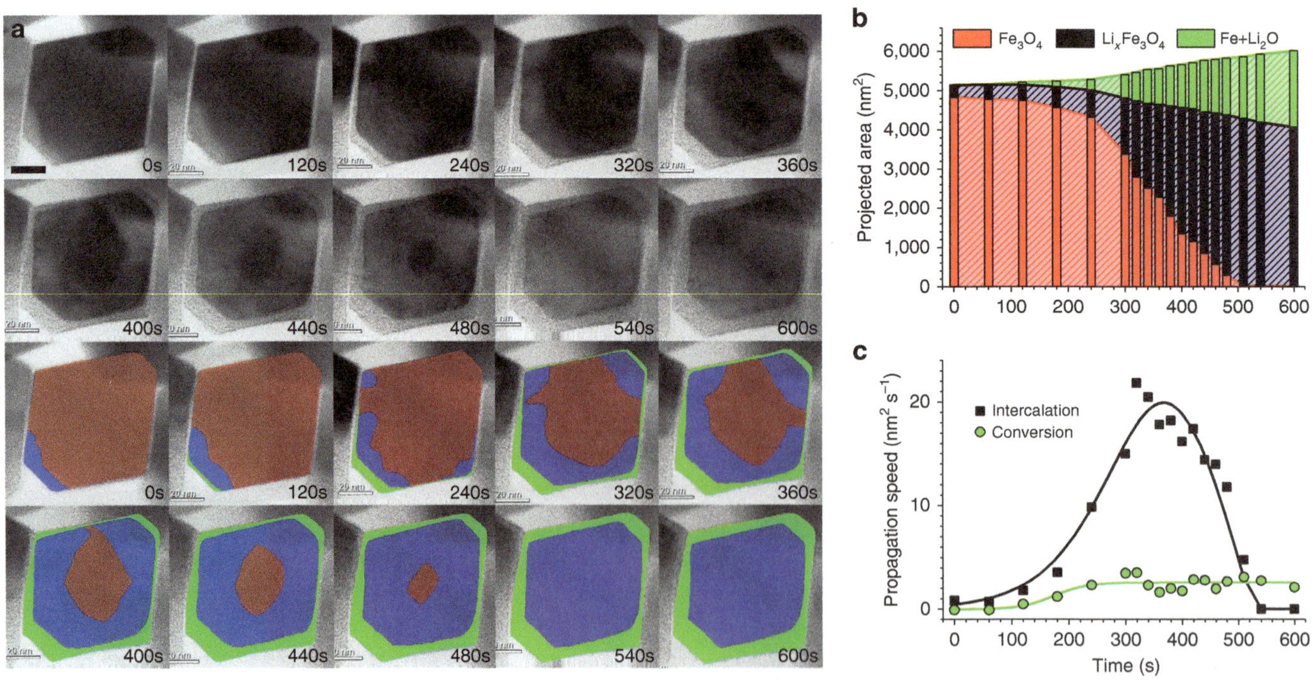

Figure 3 | *In situ* observation of two-step phase transformation during lithiation. (**a**) BF-STEM image series showing phase evolution during lithiation. The overlaid false colours indicate different phases: pristine Fe_3O_4 (red), Li-inserted $Li_xFe_3O_4$ (blue) and $Fe + Li_2O$ composite after conversion (green). Scale bar, 20 nm. The sample was lithiated at a rate of ~2,700 mA g^{-1}. (**b**) Projected areas of the three phases as a function of time. (**c**) Propagation speed of intercalation and conversion as a function of time.

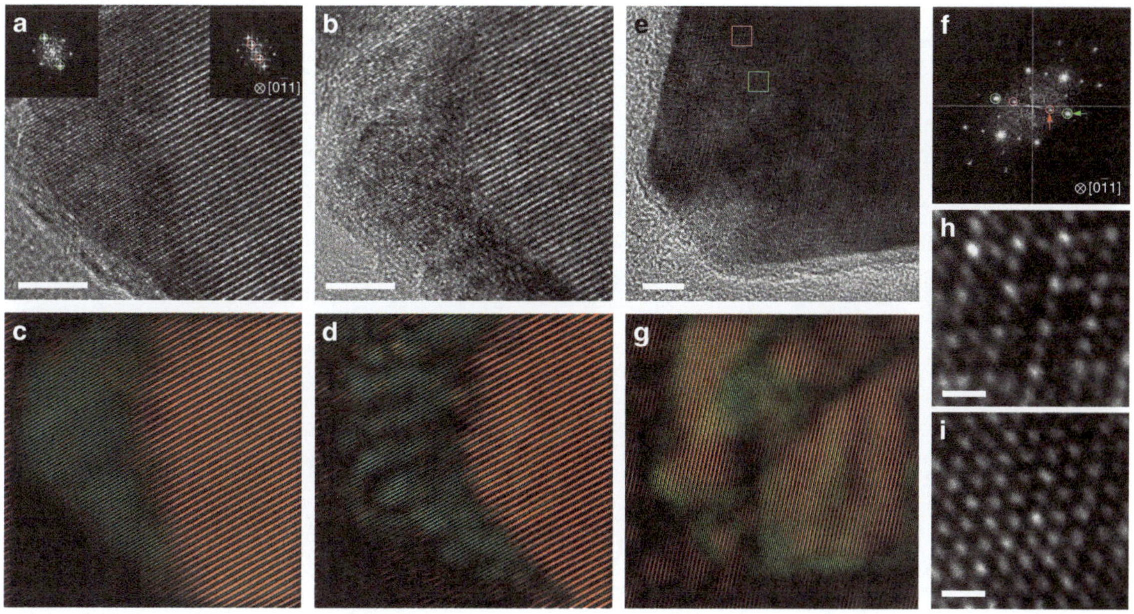

Figure 4 | Atomic structural evolution of phase transformation. (**a**,**b**) HRTEM images of a Fe_3O_4 single crystal showing the spinel and rocksalt phases during *in situ* lithium intercalation. Insets showing the FFT of the spinel (top-right of panel) and rocksalt (top-left of panel) structures along [0-11] zone axis. (**e**) HRTEM image of a partially lithiated Fe_3O_4 single crystal from *ex situ* lithiation. (**f**) The FFT of **e**. (**h**,**i**) The enlarged HRTEM images of boxed regions in **e** showing spinel and rocksalt structures, respectively. (**c**,**d**,**g**) Filtered images of (**a**,**b**,**e**) using two sets of spinel (red) and rocksalt (green) diffraction spots to indicate the corresponding phase distribution. The samples were lithiated at $\sim$150 mA g^{-1} for *in situ* lithiation, and 92.6 mA g^{-1} for *ex situ* lithiation. Scale bars: (**a**,**b**,**e**) 5 nm; (**h**,**i**) 5 Å.

reacting Fe_3O_4 crystal (Fig. 4c,d). Similarly, in an *ex situ* lithiated sample that was discharged to 0.9 V, we reproduced this result, and found a similar mixture of spinel and rocksalt phases (Fig. 4e,g). In the pristine spinel Fe_3O_4 crystals, Fe cations occupy the tetrahedral 8a sites and octahedral 16d sites, which give the uneven TEM contrast (dim at the diamond centre) shown in Fig. 4h. Upon lithium insertion, all the cations migrate to the octahedral sites (16c and 16d), leading to a uniform contrast (Fig. 4i). This means that the inserted Li ions occupy the 16c sites and also repel the adjacent Fe cations from the 8a to the 16c sites to form $Li_xFe_3O_4$ crystals with the rocksalt structure, with lithium composition up to $x=1$. It is also noted from previous reports that the excess Li ions (up to $x=2$) could possibly insert into the 8b and 48f sites to form $Li_2Fe_3O_4$ phase without Fe extrusion[16,17]. However, this particular phase was not experimentally identified from TEM observations. Nevertheless, the intercalation process goes through a two-phase mode and forms an intermediate phase of $Li_xFe_3O_4$ (refs 16–18). Thereafter, the conversion reaction occurs and the $Li_xFe_3O_4$ phase is decomposed into Li_2O and metallic Fe nanocrystals, which can be observed as the speckles in Fig. 4b,d. Combining both *in situ* TEM imaging and *ex situ* SXRD results, we now can conclude that the intercalation proceeds in a two-phase 'nucleation and growth' manner, resulting in sharp interfacial boundaries between the spinel and rocksalt phases, as expressed in equation (2).

First principles calculations and phase-field modelling. DFT computation has been performed to calculate the voltage profile of the lithiation process under the thermodynamic equilibrium condition[48–52]. The calculated reaction pathway has multiple voltage plateaus (green dashed curve in Fig. 5a) with intermediate phases, such as $LiFeO_2$, Li_2FeO_2 and Li_5FeO_4 (Fig. 5b). Given that the $LiFe_3O_4$ is an intermediate phase experimentally observed from SXRD, our DFT calculations also considered a two-step reaction route (red curve in Fig. 5a), including the intercalation plateau ($Li+Fe_3O_4 \rightarrow LiFe_3O_4$) and the conversion plateau ($LiFe_3O_4 \rightarrow Li_2O+Fe$). The ground-state $LiFe_3O_4$ structure as illustrated in the inset crystal models of Fig. 5a possesses a similar structural framework (that is, $[Fe]_{16d}O_2$ octahedron) that was inherited from the spinel Fe_3O_4 structure. Consistent with the ground-state $LiFe_3O_4$ structure calculated in the DFT computation, Li ions insert into the octahedral 16c openings, and repel the nearby 8a Fe ions to the other surface-sharing 16c site, which forms the structure identical to that observed in the HRTEM images in Fig. 4. In addition, since the octahedral 16c sites are interconnected as a 3D channel, the initial intercalation would proceed in a disordered pattern, which explains the irregular patterns that are observed during the shrinking-core process in Fig. 3. After all of the octahedral 16c sites are filled by Li, the rocksalt lattice becomes more isotropic to the incoming Li ions, which makes the subsequent conversion most likely to propagate across the outline of crystal, again as observed in Fig. 3.

Discussion

Given the basics of chemical thermodynamics, the equilibrium lithiation should go through the $LiFe_3O_4$ phase and the subsequent $Fe+Li_2O$ composite in two non-overlapping processes, as proposed by Thackeray *et al.*[16–18]. However, our *in situ* results show a clear overlap between the two steps, suggesting that the kinetic effects play an important role during lithiation. On the other hand, the phase-field theory, taking into account the effect of overpotentials, has succeeded in explaining the electrochemical kinetics in lithium-ion battery systems[53–55]. We performed the phase-field simulation in the frame of non-equilibrium thermodynamics based on the Butler–Volmer equation[55,56] and the Cahn–Hilliard equation[23,57]. To simulate the two-step reaction, we constructed a homogeneous free-energy function with three local minima corresponding to the pristine, intercalation and conversion phases, respectively. In addition, the kinetic contributions from overpotential and volumetric strain

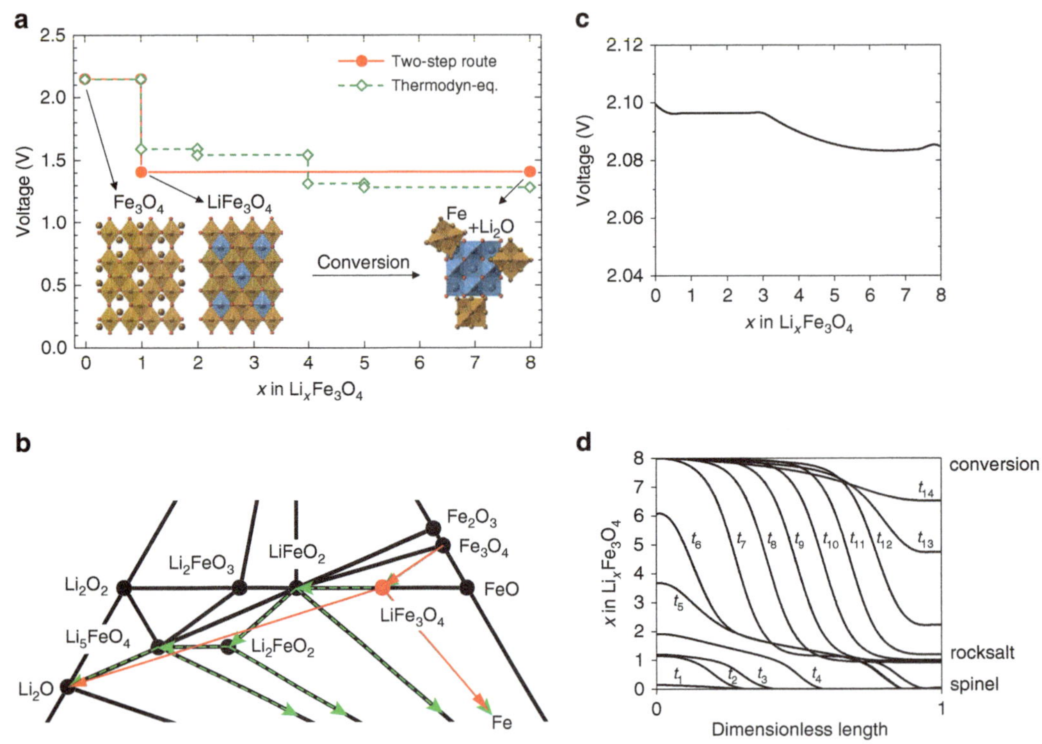

Figure 5 | DFT calculation and phase-field simulation of lithiation process in Fe_3O_4 crystals. (**a**) The voltage profiles as a function of Li composition (x = 0–8) calculated by DFT showing both thermodynamic equilibrium (green dashed lines) and two-step reaction (red solid lines) routes for an entire lithiation process. The inset models showing crystal structures of Fe_3O_4 and $LiFe_3O_4$ phases during the intercalation process and a schematic of Li_2O plus metallic Fe after the complete conversion reaction. Colour spheres representing Fe (yellow), Li (blue) and O (red). (**b**) Li-Fe-O phase diagram with overlaid arrow lines showing lithiation pathways through thermodynamic equilibrium multistep route (green dashed lines) or experimentally observed two-step route (red solid lines). (**c**) Discharge voltage profiles calculated from the phase-field modelling. (**d**) Li composition profiles along one-dimensional length as a function of reaction time simulated by the phase-field theory (also see Supplementary Movie 3). The time intervals are uniform between adjacent time stamps.

have also been included (details in Supplementary Methods). The calculated discharge voltage profile is shown in Fig. 5c, which qualitatively agrees with the experimental curves in Fig. 1e. Figure 5d shows the calculated lithium compositional profiles as a function of time (x-axis is the dimensionless lithiation direction and y-axis is the Li composition, details shown in Supplementary Movie 3). It is found that the formation of $LiFe_3O_4$ phase is predominant in the early stage of lithiation, whereas the conversion reaction initiates immediately afterward and propagates before the complete intercalation, resulting in a mixed lithiation behaviour, which is in good agreement with the *in situ* STEM observation (Fig. 3a). It is well-known that the lithiation process (redox stage and reaction speed) in a real battery is heterogeneous due to the fluctuation of local environments[22]. Here we reasonably assume that the local electrochemical conditions are identical for these two reactions within one nanoparticle. By comparing the *in situ* STEM results with phase-field simulation, we believe that the coexistence of Fe_3O_4, $Li_xFe_3O_4$, and Fe + Li_2O phases are due to the competition between the intercalation and conversion reactions. The overall electrochemical kinetics dictates to the applied C-rate, which in turn determines the propagation speeds of the interfaces of $Fe_3O_4/Li_xFe_3O_4$ and $Li_xFe_3O_4$/Fe + Li_2O. Although the reaction speed of intercalation is much faster, the conversion reaction accommodates more Li ions. Therefore, in terms of total lithium insertion capacity, both reactions give important contributions to the overall energy storage rate.

The kinetic lithiation mechanism we proposed here can accommodate the following phenomena which could not be interpreted by the equilibrium theory. (1) As for the discharge profiles in Fig. 1e and in literature[9–11], even the intercalation reaction has not completed, the occurrence of conversion reaction on the surface will reduce the apparent discharge voltage, which consequently flattens the first plateau. (2) As for the previous debate on the existence of metallic Fe extrusion[16–19], due to the kinetic effect, the conversion reaction can happen at the same time of the intercalation reaction, which induces the formation of metallic Fe. For the discharge process at a low rate, the extrusion of metallic Fe is not expected to be observed.

In summary, we have investigated the lithiation mechanism of spinel magnetite using an *in situ* strain-sensitive, BF-STEM technique, as well as DFT computation and phase-field simulation. By explicit visualization of the two-step intercalation-conversion process of the lithiation in Fe_3O_4 nanocrystals, we found that the initial lithium intercalation leads to formation of the rocksalt $LiFe_3O_4$ phase in a two-phase reaction mode, and the lithium intercalation process significantly overlaps with the subsequent conversion reaction within a single nanoparticle, leading to indistinctness of the discharge profiles. This scenario aslo clarifies the mechanism of metallic Fe extrusion during the intercalation process. Our findings have elucidated the ionic occupancy on the atomic level and revealed how rate-dependent kinetic effects can affect the reaction pathway at the single-particle scale. These findings highlight the importance of advanced *in situ* electron microscopy techniques in the field of lithium-ion batteries and provide valuable insights to improve the electrochemical performances of other spinel lithium metal oxide cathode materials.

Methods

Materials. Iron acetylacetonate ($Fe(acac)_3$, 99%), oleic acid (OAc, 90%), benzyl ether (BE, 99%) and nitrosonium tetrafluoroborate ($NOBF_4$, 95%) were purchased from Sigma-Aldrich and used without any further purification. All solvents, including hexane, isopropanol, toluene and dimethylformamide (DMF), were in ACS reagent grade and were purchased from Fisher Scientific and used without further purification. Nanoparticle synthesis was performed using standard Schlenk techniques.

Sample preparation. The Fe_3O_4 nanoparticles were synthesized by the decomposition of $Fe(acac)_3$ in the presence of OAc, as modified from the approach in ref 58. In a typical synthesis, 2 mmol of $Fe(acac)_3$ was mixed with 1.28 ml of OAc and 10 ml of BE. The mixture was magnetically stirred and was kept under vacuum at 100 °C to generate a dark red solution. After evacuation at 100 °C for 1 h to remove all impurities and moisture, the system was filled with N_2 and the solution was further heated to 290 °C at a rate of about 15 °C min^{-1} and kept at this temperature for 20 min. After cooling to room temperature, the NPs were separated by adding isopropanol, followed by centrifugation (8,500 r.p.m., 5 min). The NPs were further purified by sequential operations of dispersing in hexane, precipitation by adding isopropanol and centrifugation. The as-synthesized Fe_3O_4 nanoparticles were surface-passivated by OAc, which can be ligand-exchanged by $NOBF_4$ according to literature[59]. About 0.1 g of $NOBF_4$ was added into 10 ml DMF, resulting in immediately a light green solution. This $NOBF_4$ DMF solution is added to the hexanes solution (10 ml) of Fe_3O_4 NPs (100 mg). The mixture was shaken for 10 min, followed by the addition of 25 ml of toluene. The $NOBF_4$-modified Fe_3O_4 NPs were collected through centrifugation (7,000 r.p.m., 8 min). The Fe_3O_4 NPs after ligand exchange can be solubilized in DMF with sonication. To remove residual organics, the NPs were further purified by two cycles of re-dispersing in DMF and precipitation by adding toluene centrifugation. The NPs were either stored in DMF or evacuated at the room temperature overnight to make dry powder for further experiments.

Electrochemical measurements. Composite electrodes were prepared with 80 wt% active material, 10 wt% polyvinylidene fluoride and 10 wt% acetylene carbon black in NMP (*N*-methyl-2-pyrrolidone) and cast onto copper current collectors. 2032-type coin cells were assembled in an argon-filled glove box using the composite electrode as the positive electrode and Li metal as the negative electrode. A Celgard separator 2400 and 1 M $LiPF_6$ electrolyte solution in 1:1 w/w ethylene carbonate/dimethyl carbonate were used to fabricate coin cells. Battery testing was performed on computer controlled systems (Arbin BT2000 and BioLogic VMP3) at 21 °C.

TEM characterization. The *in situ* TEM electrochemical cell was incorporated into a Nanofactory TEM-STM specimen holder, in which Fe_3O_4 nanoparticles dispersed onto a TEM half-grid with amorphous carbon support are analogous to the Fe_3O_4-C composite electrode, Li metal is coated onto a piezo-driven W probe as the counter electrode, with a thin layer of Li_2O formed on Li metal as the solid electrolyte[36–38]. The Li and Fe_3O_4 were loaded onto the holder in an Ar-filled glove box and then transferred to TEM column using a sealed Ar bag to avoid air exposure. During the *in situ* TEM tests, a constant negative DC potential was applied to Fe_3O_4 electrode against the Li source during the lithiation process, and the lithiation processes were captured by real-time imaging in either TEM or STEM mode. We estimated the average discharge current density across the entire particle surface to be 0.017 mA cm^{-2}. The *ex situ* samples after discharge in coin cells were examined accordingly. The *in situ* BF-STEM measurements were performed on a JEOL 2100F TEM operated at 200 kV. We have used a convergence angle of 12 mrad and a collection angle cutoff at ~20 mrad. The high-resolution STEM imaging and analytical EELS were conducted on a Hitachi HD2700C STEM operated at 200 kV and equipped with a probe aberration corrector (spatial resolution <1 Å, energy resolution 0.35 eV).

Theoretical calculations. The DFT calculations were performed using the Vienna *Ab initio* Simulation Package (VASP) within the projector augmented-wave approach with the Perdew – Burke – Ernzerhof generalized-gradient approximation. The DFT parameters were consistent with the parameters used in Materials Project (MP)[48]. A Hubbard U term of 5.3 eV is adopted for Fe^{2+} and Fe^{3+} in accordance with the MP and the previous testing[51,52]. The ground-state crystal structures of $LiFe_3O_4$ were obtained by enumerating 380 symmetrically distinctive configurations of Li/Fe disordering at the 16c sites using pymatgen[49]. Several magnetic orderings in ferromagnetic, anti-ferromagnetic and ferrimagnetic structures are tested for iron oxide compounds, such as Fe_3O_4, FeO, $LiFe_3O_4$ and $LiFeO_2$, to identify the lowest energy state (more details provided in Supplementary Fig. 8 and Supplementary Table 1). The voltage plateaus are obtained using the DFT energies of all relevant compounds in the Li–Fe–O ternary space from the MP[50].

The phase-field simulation was performed using the electrochemistry theory based on non-equilibrium thermodynamics developed by Bazant *et al.*[53–55]. The standard phenomenological model based on the Butler–Volmer equation[55,56] and the Cahn–Hilliard equation[23,57] was used for modelling the electrochemical kinetics. A homogeneous free-energy was comprised using piecewise functions of polynomial with continuity and differentiability at the segment points. The overpotential and strain energy are also considered in the simulation. More details are shown in Supplementary Figs 9–11, Supplementary Table 2 and Supplementary Methods.

References

1. Poizot, P., Laruelle, S., Grugeon, S., Dupont, L. & Tarascon, J. M. Nano-sized transition-metal oxides as negative-electrode materials for lithium-ion batteries. *Nature* **407,** 496–499 (2000).
2. Tarascon, J. M. & Armand, M. Issues and challenges facing rechargeable lithium batteries. *Nature* **414,** 359–367 (2001).
3. Whittingham, M. S. Lithium Batteries and Cathode Materials. *Chem. Rev.* **104,** 4271–4301 (2004).
4. Goodenough, J. B. & Kim, Y. Challenges for rechargeable Li batteries. *Chem. Mater.* **22,** 587–603 (2010).
5. Ferg, E., Gummow, R. J., de Kock, A. & Thackeray, M. M. Spinel anodes for lithium-ion batteries. *J. Electrochem. Soc.* **141,** L147–L150 (1994).
6. Cabana, J., Monconduit, L., Larcher, D. & Palacin, M. R. Beyond intercalation-based Li-ion batteries: the state of the art and challenges of electrode materials reacting through conversion reactions. *Adv. Mater.* **22,** E170–E192 (2010).
7. Malini, R., Uma, U., Sheela, T., Ganesan, M. & Renganathan, N. G. Conversion reactions: a new pathway to realise energy in lithium-ion battery—review. *Ionics* **15,** 301–307 (2009).
8. Taberna, P. L., Mitra, S., Poizot, P., Simon, P. & Tarascon, J. M. High rate capabilities Fe_3O_4-based Cu nano-architectured electrodes for lithium-ion battery applications. *Nat. Mater.* **5,** 567–573 (2006).
9. Mitra, S., Poizot, P., Finke, A. & Tarascon, J. M. Growth and electrochemical characterization versus lithium of Fe_3O_4 electrodes made via electrodeposition. *Adv. Funct. Mater.* **16,** 2281–2287 (2006).
10. Zhou, G. *et al.* Graphene-wrapped Fe_3O_4 anode material with improved reversible capacity and cyclic stability for lithium ion batteries. *Chem. Mater.* **22,** 5306–5313 (2010).
11. Ji, L. *et al.* Fe_3O_4 nanoparticle-integrated graphene sheets for high-performance half and full lithium ion cells. *Phys. Chem. Chem. Phys.* **13,** 7170–7177 (2011).
12. Zhang, W. M., We, X. L., Hu, J. S., Guo, Y. G. & Wan, L. J. Carbon coated Fe_3O_4 nanospindles as a superior anode material for lithium-ion batteries. *Adv. Funct. Mater.* **18,** 3941–3946 (2008).
13. He, Y., Huang, L., Cai, J. S., Zheng, X. M. & Sun, S. G. Structure and electrochemical performance of nanostructured Fe_3O_4/carbon nanotube composites as anodes for lithium ion batteries. *Electrochim. Acta* **55,** 1140–1144 (2010).
14. Xiong, Q. Q. *et al.* Synthesis of hierarchical hollow-structured single-crystalline magnetite (Fe_3O_4) microspheres: the highly powerful storage versus lithium as an anode for lithium ion batteries. *J. Phys. Chem. C* **116,** 6495–6502 (2012).
15. Hariharan, S., Saravanan, K., Ramar, V. & Balaya, P. A rationally designed dual role anode material for lithium-ion and sodium-ion batteries: case study of eco-friendly Fe_3O_4. *Phys. Chem. Chem. Phys.* **15,** 2945–2953 (2013).
16. Thackeray, M. M., David, W. I. F. & Goodenough, J. B. structural characterization of the lithiated iron oxides $Li_xFe_3O_4$ and $Li_xFe_2O_3$ ($0<x<2$). *Mater. Res. Bull.* **17,** 785–793 (1982).
17. Thackeray, M. M. *et al.* Spinel electrodes for lithium batteries—a review. *J. Power Sources* **21,** 1–8 (1987).
18. Thackeray, M. M. Spinel electrodes for lithium batteries. *J. Am. Ceram. Soc.* **82,** 3347–3354 (1999).
19. Fontcuberta, J., Rodriguez, J., Pernet, M., Longworth, G. & Goodenough, J. B. Structural and magnetic characterization of the lithiated iron oxide $Li_xFe_3O_4$. *J. Appl. Phys.* **59,** 1918–1926 (1986).
20. Sivakumar, V., Kumar, S., Ross, C. A. & Shao-Horn, Y. Electrochemical lithium insertion of into iron oxide spinel thin films and nanoparticles: changes in structure and magnetic properties. *ECS Trans.* **2,** 1–11 (2007).
21. Islam, M. S. & Catlow, C. R. A. Lithium insertion into Fe_3O_4. *J. Solid State Chem.* **77,** 180–189 (1988).
22. Liu, H. *et al.* Capturing metastable structures during high-rate cycling of $LiFePO_4$ nanoparticle electrodes. *Science* **344,** 1252817 (2014).
23. Li, Y. *et al.* Current-induced transition from particle-by-particle to concurrent intercalation in phase-separating battery electrodes. *Nat. Mater.* **13,** 1149–1156 (2014).
24. Arico, A. S., Bruce, P., Scrosati, B., Tarascon, J. M. & van Schalkwijk, W. Nanostructured materials for advanced energy conversion and storage devices. *Nat. Mater.* **4,** 366–377 (2005).
25. Jamnik, J. & Maier, J. Nanocrystallinity effects in lithium battery materials: Aspects of nano-ionics Part IV. *Phys. Chem. Chem. Phys.* **5,** 5215–5220 (2003).
26. Malik, R., Burch, D., Bazant, M. & Ceder, G. Particle size dependence of the ionic diffusivity. *Nano Lett.* **10,** 4123–4127 (2010).
27. Malik, R., Zhou, F. & Ceder, G. Kinetics of non-equilibrium lithium incorporation in $LiFePO_4$. *Nat. Mater.* **10,** 587–590 (2011).
28. Huang, J. Y. *et al.* *In situ* observation of the electrochemical lithiation of a single SnO_2 nanowire electrode. *Science* **330,** 1515–1520 (2010).
29. Liu, X. H. *et al.* *In situ* atomic-scale imaging of electrochemical lithiation in silicon. *Nat. Nanotechnol.* **7,** 749–756 (2012).

30. Liu, Y. *et al.* *In situ* transmission electron microscopy observation of pulverization of aluminum nanowires and evolution of the thin surface Al_2O_3 layers during lithiation-delithiation cycles. *Nano Lett.* **11,** 4188–4194 (2011).
31. McDowell, M. T. *et al.* Studying the kinetics of crystalline silicon nanoparticle lithiation with *in situ* transmission electron microscopy. *Adv. Mater.* **24,** 6034–6041 (2012).
32. McDowell, M. T. *et al.* *In situ* observation of divergent phase transformations in individual sulfide nanocrystals". *Nano Lett.* **15,** 1264–1271 (2015).
33. Gu, M. *et al.* *In situ* tem study of lithiation behaviour of silicon nanoparticles attached to and embedded in a carbon matrix. *ACS Nano* **6,** 8439–8447 (2012).
34. Gu, M. *et al.* Probing the failure mechanism of SnO_2 nanowires for sodium-ion batteries. *Nano Lett.* **13,** 5203–5211 (2013).
35. Nie, A. *et al.* Atomic-scale observation of lithiation reaction front in nanoscale SnO_2 materials. *ACS Nano* **7,** 6203–6211 (2013).
36. He, K. *et al.* Sodiation via heterogeneous disproportionation in FeF_2 electrodes for sodium-ion batteries. *ACS Nano* **8,** 7251–7259 (2014).
37. He, K. *et al.* Transitions from near-surface to interior redox upon lithiation in conversion electrode materials. *Nano Lett.* **15,** 1437–1444 (2015).
38. He, K. *et al.* Sodiation kinetics of metal oxide conversion electrodes: a comparative study with lithiation. *Nano Lett.* **15,** 5755–5763 (2015).
39. Zhu, Y. *et al.* *In situ* atomic-scale imaging of phase boundary migration in $FePO_4$ microparticles during electrochemical lithiation. *Adv. Mater.* **25,** 5461–5466 (2013).
40. Niu, J. *et al.* *In situ* observation of random solid solution zone in $LiFePO_4$ electrode. *Nano Lett.* **14,** 4005–4010 (2014).
41. Sun, Y. *et al.* Direct atomic-scale confirmation of three-phase storage mechanism in $Li_4Ti_5O_{12}$ anodes for room-temperature sodium-ion batteries. *Nat. Commun.* **4,** 1870 (2013).
42. Lu, X. *et al.* Lithium storage in $Li_4Ti_5O_{12}$ spinel: the full static picture from electron microscopy. *Adv. Mater.* **24,** 3233–3238 (2012).
43. Laruelle, S. *et al.* On the origin of the extra electrochemical capacity displayed by MO/Li cells at low potential. *J. Electrochem. Soc.* **149,** A627–A634 (2002).
44. Ponrouch, A., Taberna, P. L., Simon, P. & Palacin, M. R. On the origin of the extra capacity at low potential in materials for Li batteries reacting through conversion reaction. *Electrochim. Acta* **61,** 13–18 (2012).
45. Hu, Y. Y. *et al.* Origin of additional capacities in metal oxide lithium-ion battery electrodes. *Nat. Mater.* **12,** 1130–1136 (2013).
46. Weppner, W. & Huggins, R. A. Determination of the kinetic parameters of mixed-conducting electrodes and application to the system Li_3Sb. *J. Electrochem. Soc.* **124,** 1569–1578 (1977).
47. Pennycook, S. J. & Nellist, P. D. *Scanning Transmission Electron Microscopy Imaging and Analysis* 97–101 (Springer Science & Business Media, 2011).
48. Jain, A. *et al.* A high-throughput infrastructure for density functional theory calculations. *Comp. Mater. Sci.* **50,** 2295–2310 (2011).
49. Ong, S. P. *et al.* Python materials genomics (pymatgen): a robust, open-source python library for materials analysis. *Comp. Mater. Sci.* **68,** 314–319 (2013).
50. Jain, A. *et al.* The materials project: a materials genome approach to accelerating materials innovation. *APL Mater.* **1,** 011002 (2013).
51. Jain, A. *et al.* Formation enthalpies by mixing GGA and GGA + U calculations. *Phys. Rev. B* **84,** 045115 (2011).
52. Wang, L., Maxisch, T. & Ceder, G. Oxidation energies of transition metal oxides within the GGA + U framework. *Phys. Rev. B* **73,** 195107 (2006).
53. Bazant, M. Z. Theory of chemical kinetics and charge transfer based on nonequilibrium thermodynamics. *Acc. Chem. Res.* **46,** 1144–1160 (2013).
54. Bai, P., Cogswell, D. A. & Bazant, M. Z. Suppression of phase separation in $LiFePO_4$ nanoparticles during battery discharge. *Nano Lett.* **11,** 4890–4896 (2011).
55. Biesheuvel, P. M., van Soestbergen, M. & Bazant, M. Z. Imposed currents in galvanic cells. *Electrochim. Acta* **54,** 4857–4871 (2009).
56. Newman, J. *Electrochemical Systems* 2nd edn (Prentice Hall, 1991).
57. Cahn, J. W. & Hilliard, J. E. Free energy of a nonuniform system. I. interfacial free energy. *J. Chem. Phys.* **28,** 258–267 (1958).
58. Kim, D. *et al.* Synthesis of uniform ferrimagnetic magnetite nanocubes. *J. Am. Chem. Soc.* **131,** 454 (2009).
59. Dong, A. *et al.* A generalized ligand-exchange strategy enabling sequential surface functionalization of colloidal nanocrystals. *J. Am. Chem. Soc.* **133,** 998–1006 (2011).

Acknowledgements

This research used resources of the Center for Functional Nanomaterials, which is a US DOE Office of Science Facility, at Brookhaven National Laboratory under Contract No. DE-SC0012704. X.Y., E.H., K.S., H.G. and X.-Q.Y. were supported by the US DOE, the Assistant Secretary for Energy Efficiency and Renewable Energy, Office of Vehicle Technologies under Contract No. DE-SC00112704. We acknowledge the technical support by Dr Wenqian Xu from Beamline 17-BM-B of Advanced Photon Source at Argonne National Laboratory. S.Z., H.Y. and C.B.M. acknowledge the support of MRSEC award No. DMR-1120901. S.Z. also acknowledges the support of the NatureNet Science Fellowship. Q.M. and Yimei Z. are supported by DOE/BES, Division of Materials Science and Engineering, under Contract No. DE-SC0012704. Those *in situ* and analytical TEM experiments performed by J.L. and E.A.S. were supported as part of the Center for Mesoscale Transport Properties, an Energy Frontier Research Center supported by the US Department of Energy, Office of Science, Basic Energy Sciences, under award #DE-SC0012673. Yizhou Z. and Y.M. acknowledge the support of the Minta Martin award at University of Maryland, and the computational resources from Extreme Science and Engineering Discovery Environment (XSEDE) supported by the National Science Foundation Grant No. TG-DMR130142 and from the University of Maryland supercomputing resources. We thank Dr Peng Bai for helpful discussions on phase-field simulaitons.

Author contributions

K.H., S.Z., D.S. and C.B.M. conceived and designed the experiments. K.H., D.S. and J.L. performed *in situ* and *ex situ* TEM experiments. S.Z., H.Y. and C.B.M. synthesized the Fe_3O_4 samples. J.L., X.Y., E.H., X.-Q.Y., K.S. and H.G. conducted electrochemical tests. Yizhou Z. and Y.M. performed DFT calculation. Q.M. performed phase-field simulation. K.H. prepared the figures and wrote the manuscript with D.S., S.Z., E.A.S., Yimei Z. and C.B.M. D.S. supervised the project. All the authors participated in discussion of the results.

Additional information

Competing financial interests: The authors declare no competing financial interests.

17

Gas-solid interfacial modification of oxygen activity in layered oxide cathodes for lithium-ion batteries

Bao Qiu[1,*], Minghao Zhang[2,*], Lijun Wu[3,*], Jun Wang[4], Yonggao Xia[1], Danna Qian[2], Haodong Liu[2], Sunny Hy[2], Yan Chen[5], Ke An[5], Yimei Zhu[3], Zhaoping Liu[1] & Ying Shirley Meng[2]

Lattice oxygen can play an intriguing role in electrochemical processes, not only maintaining structural stability, but also influencing electron and ion transport properties in high-capacity oxide cathode materials for Li-ion batteries. Here, we report the design of a gas-solid interface reaction to achieve delicate control of oxygen activity through uniformly creating oxygen vacancies without affecting structural integrity of Li-rich layered oxides. Theoretical calculations and experimental characterizations demonstrate that oxygen vacancies provide a favourable ionic diffusion environment in the bulk and significantly suppress gas release from the surface. The target material is achievable in delivering a discharge capacity as high as 301 mAh g^{-1} with initial Coulombic efficiency of 93.2%. After 100 cycles, a reversible capacity of 300 mAh g^{-1} still remains without any obvious decay in voltage. This study sheds light on the comprehensive design and control of oxygen activity in transition-metal-oxide systems for next-generation Li-ion batteries.

[1] Ningbo Institute of Materials Technology and Engineering (NIMTE), Chinese Academy of Sciences, Zhejiang 315201, China. [2] Department of Nano Engineering, University of California San Diego (UCSD), La Jolla, California 92093, USA. [3] Department of Condensed Matter Physics and Materials Science, Brookhaven National Laboratory, Upton, New York 11973, USA. [4] MEET Battery Research Center/Institute of Physical Chemistry, University of Müenster, Corrensstrasse 46, Müenster 48149, Germany. [5] Chemical and Engineering Materials Division, Oak Ridge National Laboratory, Oak Ridge, Tennessee, 37830, USA. * These authors contributed equally to the work. Correspondence and requests for materials should be addressed to Z.L. (email: liuzp@nimte.ac.cn) or to Y.X. (email: xiayg@nimte.ac.cn) or to Y.S.M. (email: shirleymeng@ucsd.edu).

The functionality of many transition metal oxides can be significantly altered by oxygen vacancies on the surface. Oxygen vacancies can behave as charge carriers for solid-oxide fuel cells[1], as well as important adsorption sites and as active sites for electro-photocatalysts[2]. In Li-ion cathode materials, these vacancies play a vital role in determining the material's electron and ion transport properties[3–5]. The influence of oxygen vacancies at the surface on electrochemical performance can be completely different depending on the type of Li-ion cathode material[6–8]. Li-rich layered oxides, either as a solid solution or as a nano-composite of layered Li_2MnO_3 and $Li(TM)O_2$ (TM = Ni, Co, Mn), are drawing attention as next-generation cathode materials for high-energy-density Li-ion batteries in electric vehicles[9–11]. Over the past 20 years, the discharge capacity at room temperature of these cathode materials[9,12–14] has been improved, from 200 mAh g^{-1}, given in the first report[12], to over 320 mAh g^{-1} (ref. 14) today as summarized by Hy *et al.*[15], and even higher at elevated temperatures[16]. While research has continued to push the limit of the available capacity of the materials over the years, debates on the origins of the ultrahigh capacity beyond the redox of transition metal have intensified recently.

Numerous studies[10,17–23] exemplify the contributions of bulk and surface oxygen on high charge–discharge capacity in Li-rich layered oxides, although this hypothesis has not been fully verified. Tarascon and co-workers[21,22] concluded that the extra-high capacity is mainly attributable to reversible anionic redox processes (O^{2-}/O_2^{2-} or O^{2-}/O_2^{n-}, where $3 > n > 1$) within the bulk of Li-rich layered oxides. Delmas and co-workers[19,20] proposed that surface oxygen is oxidized to O_2 gas and irreversibly lost from the structure, leaving some oxygen vacancies on the surface or in the sub-surface layers. Our previous work[24–26] illustrated that oxygen vacancies generated during the high-potential electrochemical process can facilitate the transition metal ion migration and surface structural transformation, finally leading to their potential degradation during extended cycles.

To reduce oxygen gas generation and utilize reversible oxygen redox activity during charging and discharging, oxygen vacancies have been proposed to form on the surface of the as-synthesized Li-rich layered oxides before electrochemical processing. The influence of the surface oxygen vacancies on the oxygen gas generation is discussed in detail in Supplementary Note 1. The creation of surface oxygen vacancies was previously attempted utilizing a reducing atmosphere[27,28] and leaching with acid accompanied by heat treatment[29,30]. However, the bulk structure reported in previous work easily transforms from a pure layered phase to spinel- and/or rock-salt phases[28–33], which diminishes the rate capability and cycling stability.

Motivated by the considerations above, we propose a strategy based on a gas–solid interface reaction (GSIR) between Li-rich layered oxides and carbon dioxide gas to create oxygen vacancies on the particles' surface. The schematic of GSIR process between Li-rich layered oxides and carbon dioxide is displayed in Fig. 1a, and the procedure is detailed in the 'Methods' section. Oxygen vacancies in the surface regions up to 20-nm thick are created on the particles without affecting structural integrity. After the surface modification, the target material $Li[Li_{0.144}Ni_{0.136}Co_{0.136}Mn_{0.544}]O_2$ exhibits discharge capacity as high as 300 mAh g^{-1} with no obvious voltage degradation after 100 cycles at a current density of 25 mA g^{-1}. As confirmed by both theoretical calculations and experimental characterization, the improved electrochemical performance is ascribed to the full utilization of oxygen activity through the creation of oxygen vacancies on the surface of the as-synthesized Li-rich layered oxides.

Results

Characterizations of oxygen vacancies after GSIR process. Li-rich layered oxide $Li[Li_{0.144}Ni_{0.136}Co_{0.136}Mn_{0.544}]O_2$ (denoted as LR-NCM) was prepared by a co-precipitation method (see the 'Methods' section). After the GSIR process is completed, the general morphology remains the same as in the pristine LR-NCM (Supplementary Fig. 1). Both synchrotron X-ray diffraction (SXRD) refinements and transition metal K-edge X-ray absorption near-edge structure show no major difference for the pristine and GSIR LR-NCM samples (Supplementary Figs 2 and 3). These results preliminarily imply that the GSIR process does not heavily influence the average crystal structures and bulk electronic environments of the LR-NCM sample.

To investigate the GSIR reaction mechanism of the formation of surface oxygen vacancies, we applied Fourier-transformed infrared spectroscopy (FTIR) and X-ray photoelectron spectroscopy (XPS) to three samples (Fig. 1b,c). The FTIR results (Fig. 1b) indicate that carbonate formed on their surfaces after the GSIR. The lithium concentration of the GSIR LR-NCM is slightly lower (6.4%) than that of the pristine LR-NCM, as was determined by inductive coupled plasma-atomic emission spectrometry. On the other hand, in the oxygen 1s XPS spectra (Fig. 1c), the intensity of the binding energy at 529.5 eV representing TM-O covalency for the GSIR LR-NCM after washing is much lower than that of the pristine LR-NCM. In contrast, the intensity of the binding energy at 531.5 eV representing the carbonate group ($-CO_3$) for the GSIR LR-NCM without washing is much larger[34,35]. Both changes indicate some lattice oxygen was extracted by CO_2, forming oxygen vacancies on the sub-surfaces.

The amount of oxygen vacancies in the GSIR LR-NCM sample was determined by neutron diffraction (ND), a technique with high sensitivity for detecting light elements[36], such as Li and O. Figure 2a,b demonstrates the as-collected time-of-flight (TOF) ND patterns with 'Rietveld' refinement for both samples (patterns refined with a solid solution of '*R-3m*' symmetry; the normalized TOF ND patterns are shown in Supplementary Fig. 4). The lattice parameters of the pristine layer are a = 2.8445(2) Å, and c = 14.2113(9) Å. The oxygen occupancy of the pristine LR-NCM is 99.97 ± 1.10%. In comparison, the lattice parameters of the GSIR LR-NCM are a = 2.8526(8) Å and c = 14.2503(7) Å, which are slightly larger than those of the pristine LR-NCM. The oxygen occupancy is reduced to 96.21 ± 1.20% after surface modification. The amount of Ni^{2+} in the Li layer, and the occupancy of Li in the Li layer are also relatively larger than that of in the pristine LR-NCM. In addition, the data are also analysed using the two-phase model (the as-collected TOF ND patterns are shown in Supplementary Fig. 5), that is, a combination of $LiTMO_2$ (*R-3m*) phase, and Li_2MnO_3 (*C/2m*) phase. The results show that no oxygen vacancies (V_O) are observed in the pristine LR-NCM sample, while the GSIR LR-NCM sample exhibits 7.20 ± 3.3% V_O on the 4i site and 3.8 ± 3.0% V_O on the 8j site in Li_2MnO_3 phase and 2.96 ± 1.86% V_O in the overall composition. Both models are consistent with the observed oxygen vacancies in the GSIR LR-NCM. The lower oxygen occupancy in the GSIR LR-NCM proves that the oxygen vacancies are successfully created after applying the GSIR process.

High-resolution transmission electron microscopy (TEM; Fig. 2c,f) was used to further illustrate the changes between the pristine and GSIR LR-NCM. A clean surface with Li and TM layers extending to the very edge of atomic planes is observed in the pristine LR-NCM (Fig. 2c). Meanwhile, a non-uniform surface (varying in thickness from 10 to 20 nm) is detected on the GSIR LR-NCM sample. In addition, high-angle annular-dark-field scanning-TEM images (Fig. 2d,g) for the pristine and GSIR LR-NCM are compared to demonstrate the lack of any obvious

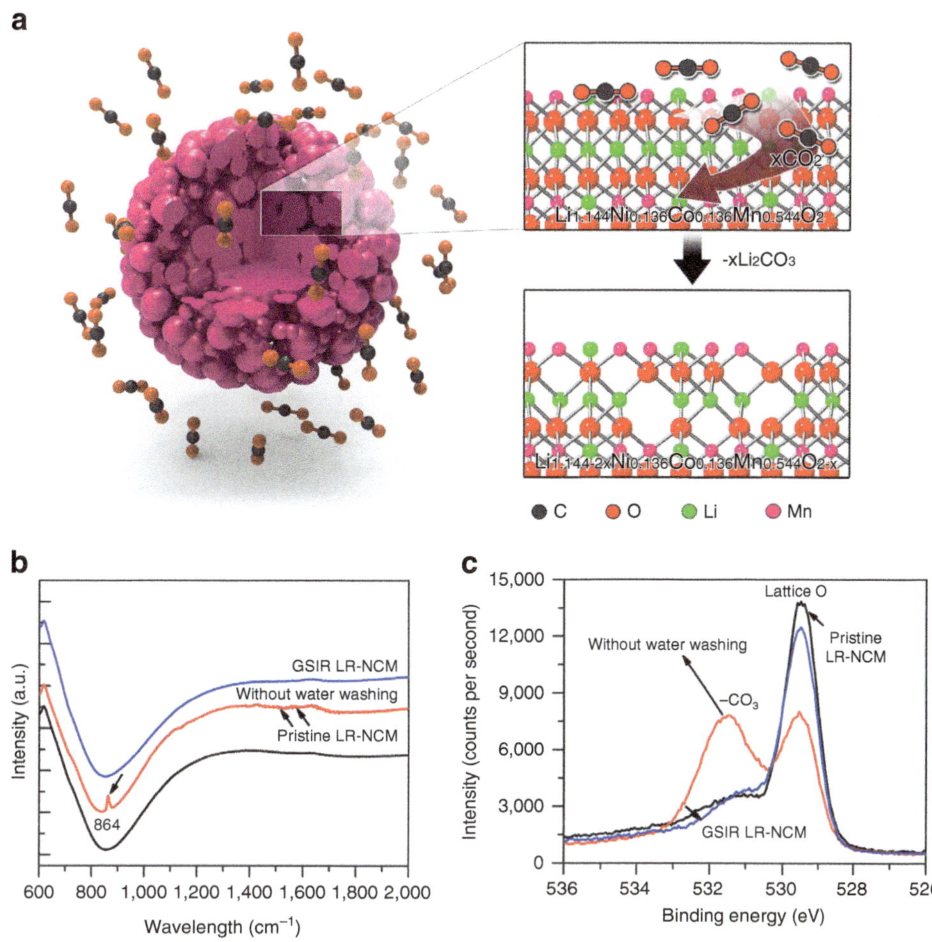

Figure 1 | GSIR. (**a**) Schematic of GSIR between Li-rich layered oxides and carbon dioxide. (**b**) FTIR spectrum. The GSIR LR-NCM without water washing (red line), compared with the pristine LR-NCM, has three additional peaks, corresponding to the IR-active modes of $-CO_3$. The sharp peak ($864\,cm^{-1}$) is the CO_3 group bending mode, while the other two peaks (1,438 and $1,500\,cm^{-1}$) are from the C-O asymmetric and symmetric stretching modes; (**c**) XPS spectra of the O 1s. It is clear that the GSIR LR-NCM has lower intensity than that of the pristine LR-NCM at lower-banding energy, which indicates some lattice O has been extracted by CO_2.

structural change. Although electron diffraction from some particles shows spinel/rock-salt-like phase in the GSIR LR-NCM compared with the pristine LR-NCM sample (Fig. 2e,h), it should be noted this phase transformation only occurs on the surface of GSIR LR-NCM within several atomic layers, as indicated in Fig. 2f, while the oxygen vacancy presence are 10–20-nm deep. These results are consistent with our SXRD and ND refinements, which verify that the GSIR process does not significantly alter the bulk structure.

Electron energy-loss spectroscopy (EELS) mapping was applied to compare relative surface composition changes for the pristine and GSIR LR-NCM sample (Fig. 3a,b). Oxygen reduction at mixed colour images is well-demonstrated on the edge of GSIR LR-NCM in comparison with the pristine LR-NCM. To study the electronic states of oxygen and transition metal from the particle, EELS was collected at different positions from bulk to surface of the GSIR LR-NCM (represented by colours in Fig. 3c,d). The spectra of the O-K-edge main peaks in Fig. 3e are normalized with the Mn L_3 peak. Interestingly, it is observed that reduced O pre-peak intensities are from the surface in the GSIR LR-NCM sample. The weak pre-peak intensities can be even seen at 20 nm for the GSIR LR-NCM sample although the difference is small. The reduced ratio of the O-K-edge pre-peak intensities to the main peak can result from a change in the local environment of oxygen, especially from the oxygen vacancies formed on the surface[26,37]. The relative distribution of chemical composition is plotted in Fig. 3f. The content of O on average is 55% in the first 10 nm surface layer compared with the bulk value of 65%. The Mn $L_{3,2}$ white lines provide information on Mn valence (oxidation); The Mn $L_{3,2}$ white lines for the GSIR LR-NCM sample display a significant energy shift, and their L_3/L_2 intensity ratio remains low until 20 nm away from the surface of the GSIR LR-NCM sample (Fig. 3g), indicating a lower Mn valence on the particles' surface. When oxygen vacancies are created on the particles' surface, the valence of Mn is reduced simultaneously with the extraction of Li to compensate for the changes in charge. Therefore, as confirmed by both the ND and STEM/EELS results, oxygen vacancies have been successfully introduced on the 10–20 nm of the surface without a noticeable interruption in the bulk structure through the GSIR approach.

Electrochemical properties of the pristine and GSIR LR-NCM. CR2032 coin cells with metallic Li as the counter electrode were assembled to investigate the electrochemical performance of the pristine and GSIR LR-NCM samples. Similar to other Li-rich layered oxides, both of them (Fig. 4a) exhibit a long plateau region at ~ 4.5 V versus Li^+/Li^0, which usually is ascribed to the

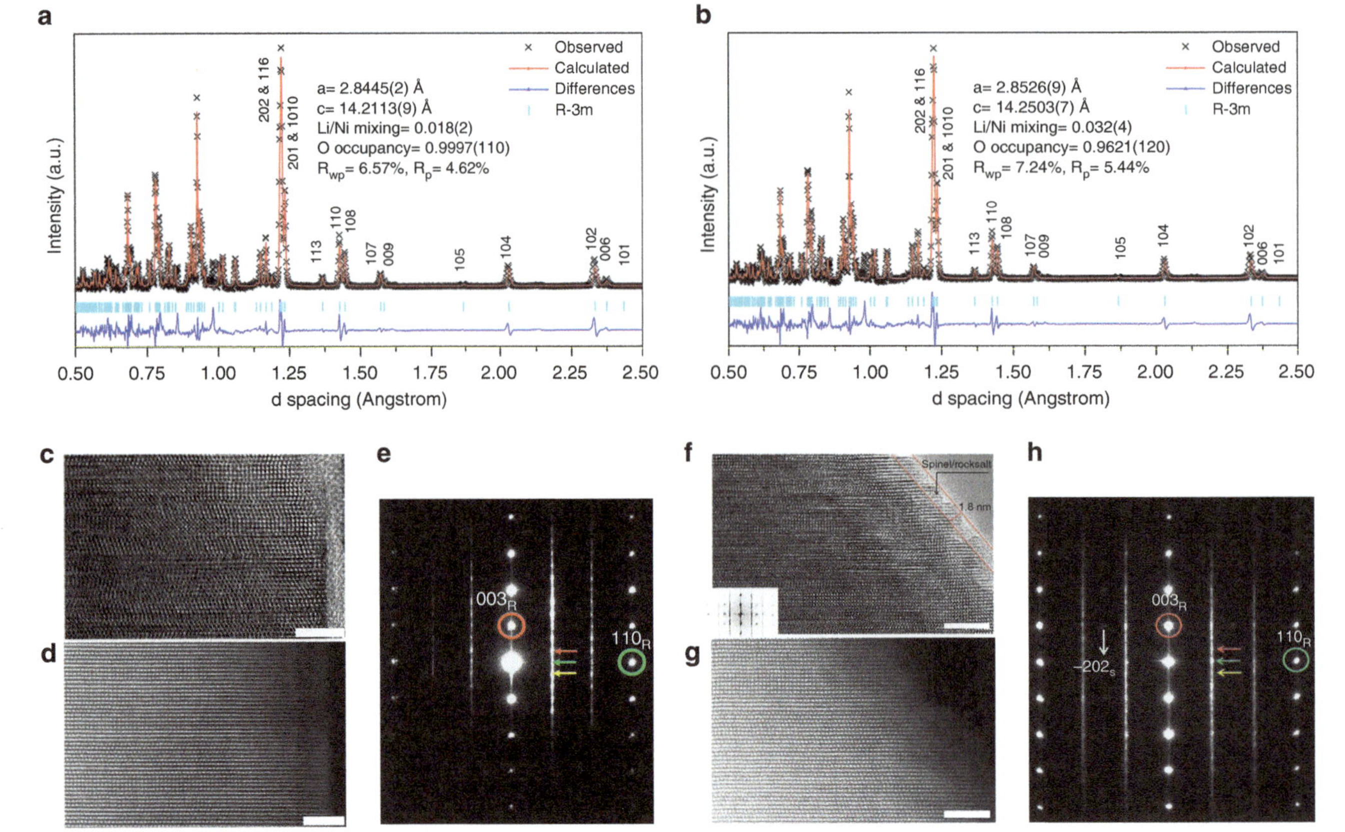

Figure 2 | Structural characterizations of the pristine and GSIR LR-NCM. (**a**,**b**) TOF ND patterns for the pristine and GSIR LR-NCM. On the basis of these results, the oxygen vacancy for the GSIR LR-NCM is around three per cent higher than that of the pristine LR-NCM sample; (**c**) high-resolution transmission electron microscopy (HRTEM) image and, (**d**) high-angle annular dark-field—STEM (HAADF-STEM) image for the pristine LR-NCM; (**e**) electron-diffraction (ED) pattern for the pristine LR-NCM. The elongation of these spots are along $[001]_R{}^*$ as well as $[001]_M{}^*$ direction. The spots indicated with red, green and yellow arrows represent (110), (020) and (− 110) of the monoclinic Li_2MnO_3 (donated as M) variants $[1\text{-}10]_M$, $[100]_M$ and $[110]_M$, respectively. The spot marked by a red circle is shared by $(003)_R$ of the rhombohedral (donated as R), as well as the structure $(001)_M$ of all monoclinic variants. The spot marked by a green circle is shared by $(110)_R$ of the rhombohedral, as well as the structure $(33\text{-}1)_M$, $(060)_M$ and $(\text{-}331)_M$ of the monoclinic. (**f**) HRTEM image with fast Fourier transform (FFT) about the surface and (**g**). HAADF-STEM image for the GSIR LR-NCM; (**h**) ED for the GSIR LR-NCM. The weak spot indicated by the white arrow can only be indexed as the structure $(-202)_S$ of the spinel (donated as S). Scale bars, 5 nm.

electrochemical activation of the Li_2MnO_3 component during the initial charge process[38–40], although an electrolyte/electrode side reaction is expected in this region. Interestingly, the value of the initial discharge capacity for the GSIR LR-NCM can reach as high as 301 mAh g^{-1} compared with that of 276 mAh g^{-1} for the pristine LR-NCM. The initial coulombic efficiency also increases from 83.8 to 93.2%. The corresponding differential capacity versus voltage (dQ/dV) curves are plotted in Fig. 4b. A lower shift in the oxidation peak after the GSIR process is due to the decrease in electrochemical impedance (Supplementary Fig. 6 and Supplementary Table 1). In addition, the decline of the oxidation peak around 4.5 V versus Li^+/Li^0 observed in the enlarged dQ/dV curves (inset in Fig. 4b) indicates that the GSIR process has pre-activated the Li_2MnO_3 component responsible for the 4.5 V plateau. The rate capability and cycling stability further highlight the advantages of our GSIR LR-NCM sample (Fig. 4c). At all tested rates, the GSIR LR-NCM exhibits a higher capacity than that of the pristine LR-NCM. The unique characteristic at different rates for the GSIR LR-NCM is that the additional discharge capacity (Supplementary Fig. 7a–f) results only from the lower-potential region (<3.5 V versus Li^+/Li^0). It is remarkable that the GSIR LR-NCM delivers a higher discharge capacity of 298 mAh g^{-1} when it returns to the 0.1 C-rate, compared with that of 288 mAh g^{-1} for the pristine LR-NCM. More importantly, the charge–discharge plots at subsequent cycles (Supplementary Fig. 7g,h) for the GSIR LR-NCM demonstrate a slight degradation in potential after 100 cycles, even for a discharge capacity as high as 300 mAh g^{-1}. To track the origin of the additional capacity after the GSIR, the energy density and discharge capacity below and above 3.5 V versus Li^+/Li^0 are plotted as a function of the cycle number, as depicted in Supplementary Fig. 8i,j. It proves that the additional capacity in discharge capacity also comes from the lower-potential region (<3.5 V versus Li^+/Li^0).

To evaluate the stability of the GSIR LR-NCM, a more challenging measurement was selected (Fig. 4d,e). Cells based on the GSIR LR-NCM show a higher initial capacity of 306 mAh g^{-1} (0.5 C-rate) and 280.9 mAh g^{-1} (1.0 C-rate), compared with that of 287 mAh g^{-1} (0.5 C-rate) and 269 mAh g^{-1} (1.0 C-rate) for the pristine LR-NCM at the elevated temperature of 55 °C. The initial charge–discharge curves (Supplementary Fig. 8a,b) are similar to the results at room temperature (Fig. 4a). In addition, the cells based on the pristine LR-NCM display only 223 mAh g^{-1} (at 0.5 C-rate) and 179 mAh g^{-1} (at 1.0 C-rate) after 100 cycles and 150 cycles, respectively, whereas the GSIR LR-NCM shows an excellent capacity of about 290 mAh g^{-1} (at 0.5 C-rate) and 262 mAh g^{-1} (at 1.0 C-rate) during the same cycling period. Moreover, the charge–discharge plots of

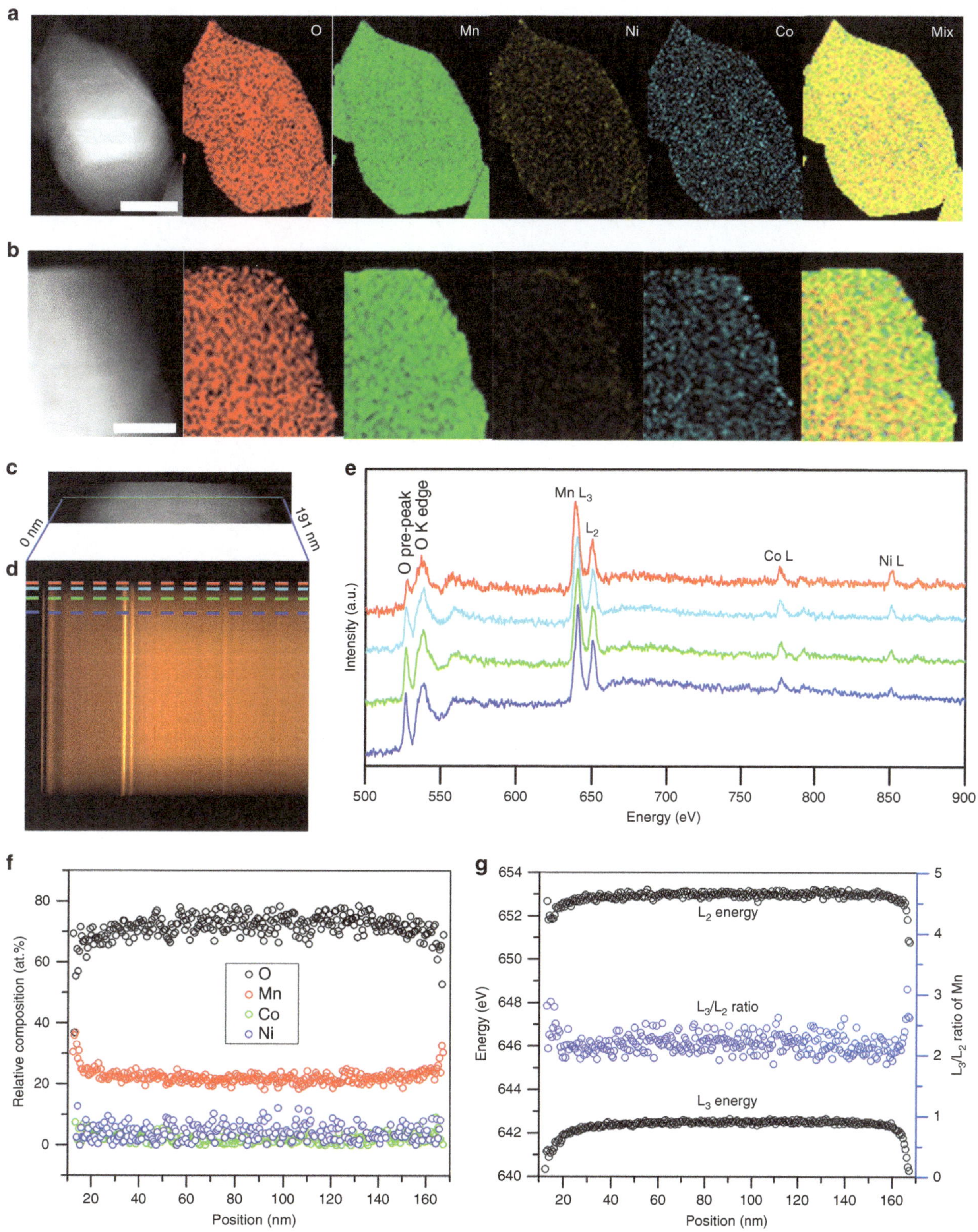

Figure 3 | Relative surface composition changes for the pristine and GSIR LR-NCM. (**a**) STEM-EELS spectrum images for the pristine LR-NCM. Scale bar, 50 nm. (**b**) The GSIR LR-NCM. Scale bar, 25 nm. Oxygen reduction at mixed colour images is well-demonstrated on the edge in comparison with the pristine LR-NCM (**a**). (**c**) STEM image of a particle of the GSIR LR-NCM. (**d**) EELS spectrum image from the vertical green line (0–191 nm) in the left STEM image (**c**), showing O K, Mn $L_{2,3}$, Co $L_{2,3}$, Ni $L_{2,3}$ edges. The dispersion is 0.25 eV/channel. The scan step is 0.127 nm. (**e**) EELS spectrum profiles from the surface to the interior as marked by the horizontal dashed lines with the same colour as in **d**. The spectra are normalized with Mn L_3 peak, and each spectrum was averaged vertically with 10 individual spectra to improve the signal noise ratio. Weak O pre-peak on the surface indicates the oxygen vacancies are formed. (**f**) Relative atomic composition of O (black), Mn (red), Co (blue) and Ni (green) as a function of position calculated based on the integrated EELS peak intensity in **e**. (**g**) Mn L_3, and L_2 energy (top and bottom of panel) from refinement, and L_3/L_2 intensity ratio.

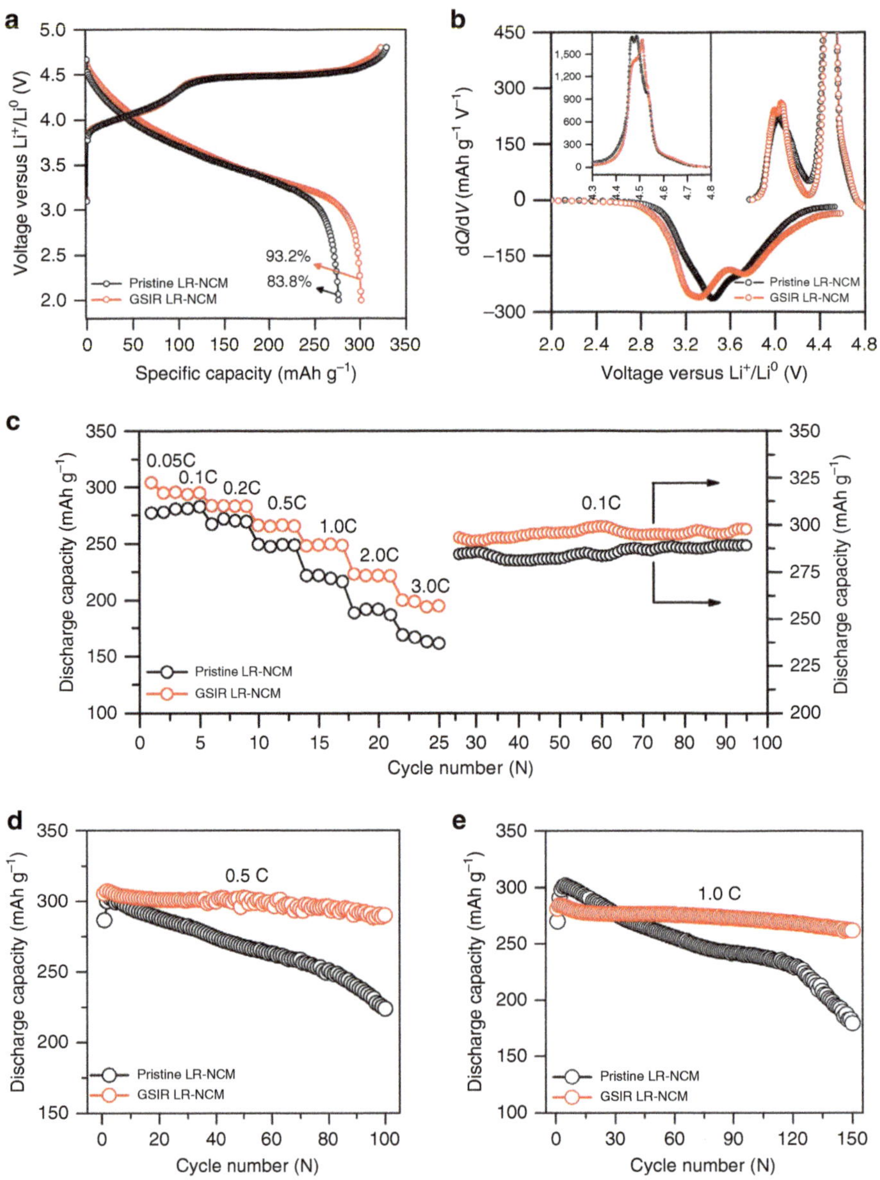

Figure 4 | Charge-discharge characteristics of the pristine and GSIR LR-NCM. (**a**) First charge-discharge profiles of the pristine and GSIR LR-NCM obtained from a 2032-type coin cells at 0.05 C-rate, where 1.0 C-rate corresponds to the current density of 250 mA g^{-1}. (**b**) Differential capacity (d*Q*/d*V*) plots of the initial cycle curves of the pristine and GSIR LR-NCM. The inset shows the enlarged d*Q*/d*V* curves about 4.5 V. (**c**) Discharge-rate capacity after charging galvanostatically at 0.1 C-rate before each discharge. The capacity retention when performing charge-discharge cycles at constant 0.1 C-rate for 70 cycles after all rates tested. (**d,e**) Cycling performance of the pristine and GSIR LR-NCM at 55 °C by applying a constant current density of 0.5 C-rate and 1.0 C-rate (250 mA g^{-1}), respectively. The loading density of the active material on the electrode was around 5.5 mg cm^{-2}. It is noticeable that the electrochemical data are derived from the average value of at least five coin cells, and the errors on the specific capacity are about 3 mAh g^{-1}.

subsequent cycles (Supplementary Fig. 8d,f) for the GSIR LR-NCM exhibit much slower potential degradation profiles at different rates than those of the pristine LR-NCM (Supplementary Fig. 8c,e). The difference clearly signifies that surface oxygen vacancies introduction in Li-rich layered oxides without severe structural destruction has a considerable effect on improving electrochemical performance.

Discussion

The influence of surface oxygen vacancies introduced by the GSIR process on the electrochemical performance can be discussed in two aspects. (1) The generation of highly reactive oxygen radicals on the particles' surface during the electrochemical process, which determines the electrochemical resistance at the electrode/electrolyte interface; (2) the participation of the oxygen reaction in the form of O^{2-}/O^{-} or O_2^{2-} in the bulk of the material, which are the dominant factors in compensating for extra electrons on the superior capacity of Li-rich layered oxides.

First, *operando* differential electrochemical mass spectrometry (DEMS) experiments were performed to evaluate the gas evolution during the initial charge–discharge process (Fig. 5a), which is largely related to the activity of oxygen on the particles' surface[18,41,42]. Irrespective of the cycling profile, as shown in Fig. 5b, O_2 is detected near 4.5–4.6 V versus Li^+/Li^0 for both the pristine and GSIR LR-NCM; however, the O_2 gas released from the GSIR LR-NCM electrode is much less than that of the pristine electrode. It is suggested that the formation of oxygen vacancies

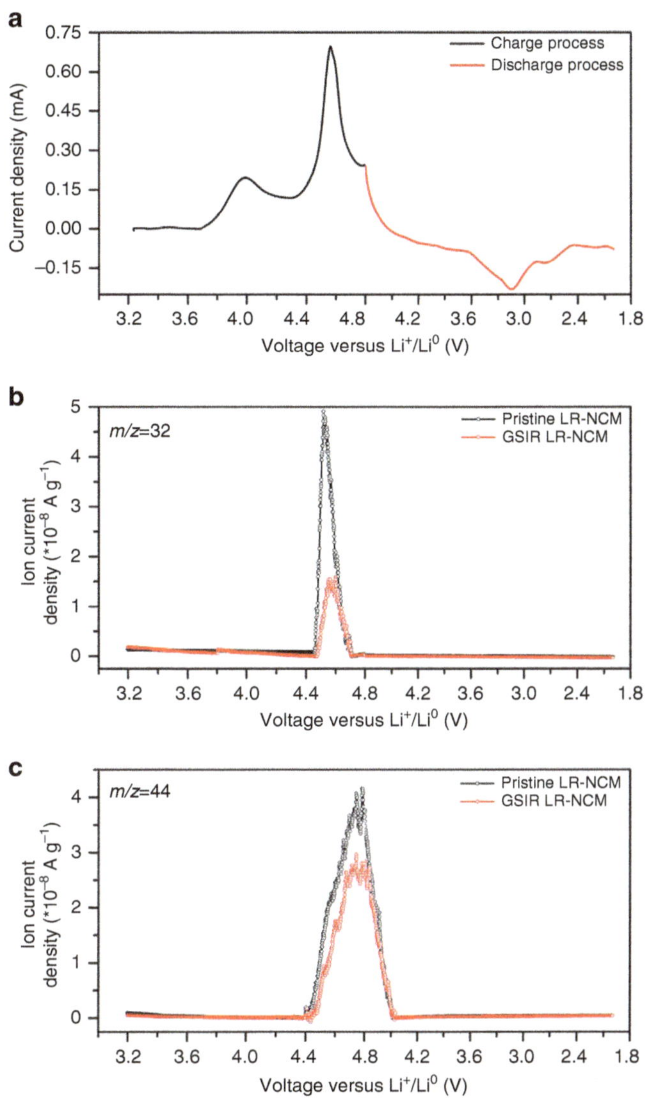

Figure 5 | *Operando* DEMS. (**a**) Initial cyclic voltammetry (CV) curve of the pristine and GSIR LR-NCM electrode. (**b**) Oxygen gas profile of the initial charge-discharge process for the pristine and GSIR LR-NCM electrode. (**c**) Carbon dioxide gas profile of the initial charge-discharge process as the above electrodes. The loading density mass for these two electrodes are the same about 7.76 mg. The scan rate was $\sim 0.1\,mV\,s^{-1}$, and the potential range was between 2.0 and 4.8 V versus Li^+/Li^0 at an elevated temperature of 55 °C.

through GSIR could greatly suppress the evolution of O_2 gas. According to the equation $P_{O_2} = K\left[V_{\ddot{O}}\right]^{-6}/16$ in the Supplementary note 1, less O_2 gas will be released due to the reduced partial pressure of oxygen on the surface of the GSIR LR-NCM electrode after oxygen vacancies are created. At the same time, the released CO_2 gas for the GSIR LR-NCM (Fig. 5c), which is usually considered as the source of decomposition of the electrolyte at high potentials[42], is smaller than that of the pristine LR-NCM. Therefore, the oxygen vacancies on the surface can hinder the highly reactive oxygen radicals from generating during the electrochemical process.

Second, the GSIR LR-NCM sample exhibits almost the same capacity in the plateau region as in the pristine sample with much less oxygen redox participation on the surface. To compensate for ionic changes during the charge process, more oxygen reactions in the form of O^{2-}/O^- or O^{2-} (instead of the O^{2-}/O_2 redox couple) are triggered in the bulk of the GSIR LR-NCM. The influence of the bulk oxygen activity in the diffusivity of lithium ions was investigated by first-principles calculations. For simplicity, the supercell was defined as $Li_{28}Ni_6Mn_{14}O_{48}$, which was adopted as a computational model in the previous work[24]. A specific $Li_{20/28}$ concentration was chosen to simulate the last stage of discharge (below ~3.2 V), in which the GSIR LR-NCM demonstrates the biggest difference in electrochemical performance compared with the pristine LR-NCM. This stage is also generally acknowledged as a rate-determining step of the whole discharge process in Li-rich layered oxides[43]. In this stage, the Li-ion migrates from the octahedral site of the Li layer into the octahedral site of the adjacent transition metal layer through an empty tetrahedral site in the former (as shown in Fig. 6a). When no oxygen vacancy is introduced, the Li tetrahedron site, which is face-sharing with the TM layer Li, is most stable. There is a high chance of trapping the Li-ion in the tetrahedral site (so-called 'Li–Li dumbbell') without diffusing into the adjacent octahedral site in the transition metal layer[24]. These dumbbells finally block the pathways of lithium diffusion, thereby decreasing ionic conductivity. In contrast, when oxygen vacancies are introduced the under-coordinated site is no longer stable, and the migration barrier is either zero or only around 170–210 meV (ref. 44). The trapped Li-ion has a higher chance of escaping from the tetrahedral sites and continuing its diffusion process.

Overall, Fig. 6b,e summarizes the proposed reaction mechanism for the pristine and GSIR LR-NCM during charging and discharging. Before charging, uniform oxygen and lithium vacancies are created on the particle surface of the GSIR LR-NCM, accompanied by the pre-activation of certain amount of the Li_2MnO_3 component. The sloped region (<4.4 V) of the charging curve is attributed to the extraction of Li^+ ions from the lithium layer. No difference is expected between the pristine and GSIR LR-NCM material during this process. During the charging plateau region, more lithium ions can be extracted from the Li_2MnO_3 component, together with the oxidation of lattice oxygen[19,20,22,23]. The pre-activated surface layer of the GSIR sample with oxygen vacancies reduces the partial pressure of oxygen on the surface, and then prevents gaseous oxygen from evolving during plateau charging. A thinner SEI layer is thus expected to form on the particles' surface of the GSIR LR-NCM due to less side reactions between released oxygen and the electrolyte species, thereby decreasing the electrochemical resistance at the electrode/electrolyte interface, as supported by our electrochemical impedance-spectroscopy (EIS) results (Supplementary Fig. 6 and Supplementary Table 1). At the same time, the more reversible oxygen redox-reaction is triggered in the bulk of the GSIR LR-NCM during the high-potential-charging process. These oxygen vacancies can activate the Li trapped in the tetrahedral sites to provide a favourable environment for Li-ion diffusion during the rate-determining step (the last stage of discharging). Higher levels of Li-ion diffusivity in the bulk, together with smaller resistance to charge-transfer and larger numbers of activated Mn^{4+}/Mn^{3+} couples lead to higher capacity and better rate capability of the GSIR LR-NCM sample. Furthermore, minimal structure changes enable the superior cycling stability after GSIR modification.

In summary, a facile but effective approach to surface modification based on GSIR has been applied to create oxygen vacancies homogeneously without causing structural destruction. Oxygen vacancies in the surface regions up to 20-nm thick are found to facilitate Li diffusion by activating the Li in the tetrahedral sites and suppress the release of gas from the surface, which finally lead to higher discharge capacity and a better rate capability in the Li-rich layered oxides. To the best of our knowledge, this study also clarifies for the first time the critical

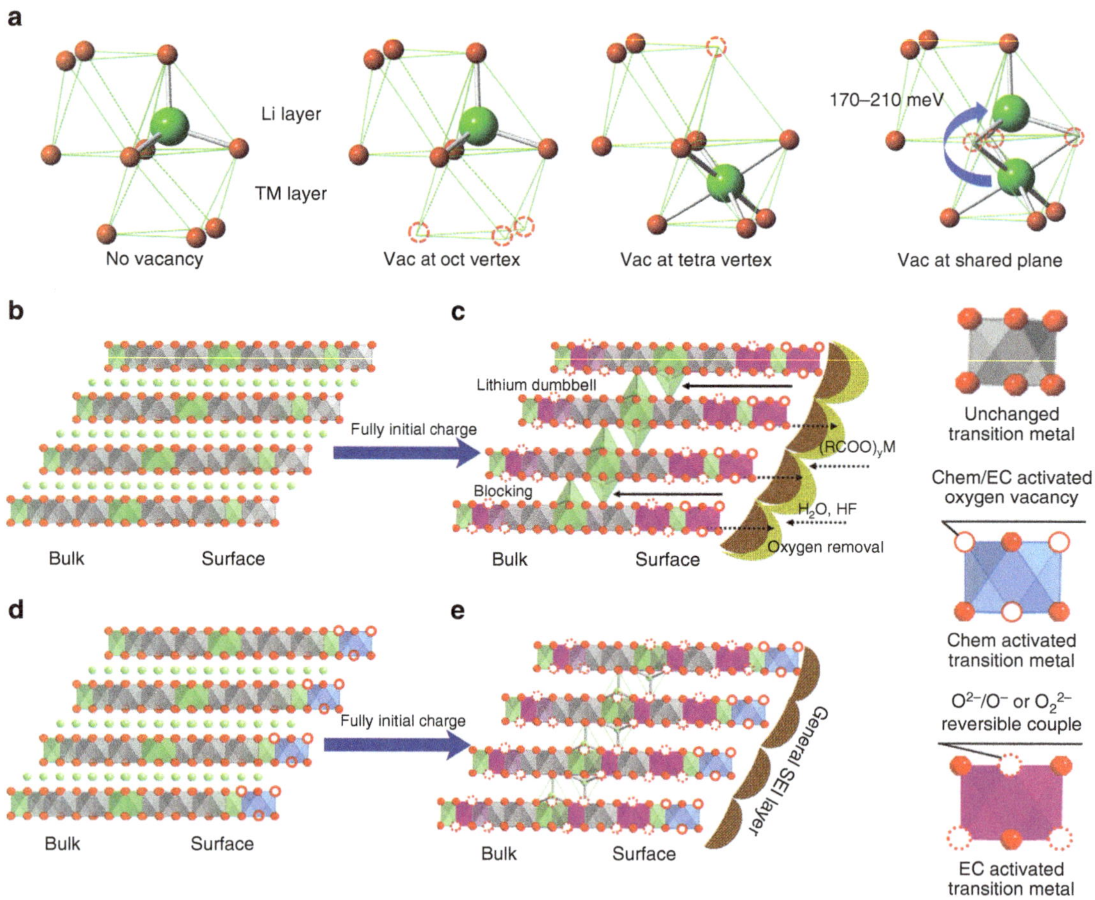

Figure 6 | Reaction mechanisms during charging and discharging. (**a**) Calculated site stability and activation barrier under conditions of no vacancy, vacancy at octahedron (Oct) vertex, vacancy (Vac) at tetrahedron (Tetra) vertex and vacancy (Vac) at shared plane. (**b**,**c**). The pristine LR-NCM before charging and after full initial charge. (**d**,**e**) The GSIR LR-NCM before charging and after full initial charge. Green, Li; red, O. The fundamental contribution of the novel surface modification is the uniform creation of lithium and oxygen vacancies on the surface of particles before electrochemical cycling. Consequently, more Mn^{4+}/Mn^{3+} redox couples in Li_2MnO_3 component are pre-activated; less oxygen gas is released from the material and less oxygen vacancies are created in the bulk during the charging process; limited electrode/electrolyte corrosion and bulk material structure transformation are formed in the battery system.

influence of the activity of surface oxygen in enhancing the electrochemical performance. We believe this facile, scalable and inexpensive approach opens a door for comprehensive designing and controlling oxygen activity in transition-metal-oxide systems to achieve excellent energy and power density in their commercial applications.

Methods

Synthesis of $Li[LO_{0.144}Ni_{0.136}Co_{0.136}Mn_{0.544}]O_2$. The synthesis procedures of $Li[LO_{0.144}Ni_{0.136}Co_{0.136}Mn_{0.544}]O_2$ were detailed as follows[45]: we pumped an aqueous solution containing $NiSO_4 \cdot 6H_2O$, $CoSO_4 \cdot 7H_2O$ and $MnSO_4 \cdot 4H_2O$ with a concentration of 2.0 mol l^{-1} into a continuously stirred tank reactor (CSTR, capacity of 50 l); at the same time, a 2.0 mol l^{-1} Na_2CO_3 solution and a 0.2 mol l^{-1} NH_4OH solution were added separately into the reactor. The co-precipitation temperature was held at 60 °C, and the pH value was fixed to 7.8. The resulting $(Ni_{1/6}Co_{1/6}Mn_{4/6})CO_3$ powders were washed several times with distilled water to remove residual Na^+, and dried in a vacuum oven at 80 °C for over 20 h. The resulting precipitates were mixed with Li_2CO_3 and the exact molar ratio between them was 0.7. The mixed powders were first pre-treated at 500 °C for 5 h in the air and then calcinated at 850 °C for 15 h in the air. Then, they were cooled to room temperature in the furnace. The as-obtained sample was labelled as pristine LR-NCM.

GSIR process. The GSIR process of the pristine LR-NCM sample with CO_2 was carried out as follows: a set amount of LR-NCM cathode materials and a given amount of NH_4HCO_3, which is as the source of CO_2 after decomposition at certain temperatures, were placed in an enclosed reactor in an argon-filled glove box ($O_2 < 0.1$ p.p.m.) with a total volume of 100 ml, and heated at 200 °C for 10 h. The optimal molar ratio between carbon dioxide and Li-rich layered oxides was about 1:5. It is important that the pristine LR-NCM cathode materials and the NH_4HCO_3 must be separated at the beginning for ensuring a homogeneous GSIR. To remove the reaction products on the surface of the LR-NCM, the sample obtained was washed with water several times and dried at 120 °C for 12 h. The as-obtained product was termed the GSIR LR-NCM. We would like to stress that the same procedures have been adopted in both NIMTE and UCSD labs, and all the electrochemical data have been completely reproduced.

Materials characterization. A field-emission scanning electron microscopy image was acquired on a Hitachi S-4800. A FTIR analysis was performed on a pellet made of active materials and KBr powder, using an FTIR spectrometer (Nicolet, 6700 series, USA). XPS measurements were collected using an AXIS Ultra DLD spectrometer with Al Kα (1,253.6 eV) radiation, to investigate the changes in the compositions on the cathode surface. The compositions of the pristine, and GSIR LR-NCM sample were confirmed quantitatively by inductive coupled plasma-atomic emission spectrometry with an emission spectrometer (Optima 2100 DV, Perkin-Elmer). The tap density of the pristine LR-NCM sample was measured and calculated by tap density instrument (FSZ-4, Ruike).

TOF powder ND. The TOF powder ND data were collected on the VULCAN beamline at the Spallation Neutron Sources (SNS) in the Oak Ridge National Laboratory[46]. Around 0.6 g of powder was packed into a vanadium sample can. An incident beam (5 mm × 12 mm) of 0.7- to 3.5-Å bandwidth allowing a 0.5–2.5 Å d-spacing in the diffracted patterns in the ± 90° 2θ detector banks was selected, using the double-disk choppers at a speed of 30 Hz. A high-resolution mode was used with $\Delta d/d \sim 0.25\%$. The SNS was at a nominal 1,100 kW. Powder-neutron-diffraction data were collected at a high-resolution mode for 3 h and reduced by the

VDRIVE software. Full-pattern Rietveld refinement was performed using the GSAS programs with the EXPGUI interface[47].

TEM characterization. TEM images and EELS spectra were carried out using the JEM-ARM200CF scanning/transmission electron microscope at Brookhaven National Laboratory. The instrument is equipped with a cold field-emission source, two aberration-correctors and a high-resolution dual electron energy-loss spectrometer (Quantum GIF). The optimal energy-resolution was ~0.35 eV, as judged by the full-width at half-maximum of the zero-loss peak. Spectra were recorded either using a convergent-beam in STEM mode, or parallel-beam in TEM mode, for diffraction, imaging and spectroscopy. The refinement Mn L_3, and L_2 energy, and L_3/L_2 intensity ratio were based on Pearson's method[48].

X-ray diffraction and X-ray absorption spectroscopy. SXRD data were collected using beamline BL14B1 at the Shanghai Synchrotron Radiation Facility (SSRF). These data were recorded in the 2θ range between 10° and 85°, using a fixed wavelength of 1.2398 Å. The structural composite study was carried out by Rietveld refinement analysis, using the GSAS program. For X-ray absorption fine-structure spectroscopy studies, we performed measurements at beamline BL14W1 of the SSRF. The incident beam was mono-chromatized using a Si (111) fixed-exit, double-crystal monochromator. Foils of nickel, manganese and cobalt were used for energy calibration with zero energy (E_0), defined according to Kraft *et al.*[49]. Spectra were acquired in transmission mode, utilizing gas-ionization chambers as detectors. Data on X-ray absorption near-edge structure was extracted with established methods using the ATHENA software package.

Electrochemical measurements. The negative electrodes were prepared by casting the slurry with a mixture of 80 wt.% the pristine or GSIR LR-NCM, 10 wt.% acetylene black and 10 wt.% polyvinylidene fluoride binder on an aluminium foil. Electrode discs of 13-mm diameter were punched from the negative electrode, pressed at 8 Mpa per disc and dried at 80 °C for 12 h in vacuum to remove any residual NMP and traces of water. The prepared electrodes then were assembled in an Ar-filled glove box ($H_2O < 0.1$ p.p.m. and $O_2 < 0.1$ p.p.m.). Metallic Li was used as the counter electrode, and Celgard 2502 as a separator. The electrolyte solution was 1 M $LiPF_6$ in a 3:7 (volume ratio) mixture of ethylene carbonate–dimethyl carbonate (Zhangjiagang Guotai-Huarong New Chemical Materials Co., Ltd.). The cells were galvanostatically charged and discharged on a LAND-CT2001A battery-test system. Unless otherwise specified, the cells typically cycled between 2.0 and 4.8 V versus Li^+/Li^0, and the temperature for testing was about 28 °C. EIS measurements were performed on an Autolab83710 impedance analyser with a.c. voltage amplitude of 5 mV, and frequency range of 0.01–100,000 Hz. The metallic Li served as both the counter and reference electrodes during the EIS measurements.

***Operando* DEMS.** *Operando* DEMS analysis was carried out to detect the gases generated during the initial charge–discharge process. The DEMS cell was assembled in an Ar-filled glove box. In all, 1 M $LiPF_6$ in 1:1 ethylene carbonate:-dimethyl carbonate was used as the electrolyte, and a glass-fiber filter (Whatman GF/D) was used as a separator. Electrodes were prepared by casting a slurry with a composition of 80 wt. % active materials, 10 wt. % Super C65 (Timcal) and 10 wt.% PVDF (Kynar FLEX 761A, Arkema Group) on to an aluminium current-collector foil. Electrode tapes then were punched into 22-mm diameter discs, and dried at 120 °C in an under vacuum oven for 24 h. The electrodes were assembled into Swagelok-type cells, using metallic Li as both the counter and the reference. Thereafter, the cell was connected to the mass spectrometer. The cell was purged continuously with gaseous Ar, which flowed from the cell into the mass spectrometer carrying the evolved gases for MS analysis. *Operando* DEMS cells were conducted on a VSP electrochemical workstation (Biologic, France). The mass signals were recorded as a function of time and the cell voltage. The time resolution of ion-current intensity is optimized by selectively scanning the *m/z* 32 and 44 signals. In line with previous studies[41,42], only O_2 and CO_2 were evolved during the initial cycling of Li-rich layered-oxide electrodes.

Computation methodology. First-principles calculations were performed in the spin-polarized GGA + U approximations to the density functional theory. The core electron states were represented by the projector-augmented-wave method[50] as implemented in the Vienna *ab initio* simulation package[51–53]. We used the Perdew-Burke-Ernzerhof[54] exchange correlation, and a plane-wave representation for the wave function with a cutoff energy of 450 eV. The Brillouin zone was sampled with a dense k-points mesh by Gamma packing. Our supercell, composed of 24 formula units of $Li[Ni_{1/4}Li_{1/6}Mn_{7/12}]O_2$, used in our previous work was used again to obtain the site's stability and activation barrier. The Hubbard U correction was introduced to describe the effect of localized d electrons of the transition-metal ions. The applied effective U value given to Mn ions is 5 eV, and that to Ni ions is 5.96 eV (ref. 55). In this calculation, a concentration of Li20/28 was applied, in which the GSIR LR-NCM exhibits a difference in electrochemical performance compared with pristine LR-NCM. The nudged elastic band method was used to find the minimum-energy path and the energy barrier for Li diffusion inside the materials.

References

1. Adler, S. B. Factors governing oxygen reduction in solid oxide fuel cell cathodes. *Chem. Rev.* **104,** 4791–4843 (2004).
2. Liao, P. & Carter, E. A. New concepts and modeling strategies to design and evaluate photo-electro-catalysts based on transition metal oxide. *Chem. Soc. Rev.* **42,** 2401–2422 (2013).
3. Huang, R. C. *et al.* Oxygen-vacancy ordering at surfaces of lithium manganese (III, IV) oxide spinel nanoparticles. *Angew. Chem. Int. Ed.* **50,** 3053–3057 (2011).
4. Okamoto, Y. Ambivalent effect of oxygen vacancies on Li_2MnO_3: a First-Principles study. *J. Electrochem. Soc.* **159,** A152–A156 (2012).
5. Sushko, P. V., Rosso, K. M., Zhang, J.-G., Liu, J. & Sushko, M. L. Oxygen vacancies and ordering of d-levels control voltage suppression in oxide cathodes: the case of spinel $LiNi_{0.5}Mn_{1.5}O_{4-\delta}$. *Adv. Funct. Mater.* **23,** 5530–5535 (2013).
6. Kim, J.-H., Myung, S.-T., Yoon, C. S., Kang, S. G. & Sun, Y.-K. Comparative study of $LiNi_{0.5}Mn_{1.5}O_{4-\delta}$ and $LiNi_{0.5}Mn_{1.5}O_4$ cathodes having two crystallographic structures: Fd-3m and $P4_332$. *Chem. Mater.* **16,** 904–916 (2007).
7. Jung, S.-K. *et al.* Understanding the degradation mechanisms of $LiNi_{0.5}Co_{0.2}Mn_{0.3}O_2$ cathode material in lithium ion batteries. *Adv. Energy Mater.* **4,** 201300787 (2014).
8. Lin, F. *et al.* Surface reconstruction and chemical evolution of stoichiometric layered cathode materials for lithium-ion batteries. *Nat. Commun.* **5,** 3529 (2014).
9. Yu, H. *et al.* Direct atomic-resolution observation of two phases in the $Li_{1.2}Mn_{0.567}Ni_{0.166}Co_{0.067}O_2$ cathode material for lithium-ion batteries. *Angew. Chem. Int. Ed.* **52,** 5969–5973 (2013).
10. Johnson, C. S. *et al.* The significance of the Li_2MnO_3 component in 'composite' $xLi_2MnO_3 \cdot (1-x)LiMn_{0.5}Ni_{0.5}O_2$ electrodes. *Electrochem. Commun.* **6,** 1085–1091 (2004).
11. Thackeray, M. M. *et al.* Li_2MnO_3-stabilized $LiMO_2$ (M = Mn, Ni, Co) electrodes for lithium-ion batteries. *J. Mater. Chem.* **17,** 3112–3125 (2007).
12. Rossouw, M. H. & Thackeray, M. M. Lithium manganese oxides from Li_2MnO_3 for rechargeable lithium battery applications. *Mater. Res. Bull.* **26,** 463–473 (1991).
13. Lu, Z., MacNeil, D. D. & Dahn, J. R. Layered cathode materials $Li[Ni_xLi_{1/3-2x/3}Mn_{2/3-x/3}]O_2$ for lithium-ion batteries. *Electrochem. Solid-State Lett.* **4,** A191–A194 (2001).
14. Xia, Q. *et al.* A Li-rich layered@spinel@carbon heterostructured cathode material for high capacity and high rate lithium-ion batteries fabricated via an *in situ* synchronous carbonization-reduction method. *J. Mater. Chem. A* **3,** 3995–4003 (2015).
15. Hy, S. *et al.* Performance and design considerations for lithium excess layered oxide positive electrode materials for lithium ion batteries. *Energy Environ. Sci.* **9,** 1931–1954 (2016).
16. Ohzuku, T., Nagayama, M., Tsuji, K. & Ariyoshi, K. High-capacity lithium insertion materials of lithium nickel manganese oxides for advanced lithium-ion batteries: toward rechargeable capacity more than 300 mAh g^{-1}. *J. Mater. Chem.* **21,** 10179–10188 (2011).
17. Lu, Z. & Dahn, J. R. Understanding the anomalous capacity of Li/ $Li[Ni_xLi_{1/3-2x/3}Mn_{2/3-x/3}]O_2$ cells using *in situ* X-ray diffraction and electrochemical studies. *J. Electrochem. Soc.* **149,** A815–A822 (2002).
18. Armstrong, A. R. *et al.* Demonstrating oxygen loss and associated structural reorganization in the lithium battery cathode $Li[Ni_{0.2}Li_{0.2}Mn_{0.6}]O_2$. *J. Am. Chem. Soc.* **128,** 8694–8698 (2006).
19. Koga, H. *et al.* Reversible oxygen participation to the redox processes revealed for $Li_{1.2}Mn_{0.54}Co_{0.13}Ni_{0.13}O_2$. *J. Electrochem. Soc.* **160,** A786–A792 (2013).
20. Koga, H. *et al.* Different oxygen redox participation for bulk and surface: a possible global explanation for the cycling mechanism of $Li_{1.20}Mn_{0.54}Co_{0.13}Ni_{0.13}O_2$. *J. Power Sources* **236,** 250–258 (2013).
21. Sathiya, M. *et al.* High performance $Li_2Ru_{1-y}Mn_yO_3$ ($0.2 \leq y \leq 0.8$) cathode materials for rechargeable lithium-ion batteries: their understanding. *Chem. Mater.* **25,** 1121–1131 (2013).
22. Sathiya, M. *et al.* Reversible anionic redox chemistry in high-capacity layered-oxide electrodes. *Nat. Mater.* **12,** 827–835 (2013).
23. Oishi, M. *et al.* Direct observation of reversible charge compensation by oxygen ion in Li-rich manganese layered oxide positive electrode material, $Li_{1.16}Ni_{0.15}Co_{0.19}Mn_{0.50}O_2$. *J. Power Sources* **276,** 89–94 (2015).
24. Xu, B., Fell, C. R., Chi, M. & Meng, Y. S. Identifying surface structural changes in layered Li-excess nickel manganese oxides in high voltage lithium ion batteries: a joint experimental and theoretical study. *Energy Environ. Sci.* **4,** 2223–2233 (2011).

25. Fell, C. R. *et al.* Correlation between oxygen vacancy, microstrain, and cation distribution in lithium-excess layered oxides during the first electrochemical cycle. *Chem. Mater.* **25,** 1621–1629 (2013).
26. Qian, D., Xu, B., Chi, M. & Meng, Y. S. Uncovering the roles of oxygen vacancies in cation migration in lithium excess layered oxides. *Phys. Chem. Chem. Phys.* **16,** 14665–14668 (2014).
27. Kubota, K. *et al.* Direct synthesis of oxygen-deficient Li_2MnO_{3-x} for high capacity lithium battery electrodes. *J. Power Sources* **216,** 249–255 (2012).
28. Song, B. *et al.* High rate capability caused by surface cubic spinels in Li-rich layer-structured cathodes for Li-ion batteries. *Sci. Rep.* **3,** 3094–3105 (2013).
29. Kang, S. H., Johnson, C. S., Vaughey, J. T., Amine, K. & Thackeray, M. M. The effects of acid treatment on the electrochemical properties of $0.5Li_2MnO_3$-$0.5LiNi_{0.44}Co_{0.25}Mn_{0.31}O_2$ electrodes in lithium cells. *J. Electrochem. Soc.* **153,** A1186–A1192 (2006).
30. Zhao, J. *et al.* An ion-exchange promoted phase transition in a Li-excess layered cathode material for high-performance lithium ion batteries. *Adv. Energy Mater* **5,** 201401937 (2015).
31. Yu, D. Y. W., Yanagida, K. & Nakamura, H. Surface modification of Li-excess Mn-based cathode materials. *J. Electrochem. Soc.* **157,** A1177–A1182 (2010).
32. Sun, Y. K. *et al.* The role of AlF_3 coatings in improving electrochemical cycling of Li-enriched nickel-manganese oxide electrodes for Li-ion batteries. *Adv. Mater.* **24,** 1192–1196 (2012).
33. Villevieille, C., Lanz, P., Bünzli, C. & Novák, P. Bulk and surface analyses of ageing of a 5V-NCM positive electrode material for lithium-ion batteries. *J. Mater. Chem. A* **2,** 6488–6493 (2014).
34. Yabuuchi, N., Yoshii, K., Myung, S. T., Nakai, I. & Komaba, S. Detailed studies of a high-capacity electrode material for rechargeable batteries, Li_2MnO_3-$LiCo_{1/3}Ni_{1/3}Mn_{1/3}O_2$. *J. Am. Chem. Soc.* **133,** 4404–4419 (2011).
35. Qiu, B. *et al.* Enhanced electrochemical performance with surface coating by reactive magnetron sputtering on lithium-rich layered oxide electrodes. *ACS Appl. Mater. Interfaces* **6,** 9185–9193 (2014).
36. Liu, H., Fell, C. R., An, K., Cai, L. & Meng, Y. S. *In situ* neutron diffraction study of the $xLi_2MnO_3 \cdot (1-x)LiMO_2$ ($x = 0$, 0.5; M = Ni, Mn, Co) layered oxide compounds during electrochemical cycling. *J. Power Sources* **240,** 772–778 (2013).
37. Wu, L. J. *et al.* Structural defects of silver hollandite, $Ag_xMn_8O_y$, nanorods: dramatic impact on electrochemistry. *ACS Nano* **9,** 8430–8439 (2015).
38. Arunkumar, T. A., Wu, Y. & Manthiram, A. Factors influencing the irreversible oxygen loss and reversible capacity in layered $Li[Li_{1/3}Mn_{2/3}]O_2$-$Li[M]O_2$ ($M = Mn_{0.5-y}Ni_{0.5-y}Co_{2y}$ and $Ni_{1-y}Co_y$) solid solutions. *Chem. Mater.* **19,** 3067–3073 (2007).
39. Wu, Y. & Manthiram, A. High capacity, surface-modified layered $Li[Li_{(1-x)/3}Mn_{(2-x)/3}Ni_{x/3}Co_{x/3}]O_2$ cathodes with low irreversible capacity loss. *Electrochem. Solid-State Lett.* **9,** A221–A224 (2006).
40. Liu, W. *et al.* Countering voltage decay and capacity fading of lithium-rich cathode material at 60 °C by hybrid surface protection layers. *Adv. Energy Mater.* **5,** 201500274 (2015).
41. Lanz, P., Sommer, H., Schulz-Dobrick, M. & Novák, P. Oxygen release from high-energy xLi_2MnO_3-$(1-x)LiMO_2$ (M = Mn, Ni, Co): electrochemical, differential electrochemical mass spectrometric, *in situ* pressure, and *in situ* temperature characterization. *Electrochim. Acta* **93,** 114–119 (2013).
42. Castel, E., Berg, E. J., El Kazzi, M., Novák, P. & Villevieille, C. Differential electrochemical mass spectrometry study of the interface of $xLi_2MnO_3 \cdot (1-x)LiMO_2$ (M = Ni, Co, and Mn) material as a positive electrode in Li-ion batteries. *Chem. Mater.* **26,** 5051–5057 (2014).
43. Yu, H. *et al.* Electrochemical kinetics of the $0.5Li_2MnO_3$-$0.5LiMn_{0.42}Ni_{0.42}Co_{0.16}O_2$ 'composite' layered cathode material for lithium-ion batteries. *RSC Adv.* **2,** 8797–8807 (2012).
44. Kang, K., Meng, Y. S., Breger, J., Grey, C. P. & Ceder, G. Electrodes with high power and high capacity for rechargeable lithium batteries. *Science* **311,** 977–980 (2006).
45. Qiu, B. *et al.* Electrochemical investigation of Li-excess layered oxide cathode materials/mesocarbon microbead in 18650 batteries. *Electrochim. Acta* **123,** 317–324 (2014).
46. An, K., Wang, X. L. & Stoica, A. D. Vulcan data reduction and interactive visualization software, ORNL Report, 621 (2012).
47. Toby, B. H. EXPGUI, a graphical user interface for GSAS. *J. Appl. Crystallogr.* **34,** 210–213 (2001).
48. Pearson, D. H. *et al.* White lines and d-electron occupancies for the 3d and 4d transition metals. *Phys. Rev. B: Condens. Matter Mater. Phys.* **47,** 8471–8478 (1993).
49. Kraft, S., Stümpel, J., Becker, P. & Kuetgens, U. High resolution X-ray absorption spectroscopy with absolute energy calibration for the determination of absorption edge energies. *Rev. Sci. Instrum.* **67,** 681–687 (1996).
50. Kresse, G. & Joubert, D. From ultrasoft pseudopotentials to the projector augmented-wave method. *Phys. Rev. B* **59,** 1758–1775 (1999).
51. Kresse, G. & Furthmüller, J. Efficiency of *ab*-initio total energy calculations for metals and semiconductors using a plane-wave basis set. *Comp. Mater. Sci.* **6,** 15–50 (1996).
52. Kresse, G. & Furthmüller, J. Efficient iterative schemes for *ab* initio total-energy calculations using a plane-wave basis set. *Phys. Rev. B* **54,** 11169–11186 (1996).
53. Kresse, G. & Hafner, J. *Ab* initio molecular-dynamics simulation of the liquid-metal–amorphous-semiconductor transition in germanium. *Phys. Rev. B* **49,** 14251–14269 (1994).
54. Perdew, J. P., Burke, K. & Wang, Y. Generalized gradient approximation for the exchange-correlation hole of a many-electron system. *Phys. Rev. B* **54,** 16533–16539 (1996).
55. Hinuma, Y., Meng, Y. S., Kang, K. & Ceder, G. Phase transitions in the $LiNi_{0.5}Mn_{0.5}O_2$ system with temperature. *Chem. Mater.* **19,** 1790–1800 (2007).

Acknowledgements

NIMTE is grateful for financial support from Strategic Priority Research Program of Chinese Academy of Sciences (CAS, Grant No. XDA09010101), Key Projects in Cooperation between CAS and Department of Energy, USA (CAS-DOE, Grant No. 174433KYSB20150047), and Ningbo Science and Technology Innovation Team (Grant No. 2012B82001). UC San Diego's efforts are supported by the Assistant Secretary for Energy Efficiency and Renewable Energy, Office of Vehicle Technologies of the U.S. Department of Energy (DOE) under Contract No. DE-AC02-05CH11231, Subcontract No. 7073923, under the Advanced Battery Materials Research (BMR) Program. Work at Brookhaven National Laboratory is supported by the U.S. DOE, Office of Basic Energy Science, Division of Materials Science and Engineering, under Contract No. DE-AC02-98CH10886. The synchrotron XRD and XAS measurements were carried out at SSRF of the beamline BL14B1 and BL14W1, respectively. The neutron experiments benefit from the SNS user facilities (VULCAN beamline) sponsored by the office of Basic Energy Sciences (BES), the Office of Science of the U.S. DOE. We acknowledge Dr Ding-Yi Tong for the design of the schematic.

Author contributions

B.Q. prepared the samples. M.Z., H.L. and Y.S.M. performed the mechanism analysis. B.Q. and M.Z. tested the electrochemical performances and performed analysis. L.W. and Y.Z performed the electron microscopy imaging, diffraction and spectroscopy experiments and analyses. J.W. performed the *operando* mass spectrometry analysis. D.Q. and Y.S.M. performed the first-principles calculations. H.L., Y.C., K.A. and S.H. recorded ND experiment data and analyses. Z.L., Y.S.M., and Y.X. supervised the research. B.Q. and M.Z. co-wrote the manuscript. All authors contributed to the discussion and provided feedback on the manuscript.

Additional information

Competing financial interests: The authors declare no competing financial interests.

Operando NMR spectroscopic analysis of proton transfer in heterogeneous photocatalytic reactions

Xue Lu Wang[1,*], Wenqing Liu[2,*], Yan-Yan Yu[3], Yanhong Song[2], Wen Qi Fang[1], Daxiu Wei[2], Xue-Qing Gong[3], Ye-Feng Yao[2,4] & Hua Gui Yang[1]

Proton transfer (PT) processes in solid-liquid phases play central roles throughout chemistry, biology and materials science. Identification of PT routes deep into the realistic catalytic process is experimentally challenging, thus leaving a gap in our understanding. Here we demonstrate an approach using operando nuclear magnetic resonance (NMR) spectroscopy that allows to quantitatively describe the complex species dynamics of generated H_2/HD gases and liquid intermediates in pmol resolution during photocatalytic hydrogen evolution reaction (HER). In this system, the effective protons for HER are mainly from H_2O, and CH_3OH evidently serves as an outstanding sacrificial agent reacting with holes, further supported by our density functional theory calculations. This results rule out controversy about the complicated proton sources for HER. The operando NMR method provides a direct molecular-level insight with the methodology offering exciting possibilities for the quantitative studies of mechanisms of proton-involved catalytic reactions in solid-liquid phases.

[1] Key Laboratory for Ultrafine Materials of Ministry of Education, School of Materials Science and Engineering, East China University of Science and Technology, Shanghai 200237, China. [2] Department of Physics, Shanghai Key Laboratory of Magnetic Resonance, East China Normal University, Shanghai 200062, China. [3] Key Laboratory for Advanced Materials, Centre for Computational Chemistry, Research Institute of Industrial Catalysis, East China University of Science and Technology, Shanghai 200237, China. [4] NYU-ECNU Institute of Physics at NYU Shanghai, 3663 Zhongshan Road North, Shanghai 200062, China. * These authors contributed equally to this work. Correspondence and requests for materials should be addressed to X.-Q.G. (email: xgong@ecust.edu.cn) or to Y.-F.Y. (email: yfyao@phy.ecnu.edu.cn) or to H.G.Y. (email: hgyang@ecust.edu.cn).

Proton transfer (PT) processes in solid–liquid phases are critically important and can be encountered throughout chemistry, biology and materials science[1–8]. The routes and dynamics of PT processes dictate the efficiency of the key biosynthetic and energy conversion systems, both natural and artificial[1–4]. Hence, tracking the PT trajectories deep into the realistic reaction becomes particularly significant. Molecular hydrogen (H_2) is a clean-burning fuel that can be produced from protons in the reductive half reaction of solar water splitting where the proton-related species are involved[2,9]. Nevertheless, critically important problems including side reactions restrict the productivity of the photocatalytic hydrogen evolution reaction (HER), in which the catalytic steps involve the coupling of light-generated redox equivalents to PT. In these cases, PT in HER is multidirectional. Thus, the addition of specific molecular reagents (also called sacrificial reagents), which is thermodynamically more oxidizable than water (H_2O), into the reaction environment can suppress the recombination of photogenerated charge carriers and the corresponding reverse reactions to some extent[10]. However, from a fundamental standpoint, some molecular reagents themselves contain protons, such as methanol, ethanol and triethanolamine, which means that such molecular agent itself can be a proton source and this undoubtedly increases the complexity of HER. Therefore, the vague and ambiguous proton source, PT routes, as well as the reaction mechanisms greatly limit our understanding and in turn the development of this field.

Recently, a series of methodologies for mechanism study, such as scanning tunnelling microscopy[11–13], time-resolved two-photon photoemission[14], temperature-programmed desorption[15–17], infrared spectroscopy[18–20], solid-state nuclear magnetic resonance[21–24] and density functional theory calculations (DFT)[25,26] have been used to address the central questions in HER. These methods can simulate and observe a rich surface chemistry of the interaction between adsorbed molecules and catalyst substrates, and distinguish various possible HER mechanisms. However, few of these techniques alone permit the direct quantification of specific species in reactions, and, moreover, the vast majority of them usually operate under ideal conditions (solid–gas or in ultrahigh vacuum), albeit with exquisite resolution, and it is still a quite tough task to take into account the complexities of practical conditions such as solvent effect, surface specificity and adsorbate–adsorbate interactions. For instance, the adsorption and dissociation of methanol (CH_3OH) gas can occur easily on anatase-TiO_2 (101) surface under continuous ultraviolet illumination; however, in the liquid solution weak interactions such as hydrogen-bonding effects between CH_3OH and H_2O may significantly complicate the reaction and probably impair the conclusion/mechanism derived from the ideal reaction conditions. Furthermore, some possible hydrolysates in HER are out of ability to be observed under the gas–solid vacuum condition, and some changes that occur during the reaction may no longer be apparent when the sample returns to a non-working state. In this regard, there still exists a large gap between over-idealized conclusions and realistic heterogeneous reaction systems. Yet, to the best of our knowledge, no experimental evidence for quantitative insight into the PT mechanisms of HER in solid–liquid phases has been reported.

Here we demonstrate an approach, using operando NMR spectroscopy, that allows to directly quantify the populations of H_2/HD gases and liquid intermediates in pmol resolution, as well as the tracking of the PT routes deep into the realistic heterogeneous HER conditions, which contains Pd/TiO_2 (anatase), H_2O and CH_3OH. In this system, the effective protons for HER are mainly from H_2O, and CH_3OH evidently serves as an outstanding sacrificial agent reacting with holes. Furthermore, it also provides evidence to rule out controversy about the complicated sources of protons for HER and the role of methanol as sacrificial molecules, with the methodology offering possibilities for the quantitative mechanism studies of proton-related catalytic reactions.

Results

Experimental set-up of operando NMR. Our experimental set-up consists of a micro-reactor system based on a NMR tube, allowing solid–liquid heterogeneous mixture to be directly studied inside the NMR coil. The experimental scheme is illustrated in Fig. 1 and Supplementary Fig. 1. The light source (300 W Xe lamp) beam is directed through a focusing lens assembly to converge on the entrance face of a homemade optical fibre bundle (seven quartz fibres, diameter 3 mm). The NMR study is incident on a suspension containing solid photocatalyst together with the aqueous mixtures, and the signals originate from not only the solid photocatalyst itself but also from the different gas–liquid molecules and chemical species dissolved in the solution[27,28]. This designed scheme has provided excellent feature information among the solid, liquid and gas mixtures without the separation of products. Our NMR studies are carried out on 700-MHz Agilent NMR spectrometers at a magnetic field strength of 16.4 T. An Agilent 5 mm z-axis pulsed field gradient triple resonance probe (^{1}H {^{13}C, ^{15}N}) is used (Detailed information can be found in Supplementary Fig. 2 and Supplementary Note 1).

To operando probe molecular species qualitatively and quantitatively in the solid–liquid phases, we measured some benchmark experiments (Supplementary Figs 3–10). First, the ^{1}H NMR signals of aqueous suspensions containing TiO_2, CH_3OH and D_2O were examined (Supplementary Figs 7 and 8). Two resonance peaks were clearly observed, corresponding to the methyl group (CH_3-) of CH_3OH (Supplementary Fig. 8b) and the

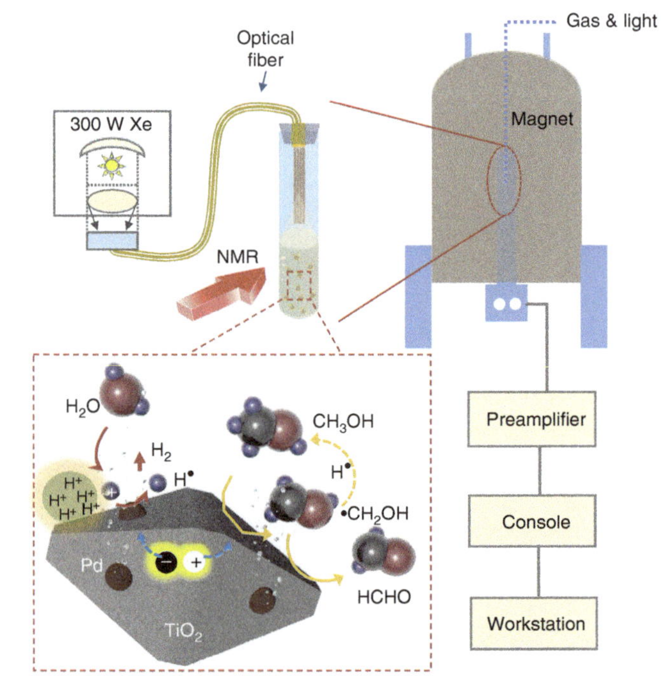

Figure 1 | Schematic layout of set-up for operando NMR studies. The set-up consists of a micro-reactor system based on a homemade NMR tube, allowing solid–liquid heterogeneous mixture to be directly studied inside the NMR coil while maintaining good uniformity. The light source (300-W Xe lamp) beam was directed through a focusing lens assembly to converge on the entrance face of a homemade optical fibre bundle (seven quartz fibres, diameter 3 mm).

residual protons of D_2O (Supplementary Fig. 8a)[29]. Interestingly, the signal of H_2 (singlet, 4.57 p.p.m.) was also remarkably detected in the spectrum of the suspension when the H_2 gas was injected into it (Supplementary Fig. 9). This could provide the way to quantitatively evaluate the amount of H_2 in the aqueous solution by comparing with the signal from the solution of saturated dissolved H_2. Second, the signal sensitivity was evaluated. Take 3-(trimethylsilyl)-1-propanesulfonic acid sodium salt (DSS) for example; when we decreased the content of DSS to nearly 50.7 pmol, the signal still had a relatively good signal-to-noise ratio (Supplementary Fig. 10). Even though by unitary sampling (number of scans is 1) of the heterogeneous mixture, the signals still had a good signal-to-noise ratio and thus could be distinctly distinguished. Therefore, the signals measured by the operando NMR could have pmol sensitivity, indicating a good capability to operando probe the gas–liquid intermediate products quantitatively deep into the solid–liquid working conditions at the very early stage of the reaction.

Operando NMR spectroscopy measurements. In a photocatalysis system, electrons are shuttled to the reductive side where they reduce protons to H_2. However, because of the complexity in building and optimizing such a complete system, when studying the reductive half reaction, it is common for a sacrificial electron donor to be used to optimize catalysts for H_2 production. Thus, for a clear recognition of the PT routes, it is significant to clarify the roles of the sacrificial agents. Widely accepted explanations of the roles for sacrificial agents (for example, methanol) are the thermodynamically more oxidizable ability than water, which work as an external driving force for the surface chemical reactions[12]. Their effects have been attributed to the consumption of holes, with the aim of suppressing the recombination of photogenerated charge carriers and the surface back reaction, which results in an increased H_2 evolution[30]. Yet, no experimental evidence for this mechanism was reported. On the other hand, some groups (for example, Xu *et al.*[15] and Highfield *et al.*[19]) pointed out that photo-reforming of methanol (gas) over the photocatalyst may also produce pure H_2, which makes it plausible to assume that the protons from methanol sacrificial agent have participated in the HER. However, these experimental simulations under ideal conditions depend on assumptions and simplifications, and many questions concerning details of the mechanisms in real conditions without the separation of products remain unanswered. As a consequence, the vague and ambiguous roles of sacrificial agents greatly limit our understanding of the PT routes.

To clarify these issues, NMR was applied on two deliberately designed reaction systems, namely System 1 ($H_2O/CD_3OD/$catalyst) and System 2 ($CH_3OH/D_2O/$catalyst), to provide some qualitative preliminary results. The virtue of those systems lays on the variation of the protonated agent of the similar reaction, which allows selective monitoring of the PT routes in HER. Figure 2a shows ^1H NMR spectra of these two systems before and after the irradiation with ultraviolet/visible (UV/Vis) light. No H_2/HD NMR signal is observed for these samples in the dark. Upon irradiation for 2 h, the H_2 signal (single, 4.56 p.p.m.) and the HD signal (triplet, 4.46, 4.52 and 4.57 p.p.m.) resonances are clearly observed for System 1. Whereas only trace amount of HD gas (triplet, 4.47, 4.53 and 4.58 p.p.m.) is found for System 2, although this system has been irradiated for 24 h before the NMR study. Considering the solution compositions of System 1, we conclude that the proton source of the H_2 gas in the photocatalysis system, $H_2O/CH_3OH/$catalyst, is at least partial from H_2O definitely.

For a deeper understanding, we changed the concentration of H_2O of System 1. As can be seen from Fig. 2b and Supplementary

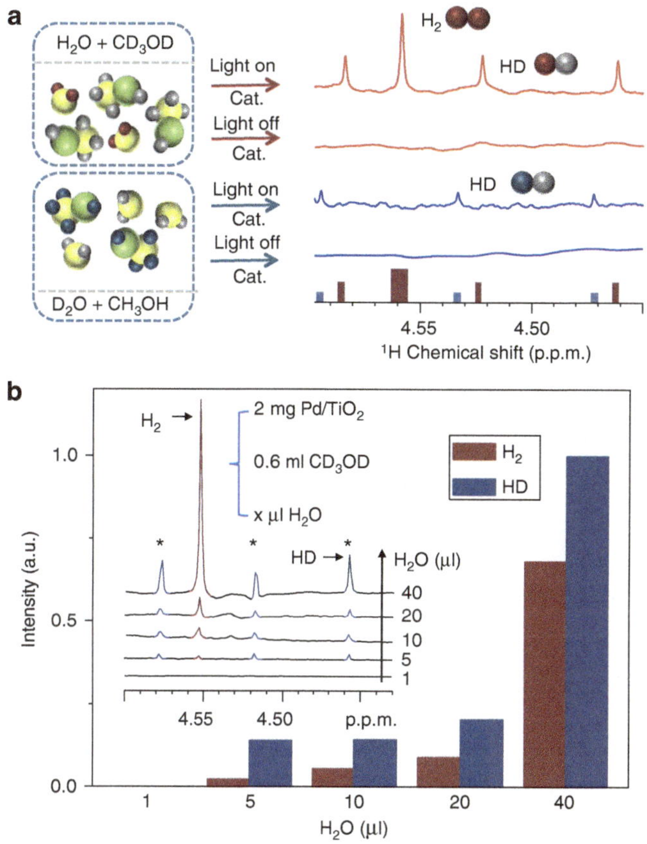

Figure 2 | ^{1}H NMR spectra of different systems. (**a**) ^1H NMR spectra of H_2/HD gases produced from heterogeneous systems before and after the irradiation with UV/Vis light. The spectra were acquired from samples containing (A,B) CH_3OH (20 μl), D_2O (500 μl) and Pd/TiO_2 (2 mg) at 298 K with a 5 s recycle delay and 16 scans before (A) and after (B) the UV/Vis irradiation for 24 h; (C, D) H_2O (20 μl), CD_3OD (500 μl) and Pd/TiO_2 (2 mg) at 298 K with a 5 s recycle delay and 32 scans before (C) and after (D) the UV/Vis irradiation (300-W Xe lamp) for 2 h. The shifts of the H_2/HD peaks are caused by the different deuterated reagents. (**b**) ^1H NMR spectra of the correlations between produced H_2/HD gas and added H_2O. The NMR experiments were acquired from samples containing Pd/TiO_2 (2 mg), CD_3OD (600 μl), D_2O (1% DSS as the internal reference) and H_2O (x μl, $x = 1, 5, 10, 20, 40$) at 298 K with a 5 s recycle delay and 32 scans after the UV/Vis irradiation for 2 h.

Fig. 11, it is observed that the signal intensity of H_2 gas in the NMR spectra increases with the content of the adding H_2O. This is another direct evidence for its involvement in PT (from H_2O to H_2), which can support our deduction for the origin of H_2 in System 1. Note that the H/D isotope effect has been realized in literature[31,32]. The presence of the H/D isotope effect can decrease the reaction rate of water splitting; however, in principle will not harm our deduction for the origin of H_2 in the studied reaction system.

The origin of HD gas is an intriguing question and also related to the photocatalytic mechanism. In a mixture of H_2O and CD_3OD, few HDO and CD_3OH species can be generated because of the H/D exchange reactions[33], which further increases the complexity of the origin of HD gas. For System 1, there are two possibilities for the origin of HD gas: the product of HDO or the reforming product of CD_3OH. As inspired by the work of Yang *et al.*[17], the methyl group H of CH_3OH would transfer to the O_{BBO} sites of TiO_2 (anatase) when it is photocatalytically dissociated, while proton reduction reaction prefers to occur on the surface of cocatalyst rather than bulk TiO_2. Consistent with this interpretation, a control experiment without loaded Pd cocatalyst cannot generate any H_2/HD product (Supplementary Fig. 13). Thus, the origin of HD gas is likely from HDO. In order to verify our hypothesis, some compared experiments for the system of $D_2O/CH_3OH/TiO_2$ were carried out (Supplementary Fig. 14). The volume ratios selected for CH_3OH are from 5 to 20%. In these samples, few HDO and CH_3OD species would be generated. After the irradiation with UV/Vis light for 7 h, no H_2/HD NMR signal is observed. Thus, we can safely infer that the reduction process for CH_3OH/CH_3OD to produce H_2/HD is quite limited within 7 h in this system. That is to say, the origin of HD gas in System 1 is mainly from HDO. Similarly, in System 2, we prolonged the illumination time to 24 h; only trace amount of HD gas was found (this might indicate that the concentration of HDO is much more smaller than D_2O in this sample). The results provide a compelling evidence that the dominating proton source of H_2 gas in the HER of CH_3OH/H_2O/catalyst mixture (CH_3OH Vol % < 20%) is H_2O species, and H_2 gas from the photo-reforming process of CH_3OH in the CH_3OH/H_2O mixture solution is negligible.

Note that in a mixture of H_2O and CD_3OD, some H_3O^+ and $CH_3OH_2^+$ (as well as OH^- and CH_3O^-) species may also be generated by autoprotolysis. In this study, the pH value of the system is 7. The influence of the concentration of H_3O^+/OH^- and $CH_3OH_2^+/CH_3O^-$ on the photocatalytic reaction is the ongoing study in our laboratory.

To get a better understanding of the role of methanol, a series of ^{13}C-labelled methanol ($^{13}CH_3OH$) instead of the naturally abundant component was added in the reaction system ($D_2O/^{13}CH_3OH$/catalyst). Figure 3 shows the $^1H/^{13}C$ NMR spectra of methanol intermediates of the reaction system containing Pd/TiO_2, D_2O and CH_3OH. In this system, CH_3OD would be generated because of the fast H/D exchange reactions. Interestingly, after irradiating for 40 h, the ^1H NMR spectrum shows a well-resolved triplet signal with equal intensities (1:1:1 triplet, $J = 1.4$ Hz) at the right side of the CH_3 peak of $CH_3OH(D)$ (Fig. 3a). While similar triplet signals can also be observed from the ^{13}C NMR spectrum (Fig. 3b). According to the chemical shift and the characteristic J coupling value, we have assigned these two signals to CH_2D- of $CH_2DOH(D)$. The possibility of -CHD- can be easily ruled out by the observation in the 1D ^{13}C DEPT-135 (Distortionless Enhancement by Polarizition Transfer) spectrum where the triplet signals show clear negative intensities (Fig. 3b, inset).

For the formation of $CH_2DOH(D)$, two processes can be proposed: one is through an addition reaction of HCHO and HD, while the other is through the reaction of $\bullet CH_2OH(D)$ with D radicals ($\bullet$D). It has been reported that the average adsorption energy of HCHO is much lower than that of CH_3OH; therefore, it may desorb from the TiO_2 surface quickly rather than recombine with an H/D atom[19]. Furthermore, as will be discussed in a further publication, the HCHO may transform into methanediol species quickly in the aqueous solution, which accelerates the desorption rate of HCHO. It is therefore reasonable to conclude that the $CH_2DOH(D)$ is likely formed through a coupling reaction of $\bullet CH_2OH(D)$ and $\bullet$D, providing an indirect evidence of the existence of $\bullet CH_2OH(D)$ radicals during the HER. Meanwhile, $\bullet CH_2OH(D)$ is the inevitable intermediate when $CH_3OH(D)$ reacts with a hole (h^+), which gives a definitive evidence that CH_3OH serves as a sacrificial agent and plays a crucial role in mediating the capture of holes to form $\bullet CH_2OH$ radicals. To our knowledge, $\bullet CH_2OH$ species only exists in the liquid (may not be present in the ideal conditions) and is hard to be captured when it is in an isolated state, while it would also rapidly convert to HCHO (as also confirmed by DFT), thus resulting in a lacking experimental evidence and unconfirmed mechanisms for this important process.

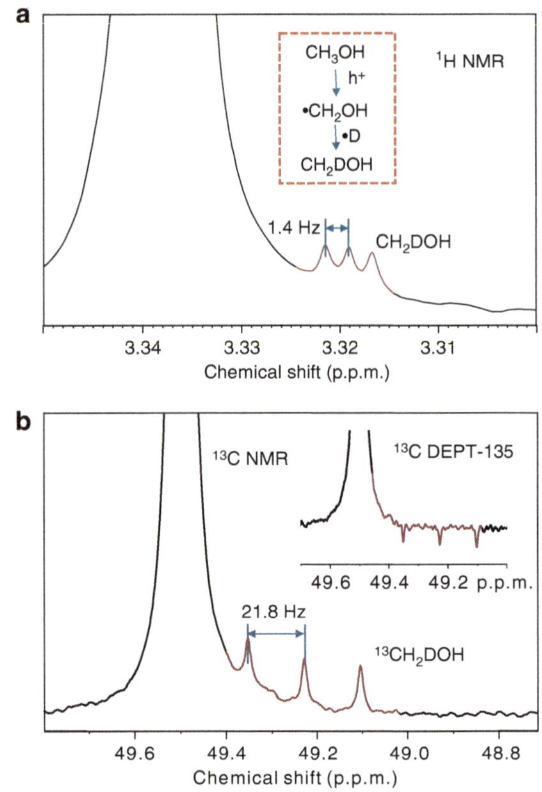

Figure 3 | High resolution NMR spectra of as-prepared samples. (**a**) ^1H NMR spectrum of the methanol intermediates after the UV/Vis irradiation. The ^1H NMR experiment was acquired from samples containing Pd/TiO_2 (2 mg), CH_3OH (20 μl) and D_2O (500 μl) at 298 K with a 5 s recycle delay and 16 scans after the UV/Vis irradiation for 40 h. (**b**) ^{13}C NMR spectrum of the methanol intermediates after the UV/Vis irradiation. The ^{13}C NMR experiment was acquired from samples containing Pd/TiO_2 (2 mg), $^{13}CH_3OH$ (20 μl) and D_2O (500 μl) at 298 K with a 2 s recycle delay and 32 scans after the UV/Vis irradiation for 40 h. Inset is the DEPT-135 NMR spectrum of the as-prepared samples. The DEPT-135 spectrum was acquired from samples containing Pd/TiO_2 (2 mg), $^{13}CH_3OH$ (20 μl) and D_2O (500 μl) at 298 K with a 1 s recycle delay and 32 scans after the UV/Vis irradiation for 40 h. In the DEPT-135 spectrum, the negative ^{13}C signals indicate that the signals are from the partially deuterated methyl ($-CH_2D$).

DFT studies. For a better understanding of the reaction mechanism, we have also conducted systematic DFT calculations concerning the reactions of CH_3OH at both ground and excited states on the anatase-TiO_2 (101) surface[34,35]. The results are illustrated in Fig. 4a (ground state) and Fig. 4b (excited state). As one can see, at the ground state, the dissociation of methanol through breaking of O–H or C–H bond is obviously an endothermic process. Therefore, the CH_3OH may stay intact at the surface under dark conditions. Interestingly, with the help of h^+, both C–H and O–H bond-breaking reactions become exothermic, and the C–H dissociation is more favourable both thermodynamically and dynamically compared with the O–H dissociation. It indicates that CH_3OH may prefer to react with h^+ to form $\bullet CH_2OH$ species under excited conditions. Then, $\bullet CH_2OH$ species may react with $\bullet$H ($\bullet$D) back to methanol or further react with h^+ to HCHO. Both reactions have also been studied in our calculations, and it turns out that, at the ground

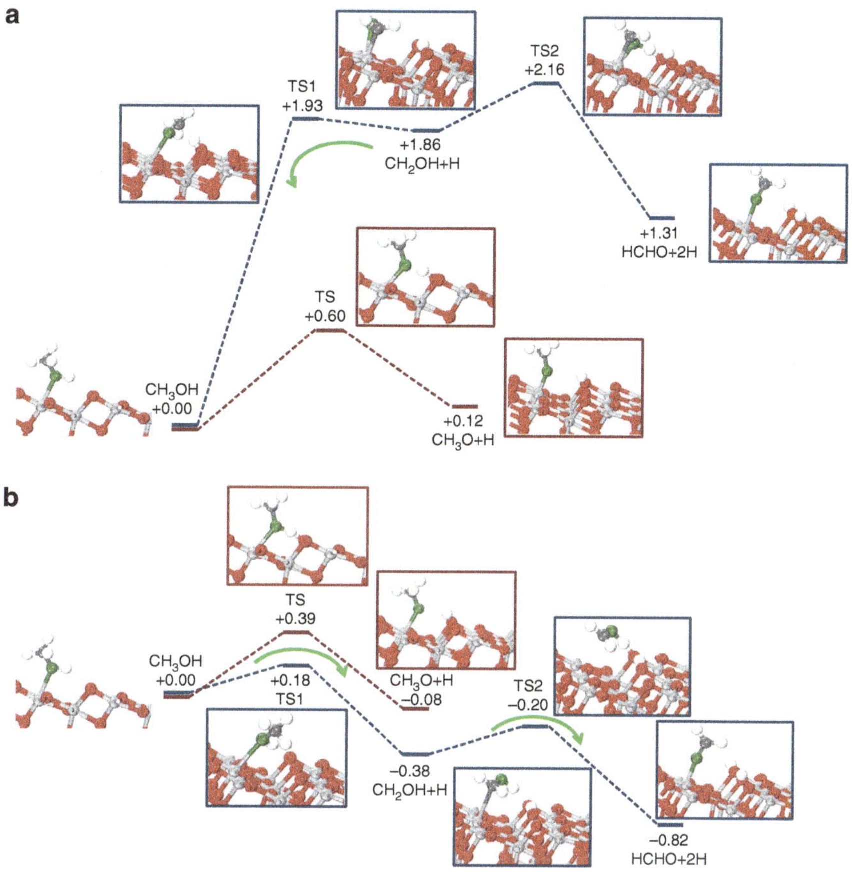

Figure 4 | DFT calculations. (**a**) DFT of the dissociation of CH_3OH at the ground state. (**b**) Photo-oxidation of CH_3OH begins with C-H broken (blue line) and O-H broken (red line). The calculated structures are displayed in the profile. The Ti atoms are in light grey and O in red, while the C atoms are in dark grey and H in white, and O atom of CH_3OH are in green. This notation is used throughout this paper.

state, the •CH_2OH may readily convert back to methanol with the barrier of only 0.07 eV—much lower than the further oxidation to HCHO (0.30 eV). By contrast, at the excited state, CH_2OH prefers the further oxidation to HCHO with a barrier of 0.18 eV rather than returning to methanol, which has a much higher barrier of 0.56 eV.

Quantitative operando NMR spectra. Operando NMR can not only provide qualitative insight into the catalytic mechanism but also be fully quantitative deep into the real solid–liquid phases. In our system, H_2 gas is a very important product, and its variation tendency may reflect some important microcosmic catalytic processes. Although some other characterization techniques such as gas chromatography (GC) and mass spectrometer (MS) can also be used to measure the gas products, they can only detect the gases that escape from the solution (after the dissolved gas reaches the saturation concentration), whereas in fact quite amount of gas molecules might be already present in the solution with partial dissolution and adsorption at the early stage of the reaction. Furthermore, the microcosmic reaction processes and trends before the dissolved gas reaches the saturation concentration will be missing, and the processes could last for more than 10 min for the 40 nmol H_2 dissolving in 100 ml H_2O solution. Meanwhile, it is also impossible for GC or MS to probe some intermediates that cannot escape from the reaction solution. In this context, the operando NMR approach is superior to GC or MS in studying the HER.

In our system, assuming that the H_2 gas does not reach the saturation state at the early stage of the reaction, the observed H_2 signals in the NMR spectra could reflect the whole H_2 gas from HER. H_2 gas can be quantified by using the intensity of the H_2 signal in the H_2–D_2O saturation solution (estimate 0.018 ml H_2

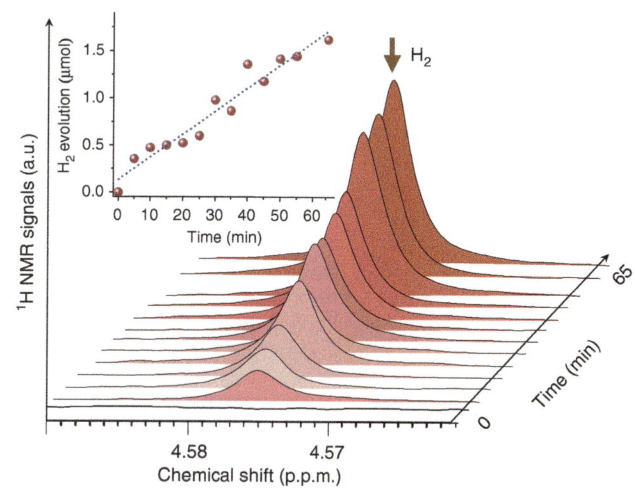

Figure 5 | Quantitative operando ^1H NMR spectra. Quantitative time-related ^1H NMR spectra of Pd/TiO_2 at 298 K in H_2O/CD_3OD mixtures under UV/Vis light irradiation. The operando NMR experiments were acquired from samples containing Pd/TiO_2 (1 mg), CD_3OD (300 μl), D_2O (1% DSS as the internal reference) and H_2O (300 μl) at 298 K with a 5 s recycle delay and eight scans under the UV/Vis irradiation. The first 1,000 points of the FID was cut off to obtain the final separable H_2 signals.

per 1 ml D_2O) at the same temperature (Supplementary Figs 15 and 16). Through comparison of the H_2 resonance intensities with calibration samples, it is possible to quantify the number of H_2 species in absolute terms at the early stage and gain insight into the PT behaviour properties in working state. Figure 5 shows the quantitative operando ^1H NMR spectrum of $H_2O/CD_3OD/$ catalyst system. For a suitable comparison with the results obtained from the conventional GC detector (Supplementary Fig. 12), the sample was irradiated directly by an Xe light source. The H_2 signal intensity increased linearly on continuous illumination. Under this condition, no HD signal can be detected, and this can be because of the few exchanged HDO products compared with H_2O, consistent with the previous results. The quantification result in Fig. 5 shows that the H_2 evolution rate at the early stage from the NMR measurement is $\sim 1.560\ \mu mol\, h^{-1}\, mg^{-1}$. The above result may be far-reaching because it indicates that the widely used methods have the intrinsic shortcoming on the detection of the reaction especially at the early stage. As a consequence, the reaction models derived from the *ex situ* measurement, especially the models for the reaction kinetics, will deviate from the actual reaction mechanisms. Thus, this operando NMR spectroscopic method provides an approach to guide the design of HER systems, offering better measures deep into the realistic reaction.

Discussion

By using an operando NMR method, we successfully quantified the generation of H_2/HD gases and liquid intermediates in pmol resolution without the separation of products, and provided evidence for the PT routes during the HER. On the basis of the photocatalytic system containing anatase-TiO_2, methanol and water, we find that the effective protons for HER are mainly from water, and methanol evidently serves as an outstanding electronic sacrificial agent reacting with holes. However, for the sacrifice reforming process of methanol, a small amount of released methyl group protons still occur, which are likely bound to the surface of TiO_2, and their migration to cocatalyst to take part in the HER would consume more time and energy. This indicates that complicated synergistic effect between water and methanol exists under practical HER conditions. The picture in Supplementary Fig. 17 illustrates the paramount PT processes together with the reaction mechanism of HER. Carrier generation and recombination occur when an electron transfers from the valence band to conduction band in a semiconductor. Once the photoexcited charges reach the surface of the particle, they can take part in surface redox reactions with adsorbed donor or acceptor molecules[34–37]. In our system, the induced electron transfers from the TiO_2 conduction band to the metal particles, and the attached proton would be reduced by an electron to its monoatomic radical species (•H), which is the main predecessor of H_2 (Supplementary Fig. 17, (equation 5)). On the other side, the holes would be consumed by CH_3OH, extracting the charges competitively with their recombination and then improving the formation rate of H_2. On the oxidative side, two sequential hole consumptions convert the adjacent CH_3OH species to •CH_2OH (equation 2) and HCHO (equation 3), respectively, accompanying with the release of protons. Subsequent spontaneous (dark) coupling reaction between •CH_2OH and •H radicals results in CH_3OH species re-formation (equation 4). On the reductive side, protons would be reduced to hydrogen by the shuttled electrons on the surface of cocatalyst (equation 5). In H_2O/CH_3OH (CH_3OH vol % <20%) mixtures, H_2O molecules would provide a vast amount of free protons in solution (equation 1). Although the photocatalytically dissociated protons from CH_3OH are constrained by the O_{BBO} nearby, only a slim chance for those to be reduced to H_2 exists. As a consequence, the current NMR spectroscopic studies provide a unique viewing angle for us to understand the predominant PT processes of HER in solution.

The results in this work deepen our understanding of interfacial heterogeneous redox reactions, and the operando NMR method will be applicable extensively with the methodology offering exciting possibilities for the microcosmic mechanism study (pmol resolution) in catalysis, photoelectrocatalysis and electrocatalysis of reactions such as CO_2 reduction, waste-water remediation and methanol reforming.

Methods

Synthesis of catalyst. The Pd(1.0 wt %)/TiO_2 photocatalyst powder was prepared by the co-precipitation method. Briefly, appropriate amount of $PdCl_2$ aqueous solution (1 wt%) was added into the TiO_2 (100 mg, Anatase, Sigma) powder and maintained at 80 °C for 1 h. After being dried, the products were calcined at 300 °C for 2 h. Before being characterized and tested, the samples were reduced in 20% H_2/Ar at 300 °C for 1 h.

NMR experiment. All NMR experiments were acquired on 700 MHz Agilent NMR spectrometers at a magnetic field strength of 16.4 T. An Agilent 5 mm z-axis pulsed field gradient triple-resonance static probe was used in all ^1H and ^{13}C NMR experiments. All ^1H NMR measurements were carried out at 298 K with a spectral width of 20 p.p.m., pulse width of 4 µs (45°), a recycle delay of 5 s and 8, 16 or 32 scans. To suppress the water signal, the first 1,000 points of the free induction decay (FID) were cut off before FT. ^{13}C NMR was carried out at 298 K with a spectral width of 253 p.p.m., pulse width of 7.3 µs (45°), a recycle delay of 2 s and 32 scans. All experiments had DSS as the internal reference. To enhance the NMR signals, the examples could also be directly
irradiated by an Xe light source during the irradiation accumulation process.

Operando ^1H NMR studies. A 300-W Xe lamp was used as the light source. The light source is outside the magnet. The light source beam was directed through a focusing lens assembly to converge on the entrance face of a homemade optical fibre bundle (seven quartz fibres, diameter 3 mm). The cap of the tube was specially designed to allow the input of the light and the H_2 gas. Note that the fibre may lead to an attenuation of the light intensity and a lower reaction rate (see Fig. 5 and Supplementary Fig. 18). In order to simulate the HER process similar to the conventional GC, the samples might be irradiated directly by an Xe light source.

Calculation details. All the calculations were carried out using the Vienna *ab initio* simulation package[38,39], and the exchange-correlation term was described by the Perdew, Burke and Ernzerhof version within the generalized gradient approximation[40]. The project-augmented wave[41,42] method was used to represent the core-valence electron interaction. The titanium 3*s*, 3*p*, 3*d*, 4*s* and the carbon and oxygen 2*s*, 2*p* electrons were treated as valence electrons and an energy cutoff of 400 eV for the basis-set expansion was used.

The anatase-TiO_2 (101) surface was modelled by a five trilayer slab with only the centre layer fixed, and other atoms were allowed to relax until atomic forces reached below 0.05 eV $Å^{-1}$. As suggested by Luo and co-workers[43], the hole was introduced by using the triplet state to mimic the singlet excited state[43–47]. A 3 × 1 surface cell and a >15 Å vacuum gap was used. Different k-point meshes were tested, and it was found the k-point sampling restricted to the Γ point only can already provide reliable results regarding adsorption energies.

The transition states in reactions were located with a constrained optimization scheme[48], and were verified when (i) all forces on atoms vanish and (ii) the when total energy is maximum along the reaction coordinate but minimum with respect to the rest of the degrees of freedom.

References

1. Han, Z., Qiu, F., Eisenberg, R., Holland, P. L. & Krauss, T. D. Robust photogeneration of H_2 in water using semiconductor nanocrystals and a nickel catalyst. *Science* **338,** 1321–1324 (2012).
2. Liu, J. *et al.* Metal-free efficient photocatalyst for stable visible water splitting via a two-electron pathway. *Science* **347,** 970–974 (2015).
3. Hundt, P. M., Jiang, B., Reijzen, M. E., Guo, H. & Beck, R. D. Vibrationally promoted dissociation of water on Ni (111). *Science* **334,** 504–507 (2014).
4. Nielsen, M. *et al.* Low-temperature aqueous-phase methanol dehydrogenation to hydrogen and carbon dioxide. *Nature* **495,** 85–90 (2013).
5. McLaren, A. D. The beckmann rearrangement of aliphatic ketoximes. *Science* **103,** 503 (1946).
6. Corma, A. & García, H. Lewis acids: from conventional homogeneous to green homogeneous and heterogeneous catalysis. *Chem. Rev.* **103,** 4307–4365 (2003).

7. Clarke, H. T., Gillespie, H. B. & Weisshaus, S. Z. The action of formaldehyde on amines and amino acids. *J. Am. Chem. Soc.* **446,** 4571–4587 (1933).
8. Albrecht, Ł. *et al.* Asymmetric organocatalytic formal [2 + 2]-cycloadditions via bifunctional H-bond directing dienamine catalysis. *J. Am. Chem. Soc.* **134,** 2543–2546 (2012).
9. Heyduk, A. F. & Nocera, D. G. Hydrogen produced from hydrohalic acid solutions by a two-electron mixed-valence photocatalyst. *Science* **293,** 1639–1641 (2001).
10. Chen, X., Shen, S., Guo, L. & Mao, S. S. Semiconductor-based photocatalytic hydrogen generation. *Chem. Rev.* **110,** 6503–6570 (2010).
11. Onishi, H. & Iwasawa, Y. Dynamic visualization of a metal-oxide-surface/gas-phase reaction: time-resolved observation by scanning tunneling microscopy at 800 K. *Phys. Rev. Lett.* **76,** 791–794 (1996).
12. Scheiber, P., Riss, A., Schmid, M., Varga, P. & Diebold, U. Observation and destruction of an elusive adsorbate with STM: O_2/TiO_2 (110). *Phys. Rev. Lett.* **105,** 216101 (2010).
13. Zhang, Z., Bondarchuk, O., White, J. M., Kay, B. D. & Dohnálek, Z. Imaging adsorbate O-H bond cleavage: methanol on TiO_2 (110). *J. Am. Chem. Soc.* **128,** 4198–4199 (2006).
14. Li, B. Ultrafast interfacial proton-coupled electron transfer. *Science* **311,** 1436–1440 (2006).
15. Xu, C. *et al.* Molecular hydrogen formation from photocatalysis of methanol on anatase-TiO_2 (101). *J. Am. Chem. Soc.* **136,** 602–605 (2014).
16. Xu, C. *et al.* Molecular hydrogen formation from photocatalysis of methanol on TiO_2 (110). *J. Am. Chem. Soc.* **135,** 19039–19045 (2013).
17. Guo, Q. *et al.* Stepwies photocatalytic dissociation of methanol and water on TiO_2 (110). *J. Am. Chem. Soc.* **134,** 13366–13373 (2012).
18. Zhang, M., Respinis, M. & Frei, H. Time-resolved observations of water oxidation intermediates on a cobalt oxide nanoparticle catalyst. *Nat. Chem.* **6,** 362–367 (2014).
19. Highfield, J. G., Chen, M. H., Nguyen, P. T. & Chen, Z. Mechanistic investigations of photo-driven processes over TiO_2 by in-situ DRIFTS-MS: part 1. Platinization and methanol reforming. *Energ. Environ. Sci.* **2,** 991–1002 (2009).
20. Chen, T. *et al.* Mechanistic studies of photocatalytic reaction of methanol for hydrogen production on Pt/TiO_2 by *in situ* fourier transform IR and time-resolved IR spectroscopy. *J. Phys. Chem. C* **111,** 8005–8014 (2007).
21. Blanc, F., Leskes, M. & Grey, C. P. In situ solid-state NMR spectroscopy of electrochemical cells: batteries, supercapacitors, and fuel cells. *Acc. Chem. Res.* **46,** 1952–1963 (2013).
22. Cattaneo, A. S. *et al.* Operando electrochemical NMR microscopy of polymer fuel cells. *Energy Environ. Sci.* **8,** 2383–2388 (2015).
23. Chan, K. W. H. & Wieckowski, A. Probing adsorbates on Pt electrode surfaces by the use of ^{13}C spin-echo NMR. Studies of CO generated from methanol electrosorption. *J. Electrochem. Soc.* **137,** 367–368 (1990).
24. Tong, Y. Y., Wieckowski, A. & Oldfield, E. NMR of electrocatalysts. *J. Phys. Chem. B* **106,** 2434–2446 (2002).
25. Sánchez, V. M., Cojulun, J. A. & Scherlis, D. A. Dissociation free energy profiles for water and methanol on TiO_2 surfaces. *J. Phys. Chem. C* **114,** 11522–11526 (2010).
26. Setvin, M. *et al.* Reaction of O_2 with subsurface oxygen vacancies on TiO_2 anatase (101). *Science* **341,** 988–991 (2013).
27. Huang, J., Jiang, Y., Vegten, N., Hunger, M. & Baiker, A. Tuning the support acidity of flame-made Pd/SiO_2-Al_2O_3 catalysts for chemoselective hydrogenation. *J. Catal.* **281,** 352–360 (2011).
28. Kolmer, A. *et al.* The influence of electronic modifications on rotational barriers of bis-NHC-complexes as observed by dynamic NMR spectroscopy. *Magn. Reson. Chem.* **51,** 695–700 (2013).
29. Suleimanov, N. M. *et al.* In situ muSR and NMR investigation of methanol dissociation on carbon-supported nanoscaled Pt-Ru catalyst. *J. Solid State Electrochem.* **17,** 2115–2121 (2013).
30. Schrauben, J. N. *et al.* Titanium and zinc oxide nanoparticles are proton-coupled electron transfer agents. *Science* **336,** 1298–1301 (2012).
31. Krishtalik, L. I. Kinetic isotope effect in the hydrogen evolution reaction. *Electrochim. Acta* **46,** 2949–2960 (2001).
32. Urey, H. C., Brickwedde, F. G. & Murphy, G. M. A hydrogen isotope of mass 2 and its concentration. *Phy. Rev.* **40,** 1–15 (1932).
33. Fenby, D. V. & Chand, A. Thermodynamic study of deuterium exchange in water + methanol systems. *J. Chem. Soc. Faraday Trans.* **74,** 1768–1775 (1978).
34. Lin, K., Zhou, X., Luo, Y. & Liu, S. The microscopic structure of liquid methanol from raman spectroscopy. *J. Phys. Chem. B* **114,** 3567–3573 (2010).
35. Valero, M. C., Raybaud, P. & Sautet, P. Nucleation of Pd_n (n = 1-5) clusters and wetting of Pd particles on γ-Al_2O_3 surfaces: a density functional theory study. *Phys. Rev. B* **75,** 045427 (2007).
36. Gerritzen, D. & Limbach, H. H. Kinetic and equilibrium isotope effects of proton exchange and autoprotolysis of pure methanol studies by dynamic NMR spectroscopy. *Ber. Bunsenges. Phys. Chem.* **85,** 527–535 (1981).
37. Galińska, A. & Walendziewski, J. Photocatalytic water splitting over Pt-TiO_2 in the presence of sacrificial reagents. *Energy and Fuels* **19,** 1143–1147 (2005).
38. Kresse, G. & Furthmüller, J. Efficiency of *ab-initio* total energy calculations for metals and semiconductors using a plane-wave basis set. *Comput. Mater. Sci.* **6,** 15–50 (1996).
39. Kresse, G. & Furthmüller, J. Efficient iterative schemes for *ab initio* total-energy calculations using a plane-wave basis set. *Phys. Rev. B* **54,** 11169–11186 (1996).
40. Perdew, J. P., Burke, K. & Ernzerhof, M. Generalized gradient approximation made simple. *Phys. Rev. Lett.* **77,** 3865–3868 (1996).
41. Kresse, G. & Joubert, D. From ultrasoft pseudopotentials to the projector augmented-wave method. *Phys. Rev. B* **56,** 1758–1775 (1999).
42. Blöchl, P. E. Projector augmented-wave method. *Phys. Rev. B* **50,** 17953–17979 (1994).
43. Ji, Y., Wang, B. & Luo, Y. A comparative theoretical study of proton-coupled hole transfer for H_2O and small organic molecules (CH_3OH, HCOOH, H_2CO) on the anatase TiO_2 (101) Surface. *J. Phys. Chem. C* **118,** 21457–21462 (2014).
44. Ji, Y., Wang, B. & Luo, Y. Location of trapped hole on rutile-TiO_2 (110) surface and its role in water oxidation. *J. Phys. Chem. C* **116,** 7863–7866 (2012).
45. Di, V. C. & Fittipaldi, D. Hole scavenging by organic adsorbates on the TiO_2 surface: a DFT model study. *J. Phys. Chem. Lett.* **4,** 1901–1906 (2013).
46. Ji, Y., Wang, B. & Luo, Y. First principles study of O_2 adsorption on reduced rutile TiO_2 (110) surface under UV illumination and its role on CO oxidation. *J. Phys. Chem. C* **117,** 956–961 (2013).
47. Jedidi, A., Markovits, A., Minot, C., Bouzriba, S. & Abderraba, M. Modeling localized photoinduced electrons in rutile-TiO_2 using periodic DFT + U methodology. *Langmuir* **26,** 16232–16238 (2010).
48. Alavi, A., Hu, P., Deutsch, T., Silvestrelli, P. L. & Hutter, J. CO oxidation on Pt (111): an *ab initio* density functional theory study. *Phys. Rev. Lett.* **80,** 3650–3653 (1998).

Acknowledgements

This work was financially supported by the National Natural Science Foundation of China (21373083, 21421004, 21322307 and 21574043), the Fundamental Research Funds for the Central Universities (WD1313009 and WD1514303), SRF for ROCS, SEM, SRFDP, Program of Shanghai Subject Chief Scientist (15XD1501300), National High-tech R&D Program of China (863 Program; 2014AA123400, 2014AA123401).

Author contributions

H.G.Y. conceived the project and contributed to the design of the experiments and analysis of the data. X.L.W. performed the catalyst preparation, characterizations and wrote the paper. Y.-F.Y., Y.S., W.Q.F., D.W. and W.L. conducted the NMR examination and contributed to writing the NMR section. X.-Q.G. and Y.-Y.Y. conducted DFT calculations and wrote part of the paper (calculations). All the authors discussed the results and commented on the manuscript.

Additional information

Competing financial interests: The authors declare no competing financial interests.

19

Electron-vibration coupling induced renormalization in the photoemission spectrum of diamondoids

Adam Gali[1,2], Tamás Demján[1,3], Márton Vörös[4,5], Gergő Thiering[1,2], Elena Cannuccia[6] & Andrea Marini[7]

The development of theories and methods devoted to the accurate calculation of the electronic quasi-particle states and levels of molecules, clusters and solids is of prime importance to interpret the experimental data. These quantum systems are often modelled by using the Born-Oppenheimer approximation where the coupling between the electrons and vibrational modes is not fully taken into account, and the electrons are treated as pure quasi-particles. Here, we show that in small diamond cages, called diamondoids, the electron-vibration coupling leads to the breakdown of the electron quasi-particle picture. More importantly, we demonstrate that the strong electron-vibration coupling is essential to properly describe the overall lineshape of the experimental photoemission spectrum. This cannot be obtained by methods within Born-Oppenheimer approximation. Moreover, we deduce a link between the vibronic states found by our many-body perturbation theory approach and the well-known Jahn-Teller effect.

[1] Institute for Solid State Physics and Optics, Wigner Research Centre for Physics, Hungarian Academy of Sciences, PO Box 49, H-1525 Budapest, Hungary. [2] Department of Atomic Physics, Budapest University of Technology and Economics, Budafoki út 8, H-1111 Budapest, Hungary. [3] Institute of Physics, Loránd Eötvös University, Pázmány Péter sétány 1/A, H-1117 Budapest, Hungary. [4] Materials Science Division, Argonne National Laboratory, Lemont, Illinois 60439, USA. [5] Institute for Molecular Engineering, University of Chicago, Chicago, Illinois 60637, USA. [6] Aix-Marseille Université, CNRS, PIIM UMR 7345, 13397 Marseille, France. [7] Institute for Material Science (ISM) of the National Research Council (CNR), Via Salaria Km 29.3, CP 10, 00016 Monterotondo Stazione, Italy. Correspondence and requests for materials should be addressed to A.G. (email: gali.adam@wigner.mta.hu).

Molecular-sized nanoparticles are applied in diverse areas of basic and applied science, from drug delivery across photovoltaics and quantum information processing, harnessing their unique response to electromagnetic fields[1–3]. Understanding the electronic structure of these systems is an inevitable key step to determine their optical, magnetic and other related properties. One of the most widely used experimental techniques to measure the electronic structure of small nanoparticles and molecules is photoemission spectroscopy. The photoemission spectrum (PES) gives detailed information about the binding energy of the (valence) electrons. Nevertheless, the interpretation of the PES of molecules and nanoparticles may be not as straightforward as for atoms whose spectrum is dominated by sharp peaks[4]. In molecules and nanoparticles, vibrations may play a major role in the ionization process resulting in broad features in PES.

The effect of vibrations in the ionization is often considered within the Born–Oppenheimer approximation where the electrons are decoupled from the vibrations and assumed to be almost perfect quasi-particles (QPs). An approach based on a many-body perturbation theory (MBPT) that goes beyond the Born-Oppenheimer approximation has been recently proposed[5–7]. In this approach both the static and dynamical coupling between electrons and vibrational modes is considered.

Here we show by means of this MBPT approach that in small diamond cages, called diamondoids[8–10], the electron–vibration coupling leads to the breakdown of the QP picture. As a consequence, the lineshape of their experimental PES can be only reproduced by taking the resultant vibronic states into account. Our MBPT approach is able to predict the Jahn–Teller nature of the corresponding electron states even for deep energy regions. These findings demonstrate the ground breaking impact of our theory in the correct understanding of the most basic photoionization phenomenon in small carbon-based nanoclusters.

Results

Photoionization spectrum of diamondoids. We demonstrate that the interpretation of the PES of selected diamondoids[8,11,12], for which the experimental data are available[13,14], in terms of atomic-like, sharp electronic levels is bound to fail. As the geometries of these nano-objects are exactly known, they serve as testbeds to study the accuracy of the standard theoretical tools that aim at describing, from first principles, the electronic structure. Indeed, several density functional theory (DFT)[15–17], MBPT and Quantum Monte Carlo[18–22] studies have focused on understanding the opto-electronic properties of diamondoids. We consider here the smallest diamondoid with a single diamond cage, called adamantane, characterized by a T_d symmetry and consisting of only carbon (C) and hydrogen (H) atoms. We also study diamantane, which is composed of two diamond cages with D_{3d} symmetry, and a derivative called urotropine with T_d symmetry, where the C–H units of adamantane are substituted by nitrogen (N) atoms (see Fig. 1**c**).

The measured and calculated PES of these structures are compared in Fig. 1 at different levels of theory. We start from Kohn–Sham (KS) density functional theory within the local density approximation (LDA). Local (LDA) and other (semi)local DFT functionals provide appropriate zeroth-order approximation for the ground state of the system including the single-particle KS states and energy levels[18,17,23]. The electron QP levels are calculated within the single-shot GW-method. We note here that fairly convergent QP levels have been obtained for energies down to -15 eV for adamantane and urotropine, and -11 eV for diamantane[22]. Thus, we study the corresponding PES in these photoionization energy ranges.

The effect of vibrations on the PES is treated semiclassically, without dynamical effects, and by including dynamical effects induced by the electron–vibration coupling via MBPT. In the former method, the broadening, or lineshape, of the electron QP levels is described with the Huang–Rhys (HR) method (see 'Methods' section and Supplementary Methods), which simulates the geometry changes induced by ionization[14] within the Born–Oppenheimer approximation. We find that the typical total HR factor is $S \approx 3$ for the considered diamondoids (see Supplementary Fig. 1), which implies that three vibrational modes are involved in the geometry change. In the latter method, instead, the effect of vibrations on the PES is directly included in the spectral function (SF), computed within the recently implemented MBPT approach to the electron–vibration coupling (see refs 5–7 and the 'Methods' section for details). The latter method describes polaronic states derived from the potential energy surface of the neutral charge state (equation (11) in the 'Methods' section) that may be induced on ionization. In our MBPT approach, the resultant SF yields only the lowest energies of these polaronic states with combination of single vibrations belonging to different dynamic distortions of the geometry. The broadening of these polaronic states due to change of the geometry on ionization involving multiple vibration excitations is accounted for by the HR method.

Figure 1 clearly demonstrates that the HR approach of vibration-assisted ionization energies does not account for the experimental PES when the overall lineshape and the broadening of the QPs are considered (see Fig. 1a). Instead, the many-body SFs convolved by HR broadening reproduce remarkably well the overall lineshape of the experimental PES and the broadening of QPs. In particular, the agreement between the experiment and theory is excellent for the adamantane structure in a wide range of photoionization energies: all broad features are correctly described as composed of two local peaks that physically represent coherent packets of electron–vibration states (Fig. 1b). For the other two diamondoids, the dynamical electron–vibration interaction significantly improves the calculated lineshapes compared with the HR method. We emphasize that the broadening is well described not just for the ionization of the highest occupied molecular orbitals but for the deeper states too.

Analysis of the QP states. In order to understand the calculated PES, we analyse the electronic structure and vibrational modes of the structures we have considered. The lowest unoccupied molecular orbital (LUMO) and the high energy states are very diffuse Rydberg states[17]. As a consequence, their coupling to the vibrational modes is small, and their SFs (see 'Methods' section) exhibit an almost perfect QP line shape. The calculated QP shifts (see 'Methods' section) ΔE_n for $n =$ LUMO,LUMO + 1,.... due to electron–vibration coupling are in the region of 0.05–0.01 eV and the renormalization factors (see 'Methods' section) Z_n are greater than 0.95. Here, LUMO + 1 refers to the index of the state above the LUMO. The situation changes significantly in the case of the highest occupied molecular orbital (HOMO) and lower energy occupied states (HOMO-1, HOMO-2 and so on.), which are the only ones contributing to the PES. These occupied states are localized on the atoms constituting the diamond cages and are strongly coupled to the atomic vibrations. This coupling induces $Z_n < 0.7$ for all these states. Values for Z_n far from one clearly point to a potential breakdown of the electron QP picture (in the limit of $Z_n = 0$, purely electronic states do not exist). In simple terms, this is confirmed and visualized by the presence of multiple peaks with well-observable intensities appearing in the many-body SFs and in the calculated PES. These states rule out any description in terms of single and well-defined electronic energy

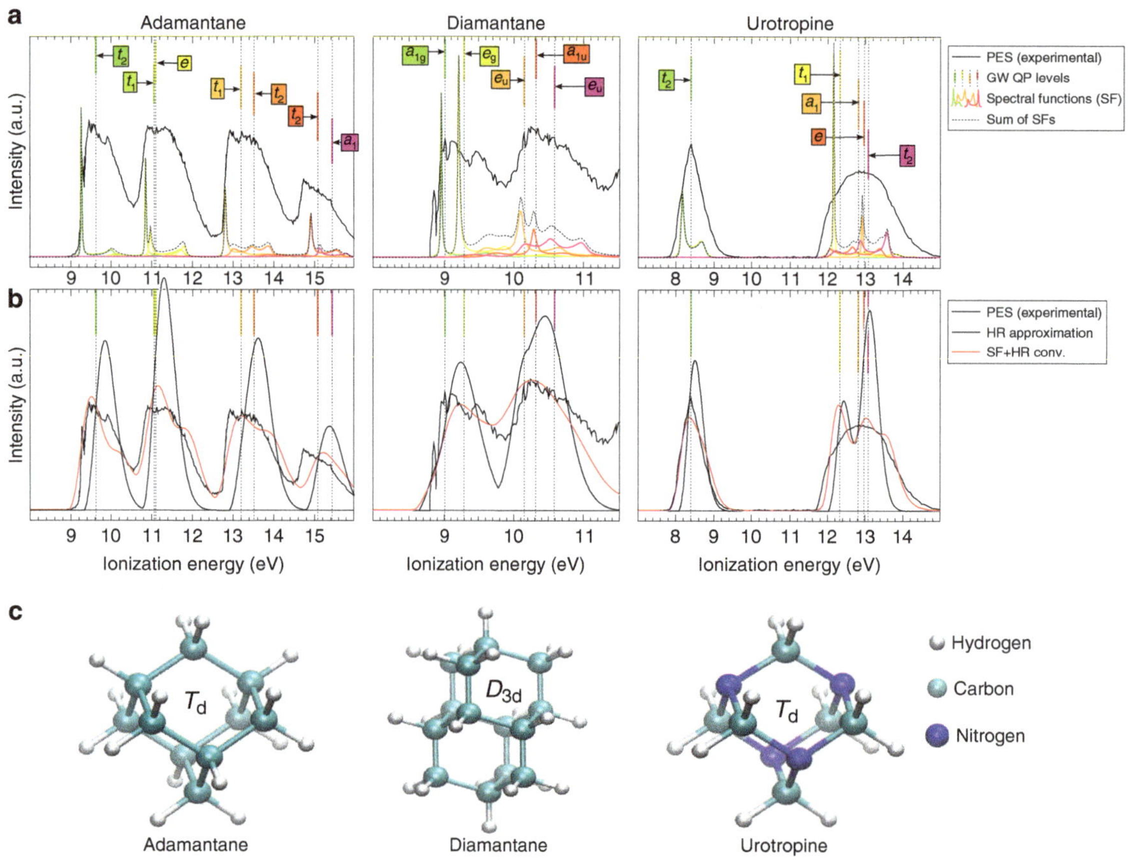

Figure 1 | Photoemission spectrum of selected diamondoids: theory and experiment. (**a**) Experimental PES (black curve, adamantane and urotropine from ref. 13, adamantane-urotropine-EXP and diamantane from ref. 14, diamantane-EXP recorded at room temperature) compared with the calculated spectral functions of the electronic QP levels (dotted columns). The different colours represent the different states labelled by their symmetry representation. The corresponding spectral functions (SFs) calculated at zero temperature (see 'Methods' section) are shown by the same colour code. The sum of the SFs is depicted as a dotted curve. (**b**) Experimental PES (black curve) compared with the broadened QP states (blue curves) and broadened SFs (red curve) where the broadening is calculated by Huang-Rhys (HR) method. The HR spectra are normalized to the area of the corresponding experimental PES and shifted by the zero-point energy difference of the ionized and neutral diamondoids that are 0.21, 0.16 and 0.05 eV for adamantane, diamantane and urotropine, respectively (see ref. 22). (**c**) Geometry of the structures considered with their point group symmetry.

states. We note that the breakdown of electron QP states has also been found in other carbon–hydrogen systems, polymers[6,7], and strong renormalization of QP levels led to a decrease of around 7% in the calculated direct gap of bulk diamond[24].

SF of adamantane and the Jahn–Teller effect. Now we analyse the nature of the calculated SFs represented by the coloured curves in Fig. 1a. We start our analysis with adamantane. The HOMO of adamantane is mostly localized on the C–C sp^3 orbitals forming a triple degenerate t_2 state within T_d symmetry. Its SF shows two peaks (see Fig. 2a): the main peak is at around 9.25 eV and a second satellite peak at around 10.0 eV. This contributes to around 2 eV broadening of the HOMO, which well reproduces the broadening observed in the experiment (see Fig. 1). The main peak might be associated with the electron QP level that would result in a 0.24 eV shift for the HOMO (for example, the shift between the green dotted line and the maximum peak in the green curve at around 9.5 eV in Fig. 1a). This is a large correction and, more importantly, it corresponds to a QP charge $Z_n = 0.64$. This clearly points to the presence of strong dynamical corrections that break the QP picture. Indeed the smaller the QP charge, the larger the participation of the particle in states where the electronic component is minor compared with the vibration excitations. In this case, the state loses a large fraction of its electronic character. Next, we demonstrate the strength of the MBPT approach in the analysis of the complex photoemission spectrum.

The main peak of the HOMO SF comes from the coupling to 36 different vibrational modes in the region of 124–162 meV corresponding mostly to C-C stretch modes. The largest contribution is due to a T_2 vibrational mode at 159.8 meV (see Supplementary Note 1, Supplementary Table 1 and Supplementary Movies 4–6). Most of the other coupled vibration modes are degenerate and an A_1 fully symmetry mode is only weakly coupled. The degenerate vibrational modes coupled to the HOMO of the neutral adamantane move the atoms out of the high symmetry geometry and include the 159.8 meV T_2 mode, which dynamically distorts the high T_d symmetry geometry to a C_{3v} symmetry as demonstrated in Fig. 3a. The DFT calculations of the ionized adamantane in the quasi-static limit result in C_{3v} distortion that was partially discussed previously[22,25]: the T_2 HOMO state is a Jahn–Teller (JT) system on ionization leaving a hole in the triple degenerate state that may be described as a $t_2 \otimes T_2$ vibronic state. We find that the 159.8 meV T_2 vibrational mode that mostly leads to formation of coherent packets of electrons and breaks the QP picture is the same that drives the JT distortion of the ionized adamantane. This finding indicates that the vibronic states found by MBPT induce the JT effect. The link between these effects can be explained by the fact that the linear

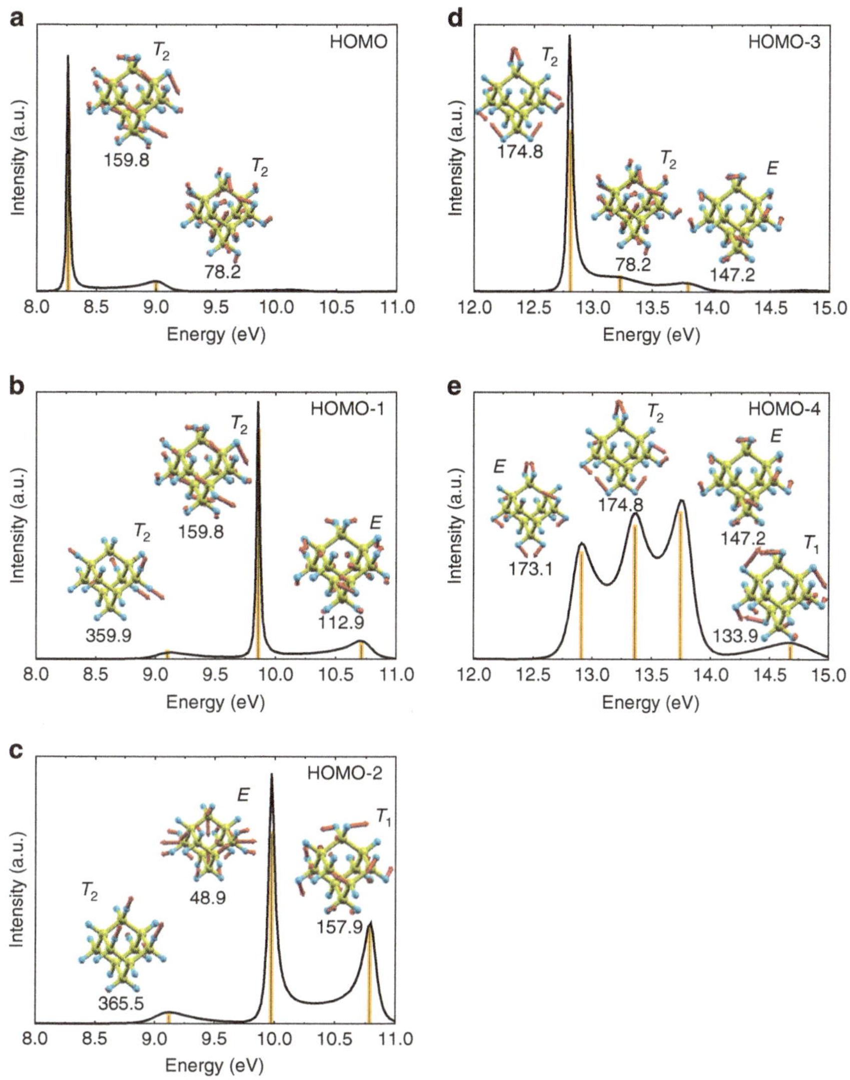

Figure 2 | Calculated electron-vibration spectral functions for the labelled occupied states of adamantane. The vibrational modes contributed with the largest extent to the peaks are given in milli-electronvolt unit where the corresponding motion of atoms are depicted by red arrows. We also label the symmetry of vibrational modes. We only show one of the eigenstates of the vibrations in the case of degenerate modes. (**a**) HOMO state with t_2 symmetry. (**b**) HOMO-1 state with t_1 symmetry. (**c**) HOMO-2 state with e symmetry. (**d**) HOMO-3 with t_1 symmetry. (**e**) HOMO-4 with t_2 symmetry.

JT coupling (see Fig. 2 of ref. 26 and references therein) corresponds to the absolute square of the electron–vibration coupling (g in Equation 4 in the 'Methods' section) which is strong for the t_2 HOMO in the neutral adamantane according to our calculations. This strong electron–phonon coupling is manifested in the PES via the calculated SF that contains the amplitude of the transition between the ionized and neutral systems in the MBPT approach (see Lehmann representation of the SF in the 'Methods' section). This shows that our MBPT approach indicates the JT nature of the ionized systems. This is a particularly important finding as this can be extended to the degenerate states lying below the HOMO level.

By following this argument other important conclusions can be drawn in the interpretation of the calculated SF and the experimental PES. The second peak of the HOMO SF corresponds to a satellite that originates from the coupling to 43 different degenerate vibrational modes with energy of 78–158 meV. These states correspond to the twist and bending modes of C atoms. The largest contribution belongs to the 78.2-meV T_2 twist mode (see Supplementary Note 1, Supplementary Table 1 and Supplementary Movies 1–3). These twist modes do not appear in bulk diamond but only in the quasi zero-dimensional diamond cages and lead to a dynamical symmetry distortion that is peculiar for these systems. Analogue to the main HOMO SF peak, this satellite HOMO SF peak can be associated with the motion of atoms in the ionized system that distorts the T_d symmetry due to JT effect. We refer to it as anomalous JT distortion by following Patrick and Giustino[25]. Remarkably, our MBPT approach is able to identify the origin of the satellite HOMO peak in PES and implies a link to the JT effect. We further note that these twist modes contribute to the second local maximum in the first broad feature of the PES of adamantane at around 10 eV (for example, Figs 1a and 2a).

Comparison of adamantane and urotropine. At this point, it is intriguing to compare the t_2 HOMO states of adamantane and urotropine that share the same T_d symmetry but show distinct shape of PES for HOMO (see Fig. 1b). We compare the HOMO SFs for adamantane and urotropine in Fig. 3. The HOMO in urotropine is strongly localized on the four N atoms. It couples to 39 (20) different vibrational modes in the energy region of 56 (56) meV to 93 (129) meV for the main (satellite) peak (see

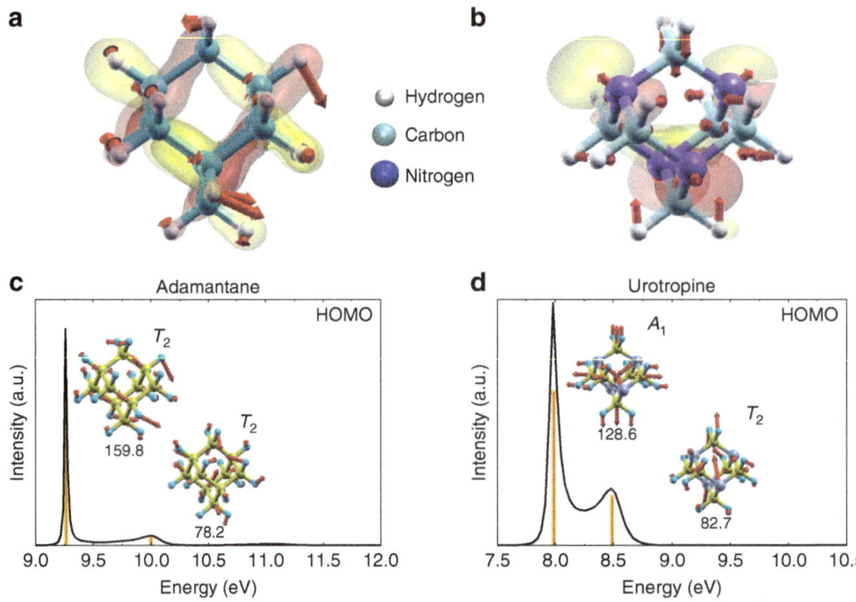

Figure 3 | Comparison of the HOMO spectral function for adamantane and urotropine sharing the same T_2 symmetry. The isosurface of the $t_{2(x)}$ wave functions with an isovalue of ±0.0028 where positive/negative isosurfaces are depicted by red/yellow lobes on the ball-and-stick model of (**a**) adamantane and (**b**) urotropine. The arrows show the displacement of atoms (multiplied by a factor of five for the sake of clarity) going from the the neutral to the ionized state as obtained by KS-DFT LDA calculation (see ref. 22). The corresponding average change of the coordinates of the atoms is larger for adamantane (0.04 Å per atom) than for urotropine (0.02 Å per atom). The vibrational modes contributed dominantly to the given SF are indicated in milli-electronvolt unit and the corresponding motion of the atoms are shown by red arrows in (**c**) adamantane and (**d**) urotropine. We also label the symmetry of vibrational modes. Note the same type of displacements between the dominant vibrational mode in the main peak of SF and the corresponding change in the optimized geometry of the neutral and ionized states. The motion of atoms in these vibration modes for adamantane and urotropine is shown in the Supplementary Movies 1-6 and 7-10, respectively.

Supplementary Note 1 and Supplementary Table 3). This results in about 1.5 eV broadening of HOMO in urotropine that is about 0.5 eV smaller than that in adamantane. Furthermore, the main and the second peaks originate from a coupling with a fully symmetric A_1 vibrational mode of 128.6 meV (see Supplementary Movie 10). In the second peak, the contribution of a T_2 vibrational mode of 82.7 meV may imply a JT distortion (see Supplementary Movies 7–9). In the main peak, the coupling to many degenerate vibrational modes appears but only with a minor weight. We find by quasi-static DFT calculations on the ionized urotropine that the JT distortion is minor and the atoms move with a smaller amplitude upon ionization than those in adamantane. In other words, the JT effect is smaller in urotropine than in adamantane. The proposed link between MBPT SF of the neutral system and the JT effect of the ionized system is in line with this finding: a stronger JT effect results in a wider electron–vibration-related broadening in PES. This feature is well captured by our MBPT approach.

Analysis of deep QP states of adamantane. We continue the analysis of PES for adamantane in Fig. 1. As we mentioned earlier, the second local maximum at around 10 eV can be explained by the satellite peak of the HOMO. However, it is evident from the dotted curve in Fig. 1a that low-energy satellite peaks of the closely lying HOMO-1 and HOMO-2 states also contribute to that. The HOMO-1 t_1 and HOMO-2 e states are degenerate and also subject to the JT distortion upon ionization. Indeed, the low-energy satellite peak, the main peak and the high-energy satellite peak originate mostly from the coupling to different degenerate vibrational modes (see Fig. 2b,c, respectively) that dynamically distort the T_d symmetry in both cases. The main peak and the high-energy satellite peak of HOMO-1 and HOMO-2 states contribute to the shape of the second broad feature in the PES of adamantane. Finally, similar conclusion can be drawn for the third and fourth broad features in PES. The third broad feature corresponds to the HOMO-3 t_1 and HOMO-4 t_2 states that are triple degenerate and subject to JT distortion again. Their SFs are shown in Fig. 2d,e, respectively. As the broadening of these states due to electron–vibration coupling is very large (above 2.5 eV), their SFs strongly overlap despite the relatively large QP energy gap of 0.4 eV (for example, orange dotted lines and the corresponding orange curves under the third broad feature of black curve in Fig. 1a). Particularly, the four peaks that are visible in the SF of HOMO-4 clearly demonstrate the breakdown of the electron QP picture. We found that more than 100 different vibrational modes within an energy range of 37–365 meV couple to this electronic state and contribute to the corresponding peaks in the SF. We note that the dynamic distortion of the 147.2 meV mode in the HOMO-3 SF couples to the HOMO-4 state whereas the dynamic distortion of the 173.1 meV mode in the HOMO-4 SF couples to the HOMO-3 state. The final lineshape of the calculated PES is very asymmetric and it is well represented by the calculated SFs convolved by HR method. A very similar process is responsible for the asymmetry of the fourth broad feature originating from the HOMO-5 and HOMO-6 states and the broad features in urotropine composed from HOMO-1 to HOMO-4 states.

Signature of pseudo Jahn–Teller effect in diamantane. Next, we analyse the PES of diamantane with D_{3d} symmetry. The calculated SFs of diamantane HOMO and the lower energy states are very broad and create an almost featureless PES, in good agreement with the experiment (Fig. 1b). The HOMO and the HOMO-1 states, the fully symmetric non-degenerate a_{1g} and the double degenerate e_g states are localized mainly on the C atoms.

The HOMO-2, HOMO-3 and HOMO-4 states are characterized by e_u, a_{1u} and e_u symmetry, respectively.

First, we focus our attention on the HOMO and HOMO-1 and their interaction. Although a_{1g} and e_g states are principally distinct in nature, their SFs look very similar (see the green and yellow curves with a maximum at around 9.0 eV in Fig. 1a). By recalling our proposal for a link between the strength of JT coupling and the width of broadening in PES based on the analysis of the HOMO in adamantane and urotropine, this might be surprising since the non-degenerate a_{1g} state is stable against JT distortion whereas the double degenerate e_g state is not.

We anticipate that this phenomenon is a manifestation of the so-called pseudo-Jahn–Teller effect[27]. We explain this feature in conjunction with the HOMO of adamantane and their relation to the valence band edge of bulk diamond. The valence band edge of diamond at the Γ-point is a triple-degenerate t_{2g} under O_h space group neglecting the spin–orbit interaction. This t_{2g} state is built up from the linear combination of sp^3 hybrid valence orbitals of carbon atoms. Adamantane, as a single cage of diamond, forms very similar triple-degenerate HOMO orbital with a correspondence of O_h:$t_{2g} \rightarrow T_d$:t_2. Diamantane with two diamond cages forms a lower symmetry, so O_h:$t_{2g} \rightarrow D_{3d}$:$a_{1g} + e_g$; the original triple-degenerate state of diamond 'HOMO' orbital splits to a non-degenerate and a double degenerate state. Nevertheless, this splitting does not originate from the chemical nature of the bonds but from the low symmetry of ions, thus these states have hidden connection, and they may be considered as pseudo-degenerate states. This is the reason why the SFs of adamantane's HOMO t_2 and diamantane's HOMO a_{1g} and HOMO-1 e_g orbitals have very similar SF line-shapes. The 159.8 meV T_2 mode, the most responsible for the JT distortion of adamantane HOMO (main peak in SF), appears as 162.7 meV A_{1g} mode for HOMO (see Supplementary Movie 13) and 159.0 meV E_g mode for HOMO-1 in diamantane (see Supplementary Movies 11,12) where similar correspondence can be derived between these vibrational modes (see Fig. 4) as for their electronic states. Particularly, the z-component of the triple-degenerate T_2 vibrational mode of adamantane is mimicked by the fully symmetric A_{1g} vibration mode in diamantane. These A_{1g} and E_g vibrational modes distort the diamond cages of diamantane similarly to the corresponding components of the T_2 vibrational mode of adamantane (Fig. 4b). Thus, A_{1g} vibration-related motion of atoms represents a pseudo-JT distortion of the diamond cages in diamantane upon ionization. To summarize, the vibrational modes that lead to formation of coherent packets of electrons and vibrational modes and break the QP picture are the same that drive the pseudo-JT distortion. The hidden connection between the electronic states can be analysed by the corresponding MBPT SFs.

We discuss further the PES of diamantane. The second peak in the SF of HOMO and HOMO-1 states of diamantane shows a similar anomalous JT distortion to that of the HOMO of adamantane. Going further with the group theory analysis, the HOMO-2 and HOMO-3 e_u and a_{1u} states have a correspondence to the adamantane's HOMO-1 t_1 state, while the HOMO-4 e_u state has a relation to the adamantane's HOMO-2 e state. However, as the symmetry of the two e_u states are the same in diamantane, these electronic states can mix the character of the ascendant t_1 and e states of adamantane. This also results in a much larger number of vibrational modes that couple these two e_u electronic states compared with the coupling of the t_1 and e states in adamantane (for example, Supplementary Table 1 versus Supplementary Table 2). The computed PES shows an almost featureless shape at around 10.3 eV for diamantane, in contrast to the corresponding spectrum at around 11.0 eV for adamantane (see Fig. 1b). Thus, the difference in the broadening and the lineshape of the two diamondoids could be well explained by the analysis of their orbitals, coupling of states via vibrations, and the proposed JT effects upon ionization.

Discussion

We note that the fine structure of the experimental PES consisting of sharp features requires further discussion. Our HR spectrum on the diamondoids suggests (see red curves in Supplementary

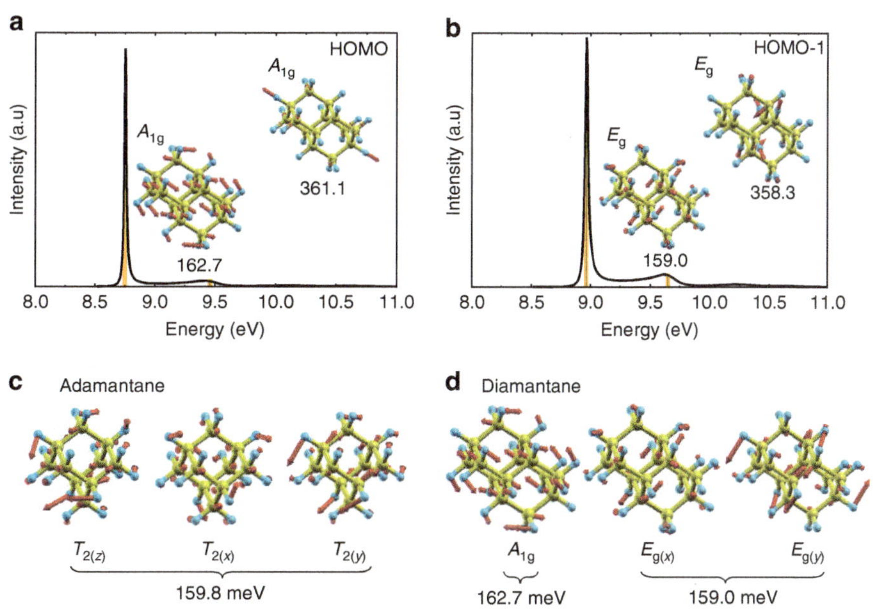

Figure 4 | Comparison of the HOMO and HOMO-1 spectral functions for diamantane possessing a_{1g} and e_g symmetry, respectively. The vibration modes contributed dominantly to the largest peak of SF are indicated in milli-electronvolt unit with the corresponding symmetry labels where the motion of the atoms are shown by red arrows for (**a**) HOMO and (**b**) HOMO-1 levels. Comparison of the vibrational modes in (**c**) adamantane and (**d**) diamantane that couples to the highest occupied molecular orbitals. The motion of atoms in these vibration modes for adamantane and diamantane is shown in the Supplementary Movies 4-6 and 11-13, respectively.

Fig. 1) that the fine structure could be related to the ionization-induced changes in the geometry. However, the accurate calculation on the strength of this contribution requires to consider the dynamical and geometry-induced effects on the same footing. This is equivalent to extend the present MBPT approach to include multiple vibrational excitations in the form of a vertex correction. Our MBPT approach has to be extended to capture multiple vibrational excitations, and we anticipate that vertex corrections can account for this. Inspired by the success of the present combined HR + MBPT approaches, we are actively pursuing the development of this more comprehensive theory.

In conclusion, we demonstrated on small nanodiamonds that the electron–vibration coupling results in the breakdown of the electron QP picture for their occupied states even at zero temperature. This has serious consequences on the features of their photoemission spectra, which cannot be predicted by semiclassical vibration broadening of electron states. We found indications in our MBPT approach for the fingerprint of different JT distortions. Our results imply that understanding of all the experimental features in the photoexcitation or photoionization of small nanoparticles needs a detailed investigation of electron–vibration coupling. We showed that our fully *ab initio* MBPT approach is capable of studying realistic systems and analyse the role of vibrations in their photoionization processes. This makes possible to design structures where the vibrations are manipulated to achieve the desired opto-electronic properties, which is of immediate importance in the field of superconductors, photovoltaic materials, colour centres in nanodiamonds[3] and related materials[28] for nanometrology.

Methods

Ground-state calculations. The ground-state geometry and electronic structure of the diamondoids were computed by plane wave supercell KS-DFT calculations as implemented in the Quantum Espresso (QE) software package[29]. We used a simple cubic simulation box and a diamondoid–diamondoid separation of about 1 nm to avoid the spurious interaction between the periodic images. For diamantane, we applied a hexagonal box to have a simulation box commensurable with its point group symmetry. The core electrons were taken into account by using norm-conserving Troullier–Martins pseudopotentials[30]. We took the standard Troullier–Martins pseudopotentials for each atom from the QE database except for nitrogen where we applied a relatively soft pseudopotential. We found that a kinetic energy cutoff of 45 Ry for plane wave expansion was sufficient to converge the band gap of the selected diamondoids. We applied *ab initio* LDA within KS-DFT to calculate the total energy and the KS spectrum[31,32]. The geometry of the diamondoids was optimized by minimizing the Hellmann–Feynmann forces below the threshold of 0.001 Ry per Bohr. In the geometry optimization procedure, we applied 80 Ry kinetic energy cutoff. We aligned single particle energies to the vacuum level by using the Martyna–Tuckerman method[33] as implemented in QE. The vibrational frequencies were computed by density functional perturbation theory[34]. The calculations were performed with a strong threshold of 10^{-16} Ry to achieve the self-consistent total energy in order to obtain accurate vibrational modes.

The implemented HR method to calculate vibration broadening. The rotations of single diamondoids and their collisions with other cages cause a broadening of about 20 meV (ref. 14) that does not explain the two orders magnitude larger broadening in the experimental PES. This is due to the relatively strong electron–vibration coupling where high energy vibration states with $n > 0$ quantum number contribute to the PES. The vibration-assisted ionization spectrum can be calculated by the HR method[35] within the Franck–Condon approximation that is explained briefly as follows. The key equation is the SF of electron–phonon coupling[36]

$$S(\hbar\omega) = \sum_{\lambda} S_{\lambda}\delta(\hbar\omega - \hbar\omega_{\lambda}), \tag{1}$$

where the sum is over all vibration modes λ with frequencies ω_{λ}, and S_{λ} is the partial HR factor for the mode λ. S_{λ} basically is the strength of the electron–vibration coupling of vibration λ. This can be defined as

$$S_{\lambda} = \omega_{\lambda} q_{\lambda}^2/(2\hbar). \tag{2}$$

where

$$q_{\lambda} = \sum_{\alpha i} m_{\alpha}^{1/2}(R_{\mathrm{I};\alpha i} - R_{\mathrm{N};\alpha i})\Delta r_{\lambda;\alpha i}. \tag{3}$$

m_{α} is the mass of atom α, $i = x,y,z$, $R_{\{\mathrm{I,N}\};\alpha i}$ is the equilibrium position in the ionized (I) and neutral (N) state and $\Delta r_{\lambda;\alpha i}$ is normalized vector that describes the displacement of the atom α along the direction i in the vibration mode λ. On ionization of the diamondoids, the optimized geometry changes from neutral to the ionized state within our Born–Oppenheimer approximation calculated by the chosen DFT. The vibration modes of the neutral diamondoids are calculated by the density functional perturbation theory method. The total HR factor S is obtained by $S = \sum S_{\lambda}$. Generally, if $S \ll 1$ then the spectrum is very sharp and mostly localized at the position of the QP energy. If $S > 10$, then the spectrum is very broad, featureless and can be well described with a symmetrical Gaussian function. We found that the typical value for diamondoids, $S \sim 3$, leads to a more complicated structure. The spectrum consists of a few sharp features lying on a broad background and the shape of the spectrum is non-symmetric. The $S \sim 3$ implies that three vibration quanta should contribute to the spectrum.
To create a simplified curve for broadening, we replaced the structured HR SF with a continuous envelope function. In the Supplementary Methods and the corresponding Supplementary Fig. 1, we give a more detailed description about this methodology and the results on the selected diamondoids.

The dynamical approach to the electron-vibration problem. Our methodology has been derived in detail in refs 6,7. Here, we only briefly summarize the basic assumptions and the key equations. Since we have finite objects within the supercell formalism, we use the $k = 0$ in the Brillouin zone and thus the momentum transfer q is also set to zero.

Using finite temperature MBPT, one can derive the self energy of an electron due to its interaction with vibrational modes. We consider the lowest-order self-energy terms: Fan-term (Σ_n^{Fan}) and DW-term (Σ_n^{DW}) (refs 6,7). Both depend on the temperature (T) through the boson occupation number of the vibration states ($B(\omega_{\lambda})$):

$$\Sigma_n^{\mathrm{Fan}}(\omega, T) = \sum_{n'\lambda} |g_{n'n}^{\lambda}|^2 \left[\frac{B(\omega_{\lambda}) + 1 - f_{n'}}{\omega - \varepsilon_{n'} - \omega_{\lambda} - i0^+} + \frac{B(\omega_{\lambda}) + f_{n'}}{\omega - \varepsilon_{n'} + \omega_{\lambda} - i0^+}\right], \tag{4}$$

$$\Sigma_n^{\mathrm{DW}}(T) = -\sum_{n'\lambda} \frac{1}{2} \frac{\Lambda_{nn'}^{\lambda}}{\varepsilon_n - \varepsilon_{n'}} (2B(\omega_{\lambda}) + 1), \tag{5}$$

Here, n and n' are the KS electron states with KS energies ε_n and $\varepsilon_{n'}$, respectively. f_n is the Fermi function of electrons. Since the diamondoids have large gaps, f_n is 1(0) for occupied (unoccupied) states, respectively. $g_{n'n}^{\lambda}$ is the electron–vibration matrix element between states n and n' due to the vibrational mode λ and energy ω_{λ}. This is a linear combination of $\langle n \mid \frac{\partial V_{\mathrm{scf}}}{\partial R_{\alpha,s}} \mid n' \rangle$ weighted by the amount they contribute to each vibrational mode, where $\frac{\partial V_{\mathrm{scf}}}{\partial R_{\alpha,s}}$ is the change in the self-consistent potential due to the displacement $R_{\alpha,s}$ of atom s along the cartesian direction α. We compute the first-order derivatives of the KS potential using density functional perturbation theory[34]. The second-order derivatives appearing in equation (5) ($\Lambda_{nn'}^{\lambda}$) can be turned into products of first-order derivatives by using translational invariance. In the summations over n', we used 200 states and checked convergence by using 300 states for selected diamondoids. In the summations over λ, we did not include the first six modes corresponding to translations and rotations of the diamondoids.

The temperature-dependent Green-function due to electron–vibration interaction can be written as

$$G_n(\omega, T) = \frac{1}{\omega - \varepsilon_n - \Sigma_n^{\mathrm{DW}}(T) - \Sigma_n^{\mathrm{Fan}}(\omega, T)}. \tag{6}$$

By Taylor expanding the Fan term around the unperturbed energies, we can also define perturbative corrections to the electron QP states:

$$\begin{aligned} E_n(T) &= \varepsilon_n + \Sigma_n^{\mathrm{DW}}(T) + \Sigma_n^{\mathrm{Fan}}(\varepsilon_n, T) + \left.\frac{\partial \Sigma_n^{\mathrm{Fan}}(\omega, T)}{\partial \omega}\right|_{\omega=\varepsilon_n} (E_n(T) - \varepsilon_n) \\ &= \varepsilon_n + Z_n(T)\left[\Sigma_n^{\mathrm{DW}}(T) + \Sigma_n^{\mathrm{Fan}}(\varepsilon_n, T)\right], \end{aligned} \tag{7}$$

where the renormalization factor $Z_n(T)$ reads

$$Z_n(T) = \left[1 - \left.\frac{\partial \Sigma_n^{\mathrm{Fan}}(\omega, T)}{\partial \omega}\right|_{\omega=\varepsilon_n}\right]^{-1}. \tag{8}$$

The QP shift $\Delta E_n(T)$ due to electron–vibration interaction may be simply defined as $\Delta E_n(T) = E_n(T) - \varepsilon_n$. We note that the PES for diamondoids was measured at room temperature. For adamantane, we checked that bosonic occupation of vibration states indeed do not change the results going from zero kelvin up to room temperature. Since N-C, C-C and C-H vibrational modes have much higher energies than the room temperature energy of 26 meV, only the ground-state vibration states have non-zero occupations, and thus the calculated properties do not show measurable temperature dependence in the region of zero kelvin and room temperature. Therefore, we omit the temperature dependence in the QP shift $\Delta E_n \equiv \Delta E_n(T)$ and the renormalization factor $Z_n \equiv Z_n(T)$.

We can also compute the SF, which is defined as

$$A_n(\omega) \equiv \pi^{-1}|\Im[G_n(\omega)]|, \tag{9}$$

where $G_n(\omega)$ is taken from equation (6) by omitting the temperature. These SFs are plotted for a given n KS state in Fig. 2 for adamantane from which the PES was derived as shown in Fig. 1b. We note here that equation (9) does not include transition matrix element effects and thus our method of computing PES corresponds to a constant, energy-independent transition matrix element approximation.

For the analysis of the contribution of the vibrations to the corresponding SFs for a given KS state, we use the Lehmann representation,

$$A_n(\omega) = \sum_I |\langle \Psi_0 \mid c_n \mid I \rangle|^2 \frac{\eta}{(\omega - E_I)^2 + \eta^2}, \tag{10}$$

where E_I is the energy of the true eigenstate ($|I\rangle$) of the system and η is a small infinitesimal number. $|\Psi_0\rangle$ is the exact ground state and c_n is the electron annihilation operator. Here, we again omit for simplicity the temperature dependence. By definition, E_I are the real poles of $G_n(\omega)$.

In ref. 7, it was shown that the states $|I\rangle$ are mixed packets of electrons and vibrations. Moreover, their internal structure was found by showing that the electron–vibration problem can be fully rewritten as the solution of an effective Hamiltonian. The solution of this problem allows to introduce an explicit expression for the interacting states $|I\rangle$:

$$|I\rangle = \sum_n A_n^I \mid n\rangle + \sum_{n'\lambda} B_{n'\lambda}^I \mid n'\rangle \otimes \mid \lambda\rangle. \tag{11}$$

$|n\rangle$ is the purely electronic state obtained by creating an electron in the n-th KS level. $|A_n^I|^2$ is the corresponding probability to find the electron–vibration state along $|n\rangle$. $|B_{n'\lambda}^I|^2$, instead, is the probability to find the fully interacting state in a mixed electronic $|n'\rangle$ and vibrational λ state (see refs 7,37). These coefficients can be found by diagonalizing the electron–vibration Hamiltonian on the basis of equation (11). This tool was applied to analyse the contribution of vibrations λ to the given peaks in the SFs, for instance, as depicted in Fig. 2 for adamantane.

From the Lehmann representation, it follows that the Green-function represents the dynamical evolution of the charged state $c_n^\dagger \mid \Psi_0\rangle$. This situation is quite similar to the physics described in the Jahn–Teller effect. The difference is that, in the present case, the atoms are allowed to move only within the harmonic approximation. Nevertheless, as done in ref. 26, the analysis of the $|B_{n'\lambda}^I|^2$ terms can be used to deduce the most intense phonon modes (and consequently, atomic displacement directions) that contribute to the formation of the coupled electron–vibration states.

The complex structure of the $|I\rangle$ state, then, induces the wealth of structures that are crucial to explain the PES and, at the same time, provides a sound and clear indication of the most intense modes that contribute to the physics of the Jahn–Teller effect.

References

1. Kovalenko, M. V. *et al.* Prospects of nanoscience with nanocrystals. *ACS Nano* **9,** 1012–1057 (2015).
2. Mochalin, V. N., Shenderova, O., Ho, D. & Gogotsi, Y. The properties and applications of nanodiamonds. *Nat. Nanotechnol.* **7,** 11–23 (2012).
3. Schirhagl, R., Chang, K., Loretz, M. & Degen, C. L. Nitrogen-vacancy centers in diamond: nanoscale sensors for physics and biology. *Annu. Rev. Phys. Chem.* **65,** 83–105 (2014).
4. Hüfner, S. *Photoelectron Spectroscopy: Principles and Applications* (Springer, 2013).
5. Marini, A. *Ab Initio* finite-temperature excitons. *Phys. Rev. Lett.* **101,** 106405 (2008).
6. Cannuccia, E. & Marini, A. Effect of the quantum zero-point atomic motion on the optical and electronic properties of diamond and trans-polyacetylene. *Phys. Rev. Lett.* **107,** 255501 (2011).
7. Cannuccia, E. & Marini, A. Zero point motion effect on the electronic properties of diamond, trans-polyacethylene and polyethylene. *Eur. Phys. J. B* **85,** 320 (2012).
8. Dahl, J. E., Liu, S. G. & Carlson, R. M. K. Isolation and structure of higher diamondoids, nanometer-sized diamond molecules. *Science* **299,** 96–99 (2003).
9. Dahl, J. E. P. *et al.* Synthesis of higher diamondoids and implications for their formation in petroleum. *Angew. Chem. Int. Ed.* **49,** 9881–9885 (2010).
10. Schwertfeger, H., Fokin, A. A. & Schreiner, P. R. Diamonds are a chemist's best friend: diamondoid chemistry beyond adamantane. *Angew. Chem. Int. Ed.* **47,** 1022–1036 (2008).
11. Mansoori, G., de Araujo, P. & de Araujo, E. *Diamondoid Molecules: With Applications in Biomedicine, Materials Science, Nanotechnology & Petroleum Science* (World Scientific Pub., 2012).
12. Clay, W. A., Dahl, J. E. P., Carlson, R. M. K., Melosh, N. A. & Shen, Z.-X. Physical properties of materials derived from diamondoid molecules. *Rep. Prog. Phys.* **78,** 016501 (2015).
13. Schmidt, W. Photoelectron spectra of diamondoid molecules: adamantane, silamantane and urotropine. *Tetrahedron* **29,** 2129–2134 (1973).
14. Zimmermann, T. *et al.* Exploring covalently bonded diamondoid particles with valence photoelectron spectroscopy. *J. Chem. Phy.* **139,** 084310 (2013).
15. Fokin, A. A., Tkachenko, B. A., Gunchenko, P. A., Gusev, D. V. & Schreiner, P. R. Functionalized nanodiamonds part i. an experimental assessment of diamantane and computational predictions for higher diamondoids. *Chemistry* **11,** 7091–7101 (2005).
16. Landt, L. *et al.* Experimental and theoretical study of the absorption properties of thiolated diamondoids. *J. Chem. Phys.* **132,** 144305 (2010).
17. Vörös, M. & Gali, A. Optical absorption of diamond nanocrystals from ab initio density-functional calculations. *Phys. Rev. B* **80** (2009).
18. Drummond, N. D., Williamson, A. J., Needs, R. J. & Galli, G. Electron emission from diamondoids: a diffusion quantum Monte Carlo study. *Phys. Rev. Lett.* **95,** 096801 (2005).
19. Nguyen, H., Pham, T. A., Rocca, D. & Galli, G. Improving accuracy and efficiency of calculations of photoemission spectra within the many-body perturbation theory. *Phys. Rev. B* **85,** 081101 (2012).
20. Marsusi, F., Sabbaghzadeh, J. & Drummond, N. D. Comparison of quantum monte carlo with time-dependent and static density-functional theory calculations of diamondoid excitation energies and stokes shifts. *Phys. Rev. B* **84,** 245315 (2011).
21. Yin, H. *et al.* Quasiparticle electronic structure and optical absorption of diamond nanoparticles from ab initio many-body perturbation theory. *J. Chem. Phys.* **140,** 214315 (2014).
22. Demján, T., Vörös, M., Palummo, M. & Gali, A. Electronic and optical properties of pure and modified diamondoids studied by many-body perturbation theory and time-dependent density functional theory. *J. Chem. Phys.* **141,** 064308 (2014).
23. Rander, T. *et al.* Electronic structure tuning of diamondoids through functionalization. *J. Chem. Phys.* **138,** 024310 (2013).
24. Giustino, F., Louie, S. G. & Cohen, M. L. Electron-phonon renormalization of the direct band gap of diamond. *Phys. Rev. Lett.* **105,** 265501 (2010).
25. Patrick, C. E. & Giustino, F. Quantum nuclear dynamics in the photophysics of diamondoids. *Nat. Commun.* **4,** 2006 (2013).
26. Manini, N., Gattari, P. & Tosatti, E. Jahn-Teller spectral fingerprint in molecular photoemission: c_{60}. *Phys. Rev. Lett.* **91,** 196402 (2003).
27. Bersuker, I. B. Pseudo-Jahn-Teller effect—a two-state paradigm in formation, deformation, and transformation of molecular systems and solids. *Chem. Rev.* **113,** 1351–1390 (2013).
28. Falk, A. L. *et al.* Electrically and mechanically tunable electron spins in silicon carbide color centers. *Phys. Rev. Lett.* **112,** 187601 (2014).
29. Giannozzi, P. *et al.* Quantum espresso: a modular and open-source software project for quantum simulation of materials. *J. Phys. Cond. Matter* **21,** 395502 (2009).
30. Troullier, N. & Martins, J. L. Efficient pseudopotentials for plane-wave calculations. *Phys. Rev. B* **43,** 1993–2006 (1991).
31. Ceperley, D. M. & Alder, B. J. Ground state of the electron gas by a stochastic method. *Phys. Rev. Lett.* **45,** 566–569 (1980).
32. Perdew, J. P. & Zunger, A. Self-interaction correction to density-functional approximations for many-electron systems. *Phys. Rev. B* **23,** 5048–5079 (1981).
33. Martyna, G. J. & Tuckerman, M. E. A reciprocal space based method for treating long range interactions in ab initio and force-field-based calculations in clusters. *J. Chem. Phys.* **110,** 2810–2821 (1999).
34. Baroni, S., de Gironcoli, S., Dal Corso, A. & Giannozzi, P. Phonons and related crystal properties from density-functional perturbation theory. *Rev. Mod. Phys.* **73,** 515–562 (2001).
35. Huang, K. & Rhys, A. Theory of light absorption and non-radiative transitions in *f*-centres. *Proc. Roy. Soc.* **204,** 407–423 (1970).
36. Miyakawa, T. & Dexter, D. L. Phonon sidebands, multiphonon relaxation of excited states, and phonon-assisted energy transfer between ions in solids. *Phys. Rev. B* **1,** 2961–2969 (1970).
37. Cannuccia, E. *Giant Polaronic Effects in Polymers: Breakdown of the Quasiparticle Picture.* (PhD thesis, Rome Tor Vergata Univ, 2011).

Acknowledgements

A.G. acknowledges the support from the Lendület program of the Hungarian Academy of Sciences, the PRACE DECI6 DIASIC project and the PRACE DECI7 DIAVIB project (FP7 RI-283493) for the resources CINECA in Italy and HUYGENS in Netherlands. The technical assistance of Andrew Emerson from CINECA and Dr Jörg Hertzer from High Performance Computing Center, Germany is gratefully acknowledged. A.G. and M.V. also acknowledge the support from the NIIF Supercomputer Center Grant No. 1090. A.M. acknowledges financial support by the *Futuro in Ricerca* grant No. RBFR12SW0J of the Italian Ministry of Education, University and Research MIUR, the European Union project MaX *Materials design at the eXascale* H2020-EINFRA-2015-1, Grant agreement No. 676598 and *Nanoscience Foundries and Fine Analysis - Europe* H2020-INFRAIA-2014-2015, Grant agreement No. 654360. This work was also supported by U.S. DOE, Office of Science under Contract No. DE-AC02-06CH11357 (M.V.). This research used resources of the National Energy Research Scientific Computing Center, a DOE Office of Science User Facility supported by the Office of Science of the U.S. Department of Energy under Contract No. DE-AC02-05CH11231.

Author contributions

A.G. proposed the idea and the strategy for the calculation on the effect of electron–vibration coupling in the QP states, and wrote the paper. T.D., G.T. and M.V. carried out the calculations whereas E.C. and A.M. developed the many-body perturbation theory and implemented into the applied scientific code. A.G. and G.T. implemented the Huang–Rhys method. All the authors discussed the results and commented on the manuscript.

Additional information

Competing financial interests: The authors declare no competing financial interests.

20

Observation of coherent delocalized phonon-like modes in DNA under physiological conditions

Mario González-Jiménez[1], Gopakumar Ramakrishnan[1], Thomas Harwood[2], Adrian J. Lapthorn[1], Sharon M. Kelly[3], Elizabeth M. Ellis[2] & Klaas Wynne[1]

Underdamped terahertz-frequency delocalized phonon-like modes have long been suggested to play a role in the biological function of DNA. Such phonon modes involve the collective motion of many atoms and are prerequisite to understanding the molecular nature of macroscopic conformational changes and related biochemical phenomena. Initial predictions were based on simple theoretical models of DNA. However, such models do not take into account strong interactions with the surrounding water, which is likely to cause phonon modes to be heavily damped and localized. Here we apply state-of-the-art femtosecond optical Kerr effect spectroscopy, which is currently the only technique capable of taking low-frequency (GHz to THz) vibrational spectra in solution. We are able to demonstrate that phonon modes involving the hydrogen bond network between the strands exist in DNA at physiologically relevant conditions. In addition, the dynamics of the solvating water molecules is slowed down by about a factor of 20 compared with the bulk.

[1] School of Chemistry, WestCHEM, University of Glasgow, Glasgow G12 8QQ, UK. [2] Institute of Pharmacy and Biomedical Sciences, University of Strathclyde, Glasgow G4 0RE, UK. [3] Institute of Molecular Cell and Systems Biology, University of Glasgow, Glasgow G12 8QQ, UK. Correspondence and requests for materials should be addressed to K.W. (email: klaas.wynne@glasgow.ac.uk).

The processes important to the biological function of DNA (replication, transcription, denaturation and molecular intercalation) have in common that they start with the breaking of the hydrogen bonds between the bases of the nucleic acid. Driven by the torsional stress of the molecule[1], the destablization of the weak bonds leads to the splitting of a section of the double helix of DNA into single strands, forming a gap in the nucleic acid known as a transcriptional bubble[2].

The DNA molecule is not a static object, but vibrates and fluctuates on timescales from femtoseconds to nanoseconds. The origin of the rupture of the hydrogen bonds in DNA is thought to be low-frequency vibrational modes propagating along its length in the form of phonon-like modes that expand and contract the space between the bases[3–5]. Because adenine–thymine and guanine–cytosine base pairs differ in their hydrogen bonding and stacking, the physical properties of DNA that influence these waves, such as helical structure[6] or elasticity[7], depend on the local DNA sequence[8]. It has been suggested that DNA has regions where specific sequences favour the resonance between low-frequency vibrational modes[9,10], promoting the temporary splitting of a significant number of bases, even at physiological temperatures[11–13]. During the short period of time that the bubbles exist[14], the bases are exposed to the surrounding solvent, which has two effects. On one hand, bubbles expose the nucleic acid to reactions of the bases with mutagens in the environment[15], while molecular intercalators can take advantage of the gap and may insert themselves between the strands of DNA[11]. On the other hand, bubbles allow helicases access to DNA to stabilize the bubble, followed by splitting the strands to start the transcription and replication process[16]. For this reason, it is believed that DNA directs its own transcription.

Despite the importance of the low-frequency vibrational modes of DNA, evidence for their existence is only indirect[17]. The relatively weak absorption of far-infrared radiation by DNA compared with the extremely strong absorption by water has limited investigation by infrared spectroscopy[18]. However, a recent report suggested a non-thermal response of cellular gene expression to terahertz radiation[19]. The presence of low-frequency vibrational modes in DNA has been shown through Raman[20,21], Brillouin[22], inelastic X-ray[23] and inelastic neutron-scattering[24] measurements. However, all of these studies have been performed in unnatural solid DNA preparations (humidified films, fibres and so on) that modify the dynamics and are likely to introduce low-frequency lattice-phonon modes associated with crystal packing. Furthermore, inelastic neutron-scattering experiments have shown great sensitivity of the speed of sound to relative humidity. No spectroscopic experiments have been carried out under relevant physiological conditions in aqueous solution except over a very narrow frequency range ($10–35\,cm^{-1}$)[25–27]. It seems highly likely that exposure of the nucleic acid to bulk liquid water will lead to strong damping of phonon-like modes, resulting in localization. Therefore, the importance of bubbles and the associated low-frequency phonon-like vibrational modes under physiological conditions remains unproven and unlikely.

In recent work, we have shown that the technique of ultrafast optical Kerr effect (OKE) spectroscopy could be used to determine the presence of delocalized phonon-like modes in a protein and prove their relevance to biological function[28]. This technique measures the low-frequency depolarized Raman spectrum in the time domain and obtains a spectrum through numerical Fourier transformation[29–31]. Spontaneous scattering techniques such as Raman spectroscopy and inelastic neutron scattering when applied to liquids and solutions become unreliable at low frequencies (<1 THz) due to a very strong Rayleigh peak from elastic scattering[32]. The spectral resolution of inelastic X-ray and neutron scattering is poor ($\sim 15\,cm^{-1}$ or 0.5 THz), prohibiting spectral characterization especially below ~ 1 THz (ref. 23). OKE spectroscopy has proven to be far superior at low frequencies, as it does not suffer from a large Rayleigh peak and its high signal to noise allows a detailed analysis of the spectra[29,30]. Furthermore, the OKE signal from liquid water is relatively weak, allowing it to be subtracted from the total signal to reveal the spectrum of the solvated biomolecule. Thus, OKE is the only technique that can reliably determine the terahertz and sub-terahertz spectra of biomolecules in physiologically relevant aqueous solution.

Here we use OKE spectroscopy to investigate the low-frequency vibrational modes of DNA to determine the presence and possible role of phonon-like modes in nucleic acids in aqueous solution. We present the OKE spectra of two DNA oligomers in phosphate buffer solution, at various degrees of denaturation as a function of increasing temperature. Two phonon-like modes associated with the hydrogen bonds of the double strand of DNA are identified. To confirm the assignments given to the observed bands, the OKE spectra of two different oligomers designed to form a double helix only when they are dissolved together are measured. Since nucleotides in solution stack even at low concentrations while not forming hydrogen bonds between them[33], the OKE spectra of cytosine are measured to investigate the influence of stacking on the nucleic acid spectra.

Results

OKE spectra of 20mers. Experiments were carried out on two relatively short palindromic DNA sequences containing mostly cytosine and guanine: $d(GGCGGCCCGCGCGGGCCGCC)_2$ (CG 20mer), and mostly adenine and thymine: $d(TTATTAAATATATTTAATAA)_2$ (AT 20mer). OKE spectra for 10 mM solutions of the CG and AT oligomers in phosphate buffer (pH = 7) were measured at 10 K temperature intervals from 298 to 358 K. The OKE spectrum of water was subtracted at each temperature (Supplementary Note 1; Supplementary Figs 1–4; Supplementary Tables 1–4) to obtain the spectra of the solvated oligomers. At the concentration employed (0.15 M), the OKE spectrum of phosphate buffer is indistinguishable from that of water.

Figure 1 shows the experimental temperature-dependent OKE spectra of the AT 20mer and fits to theoretical expressions. Below ~ 50 GHz, the contribution of the molecular orientational diffusion of the oligomer is expected to dominate. Using the Stokes–Einstein–Debye equation[29], a relaxation frequency between 0.2 GHz at 298 K and 1.6 GHz at 358 K was estimated for this process (Supplementary Note 2; Supplementary Fig. 5). Because this is at significantly lower frequency than accessible in these experiments, only the high-frequency wing of the band can be seen in the spectra. Despite this, the orientational relaxation band could be fitted using a Debye function (D in Fig. 1; Methods). The relaxation time constants obtained through curve fitting are broadly consistent with the values calculated using the Stokes–Einstein–Debye equation.

The remainder of the low-frequency portion of the spectra is caused by a very broad band that can be fitted with a Cole–Cole function (Methods) with its maximum moving from 14.2 GHz at 298 K to 27.8 GHz at 358 K. This band extends to unusually high frequency and is responsible for the increase in intensity in the region between 4 and 9 THz.

The most interesting features of the OKE spectra of the AT 20mer are the changes with temperature that appear in the high-frequency part of the spectra (>200 GHz). This portion of the spectrum can be fitted by four Brownian oscillators (Methods). There is only a very small change with temperature in the

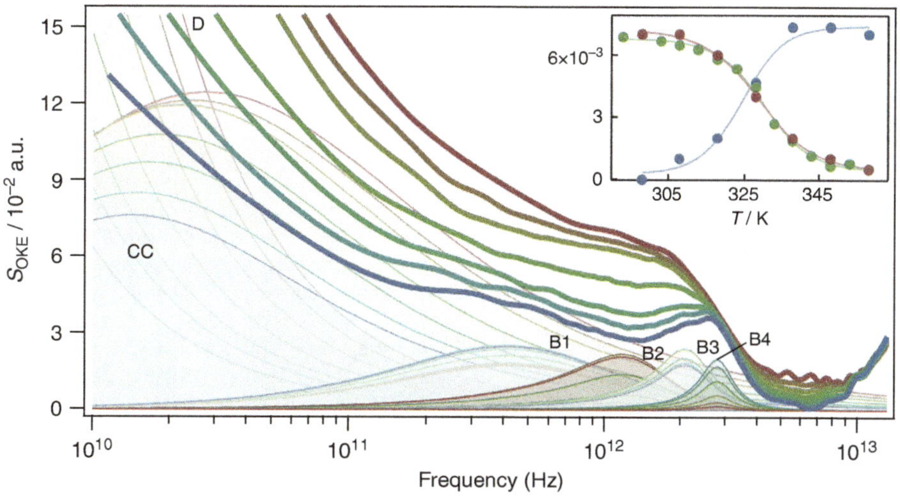

Figure 1 | Experimental optical Kerr effect spectra and the component fit functions used to fit the data for the 20-bp AT oligomer. The temperatures run from 298 (blue) to 358 K (red) in steps of 10 K. Each spectrum has been fitted using the combination of a Debye function (D), a Cole-Cole function (CC) and four Brownian oscillators (B1-B4). The features in the figure that change in an unexpected manner are the bands B2 and B4 (with emphasized colours). The intensity of B2 increases with temperature, while the intensity of B4 reduces. The inset shows the intensity of the bands B2 (blue) and B4 (red). They are compared with the scaled results of a circular dichroism experiment (ellipticity at 248 nm, green).

intensity of the first (B1 in Fig. 1, with the Brownian oscillator frequency $\omega_0/2\pi = 1.01$ THz and damping rate $\gamma/2\pi = 1.39$ THz) and third (B3, $\omega_0/2\pi = 2.19$ THz and $\gamma/2\pi = 0.70$ THz) bands. However, there are significant changes in the intensities of the second (B2) and fourth (B4) bands, which is unexpected. At room temperature (298 K), only B4 occurs in the spectrum but as the temperature increases, the intensity of B4 decreases, while B2 appears and increases in intensity. The inset of Fig. 1 shows the sigmoidal dependence of the intensity of these bands on temperature. A circular dichroism (CD) experiment to determine the proportion of denatured oligomer at each temperature was carried out (Supplementary Note 3; Supplementary Figs 6 and 7) showing the same sigmoidal dependence on temperature from double stranded at room temperature to single stranded at 358 K. This strongly suggests that B2 and B4 are associated with the single- and double-stranded conformation of the oligomer, respectively. The melting temperatures for the AT 20mer derived from a sigmoidal fit to the intensity of the B2 and B4 bands (324 ± 2 and 329 ± 1 K, respectively) and from the CD experiment (330.1 ± 0.6 K) are in good agreement as expected.

The remaining fit parameters of these bands are $\omega_0/2\pi = 1.38$ THz and $\gamma/2\pi = 0.75$ THz for B2, and $\omega_0/2\pi = 2.83$ THz and $\gamma/2\pi = 0.50$ THz for B4. Thus both vibrational modes have a damping rate smaller than their oscillator frequencies and are therefore underdamped. Furthermore, these bands (in particular the high frequency B4 band) cannot be fitted by a Gaussian function (Supplementary Note 4; Supplementary Fig. 8).

The temperature-dependent OKE spectra of the CG 20mer (Supplementary Note 5; Supplementary Fig. 9) show similar behaviour and can be fitted in a similar manner. However, the characteristic B4 band only starts changing at relatively high temperature consistent with results of CD, which shows that the melting temperature of the CG 20mer is higher than the accessible temperature range in the OKE experiment.

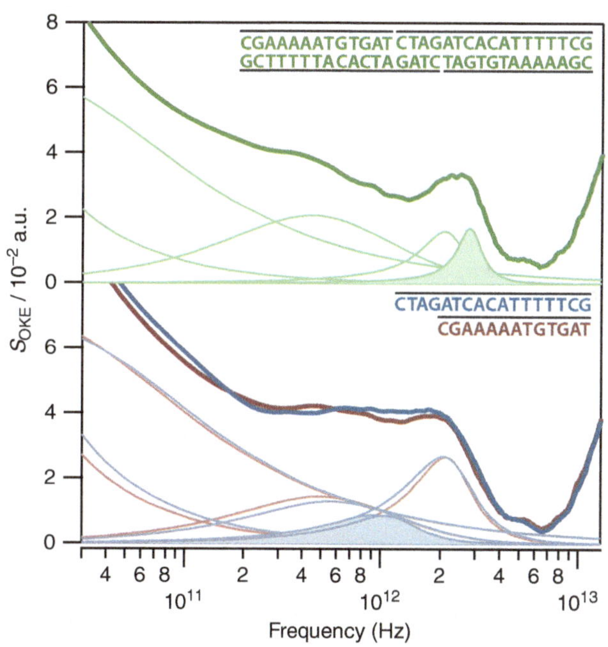

Figure 2 | OKE spectra and fits for the solutions of the complementary oligomers of 13 and 17 bases dissolved together and separately. Dissolved together (top) and separately (bottom, where the spectrum from the 13mer is shown in red and that of the 17mer in blue). The significant difference between top and bottom is the bands filled with colour. When the nucleic acids are single stranded, there is a band at 1.08 (13mer) and 1.20 THz (17mer). However, when they form a double strand, only a characteristic band at 2.83 THz appears.

Investigation of 13- and 17-base oligomers. The data in Fig. 1 strongly suggest that the B2 band corresponds to single-stranded DNA and the B4 band to double-stranded DNA. To prove this, a room-temperature experiment was carried out using two oligomers of 13 and 17 bases with sequences d(CGAAAAATGTGAT) and d(CTAGATCACATTTTTCG) that minimize the possibility of association between strands when they are dissolved separately. However, when they are dissolved together, the strands match perfectly to form a 30-bp double-stranded helix. OKE spectra for 10 mM solutions in phosphate buffer of each oligomer separately and both oligomer together were measured at 298 K (Fig. 2). The spectra of the oligomers separately and together can be fitted in a very similar manner. However, solutions of the 13mer and 17mer separately show unique bands at 1.08 and 1.20 THz, respectively,

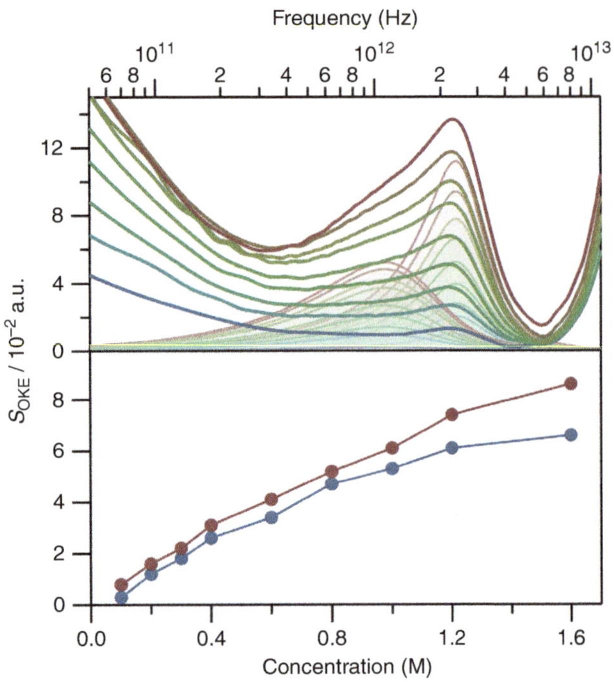

Figure 3 | Influence of concentration on the OKE spectra of cytidine monophosphate. The terahertz band of the spectra has been fitted using two Brownian oscillators at 1.48 and 2.58 THz (top). The effect of the concentration on the intensity of these bands (blue and red, respectively) has been plotted (bottom).

while a fourth band at higher frequency is absent. However, in the spectrum of the solution with both oligomers together, only a unique underdamped band at 2.83 THz appears (the B4 band, but not the B2 band), confirming our hypothesis.

Base stacking in mono-nucleotide solutions. It is now clear that the B2 and B4 bands in Fig. 1 are closely associated with the single- and double-stranded forms of DNA. The B4 band, which disappears on melting, is likely associated with hydrogen bonding between the strands. To rule out an origin in base stacking, concentration-dependent OKE experiments were carried out at room temperature on aqueous solutions of cytidine monophosphate with concentrations between 0.1 and 1.6 M. It is known that mono-nucleotides in solution tend to stack without forming intermolecular hydrogen bonds between them[33]. The OKE spectra for these solutions (Fig. 3) show a characteristic pattern with a low-frequency band (below 100 GHz) due to the orientational diffusion and a higher-frequency band between 400 GHz and 5 THz. The latter can be fitted using two Brownian oscillators, yielding the parameters $\omega_0/2\pi = 1.48$ THz and $\gamma/2\pi = 1.08$ THz, for the first oscillator, and $\omega_0/2\pi = 2.58$ THz and $\gamma/2\pi = 1.06$ THz, for the second. These vibrational modes do not coincide with the frequency of the modes observed in the oligomers and are less underdamped. The relation between intensity of the oscillators and nucleotide concentration is also plotted in Fig. 3. The curves show the same saturation behaviour that can be seen in other studies of stacking of nucleotides[33,34], that comes from the dependence on concentration of the average number of aggregated nucleotides.

Discussion

Thus, we have shown that the bands labelled B2 and B4 in Fig. 1 are associated with single- and double-stranded DNA respectively and that the B4 band reflects the dynamics of hydrogen bonding between the two strands. The B4 hydrogen-bonding band can be fitted with a Brownian oscillator function, but not by a Gaussian (Supplementary Note 4). Although spontaneous Raman scattering and related techniques such as OKE cannot determine whether a band is homogeneously or inhomogeneously broadened[35–37], the good fit to a Brownian oscillator function does strongly suggest a homogeneously broadened band or at most a very narrow distribution of phonon frequencies. Therefore, this band is not a broad inhomogeneous distribution of multiple vibrational modes, instead a single delocalized normal mode of the DNA 20mer. As the frequency of this mode is high ($\omega_0/2\pi = 2.83$ THz corresponding to a period of 350 fs) and the damping rate low ($\gamma/2\pi = 0.50$ THz corresponding to a 2-ps decay), it is underdamped undergoing approximately five oscillations before vanishing.

According to the calculations using a fully atomistic model of B-DNA[4] as well as inelastic X-ray scattering experiments on humidified films of oriented DNA[23], the acoustic phonon branch for delocalized phonons in DNA peaks at ~12 meV (3.2 THz) at the Brillouin zone edge. OKE spectroscopy has the same selection properties as Raman scattering and is therefore normally only sensitive to optical phonon modes near the zone centre where the phonon wave vector vanishes and the wavelength tends to infinity. The observed frequency of 2.83 THz is therefore consistent with an optical phonon mode with a wavelength extending throughout the entire 20mer (Brillouin zone centre). This explains the observed reduction in intensity of the B4 band with temperature, since the heating-induced denaturation disrupts the hydrogen bonds of the base pairs and thereby causes the strands to separate eliminating the phonon mode.

The phonon-like mode in double-stranded DNA observed here using OKE spectroscopy has a half-period of 177 fs, which sets the timescale for the breaking of hydrogen bonds in double-stranded DNA. However, this result is consistent with nuclear magnetic resonance experiments[2,38], fluorescence spectroscopy[39,40] and simulations[8] showing much longer timescales. These experiments measure a global conformational change in the double-stranded DNA or the probability of imino-proton exchange in an established bubble[12]. The phonon-like mode as observed here using OKE sets the approach rate in the barrier crossing that leads to bubble formation, and is therefore expected to be a much faster process.

Thus, the assumption that interaction of DNA with the surrounding water would dampen and localize the phonon mode[41] is found to be incorrect. This observation is consistent with the previous observation of similar delocalized underdamped phonon-like modes in the protein lysozyme[28].

The OKE spectrum of the AT 20mer (Fig. 1) contains a broad band peaking between 14.2 and 27.8 GHz that has been fitted to a Cole–Cole function. It is tempting to assign this CC band to some overdamped motion of the DNA insensitive to the hybridization state. However, the data analysis shows that the CC band is responsible for intensity between 4 and 9 THz, that is, above the frequency of the highest-frequency mode that can clearly be assigned to the 20mer (the B4 hydrogen-bond phonon mode). This would be unphysical (relaxational processes involving a particular molecule, by their very nature must be slower than underdamped processes of the same molecule) unless the CC band is associated with liquid water whose librational frequency is ~25 THz. In fact, OKE and Raman scattering studies on lysozyme[28,32] observed a similar band at ~20–40 GHz, which could be shown using neutron scattering to originate in the diffusive translational motion of water molecules in the solvation shell of the protein[32].

Thus, the CC band can similarly be interpreted as the diffusive translational dynamics of water molecules in the solvation shell of the DNA oligomer. In bulk water at 298 K, this band peaks at[28] 245 GHz, showing a slowdown of the dynamics by a factor of 17 consistent with ultrafast solvation experiments[42]. However, this appears inconsistent with the predictions made by the jump model of water diffusion[43,44], which only predicts a slowdown by a factor 2 near a flat non-interacting surface. The surface of DNA is much more complex though with convex and concave areas. This would give rise to a greater slowdown in addition to the extreme broadening seen here[45]. At 248 K, when DNA is almost denatured, the slowdown of the dynamics is slightly larger (it changes from 17 at 298 K to 21.5 at 348 K). This is most likely caused by hydrogen bonding with the exposed bases.

Thus, the results presented here demonstrate that the inter-strand hydrogen-bond modes are coherent delocalized phonon modes even at physiological conditions. As the hydrogen bonds need to be broken for a transcription bubble to form, this result suggests that at least the initial steps of bubble formation are coherent. This is consistent with the recently observed dynamics in enzymes and enzyme–inhibitor complexes, which were similarly shown to be delocalized and coherent (although damped much more strongly than seen here in DNA). The release of water molecules from the solvation shell of DNA is thought to provide an entropic driving force for DNA–protein and anti-cancer drug binding[43]. The unexpectedly large slowdown of water dynamics by a factor of ~20 observed here will have a large effect on the dynamics and energetics of DNA binding.

Methods

Sample preparation. DNA oligomers (salt-free and reverse-phase cartridge purified) were purchased from Eurofins and cytidine monophosphate from Merk. All the samples were prepared by directly dissolving in phosphate buffer (pH = 7, 0.15 M) and filtering using 0.2 μm hydrophilic polytetrafluoroethylene (PTFE) filters (Millipore) to remove dust. Before the measurements, the solutions of DNA oligomers were annealed by equilibrating them in a water bath at 363 K, followed by slow cooling down to the desired temperature. CD spectroscopy was used to determine the temperature-dependent denaturation of the DNA (Supplementary Note 3; Supplementary Figs 6 and 7).

OKE experimental details. The OKE data were recorded in a standard time-domain step-scan pump-probe configuration and Fourier transformed to obtain the frequency-domain reduced depolarized Raman spectrum as described previously[28–31]. A laser oscillator (Coherent Micra) provided ~10-nJ pulses (0.8 W average power) with a nominal wavelength of 800 nm at a repetition rate of 82 MHz providing a 20-fs pulse width in the sample. For the longer-timescale relaxation measurements, a second set of data was taken in a similar configuration using a higher pulse energy of typically 1 μJ provided by a regeneratively amplified laser (Coherent Legend Elite USX) at a repetition rate of 1 kHz with a pulse duration stretched to ~1 ps. Stretching the pulse enables a higher energy to be used without nonlinear effects and reduces the upper bandwidth limit allowing large step size scanning without introducing undersampling artefacts. Pump-probe OKE experiments were carried out with delay times from a few femtoseconds to a maximum of 1–4 ns, resulting in a spectral resolution (after Fourier transformation) of better than 1 GHz ($<0.033\,\mathrm{cm}^{-1}$)[28,29]. The sample was contained in a rectangular quartz cuvette (Starna, thickness: 1 mm) held in a brass block that was temperature controlled with a precision of ± 0.5 K.

OKE data analysis. The OKE spectra consist of several broad overlapping bands that are analysed through curve fitting to a number of analytical functions (Supplementary Note 4). At the lowest frequencies, one finds processes associated with diffusive orientational relaxation of the molecules. The diffusive orientational relaxation of DNA in solution has been fitted here with a Debye function

$$S_{\mathrm{D}}(\omega) = \mathrm{Im}\frac{A_{\mathrm{D}}}{1 + i\omega\tau}, \tag{1}$$

where A_{D} is the amplitude of the band, ω is the angular frequency and τ is the relaxation time. The band at slightly higher frequency, which can be assigned to the diffusive relaxation of water in the solvation shell of the DNA, cannot be fitted to a Debye function due to its much greater width. This band was modelled using a Cole–Cole function

$$S_{\mathrm{CC}}(\omega) = \mathrm{Im}\frac{A_{\mathrm{CC}}}{1 + (i\omega\tau)^{\alpha}}, \tag{2}$$

where A_{CC} is the amplitude of the band and α is a parameter that accounts for the broadness of the observed band.

In the terahertz range, one finds bands from modes that are not diffusive, but critically damped or underdamped. These originate in librations, vibrations and phonon-like modes. These have been fitted using the Brownian oscillator model[28]

$$\begin{aligned} S_{\mathrm{BO}}(\omega) &= \mathrm{Im}\frac{A_{\mathrm{BO}}\omega_0^2}{\omega_0^2 - \omega(\omega + 2i\gamma)} \\ &= \mathrm{Im}\frac{A_{\mathrm{BO}}\omega_0^2}{\omega_0^2 - \omega^2 - 2i\gamma\omega}, \end{aligned} \tag{3}$$

where ω_0 is the undamped oscillator angular frequency and γ is the damping rate.

References

1. Kouzine, F., Liu, J., Sanford, S., Chung, H.-J. & Levens, D. The dynamic response of upstream DNA to transcription-generated torsional stress. *Nat. Struct. Mol. Biol.* **11,** 1092–1100 (2004).
2. Printz, M. P. & Von Hippel, P. H. Hydrogen exchange studies of DNA structure. *Proc. Natl Acad. Sci. USA* **53,** 363–370 (1965).
3. Chou, K. C., Maggiora, G. M. & Mao, B. Quasi-continuum models of twist-like and accordion-like low-frequency motions in DNA. *Biophys. J.* **56,** 295–305 (1989).
4. Merzel, F., Fontaine-Vive, F., Johnson, M. R. & Kearley, G. J. Atomistic model of DNA: phonons and base-pair opening. *Phys. Rev. E* **76,** 031917 (2007).
5. Ares, S., Voulgarakis, N. K., Rasmussen, K. O. & Bishop, A. R. Bubble nucleation and cooperativity in DNA melting. *Phys. Rev. Lett.* **94,** 035504 (2005).
6. Packer, M. J. *et al.* Sequence-dependent DNA structure: tetranucleotide conformational maps. *J. Mol. Biol.* **295,** 85–103 (2000).
7. Lankaš, F., Šponer, J., Hobza, P. & Langowski, J. Sequence-dependent elastic properties of DNA. *J. Mol. Biol.* **299,** 695–709 (2000).
8. Knotts, T. A., Rathore, N., Schwartz, D. C. & De Pablo, J. J. A coarse grain model for DNA. *J. Chem. Phys.* **126,** 084901 (2007).
9. Dornberger, U., Leijon, M. & Fritzsche, H. High base pair opening rates in tracts of GC base pairs. *J. Biol. Chem.* **274,** 6957–6962 (1999).
10. Choi, C. H. *et al.* DNA dynamically directs its own transcription initiation. *Nucleic Acids Res.* **32,** 1584–1590 (2004).
11. Chou, K.-C., Chou, K. C., Mao, B. & Mao, B. Y. Collective motion in DNA and its role in drug intercalation. *Biopolymers* **27,** 1795–1815 (1988).
12. Cuesta-López, S., Menoni, H., Angelov, D. & Peyrard, M. Guanine radical chemistry reveals the effect of thermal fluctuations in gene promoter regions. *Nucleic Acids Res.* **39,** 5276–5283 (2011).
13. Theodorakopoulos, N. & Peyrard, M. Base pair openings and temperature dependence of DNA flexibility. *Phys. Rev. Lett.* **108,** 078104 (2012).
14. Guéron, M. *et al.* A single-mode of DNA base-pair opening drives imino proton-exchange. *Nature* **328,** 89–92 (1987).
15. Ruddon, R. W. *Cancer Biology* (Oxford Univ. Press, 2007).
16. Phelps, C., Lee, W., Jose, D., Von Hippel, P. H. & Marcus, A. H. Single-molecule FRET and linear dichroism studies of DNA breathing and helicase binding at replication fork junctions. *Proc. Natl Acad. Sci. USA* **110,** 17320–17325 (2013).
17. O'neil, M. A. & Barton, J. K. DNA charge transport: conformationally gated hopping through stacked domains. *J. Am. Chem. Soc.* **126,** 11471–11483 (2004).
18. Heilweil, E. J. & Plusquellic, D. *Terahertz Spectrocopy Principles and Applications. Optical Science and Engineering.* (ed. Dexheimer, S.) 269–297 (CRC Press, 2007).
19. Alexandrov, B. S. *et al.* Specificity and heterogeneity of terahertz radiation effect on gene expression in mouse mesenchymal stem cells. *Sci. Rep.* **3,** 1–8 (2013).
20. Painter, P. C., Mosher, L. E. & Rhoads, C. Low-frequency modes in the Raman spectrum of DNA. *Biopolymers* **20,** 243–247 (1981).
21. Tominaga, Y. *et al.* Coupled dynamics between DNA double helix and hydrated water by low frequency Raman spectroscopy. *J. Chem. Phys.* **83,** 5972 (1985).
22. Hakim, M. B., Lindsay, S. M. & Powell, J. The speed of sound in DNA. *Biopolymers* **23,** 1185–1192 (1984).
23. Krisch, M., Mermet, A., Grimm, H., Forsyth, V. T. & Rupprecht, A. Phonon dispersion of oriented DNA by inelastic x-ray scattering. *Phys. Rev. E* **73,** 061909 (2006).
24. Grimm, H., Stiller, H., Majkrzak, C., Rupprecht, A. & Dahlborg, U. Observation of acoustic umklapp-phonons in water-stabilized DNA by neutron scattering. *Phys. Rev. Lett.* **59,** 1780–1783 (1987).
25. Globus, T., Gelmont, B. & Sizov, I. *Biological Identification. DNA Amplification and Sequencing, Optical Sensing, Lab-On-Chip and Portable Systems.* (ed. Schaudies, P.Ch. 10, 281–312 (Woodhead Publishing Limited, 2014).
26. Arora, A. *et al.* Terahertz-time domain spectroscopy for the detection of PCR amplified DNA in aqueous solution. *Analyst* **137,** 575–579 (2012).
27. Wu, X., E, Y., Xu, X. & Wang, L. Label-free monitoring of interaction between DNA and oxaliplatin in aqueous solution by terahertz spectroscopy. *Appl. Phys. Lett.* **101,** 033704 (2012).

28. Turton, D. A. *et al.* Terahertz underdamped vibrational motion governs protein-ligand binding in solution. *Nat. Chem.* **5,** 3999 (2014).
29. Turton, D. A. & Wynne, K. Stokes–Einstein–Debye failure in molecular orientational diffusion: exception or rule? *J. Phys. Chem. B* **118,** 4600–4604 (2014).
30. Giraud, G., Karolin, J. & Wynne, K. Low-frequency modes of peptides and globular proteins in solution observed by ultrafast OHD-RIKES Spectroscopy. *Biophys. J.* **85,** 1903–1913 (2003).
31. Hunt, N. T., Kattner, L., Shanks, R. P. & Wynne, K. The dynamics of water-protein interaction studied by ultrafast optical Kerr-effect spectroscopy. *J. Am. Chem. Soc.* **129,** 3168–3172 (2007).
32. Perticaroli, S. *et al.* Broadband depolarized light scattering study of diluted protein aqueous solutions. *J. Phys. Chem. B* **114,** 8262–8269 (2010).
33. Ts'o, P. O. P. in *Basic Principles in Nucleic Acid Chemistry* Vol. 1 (ed. Ts'o, P. O. P.) Ch. 6, 453–584 (Elsevier, 1974).
34. Rymdén, R. & Stilbs, P. Nucleotide aggregation in aqueous solution. A multicomponent self-diffusion study. *Biophys. Chem.* **21,** 145–156 (1985).
35. Loring, R. F. & Mukamel, S. Selectivity in coherent transient Raman measurements of vibrational dephasing in liquids. *J. Chem. Phys.* **83,** 2116–2128 (1985).
36. Muller, M., Wynne, K. & Vanvoorst, J. The interpretation of echo experiments. *Chem. Phys.* **125,** 225–230 (1988).
37. Wynne, K., Muller, M., Brandt, D. & Vanvoorst, J. Diagrammatic density-matrix analysis of the Raman photon-echo. *Chem. Phys.* **125,** 211–223 (1988).
38. Guéron, M. & Leroy, J.-L. Studies of base pair kinetics by NMR measurement of proton exchange. *Methods Enzymol.* **261,** 383–413 (1995).
39. Altan-Bonnet, G., Libchaber, A. & Krichevsky, O. Bubble dynamics in double-stranded DNA. *Phys. Rev. Lett.* **90,** 138101 (2003).
40. Brauns, E., Madaras, M., Coleman, R., Murphy, C. & Berg, M. Complex local dynamics in DNA on the picosecond and nanosecond time scales. *Phys. Rev. Lett.* **88,** 158101 (2002).
41. Frank-Kamenetskii, M. D. & Prakash, S. Fluctuations in the DNA double helix: a critical review. *Phys. Life Rev.* **11,** 153–170 (2014).
42. Pal, S., Zhao, L. & Zewail, A. H. Water at DNA surfaces: Ultrafast dynamics in minor groove recognition. *Proc. Natl Acad. Sci. USA* **100,** 8113–8118 (2003).
43. Fogarty, A. C., Duboué-Dijon, E., Sterpone, F., Hynes, J. & Laage, D. Biomolecular hydration dynamics: a jump model perspective. *Chem. Soc. Rev.* **42,** 5672–5683 (2013).
44. Sterpone, F., Stirnemann, G. & Laage, D. Magnitude and molecular origin of water slowdown next to a protein. *J. Am. Chem. Soc.* **134,** 4116–4119 (2012).
45. Franck, J. M., Ding, Y., Stone, K., Qin, P. Z. & Han, S. Anomalously rapid hydration water diffusion dynamics near DNA surfaces. *J. Am. Chem. Soc.* **137,** 12013–12023 (2015).

Acknowledgements

This work was supported by EPSRC grants EP/J009733/1, EP/J00975X/1, EP/K034995/1 and EP/N508792/1.

Author contributions

All authors contributed to the study and manuscript. M.G.-J, G.R. and T.H. were responsible for data collection and numerical analysis. A.J.L., S.M.K. and E.M.E. were responsible for sample preparation and basic characterization. M.G.-J. and K.W. carried out data analysis and manuscript preparation.

Additional information

Competing financial interests: The authors declare no competing financial interests.

21

Avoiding short circuits from zinc metal dendrites in anode by backside-plating configuration

Shougo Higashi[1,2], Seok Woo Lee[1,3], Jang Soo Lee[1], Kensuke Takechi[2] & Yi Cui[1,4]

Portable power sources and grid-scale storage both require batteries combining high energy density and low cost. Zinc metal battery systems are attractive due to the low cost of zinc and its high charge-storage capacity. However, under repeated plating and stripping, zinc metal anodes undergo a well-known problem, zinc dendrite formation, causing internal shorting. Here we show a backside-plating configuration that enables long-term cycling of zinc metal batteries without shorting. We demonstrate 800 stable cycles of nickel–zinc batteries with good power rate (20 mA cm^{-2}, 20 C rate for our anodes). Such a backside-plating method can be applied to not only zinc metal systems but also other metal-based electrodes suffering from internal short circuits.

[1] Department of Materials Science and Engineering, Stanford University, 476 Lomita Mall, McCullough Building 343, Stanford, California 94305, USA. [2] Smart Design of Materials and Process Research-Domain, Toyota Central R&D Laboratories, Inc., 41-1 Nagakute, Aichi 480-1192, Japan. [3] School of Electrical and Electronic Engineering, Nanyang Technological University, 50 Nanyang Avenue, Singapore 639798, Singapore. [4] SLAC National Accelerator Laboratory, Stanford Institute for Materials and Energy Sciences, 2575 Sand Hill Road, Menlo Park, California 94025, USA. Correspondence and requests for materials should be addressed to Y.C. (email: yicui@stanford.edu).

Rechargeable batteries in various forms are critical as power sources for portable electronics, electrical vehicles and grid-scale storages[1–5]. For very large-scale deployment, battery materials ideally need to be abundant, accessible and environmentally friendly. Zn satisfies the above criteria and is also recyclable[6]. It has an attractive electrochemical potential as an anode in aqueous battery systems (1.83 V versus Li, − 1.245 V versus standard hydrogen electrode in alkaline electrolyte) and possesses high capacity (5,854 Ah l^{-1} and 820 Ah kg^{-1}). Zn-based batteries offer the highest energy density of all aqueous battery systems at low cost. In contrast to other high-energy-density metals such as lithium (Li) and sodium, Zn is chemically stable in air and non-flammable. As the invention of the Volta pile (Zn-H_2O system), which generated continuous current for the first time in early nineteenth century, many battery systems using Zn metal as anode were proposed, such as Zn-NiOOH (Ni-Zn), Zn–air, Zn-MnO_2 and Zn-Ag_2O, and some were successfully commercialized as primary battery systems[7]. However, despite past efforts dedicated to making rechargeable Zn-based batteries, stable electrode operation in practical cells has still been a challenge. Unlike the anodes of Li-ion batteries where ions are intercalated and de-intercalated from a graphite host structure, Zn metal anode is a hostless electrode in which the metal dissolves into electrolyte at discharge and plated back at charge, accompanying redistribution of the Zn metal. On charging, dendritic Zn is formed easily and can cause internal short circuits in an unpredictable manner[8–12]. Although an internal short circuit in Zn-based batteries does not result in the same hazardous situation as in Li-ion batteries, uncontrolled energy release remains a safety concern. In addition, cycle life can be shortened due to internal shorts[13]. Recently, Zn-based battery systems have been re-visited with a material design of porous Zn metal sponge, demonstrating improved cyclability in Ag-Zn cells (50 cycles with 3–5 mA cm^{-2} of current densities)[14]. A high-energy, high-power cathode for Ni-Zn rechargeable battery has also been demonstrated with nanoscale material design of the Ni hydroxide cathode[15,16]. Zn–air batteries with a metal-free bifunctional catalyst show a stable cycling at a current density of 2 mA cm^{-2} (ref. 5). Despite the above progress, the dendrite issue under a range of current densities still remains a critical concern.

Here we tackle the issue of dendrite-induced shorting by developing a concept of backside metal plating. We use half-cells to illustrate our concept as shown in Fig. 1. Figure 1a shows a conventional Zn metal foil electrode, which serves as both reference and counter electrodes, directly facing the Cu working electrode. During Zn plating onto Cu foil in the conventional frontside plating configuration, Zn dendrites can grow towards the Zn reference electrode, causing battery shorting. In our concept demonstrated here (Fig. 1b), the backside plating of Zn is realized by coating an insulating layer on the edges and the 'front' surface of Cu foil facing the Zn metal counter electrode. Thus, during Zn plating, Zn ions in the aqueous electrolyte travel over the edge and are deposited on the open back surface of Cu foil. Therefore, even if Zn dendrites form, they grow away from the counter electrode and do not short a battery. Looking into the ionic pathway in back plating configuration, Zn-related ions need to go around the insulating layer, to access the back surface of Cu foil. This configuration might seem to be a disadvantage, as it would cause a decrease of the rate capability. However, we note that the high ionic conductivity of 6 M aqueous KOH (0.6 S cm^{-1})[17], a common electrolyte for conventional Zn-based batteries, can afford sufficient ion conduction for maintaining reasonable power rates. To put this into perspective, such a high ionic conductivity is ~50–100 times of that of an organic electrolyte used in existing lithium ion batteries. Our experiments and numerical analyses demonstrate how this configuration maintains performance.

Results

Proof of concept of Zn backside plating with half-cell. To prove that our concept of backside-plating configuration can avoid shorting, we made simple half-cells for both front (Fig. 2a) and backside-plating configurations (Fig. 2b) and performed electrochemical plating–stripping cycles. We used a Zn foil (100 μm in thickness) as reference and counter electrodes, and a Cu foil (30 μm) as a current collector for a working electrode. The electrodes are both round (0.8 cm in radius) and half-immersed into electrolyte in a round, airtight polycarbonate bottle (flooded cell) after cell assembly. The conductive electrode area in contact with the electrolyte is ~1 cm^2 (given by $(0.8\ cm)^2 \times \pi/2$) and the upper half of the electrodes not in contact with electrolyte is connected to stainless steel current collectors. We use Cu foil as a current collector instead of Zn foil so that we can plate a known capacity of Zn onto Cu foil and then strip it away, allowing us to calculate meaningful Coulombic efficiency numbers. Cu also has a relatively high overpotential for the hydrogen evolution reaction (HER) and thus is used as current collector for Zn battery[13,16]. For a conventional frontside half-cell, a 0.3-mm gap between electrodes was allowed. For a backside-plating configuration half-cell, a film of insulator made of polypropylene (PP) (30 μm) is glued to Cu foil, to protect the frontside of Cu facing the cathode. To protect the edge of Cu substrate where current concentration takes place as indicated by our simulation (see Supplementary Fig. 1), an amorphous insulating carbon layer was sputtered around the edge[18]. The edge of the PP insulator extends out from the edge of Cu substrate by an additional 1.5 mm and thus immersed electrode area including the outside insulating ring is ~1.4 cm^2 (given by $(0.8\ cm + 0.15\ cm)^2 \times \pi/2$). We used 6 M KOH aqueous solution saturated with ZnO as an electrolyte

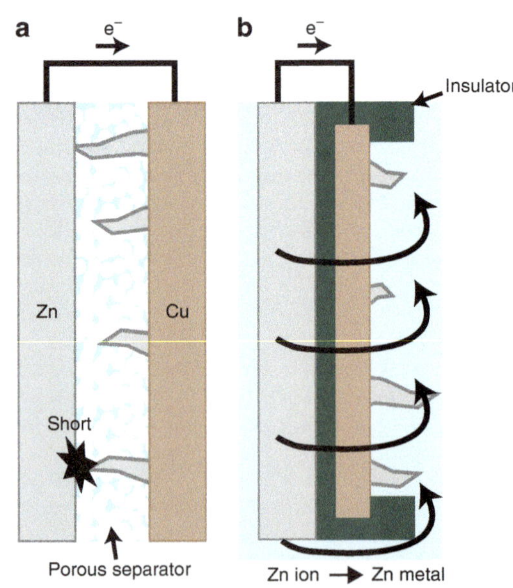

Figure 1 | Scheme of backside-plating configuration for avoiding internal shorts. Schematic representation of (**a**) conventional frontside- and (**b**) backside-plating configuration cells. In conventional configuration, electrodes face each other separated by a porous polymer separator, which causes a battery short due to the formation of dendrites under repeated cycles. In the backside-plating configuration, an insulator protects the front side and edge of Cu, allowing Zn to grow on the backside of substrate Cu. Backside cell is configured to have aqueous electrolyte between Zn and Cu back surface.

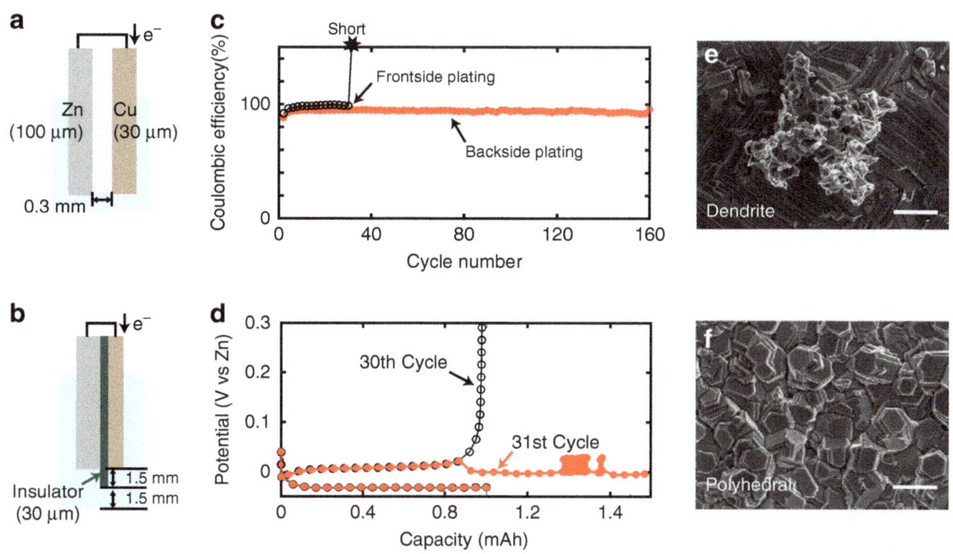

Figure 2 | Electrochemical data of half-cells. Schematics of the half-cells with (**a**) conventional frontside- and (**b**) backside-plating cell configurations. (**c**) Corresponding high rate (20 mA cm^{-2}) galvanostatic plating/stripping stability curves of the front and backside cells. (**d**) Plating/stripping curves of the conventional front-side cell at 30th and 31st cycle. (**e**) *Ex-situ* SEM image of Cu substrate surface at the 20th cycle of frontside and (**f**) backside-plating cell. Scale bars, 2 μm (**e**,**f**).

and all electrochemical measurements were done in airtight flooded cells.

In strong alkaline solutions such as 6 M KOH, $Zn(OH)_2$ is reduced to Zn metal during plating as shown in equation (1) (ref. 19). Figure 2c,d show the Coulombic efficiencies of Zn plating (charge) and stripping (discharge) versus cycle number in half-cells for frontside- and backside-plating cells, respectively. For charging, Zn was plated with constant current density of 20 mA cm^{-2} until the total charge reached 1 mAh cm^{-2}. Next, for discharging, Zn was stripped with the same constant current density until the voltage reached the cut-off voltage, 0.8 V versus Zn reference electrode for every cycle.

For the conventional frontside-plating configuration, a Coulombic efficiency of ~96% was obtained until the 30th cycle, but a short occurred at the 31st cycle during Zn stripping due to dendrite formation on Zn counter electrode where Zn plating occurs. Corresponding charge–discharge voltage curves of the 30th and 31st cycle are shown in Fig. 2c. In the 30th cycle, the plating potential of Zn was at − 30 mV versus Zn reference and the stripping was at 12 mV (*cf.* Fig. 3a,b). In the following Zn stripping at the 31st cycle, the curve showed an abrupt overpotential drop at ~1 mAh and apparent capacity exceeding plating capacity with voltage noise around 1.3 mAh due to a short. Figure 2e shows the scanning electron microscopy (SEM) image of a dendrite formed on Cu surface after the Zn plating at the 20th cycle. Most of the Zn-plated surfaces were smooth with polyhedral type of Zn but formation of dendrites ranging from 5 to 10 μm in size was also confirmed (see Supplementary Fig. 2)[20]. The small dendrites would not penetrate immediately after the formation but they grow larger in size and cause a short. This is exactly the issue that Zn-based batteries face, hindering stable operation over hundreds of charge–discharge cycles.

In contrast, for backside plating, 160 cycles of stable plating stripping with Coulombic efficiency of 92% was achieved and no electrical shorts were observed. The plating potential of Zn was at − 70 mV versus Zn reference and the stripping was at 50 mV (Fig. 3), which suggests that backside plating does not cause significant overpotential. Shown in Fig. 2f is the SEM image of the Cu surface after Zn plating at 20 cycles with backside-plating configuration. Only polyhedral Zn deposits were seen throughout the back surface of Cu. Polyhedra near the edge were larger than the ones at the terrace but there was no dendrite formation (see Supplementary Fig. 2). It should be emphasized that even if dendrites form in the back plating configuration, they do not result in an electrical short.

The reason for the slightly lower Coulombic efficiency in the backside-plating configuration compared with conventional one is HER caused by the carbon coating at the edge, which has a lower overpotential for HER than Cu. We confirmed that when the edge is not covered by carbon in backside-plating configuration, Coulombic efficiency becomes ~96% between the 1st cycle and the 30th cycle, which is similar to the data obtained for frontside cells (see Supplementary Fig. 1e). To improve the Coulombic efficiency, alternative materials need to be explored in the future.

Power capability of frontside- and backside-plating half-cells. Owing to the increased ionic pathway in the backside-plating configuration, it is essential to compare its power capability with that of the frontside. We performed galvanostatic electrochemical Zn plating and stripping at various current densities and measured the overpotentials as the key parameter for comparison. Figure 3a–d show the plating/stripping curves of 1 mAh cm^{-2} of Zn for conventional frontside- and backside-plating configuration cells, respectively, performed by half-cells at various current densities (5, 10, 20 and 30 mA cm^{-2}).

For plating, overpotential for the frontside cell was only 30 mV at 20 mA cm^{-2}. Backside-plating cells show slightly higher plating overpotential than frontside one and it was 70 mV at 20 mA cm^{-2}. For stripping, the overpotentials for frontside and backside cells were ~12 and 50 mV, respectively, at 20 mA cm^{-2} and there were no considerable difference between them. Larger overpotentials of the plating than of the stripping for both front and backside cells are presumably due to the activation energy required for Zn nucleation on Cu surface. When the current density is increased to 30 mA cm^{-2}, we found that the backside cell shows relatively high overpotential over 200 mV, which may not be desired for practical use.

Thus, we found that backside-plating configuration allows a reasonably high current density of 20 mA cm^{-2} for plating

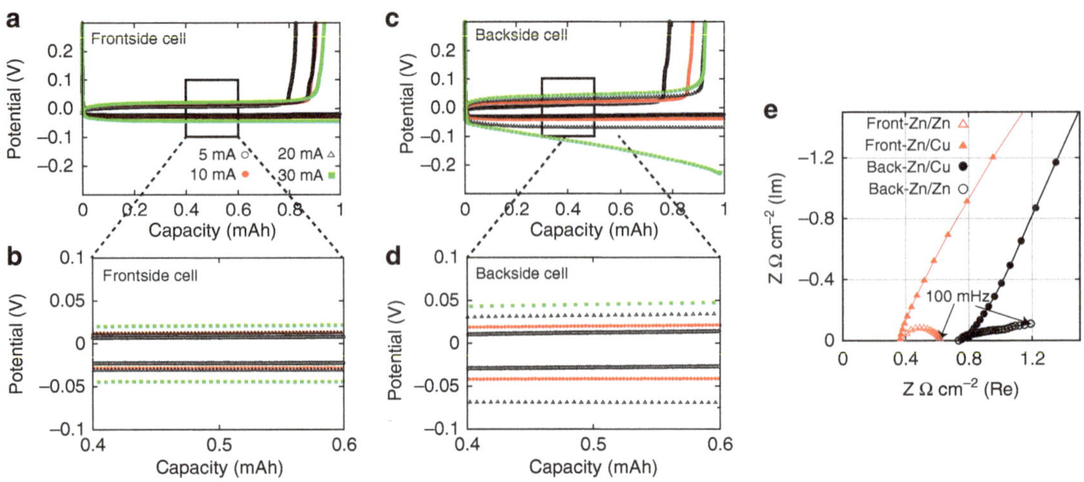

Figure 3 | Comparison of power capability. Plating/stripping curves of (**a**,**b**) the frontside and (**c**,**d**) the backside-plating half-cells in different galvanostatic currents. 5, 10, 20 and 30 mA cm^{-2} were applied for both plating and stripping. (**e**) Corresponding impedance curves for the front and backside plating cells measured at open circuit potentials (~0.9 V versus Zn). Half-cells use Zn and Cu as electrodes are denoted as Zn/Cu. In the same way, a cell using Zn for both electrodes are denoted as Zn/Zn. Frequency range for measurement is 1-100 MHz.

without causing appreciable overpotential. The low overpotential at such a high current density of 20 mA cm^{-2} for backside-plating cell can be explained by the low solution resistivity (that is, the high ionic conductivity) of 6 M KOH aqueous electrolyte. The solution resistances of frontside and backside-plating cells were ~0.4 and 0.8 Ω cm^{-2}, respectively, as seen in Fig. 3e of impedance curves. The ionic conductivity of 6 M KOH zincate saturated aqueous solution was estimated to be ~0.1 S cm^{-1} in our experiments, which is of the same order as reported values[17].

Ni-Zn full cell performance with backside-plating configuration. With the successful demonstration of backside Zn plating/stripping without dendrite formation or large overpotential at reasonably high current density, here we demonstrate a Ni-Zn full-cell with high power and long-cycle life as an example of the application of the backside-plating configuration concept. Electrochemical reactions involved here are given as follows:

$$\text{Anode.}\quad Zn + 4OH^- = Zn(OH)_4^{2-} + 2e^-, \quad E_0 = 1.245\ \text{V vs. SHE} \tag{1}$$

$$\text{Cathode.}\ 2NiOOH + 2H_2O + 2e^- = 2Ni(OH)_2 + 2OH^-, \ E_0 = 0.49\ \text{V vs. SHE} \tag{2}$$

$$\text{Overall.}\quad Zn + 2H_2O + 2NiOOH = 2Ni(OH)_2 + Zn(OH)_2, E = 1.735\ \text{V} \tag{3}$$

During charging, $Ni(OH)_2$ at the cathode releases protons and is electrochemically converted to NiOOH. At the same time, zincate ions $Zn(OH)_4^{2-}$ are reduced and plated on the Cu substrate at anode. Discharge will progress in the opposite manner. The full-cell configuration is similar to the half-cell other than changing the cathode to β-$Ni(OH)_2$ as illustrated in Fig. 4a. We confirmed the phase of β-$Ni(OH)_2$ by X-ray diffraction (XRD) (see Supplementary Fig. 3). We used β-$Ni(OH)_2$, as it has superior stability in strong alkaline solutions compared with α-$Ni(OH)_2$. The theoretical capacity of β-$Ni(OH)_2$ is 289 mAh g^{-1} and this material was successfully used as a cathode material in Ni-Cd batteries because of its high reliability. It is still used in a Ni-MH battery for hybrid vehicles. When combined with a Zn electrode, this battery offers one of the highest overall potentials (~1.7 V) in aqueous batteries. The specific energy (calculated based on active material weight only) can reach ~372 Wh kg^{-1}.

Figure 4b shows the cycle stability curves of Ni-Zn full cells with frontside- and backside-plating configurations, respectively. The battery is charged up to total charge capacity of 1 mAh cm^{-2} and discharged to cutoff voltage of 1.3 V with current density of 20 mA cm^{-2} (20 C or 3 min for charge and discharge, respectively) for every cycle. Capacity retention, which is equivalent to Coulombic efficiency, is calculated by the following equation:

Capacity retention (%) = discharge capacity/charge capacity (fixed to be 1 mAh cm^{-2}) × 100 (4)

The frontside cell shows a spike in stability curve at around 100 cycles due to a short and cycled until about 150 cycles when it shorts again. Around the 280th cycle, there is a short recovery of the capacity, presumably due to the detachment of the zinc dendrite that was bridging the two electrodes, and then the cell shorts again at later cycles.

In contrast, the backside cell shows expected highly stable cycling more than 800 times free of any indication of shorting. 88% capacity retention was maintained at 800 cycles, which has never been achieved before with conventional frontside batteries, except using a flow cell system with the electrode separation much larger than the static cells[21,22]. Galvanostatic charge–discharge curves of Ni-Zn with backside-plating configuration at the 800th cycle are shown in Fig. 4c,d. The average charge and discharge voltage were found to be 1.95 and 1.65 V, respectively, which are typical for Ni-Zn full-cell battery. Thus, the specific energy density for the full cell shown in Fig. 4 was estimated to be ~246 Wh kg^{-1} based on the active materials' weight (~4.8 mg of commercial $Ni(OH)_2$ and ~1.2 mg of Zn) for both conventional and backside-plating configuration Ni-Zn full-cell batteries tested (1 mAh × 0.9 × 1.65 V × (4.8 mg + 1.2 mg)$^{-1}$). In general, a battery suffers a decrease of capacity in real devices due to the volume of inactive components such as separator, current collectors, packaging and so on, and volumes of these parts need to be small for achieving the energy density closer to the theoretical one. In the backside-plating configuration, it is also important to reduce inactive volume for the same reason. In the following section, we discuss the energy density considering not only the active materials' weight but also the volume of electrolyte over the backside of Cu foil required for ion transport in the backside-plating configuration.

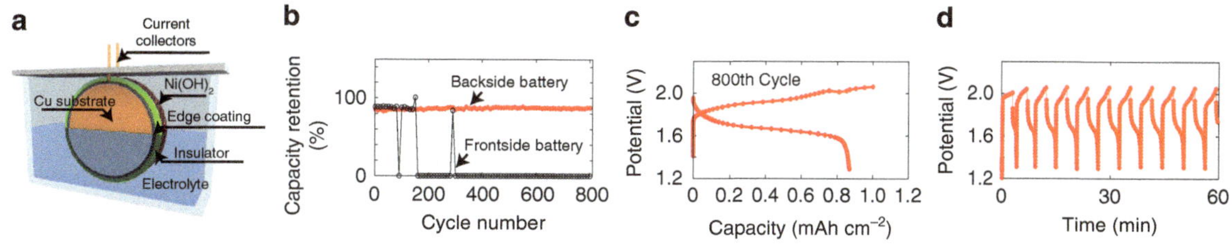

Figure 4 | Ni-Zn full cells with backside-plating configuration. (**a**) Schematic representation of the backside-plating configuration full-cell battery. (**b**) Comparison of cycle stability curves of the Ni-Zn battery with the conventional frontside and backside battery cells. (**c**) Galvanostatic charge-discharge curves of Ni-Zn with the backside configuration (**c**,**d**). A steady charge-discharge cycle is achieved in the backside-plating configuration with high current density.

Discussion

Stable operation for more than 800 cycles (as demonstrated in the Ni-Zn full-cell batteries) was achieved with the backside-plating configuration. Here we discuss the importance of insulating the Cu foil edge in avoiding dendrite formation there, as well as the importance of solution thickness above the Cu surface for avoiding dendrite formation near the centre of backside of Cu surface. In the backside cell, we covered the edge of the Cu substrate by a carbon insulating layer. The key role of this layer, which is amorphous and highly resistive[18], is to avoid preferential plating of Zn on the very edge. Metallic Cu has higher electrical conductivity than amorphous carbon and therefore Zn will be plated onto Cu substrate rather than the protected edge. Our numerical analysis results indicated a certain current concentration at the edge (see Supplementary Fig. 1a–d) and we confirmed the dendrite formation at the edge of Cu after repeated cycles of Zn plating and stripping when the edge of Cu was not protected (see Supplementary Fig. 4). Accordingly, the cycle stability curve of the half-cell without carbon coating shows small spikes and the curve became noisy (see Supplementary Fig. 1e). Although suggesting an intermittent and less disruptive process in comparison with the clear signs of shorting in the conventional cells, this unstable cycling behaviour nonetheless needs to be avoided from a practical viewpoint.

In backside-plating configuration, certain thickness of electrolyte above the Cu is necessary to supply enough ions for the whole back surface, in particular the surface of the centre far from the edge. If the thickness of the solution is not adequate for the distance from the edge to the centre of the Cu back surface, ions would be reduced at around the edge side during charging and most probably a thicker layer of Zn would be formed, resulting in depletion of ions near the centre of the surface. Such depletion of ions can also cause the formation of dendrites as reported[11,23,24]. Hence, the optimal design should allow a certain solution thickness to avoid such depletion of ions to support sufficient ion supply, yet it should not be too thick, to maintain useful energy density of the cell.

To investigate how thick the electrolyte must be for plating only smooth Zn for the whole back surface, we prepared the backside-plating half-cells with different solution thicknesses ranging from 0.5 to 2.0 mm by placing polycarbonate film above the solution and plated 1 mAh cm^{-2} of Zn with fixed current density of 20 mA cm^{-2}. We found that a majority of Zn deposits were polyhedral/boulder structures within ~2 mm from the edge of Cu for the cell with 0.5 mm of solution thickness. Beyond 2 mm from the edge to the Cu back surface centre, Zn dendrites were formed due to the depletion of Zn ions (see Supplementary Fig. 5a). We gradually increased the solution thickness so that the whole back surface can be covered by Zn polyhedral/boulder structure and found that 2 mm of solution thickness was required for plating polyhedral/boulder Zn at the centre of backside of Cu surface (see Supplementary Fig. 5b).

Next, we prepared the half-cells with a smaller solution thickness of 50 μm, which is secured by inserting the PP battery separator placed between the back surface of Cu and polycarbonate outer film. We plated Zn corresponding to the capacity of 1 mAh cm^{-2} with the different current densities as shown in Supplementary Fig. 6a (2, 5 and 10 mA cm^{-2}). At the current density of 10 mA cm^{-2}, the overpotential was ~0.15 V at the end of plating. When the current density was decreased to 2 mA cm^{-2}, the overpotential was ~0.05 V and the whole back Cu surface was plated with polyhedral/boulder Zinc (Supplementary Fig. 7). With this solution thickness and plating capacity, the energy density was calculated to be higher than 150 Wh kg^{-1} (Supplementary Fig. 6b); higher values are possible if a fraction of active materials is increased, which may be possible through engineering efforts.

When the radius of Cu foil was 0.8 cm, the solution thickness was 2 mm and the current density was 20 mA cm^{-2}; the whole back surface was plated with polyhedral/boulder Zn with no dendrites (Supplementary Fig. 5). Thus, the ratio of the radius of the Cu foil to the solution thickness for plating polyhedral/boulder Zn without a dendrite can be calculated to be 4:1. In the same way, at the current density of 2 mA cm^{-2}, the ratio is calculated to be 160 (0.8 cm) to 1 (50 μm) based on the SEM observations shown in Supplementary Fig. 7. The ratio of the radius of the Cu foil to the solution thickness for preventing dendrite formation would be helpful for designing backside-plating configuration batteries. Finally, we note that the geometric advantage of backside-plating configuration prevents shorting problem even if dendrite formation takes place during Zn plating. Thus, it may not be always necessarily required to plate only polyhedral Zn throughout the surface in practical use. Therefore, the solution thickness might be reduced further for better energy density.

In summary, we developed a backside-plating configuration for avoiding a short of Zn-based batteries. As a proof-of-concept, we demonstrated highly stable 800 charge–discharge cycles with a Ni-Zn full-cell battery under high rate (20 mA cm^{-2}) using 6 M KOH zincate aqueous solution in backside-plating configuration. We found that the backside-plating configuration requires a certain thickness of solution above the plating substrate for homogenous plating of Zn throughout the substrate. We prepared half-cells with different solution thickness (2 mm and 50 μm) and applied different current densities (20 and 2 mA cm^{-2}) to estimate the ratio of the radius of the Cu foil to the solution thickness, as an attempt to establish the design rule for the backside-plating configuration batteries. Many different approaches have been studied to avoid shorting due to dendrite formation but further improvements are needed. The backside-plating method can provide a way for turning high-capacity electrodes limited to primary batteries by dendrite formation into reliable components of rechargeable systems.

Methods

Material characterization by uses of *ex situ* SEM and XRD. SEM images were taken using a FEI XL30 Sirion SEM (accelerating voltage 5 kV). XRD patterns were recorded on a PANalytical X 'Pert Diffractometer operated at 45 kV and 40 mA at the wavelength of Cu Kα radiation ($\lambda = 0.15406$ nm). After the plating of Zn, substrates are detached from the insulator and outer package by ethanol, thoroughly washed by deionized water (18 MΩ) and immediately dried by blowing air. Dried substrate was glued to slide glass for XRD measurements.

Electrochemistry. Galvanostatic experiments were performed using multi-channel potentiostat, VMP3 (BioLogic). For half-cell galvanostatic experiments, 1 mAh of Zn (corresponding to ~1.2 mg of Zn) was plated/charged onto Cu substrate, followed by stripping of Zn up to 0.8 V versus Zn reference electrode for every cycle. For full cells, 1 mAh is charged and then discharged to 1.3 V of battery potential for every cycle. For both experiments, electrode surface touching the electrolyte is designed to be 1 cm^2.

KOH (6 M) zincate saturated aqueous solution was prepared by mixing 500 ml of deionized water, 168 g of KOH and 20.35 g of ZnO (Sigma-Aldrich), and stirred for 12 h at 70 °C. The mixed solution was bubbled by N_2 gas for 2 h and centrifuged before use. Zn (Alfa Aesar) and/or Mercury Oxide Electrode (Koslow 5088) were used as reference electrodes. Airtight beaker cells with 30 ml of electrolyte were used for measurements. Cathode, anode and reference electrodes are sunk into the electrolyte during electrochemical measurements.

Full-cell preparation. Cathodes are prepared as follows: 700 mg of $Ni(OH)_2$ (Sigma-Aldrich), 200 mg of Acetylene black, 100 mg of polyvinylidene difuluoride (Kynar HSV900) and 10 ml of *N*-methyl-2-pyrrolidone (Sigma-Aldrich) were mixed and stirred for 12 h at room temperature. The obtained slurry was dip coated on current collector of Ni foam (2 mm thick) and dried at vacuum oven 80 °C. Dried Ni foam was pressed in to 0.1 mm to be used as cathode. Approximately 4.8 mg commercial $Ni(OH)_2$ is loaded on the prepared cathode. Carbon coater (EMS150R ES) was used for sputtering the amorphous carbon nano layer. Carbon fibres were used as the evaporation target. The evaporation chamber was pumped down to 5×10^{-2} mbar before an out gassing current of 30 A was passed through the carbon fibres. After out-gas recovery, a pulse current was passed through the fibre to allow flash evaporation of carbon. The pulse current was set to 60 A for 20 s, with a 10-s interval between pulses. All the procedure is the same as that in ref. 18. A mask was used to coat the carbon layer on only the edge of Cu substrate. The edge-protected Cu substrate was then glued to PP film by Polyvinyl butyral (Santa Cruz Biotechnology).

Numerical analysis. COMSOL multiphysics (COMSOL Inc.) was used, to estimate current density profile in the electrolyte and on the surface of the copper electrode. The two-dimensional axisymmetric structure of simulated electrolyte consists of the surface of zinc electrode as an anode, insulating surface and the surface of copper electrode a cathode. The radius of the copper electrode is 8 mm, the thickness of zinc electrode is 1 mm and the height of electrolyte above the copper surface for zinc deposition is varied from 50 μm to 2 mm. The input current density on zinc electrode is varied from 1 to 20 mA cm^{-2}. For the estimation of current density profile of zinc ion on the copper electrode, Nernst-Planck equation is employed as a governing equation in COMSOL Multiphysics. This mode of simulation counts diffusion and migration due to the electric field as a regime of the transportation of zinc ion during electrochemical deposition. The given diffusivity of zinc ion in the aqueous electrolyte is 6×10^{-6} cm^2 s^{-1}. The simulation results visualize the colour map of current density, contour lines of overpotential and the streamlines (red) of zinc ion flow.

References

1. Goodenough, J. B. & Park, K. S. The Li-ion rechargeable battery: a perspective. *J. Am. Chem. Soc.* **135,** 1167–1176 (2013).
2. Manthiram, A., Fu, Y., Chung, S.-H., Zu, C. & Su, Y.-S. Rechargeable lithium-sulfur batteries. *Chem. Rev.* **114,** 11751–11787 (2014).
3. Skyllas-Kazacos, M., Chakrabarti, M. H., Hajimolana, S. A., Mjalli, F. S. & Saleem, M. Progress in flow battery research and development. *J. Electrochem. Soc.* **158,** R55–R79 (2011).
4. Li, Y. & Dai, H. Recent advances in zinc-air batteries. *Chem. Soc. Rev.* **43,** 5257–5275 (2014).
5. Zhang, J., Zhao, Z., Xia, Z. & Dai, L. A metal-free bifunctional electrocatalyst for oxygen reduction and oxygen evolution reactions. *Nat. Nanotechnol.* **10,** 444–452 (2015).
6. Jewell, S. & Kimball, S. M. Mineral commodity summaries 2015. *U.S. Geological Survey* **9,** 196 (2015).
7. Reddy, T. *Linden's Handbook of Batteries* 4th edn (McGraw-Hill Education, 2010).
8. McLarnon, F. R. & Cairns, E. J. The secondary alkaline zinc electrode. *J. Electrochem. Soc.* **138,** 645–656 (1991).
9. Diggle, J. W., Despic, A. R. & Bockris, J. O. The mechanism of the dendritic electrocrystallization of zinc. *J. Electrochem. Soc.* **116,** 1503–1514 (1969).
10. Wang, R. Y., Kirk, D. W. & Zhang, G. X. Effects of deposition conditions on the morphology of zinc deposits from alkaline zincate solutions. *J. Electrochem. Soc.* **153,** C357–C364 (2006).
11. López, C. M. & Choi, K.-S. Electrochemical synthesis of dendritic zinc films composed of systematically varying motif crystals. *Langmuir* **22,** 10625–10629 (2006).
12. Gallaway, J. W. *et al.* A lateral microfluidic cell for imaging electrodeposited zinc near the shorting condition. *J. Electrochem. Soc.* **157,** A1279–A1286 (2010).
13. Bonnick, P. & Dahn, J. R. A simple coin cell design for testing rechargeable zinc-air or alkaline battery systems. *J. Electrochem. Soc.* **159,** A981–A989 (2012).
14. Parker, J. F., Chervin, C. N., Nelson, E. S., Rolison, D. R. & Long, J. W. Wiring zinc in three dimensions re-writes battery performance-dendrite-free cycling. *Energy Environ. Sci.* **7,** 1117–1124 (2014).
15. Sugimoto, A., Ishida, S. & Hanawa, K. Preparation and characterization of Ni/Al-layered double hydroxide. *J. Electrochem. Soc.* **146,** 1251–1255 (1999).
16. Gong, M. *et al.* Ultrafast high-capacity NiZn battery with NiAlCo-layered double hydroxide. *Energy Environ. Sci.* **7,** 2025–2032 (2014).
17. Iwakura, C., Nohara, S., Furukawa, N. & Inoue, H. The possible use of polymer gel electrolytes in nickel/metal hydride battery. *Solid State Ionics* **148,** 487–492 (2002).
18. Zheng, G. *et al.* Interconnected hollow carbon nanospheres for stable lithium metal anodes. *Nat. Nanotechnol.* **9,** 618–623 (2014).
19. Beverskog, B. & Puigdomenech, I. Revised pourbaix diagrams for zinc at 25–300°C. *Corros. Sci.* **39,** 107–114 (1997).
20. Fukami, K. *et al.* General mechanism for the synchronization of electrochemical oscillations and self- organized dendrite electrodeposition of metals with ordered 2d and 3d microstructures. *J. Phys. Chem. C.* **111,** 1150–1160 (2007).
21. Cheng, J. *et al.* Preliminary study of single flow zinc–nickel battery. *Electrochem. Commun.* **9,** 2639–2642 (2007).
22. Ito, Y. *et al.* Zinc morphology in zincnickel flow assisted batteries and impact on performance. *J. Power Sources* **196,** 2340–2345 (2011).
23. Kahanda, G. L. M. K. S. & Tomkiewicz, M. Morphological evolution in Zn electrodeposition. *J. Electrochem. Soc.* **136,** 1497–1502 (1989).
24. Grier, D., Ben-Jacob, E., Clarke, R. & Sander, L. M. Morphology and microstructure in electrochemical deposition of zinc. *Phys. Rev. Lett.* **56,** 1264–1267 (1986).

Acknowledgements

We thank H. R. Lee, H. -W. Lee, and T. M. Hymel (Stanford University) for their technical assistance and fruitful discussions. We are grateful to T. Kobayashi and T. Sasaki (TCRDL) for critical help.

Author contributions

S.H. and Y.C. conceived and designed the experiments. S.H. performed the experiments. S.W.L. performed numerical simulations and data analysis. S.H., S.W.L., J.S.L. and Y.C. discussed, and S.H. and Y.C. wrote the paper. K.T. and Y.C. directed this study. All authors discussed the results and commented on the manuscript.

Additional information

Competing financial interests: The authors declare no competing financial interests.

22

Highly selective plasma-activated copper catalysts for carbon dioxide reduction to ethylene

Hemma Mistry[1,2], Ana Sofia Varela[3], Cecile S. Bonifacio[4], Ioannis Zegkinoglou[2], Ilya Sinev[2], Yong-Wook Choi[2], Kim Kisslinger[5], Eric A. Stach[5], Judith C. Yang[4], Peter Strasser[3] & Beatriz Roldan Cuenya[2]

There is an urgent need to develop technologies that use renewable energy to convert waste products such as carbon dioxide into hydrocarbon fuels. Carbon dioxide can be electrochemically reduced to hydrocarbons over copper catalysts, although higher efficiency is required. We have developed oxidized copper catalysts displaying lower overpotentials for carbon dioxide electroreduction and record selectivity towards ethylene (60%) through facile and tunable plasma treatments. Herein we provide insight into the improved performance of these catalysts by combining electrochemical measurements with microscopic and spectroscopic characterization techniques. Operando X-ray absorption spectroscopy and cross-sectional scanning transmission electron microscopy show that copper oxides are surprisingly resistant to reduction and copper^{+} species remain on the surface during the reaction. Our results demonstrate that the roughness of oxide-derived copper catalysts plays only a partial role in determining the catalytic performance, while the presence of copper^{+} is key for lowering the onset potential and enhancing ethylene selectivity.

[1] Department of Physics, University of Central Florida, Orlando, Florida 32816, USA. [2] Department of Physics, Ruhr-University Bochum, 44780 Bochum, Germany. [3] Department of Chemistry, Chemical Engineering Division, Technical University Berlin, 10623 Berlin, Germany. [4] Chemical and Petroleum Engineering and Physics, University of Pittsburgh, Pittsburgh, Pennsylvania 15261, USA. [5] Center for Functional Nanomaterials, Brookhaven National Laboratory, Upton, New York 11973, USA. Correspondence and requests for materials should be addressed to B.R.C. (email: beatriz.roldan@rub.de).

The development of technologies to capture and convert carbon dioxide (CO_2)—a greenhouse gas released by the burning of fossil fuels—to useful products is critically needed to mitigate global warming. The electrocatalytic reduction of CO_2 (CO_2RR) shows promise as a viable CO_2 utilization process since it can occur under ambient conditions and low cost copper (Cu) catalysts can facilitate the reduction pathway to hydrocarbons[1–3]. However, more efficient catalysts are still required because large overpotentials are required for bulk Cu to reduce CO_2 to hydrocarbon products (methane and ethylene) while competing with the hydrogen evolution side reaction[4–8]. Efficient Cu catalysts with high selectivity to ethylene are especially desirable due to the value of ethylene as a widely used chemical feedstock.

Investigations in recent years have led to the development of many interesting metal-based catalytic materials for CO_2RR. Nanostructured catalysts such as nanoparticles[9–14], nanocubes[15,16], nanofoams[17] and nanowires[18] have shown vastly improved activity or selectivity over bulk materials[19,20]. Recently, nanostructures derived from the reduction of copper oxides have shown vastly improved CO_2 reduction efficiency at lower overpotentials[21–24]. These materials were synthesized from the reduction of thermally oxidized Cu or electrodeposited copper(I) oxide (Cu_2O), and in general display improved current density, enhanced CO_2 reduction to carbon monoxide (CO) at low overpotentials, and a partial suppression of methane (CH_4) in favour of ethylene (C_2H_4) at higher overpotentials. Despite these promising results, the mechanism behind the reactivity trends of oxide-derived CO_2RR catalysts is still under dispute. Grain boundaries in oxide-derived Cu have been suggested to support unique surface sites which may be active for CO_2RR[22,25,26]. Another mechanism could be the increase in local pH due to high current densities[27] on the highly roughened surfaces, which could alter the reaction pathway in favour of ethylene[28–30]. Alternatively, the nanostructuring of the catalyst surface during the oxide reduction may also provide low-coordinated atoms as active sites[9]. $Cu^{\delta+}$ sites which may remain on the catalyst surface during the reaction have also been suggested to be the active sites[31–35]. However, no operando evidence of this claim exists up to this point, and only *ex situ* evidence has been presented. This *ex situ* evidence is complicated by the rapid formation of native oxides on metallic Cu surfaces upon removal from the reactor and exposure to air[23]. In general, it has been assumed that Cu oxides are completely reduced during the relatively high CO_2RR potentials used (that is, ~ 1 V versus reversible hydrogen electrode (RHE)), and that the reaction occurs only on metallic Cu species.

Oxygen plasma treatment is a facile and scalable technique to controllably oxidize and nanostructure metal catalysts for CO_2RR that has not yet been explored. Plasma treatment is a powerful method to activate catalysts, for example by removing capping ligands used in the synthesis of nanoparticles without sintering[36,37], by rapidly changing the chemical state of the surface at room temperature[38,39], or by creating defects or embedded atoms which can improve reactivity[40]. We have used oxygen (O_2) and hydrogen (H_2) plasmas to create novel nanostructured oxide layers and porous surfaces with tunable morphology and chemical state on polycrystalline Cu. These catalysts can lower the onset potential of CO_2 reduction to ethylene by 350 mV compared with electropolished Cu, with a maximum faradaic selectivity of >60% at −0.9 V versus RHE. Here, using a combination of characterization techniques—including operando X-ray absorption fine-structure spectroscopy (XAFS), scanning transmission electron microscopy (STEM) equipped with energy dispersive X-ray spectroscopy (EDS)—we gain insight into the catalysts in their working state and find that Cu^+ is the active species for reducing CO_2 to ethylene. Through our results, we unravel the mechanism behind the improved activity and unique ethylene selectivity of oxide-derived Cu catalysts and lay out the design principles necessary for improved ethylene-selective CO_2RR catalysts.

Results

Synthesis and characterization. To synthesize plasma-activated Cu, electropolished polycrystalline Cu foils were treated in O_2 and H_2 plasmas of varying power and duration. Figure 1 presents scanning electron microscopy (SEM) images of the Cu films treated with O_2 plasma at 20 W for 2 min (top row) and 100 W for 2 min (middle row) and 10 min (bottom row). By tuning the plasma conditions, the morphology of the surface and oxide thickness can be easily controlled. Growth of the Cu oxide begins at grain boundaries, forming micron-sized wires as shown in Fig. 1a. With increasing plasma time and power, the wires merged, forming a highly roughened surface, (c.f. Fig. 1b,e,i) and became highly porous with an observed 64% increase in the number of pores and 50% decrease in pore size (Supplementary Figs 1 and 2). Subsequently, the plasma-activated Cu samples were partially reduced *in situ* during CO_2RR, and the resulting surface morphology is shown in the last column of Fig. 1. For most samples, an increase in the porosity was observed. Interestingly, the sample oxidized at 100 W for 10 min showed the growth of hair-like nanoneedles after the reaction. To compare the effect of plasma reduction on the film morphology to the electrochemical reduction mentioned above, a H_2 plasma was used (100 W for 10 min), and is shown in Fig. 1g. The resulting structure was very similar to the electrochemically reduced sample (c.f. Fig. 1f), indicating that plasma reduction or electrochemical reduction of the Cu oxides results in a similar structure. SEM images for additional plasma treatments are shown in Supplementary Fig. 1.

To gain further information on the structure and chemical state of the plasma-treated Cu films, cross-sections of the samples were measured using STEM combined with elemental mapping using EDS. Cross-sectional EDS maps of plasma-treated Cu samples before and after CO_2RR are shown in Fig. 2 and Supplementary Fig. 3. After the O_2 plasma treatment, the EDS maps show two well-defined oxide layers over the bulk metallic Cu film. Stoichiometric analysis using EDS data (Supplementary Table 1) resulted in Cu:O ratios of 2.7:1.0 and 1.3:1.0 for the interlayer and upper layers, respectively, indicating that the interlayer is Cu_2O, while the upper layer is copper(II) oxide (CuO). It is also apparent that at grain boundaries and crystallographic domains with the fastest oxide growth, a thick 1.5-μm Cu_2O interlayer grows with a thinner 100 nm CuO top layer, while in domains with slower oxide growth, the CuO layer is thicker at 260 nm—as clearly shown in Fig. 2a. Figure 2g shows the 100 W 2 min oxidized Cu with a subsequent H_2 plasma treatment. A 100 nm layer at the surface is reduced entirely to metallic Cu, with the CuO and Cu_2O oxide layers being still observed subsurface. After H_2 plasma treatment, the sample was exposed to air for more than a week before TEM sample preparation and analysis, and the native Cu oxide layer, measured to be <3 nm thick, was insignificant compared with the thick oxide layers which formed from the plasma treatment and remained during the reaction. This indicates that the observed oxides in our samples before or after reaction are not simply due to surface oxidation in air during the *ex situ* sample transfer to the TEM. Figure 2c,f,h also shows EDS maps of these samples after 1 h of CO_2RR at −0.91 V versus RHE. Remarkably, after the reaction a single layer with O atom concentration ranging from 3 to 29 atomic % remains with porous and oxygen-rich regions

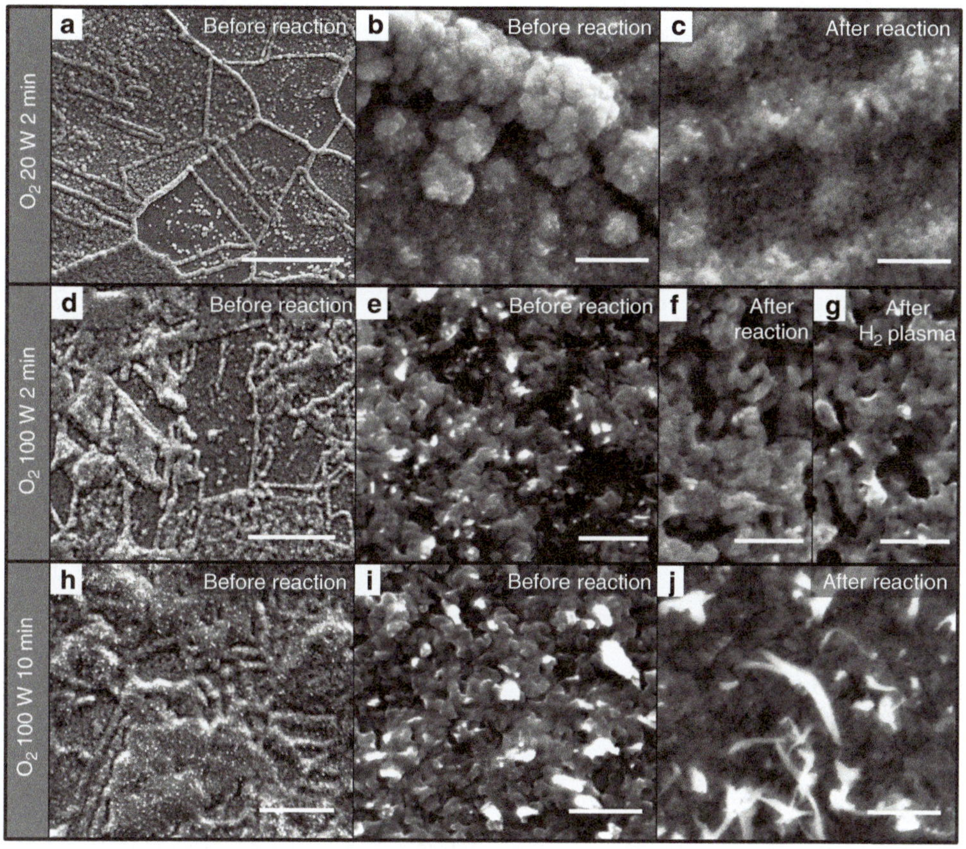

Figure 1 | Morphological characterization of plasma-treated Cu foil electrodes. SEM images of Cu foils treated with O_2 plasma for (**a–c**) 20 W 2 min; (**d–g**) 100 W 2 min; and (**h–j**) 100 W 10 min. (**a,b,d,e,h,i**) The morphology of the as-prepared foils. (**c,f,j**) The morphology after the CO_2RR reaction. (**g**) The sample plasma treated with O_2 at 100 W for 2 min after an additional H_2 plasma treatment at 100 W for 10 min. Scale bars, (**a**) 10 μm; (**b,c,e–g,i,j**) 500 nm; (**d,h**) 20 μm.

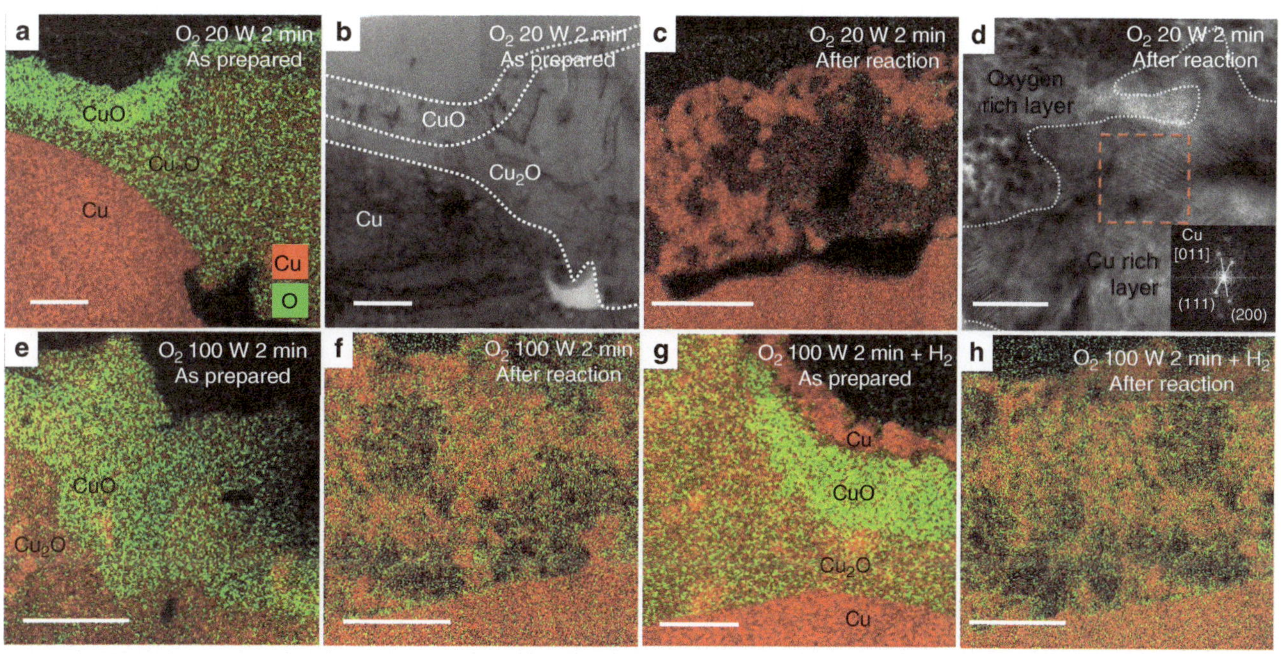

Figure 2 | Morphological and chemical analysis of plasma-activated Cu foils. EDS elemental maps of Cu foils treated with O_2 plasma for (**a–d**) 20 W 2 min; (**e–f**) 100 W 2 min; and (**g–h**) 100 W 2 min + H_2 plasma. The images labelled 'after reaction' were used as catalyst for CO_2RR for 1 h at −0.91 V versus RHE. (**b,d**) HRTEM and SAED analysis of the O_2 plasma 20 W 2 min treated sample before and after the reaction, respectively. Scale bars, (**a–c**) 300 nm; (**d**) 20 nm; (**e–h**) 200 nm.

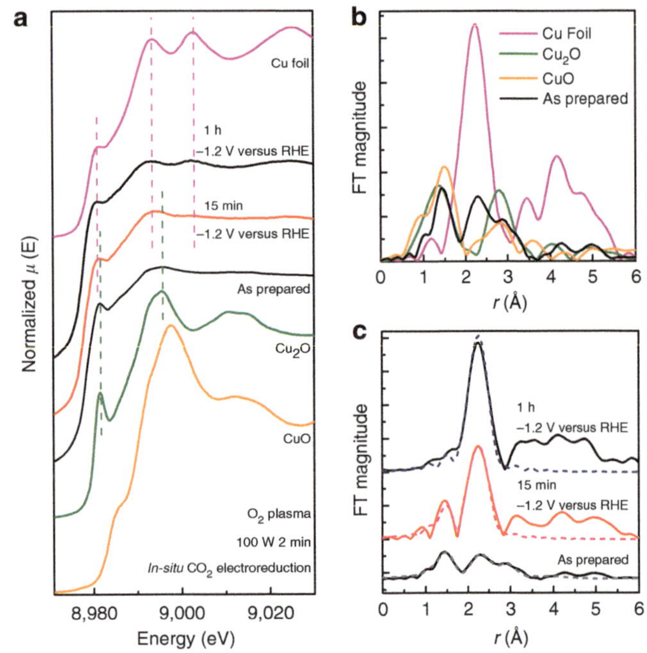

Figure 3 | Operando structural and chemical characterization during CO_2 electroreduction. (**a**) XANES spectra of the O_2 100 W 2 min treated sample measured under operando conditions in 0.1 M $KHCO_3$ during the first 15 min and after 1 h of reaction at −1.2 V versus RHE. Bulk Cu, Cu_2O and CuO spectra are plotted as reference. (**b**) EXAFS spectrum of the as-prepared sample plotted with references. (**c**) EXAFS spectra and fits (dashed lines) of the sample measured under operando conditions. Fourier transforms are k^2-weighted. Fit results are given in Supplementary Table 2.

observed. The Cu oxide species are present throughout the layer, indicating that the surface is rich in Cu^+ sites which remain stable on the surface and near surface layers during the reaction. This finding challenges the conventional assumption that only metallic Cu (Cu^0) is the active species during CO_2RR and indicates that Cu^+ plays a significant role in the unique reactivity of oxide-derived Cu.

Figure 2b,d and Supplementary Fig. 4 show high-resolution TEM (HRTEM) and selected area electron diffraction (SAED) patterns of the O_2 20 W 2 min plasma-treated sample before and after the reaction. The diffraction analysis confirms that a CuO top layer and Cu_2O interlayer exist before the reaction (Fig. 2b). After the reaction, the layer transforms into a mixture of Cu-rich regions, which contain only $3.4 \pm 0.9\%$ oxygen atoms, and oxygen-rich regions (Supplementary Fig. 4) with $19.5 \pm 2.6\%$ oxygen atoms identified as Cu_2O. It is clear that the oxygen is depleted during the reaction, but the surface layer of the oxidized Cu films is still rich in oxygen even after 1 h of reaction. It is unlikely that the Cu oxides in the sample after reaction could be a result of air exposure, since the native oxide would be much thinner due to its slow growth rate at room temperature in air and would contain Cu(II) species, which are not observed here.

Operando X-ray absorption spectroscopy. To further investigate the changes in the structure of the oxide layer during CO_2RR and gain insight into the chemical state of the active Cu species, operando X-ray absorption near-edge structure (XANES) and extended XAFS (EXAFS) were used. Figure 3 shows the Cu K-edge XANES and EXAFS spectra of the O_2 100 W 2 min plasma-treated Cu measured in fluorescence at small incidence angle (for enhanced surface sensitivity) during CO_2RR at −1.2 V versus RHE. Although the thickness of the sample caused dampening of the XAFS signals due to self-absorption, the XANES spectrum of the as-prepared plasma-oxidized Cu shows the features of Cu_2O, in particular, the prominent shoulder at the edge marked with a green dashed line at ~8,982 eV. The presence of metallic Cu and Cu oxide in the initial sample is also clear in the EXAFS data shown in Fig. 3b,c. During the first 15 min of reaction, a combination of Cu and Cu_2O features are still present, and at 1 h, only metallic Cu features are discernable. Although an oxygen content of ~25–28% is still present in the surface layer of the catalyst after 1 h of reaction according to STEM-EDS, XAFS probes further into the bulk, and when the films are partially reduced, the XAFS signals become dominated by the signal of the metallic Cu underlayer. Nevertheless, we have shown here unquestionably that for our plasma-oxidized Cu catalyst, Cu_2O species are present and can participate in the reaction. It is clear that significant changes occur in the bulk of the Cu oxide layer during the first 15 min of reaction in which the initial CuO and Cu_2O layers are reconstructed and some of the oxygen is depleted. After this initial change, it is likely that the Cu_2O species remain relatively stable in the surface layer, since the activity and selectivity measured after 10 min and 1 h are unchanged for all the oxidized samples, see Supplementary Fig. 5. Furthermore, the reactivity of the O_2 20 W 2 min sample was observed to be stable over the course of 5 h, suggesting there is no significant change on the active phase during this time (Supplementary Fig. 6). Therefore, it is likely that Cu^+ species can be supplied to the surface from the thick initial oxide surface and subsurface layers during the reaction to maintain the reactivity of these catalysts over time.

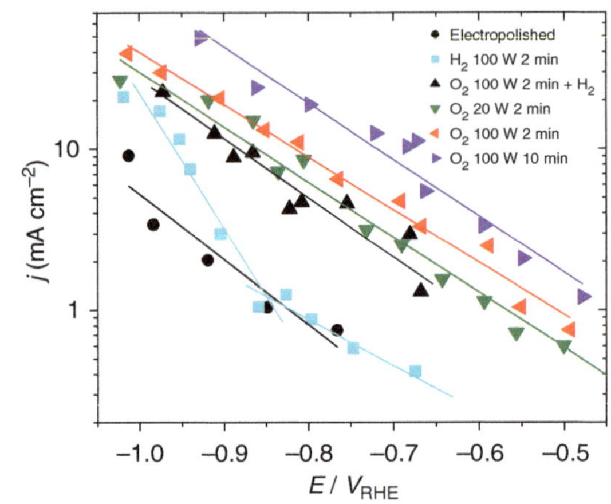

Figure 4 | Electrochemical activity during CO_2 electroreduction. Geometric reduction current density as a function of applied potential.

CO_2 electroreduction performance. The catalytic activity and selectivity of the plasma-activated copper foils were studied by performing bulk electrolysis in 0.1 M potassium bicarbonate ($KHCO_3$) at a constant potential and analysing the reaction products via chromatographic techniques. Figure 4 shows the geometric current density as a function of applied potential. The plasma treatment has a clear effect on the activity of Cu towards CO_2RR which can partially be attributed to the surface roughness. In general, longer and higher power O_2 plasma treatment increases the roughness factor (Table 1), which is an estimate of the active surface area derived by measuring the double layer capacity and normalizing by the electropolished foil (Supplementary Table 3; Supplementary Fig. 7). However, roughness effects alone cannot explain all of our experimental findings satisfactorily. In fact, the initial oxidation state of the sample also affects the current density. This is particularly evident when we compare the catalytic activity of the O_2 100 W 2 min and O_2 100 W 2 min + H_2 plasma-treated Cu. Although both samples have a similar surface roughness, as shown in Table 1, their catalytic activity is significantly different. The H_2 plasma treatment on the oxidized sample reduces the activity by about two times compared with the oxidized sample without H_2 treatment. The H_2 plasma treatment depletes the oxide species available initially in the sample, as shown in Fig. 2g, while having minimal effect on the surface roughness. This result is an indication that the Cu oxide species are critical for improved current density. As demonstrated by our STEM-EDS data, even after 1 h of CO_2RR, patches of Cu_2O are still observed near the sample surface.

Figure 5 shows the faradaic selectivity of the CO_2RR products for the plasma-activated Cu samples as a function of applied potential. The plasma treatments not only affect the activity of copper, but more importantly, they also have a remarkable effect on selectivity. The selectivity change is most drastic on the O_2 treated samples: on one hand, this treatment clearly suppresses methane formation (Fig. 5c), while on the other hand, it enhances the formation of the other products of CO_2RR (CO, formate and ethylene). Consistent with results from Kanan and co-workers[22], we observe that the onset potential towards CO and formate ($HCOO^-$) is shifted to lower overpotentials. Furthermore, for our plasma-oxidized Cu, the selectivity towards CO reaches a maximum of 60% at −0.5 V versus RHE, which is three times higher than for electropolished Cu foil. In our study, the onset of ethylene production is also shifted from approximately −0.85 V versus RHE for metallic Cu foils to as low as −0.5 V versus RHE for the most oxidized Cu foils. In addition, we observe a remarkably high selectivity towards ethylene, reaching 60% at −0.9 V versus RHE for the O_2 20 W 2 min plasma-treated sample. This value is higher than other oxide-derived Cu catalysts synthesized from thermal oxidation[22], which show <10% ethylene efficiency at these potentials, or from electrodepositon of Cu_2O[21,23,24]. In addition, the most oxidized foils in this study also produce trace amounts of ethanol (Supplementary Fig. 8). Interestingly, the 100 W O_2 plasma-treated samples, which have been exposed to a stronger oxygen plasma treatment and are characterized by the highest roughness factors, also produce trace amounts of ethane, but nonetheless exhibit lower C_2H_4 selectivities. These results indicate that there is an optimal oxidation treatment to favour ethylene formation, in agreement with results from Ren *et al*[21]. We hypothesize that this trend could be related to mass diffusion limiting ethylene formation on the most roughened surfaces where high geometric currents are achieved. This is also consistent with our observation that at high overpotentials—at which high geometric current densities are also achieved—ethylene selectivity is suppressed.

Table 1 | Roughness factors estimated through capacitance measurements.

Sample	Roughness factor
Electropolished	1.0
H_2 100 W 2 min	1.5
O_2 20 W 2 min	26.4
O_2 100 W 2 min	43.7
O_2 100 W 2 min + H_2	48.3
O_2 100 W 10 min	89.7

To confirm that the selectivity trend for the oxidized samples is due to the oxidation state, not the surface roughness, we can compare the O_2 100 W 2 min treated sample with the same sample which was exposed to an identical O_2 plasma treatment plus H_2 plasma. Despite their similar roughness (Table 1), their selectivity is drastically different. On both samples, methane formation is suppressed, however on the former sample, which

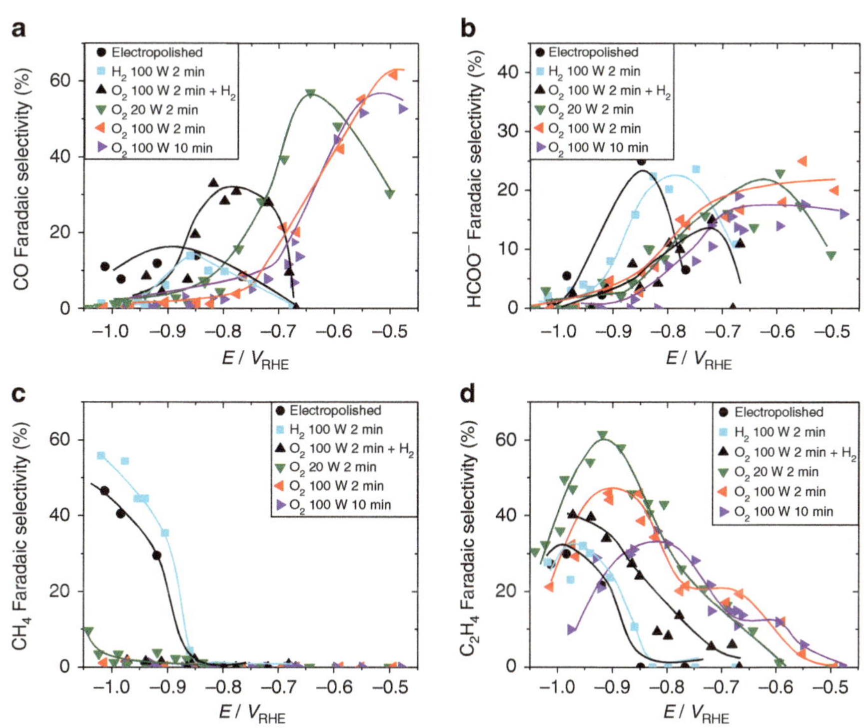

Figure 5 | Faradaic selectivities of the main CO_2RR products. Data were acquired after 60 min of CO_2 electrolysis at a constant potential in CO_2 saturated 0.1 M $KHCO_3$. **(a)** CO, **(b)** formate, **(c)** CH_4 **(d)** C_2H_4. Solid lines are guides for the eye. The remaining selectivity is due to H_2, which is shown in Supplementary Fig. 8.

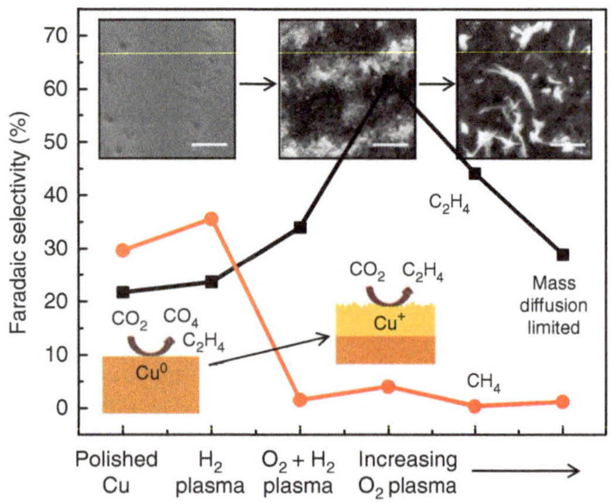

Figure 6 | Summary of hydrocarbon selectivity of plasma-treated Cu foils. From left to right, the insets show SEM images of the low surface area H_2 plasma-treated metallic Cu foil, the O_2 20 W 2 min plasma-treated Cu foil with optimal ethylene selectivity, and the high surface area nanoneedles on the O_2 100 W 10 min oxidized sample after the reaction (500 nm scale bars).

was only treated with O_2 plasma, the onset potential for the formation of CO, $HCOO^-$ and C_2H_4 is clearly shifted to less negative potentials. The main difference between these two samples is the availability of Cu^+ sites in their initial state. Thus, by depleting the Cu^+ sites using the H_2 plasma, we have reduced the number of active sites available for C_2H_4 formation, and therefore altered the selectivity.

We also observed that the H_2 100 W 2 min plasma-treated sample, which has a similar roughness to the electropolished Cu foil, has an improved catalytic performance towards CO_2RR compared with the polished foil. The H_2 plasma-treated sample exhibits an earlier onset potential for formate production and a higher faradaic selectivity towards methane. An explanation for this effect could be improved CO_2 reduction on defect sites generated by the plasma treatment, as shown in Supplementary Fig. 1. Another possible mechanism could be cleaning of contaminants or the native oxide from the Cu surface by the H_2 plasma, which may improve the CO_2RR performance in comparison to the electropolished sample. XPS shows that the small adventitious carbon 1s peak present on the electropolished foil is diminished after the H_2 plasma treatment, although the oxidation state of the surface of the foil is not changed (Supplementary Fig. 9).

Discussion

Figure 6 shows a summary of the hydrocarbon selectivity as a function of the plasma treatment of the Cu foils. Two competing effects are controlling the reactivity of the plasma-treated Cu foils: the surface roughness and the oxidation state of the surface layer. As described above, after oxidizing the surface layer with O_2 plasma, we see an almost complete suppression of methane formation. As the foils are more oxidized, the surface roughness increases, and there is a drop in the ethylene selectivity, likely due to mass diffusion limitations.

Previous studies have proposed that the catalytic behaviour of such oxygen-derived Cu catalysts can be attributed to a roughness effect or to the presence of strongly binding defect sites such as grain boundaries. Recent work in Verdaguer-Casadevall *et al.*[25] have shown that CO-binding energy on oxide-derived Cu is higher than on metallic copper. This high-binding energy could explain the early onset potential for CO_2RR, but does not explain why subsequent H_2 plasma treatment would suppress the ethylene selectivity of the 100 W 2 min oxidized sample in our study.

Another proposed mechanism for the behaviour of oxide-derived CO_2RR catalysts is a pH effect. It is expected that during the catalytic reaction on these rough surfaces there is a significant rise in the local pH, suppressing the pH-dependent CO protonation that leads to methane formation[41,42]. The pH-independent pathway via CO dimerization, however, is not affected, resulting in high ethylene selectivities[28,43,44]. Consistently, all samples with some initial oxidation treatment (rough surfaces) exhibit suppression of methane, as shown in Fig. 6, which may be attributed to the pH effect. However, the lower onset potential for ethylene production observed in the oxygen-treated samples (Fig. 5d) cannot be attributed to a local pH effect. In fact we observed that that the increasing oxidation treatment leads to a decrease in the overpotential for C_2H_4 production, indicating that high activity for ethylene can be related to the oxidation–reduction treatment and not to a local pH effect.

By comparing the catalytic performance of O_2 plasma 100 W 2 min + H_2 and O_2 plasma 100 W 2 min Cu, we have demonstrated that purely a roughness effect cannot explain the reactivity trends of our plasma-activated oxide-derived Cu catalysts, since these catalysts with nearly the same roughness, but different oxide content have significantly different reactivity and selectivity. In addition, while the high local pH could explain the suppression of CH_4, it cannot explain the earlier onset potential for CO_2RR. Since surface roughness and pH effects cannot fully explain the reactivity of our plasma-activated Cu, it is clear that another mechanism is also involved in controlling the reactivity. Based on our STEM-EDS and operando XAFS measurement we conclude that, contrary to common belief, Cu^+ species are stable under reaction conditions and play a determining role in the reactivity of oxide-derived Cu. These species may interact with negatively charged CO_2 reduction intermediates which may play an important role in determining selectivity.[41]

The mechanism behind the stability of Cu^+ species under the thermodynamically unfavourable conditions of CO_2RR is still under investigation, however, several mechanisms may explain this phenomenon. One possibility is that the highly roughened surfaces with low-coordinated sites which are formed after O_2 plasma treatment may bind more strongly to oxygen, helping to stabilize oxides during the reaction. Strain in the surface caused by the oxidation and reduction of the surface may also cause strong oxygen binding. The highly porous nature of the films formed might also serve as favourable oxygen reservoir. Another possibility is that the high local pH could help to stabilize Cu^+ by negatively shifting the overpotential for Cu_2O reduction. Recent studies of nanostructured tin oxide catalysts for CO_2 electroreduction have also suggested that oxide species are stable during the reaction and play a controlling role in their selectivity. Sn oxides and not metallic Sn were shown to be the active species for CO_2 reduction to formic acid and are metastable during the reaction at reducing potentials, even though the oxides are thermodynamically unstable under these conditions[45–47]. While similar stability of Cu^+ has been suggested for oxide-derived Cu catalysts[23,31,48], we provide here the first direct evidence that Cu^+ species are stable during CO_2RR and are key for controlling selectivity.

In summary, we have synthesized superior oxygen-derived Cu catalysts using plasma treatments and used a synergistic combination of advanced *ex situ*, *in situ* and operando techniques to reveal the key mechanisms behind the improved performance of oxide-derived copper catalysts. Plasma treatment was found to be a facile method to rapidly oxidize Cu foils, which resulted in stable oxide layers with a unique surface structure. These plasma-oxidized catalysts achieve lower onset potentials for CO, formate and ethylene, as well as outstanding ethylene selectivity of 60% at − 0.9 V versus RHE. In addition, using STEM-EDS and operando XAFS, we have found that the oxides in the surface layer are surprisingly stable against reduction, with a significant amount of oxide species and dissolved oxygen remaining after 1 h of reaction at relatively high potentials (− 0.91 V versus RHE). Furthermore, ethylene synthesis and methane suppression are due to the presence of these oxides. It is expected that the fundamental understanding extracted from our experimental study could aid the further design and optimization of oxide-derived Cu catalysts.

Methods

Catalyst synthesis. All Cu foils were initially electropolished in 85% phosphoric acid at 3 V versus a titanium foil, then thoroughly rinsed with ultra-pure water and dried with nitrogen. Plasma treatments were performed in 400 mtorr of H_2 or O_2 for the indicated plasma power and time. For the O_2 100 W 2 min + H_2-treated sample, 100 W 10 min of H_2 plasma was used.

Electron microscopy. SEM images were obtained using a Quanta 200 FEG microscope from FEI with a field emitter as electron source. A secondary electron (Everhart-Thornley) detector was used for the image acquisition. An electron acceleration voltage of 10 kV and a working distance of 10 mm were chosen for the measurements. TEM cross-section samples were prepared using a FEI Helios 600 Dual Beam Focused Ion Beam (FIB) at the Center for Functional Nanomaterials (CFN) at Brookhaven National Laboratory (BNL). The FIB sectioning was performed at 30 keV with final milling at 5 KeV. Spatially resolved elemental maps were acquired using a FEI Titan TEM/STEM microscope with ChemiSTEM technology (X-FEG and SuperX with four windowless silicon drift EDS detectors) operated at 200 kV, at the National Center for Electron Microscopy (NCEM), Molecular Foundry at Lawrence Berkeley National Laboratory (LBL). A Hitachi 9500 TEM at the Nanoscale Fabrication and Characterization Facility (NFCF), University of Pittsburgh operated at 200 keV was used to acquire HRTEM images and SAED patterns.

X-ray absorption spectroscopy. XANES and EXAFS spectra were measured at beamline 2-2 of the Stanford Synchrotron Light Source. A home-built operando electrochemical cell was used, with a platinum foil counter electrode and silver/silver chloride (Ag/AgCl) reference electrode. The samples were mounted behind an X-ray kapton window with 1 mm of electrolyte between the sample and window. The electrolyte (0.1 M $KHCO_3$) was circulated between the cell and a reservoir in which CO_2 was continuously bubbled. Measurements were performed in fluorescence at small incidence angle using a passivated implanted planar silicon (PIPS) detector. The acquisition of each spectrum took 15 min, and they were acquired during the first 15 min and after 1 h of reaction.

Data analysis was performed using the Athena and Artemis software. The FEFF8 code was used to simulate Cu, Cu_2O and CuO spectra for EXAFS fitting. Further details are given in the Supplementary Table 2.

Electrochemical measurements. Electrochemical measurements were carried out in a custom made two compartment cell, separated by a Nafion membrane. The glassware was cleaned in a 'nochromix' bath and afterwards in concentrated HNO_3 for 1 h, respectively, rinsed and sonicated with ultra-pure water several times. The working compartment was filled with 120 ml 0.1 M $KHCO_3$ (Sigma-Aldrich ≥99.95%). Before and during the electrochemical reaction the cell was purged continuously with CO_2 (30 ml min^{-1}), reaching a stable pH value of 6.8.

A platinum mesh 100 (Sigma-Aldrich 99.9%) was used as counter electrode and a leak-free Ag/AgCl electrode as reference electrode (Hugo Sachs Elektronik Harvard apparatus GmbH). The plasma-treated Cu foils were used as working electrode and contacted by a gold clamp. Every measurement was started with a linear voltammetric sweep, performed with a scan rate of − 5 mV s^{-1} between E = + 0.05 V/RHE and the working potential (between − 0.45 V and − 1.0 V/RHE) followed by a chronoamperometric step for 60 min. All reported potentials are corrected for Ohmic drop determined by electrochemical impedance spectroscopy.

Product analysis. After 10 and 60 min of bulk electrolysis at constant potential, a sample of the gas was analysed by gas chromatography (Shimadzu GC 2016) to determine the production rate and faradaic selectivity of the gaseous products. In addition, an aliquot of the electrolyte was analysed by a high-performance liquid chromatograph (Agilent 1200 series).

References

1. Hori, Y. in *Modern Aspects of Electrochemistry* Vol. 42 (eds Vayenas, C. G., White, R. E. & Gambao-Aldaco, M. E.) 89–189 (Springer, 2008).
2. Kuhl, K. P., Cave, E. R., Abram, D. N. & Jaramillo, T. F. New insights into the electrochemical reduction of carbon dioxide on metallic copper surfaces. *Energy Eviron. Sci.* **5**, 7050–7059 (2012).
3. Hori, Y., Kikuchi, K. & Suzuki, S. Production of CO and CH_4 in electrochemical reduction of CO_2 at metal electrodes in aqueous hydrogencarbonate solution. *Chem. Lett.* **14**, 1695–1698 (1985).
4. Hori, Y., Murata, A. & Takahashi, R. Formation of hydrocarbons in the electrochemical reduction of carbon dioxide at a copper electrode in aqueous solution. *J. Chem. Soc. Farad. Trans. 1* **85**, 2309–2326 (1989).
5. Peterson, A. A. & Nørskov, J. K. Activity descriptors for CO_2 electroreduction to methane on transition-metal catalysts. *J. Phys. Chem. Lett.* **3**, 251–258 (2012).
6. Hansen, H. A., Varley, J. B., Peterson, A. A. & Nørskov, J. K. Understanding trends in the electrocatalytic activity of metals and enzymes for CO_2 reduction to CO. *J. Phys. Chem. Lett.* **4**, 388–392 (2013).
7. Peterson, A. A., Abild-Pedersen, F., Studt, F., Rossmeisl, J. & Nørskov, J. K. How copper catalyzes the electroreduction of carbon dioxide into hydrocarbon fuels. *Energy Eviron. Sci.* **3**, 1311–1315 (2010).
8. Zhang, Y.-J., Sethuraman, V., Michalsky, R. & Peterson, A. A. Competition between CO_2 reduction and H_2 evolution on transition-metal electrocatalysts. *ACS Catal.* **4**, 3742–3748 (2014).
9. Reske, R., Mistry, H., Behafarid, F., Roldan Cuenya, B. & Strasser, P. Particle size effects in the catalytic electroreduction of CO_2 on Cu nanoparticles. *J. Am. Chem. Soc.* **136**, 6978–6986 (2014).
10. Mistry, H. *et al.* Exceptional size-dependent activity enhancement in the electroreduction of CO_2 over Au nanoparticles. *J. Am. Chem. Soc.* **136**, 16473–16476 (2014).
11. Mistry, H. *et al.* Tuning catalytic selectivity at the mesoscale via interparticle interactions. *ACS Catal.* **6**, 1075–1080 (2015).
12. Tang, W. *et al.* The importance of surface morphology in controlling the selectivity of polycrystalline copper for CO_2 electroreduction. *Phys. Chem. Chem. Phys.* **14**, 76–81 (2012).
13. Baturina, O. A. *et al.* CO_2 electroreduction to hydrocarbons on carbon-supported Cu nanoparticles. *ACS Catal.* **4**, 3682–3695 (2014).
14. Manthiram, K., Beberwyck, B. J. & Alivisatos, A. P. Enhanced electrochemical methanation of carbon dioxide with a dispersible nanoscale copper catalyst. *J. Am. Chem. Soc.* **136**, 13319–13325 (2014).
15. Chen, C. S. *et al.* Stable and selective electrochemical reduction of carbon dioxide to ethylene on copper mesocrystals. *Catal. Sci. Technol.* **5**, 161–168 (2015).
16. Roberts, F. S., Kuhl, K. P. & Nilsson, A. High selectivity for ethylene from carbon dioxide reduction over copper nanocube electrocatalysts. *Angew. Chem. Int. Ed. Engl.* **127**, 5268–5271 (2015).
17. Sen, S., Liu, D. & Palmore, G. T. R. Electrochemical reduction of CO_2 at copper nanofoams. *ACS Catal.* **4**, 3091–3095 (2014).
18. Zhu, W. *et al.* Active and selective conversion of CO_2 to CO on ultrathin Au nanowires. *J. Am. Chem. Soc.* **136**, 16132–16135 (2014).
19. Lu, Q., Rosen, J. & Jiao, F. Nanostructured metallic electrocatalysts for carbon dioxide reduction. *ChemCatChem* **7**, 38–47 (2015).
20. Mistry, H., Varela, A. S., Kuehl, S., Strasser, P. & Roldan Cuenya, B. Nanostructured electrocatalysts with tunable activity and selectivity. *Nat. Rev. Mater.* **1**, 16009 (2016).
21. Ren, D. *et al.* Selective electrochemical reduction of carbon dioxide to ethylene and ethanol on copper (I) oxide catalysts. *ACS Catal.* **5**, 2814–2821 (2015).
22. Li, C. W. & Kanan, M. W. CO_2 reduction at low overpotential on Cu electrodes resulting from the reduction of thick Cu_2O films. *J. Am. Chem. Soc.* **134**, 7231–7234 (2012).
23. Kim, D. *et al.* Insights into an autonomously formed oxygen-evacuated Cu_2O electrode for the selective production of C_2H_4 from CO_2. *Phys. Chem. Chem. Phys.* **17**, 824–830 (2015).
24. Kas, R. *et al.* Electrochemical CO_2 reduction on Cu_2O-derived copper nanoparticles: controlling the catalytic selectivity of hydrocarbons. *Phys. Chem. Chem. Phys.* **16**, 12194–12201 (2014).
25. Verdaguer-Casadevall, A. *et al.* Probing the active surface sites for CO reduction on oxide-derived copper electrocatalysts. *J. Am. Chem. Soc.* **137**, 9808–9811 (2015).
26. Li, C. W., Ciston, J. & Kanan, M. W. Electroreduction of carbon monoxide to liquid fuel on oxide-derived nanocrystalline copper. *Nature* **508**, 504–507 (2014).
27. Gupta, N., Gattrell, M. & MacDougall, B. Calculation for the cathode surface concentrations in the electrochemical reduction of CO_2 in $KHCO_3$ solutions. *J. Appl. Electrochem.* **36**, 161–172 (2006).
28. Varela, A. S., Kroschel, M., Reier, T. & Strasser, P. Controlling the selectivity of CO_2 electroreduction on copper: the effect of the electrolyte concentration and the importance of the local pH. *Catal. Today* **260**, 8–13 (2016).

29. Kas, R., Kortlever, R., Yılmaz, H., Koper, M. & Mul, G. Manipulating the hydrocarbon selectivity of copper nanoparticles in CO_2 electroreduction by process conditions. *ChemElectroChem* **2,** 354–358 (2015).
30. Schouten, K. J. P., Gallent, E. P. & Koper, M. T. The influence of pH on the reduction of CO and CO_2 to hydrocarbons on copper electrodes. *J. Electroanal. Chem.* **716,** 53–57 (2014).
31. Lee, S. & Lee, J. Electrode build-up of reducible metal composites toward achievable electrochemical conversion of carbon dioxide. *ChemSusChem* **9,** 333–344 (2015).
32. Shibata, H., Moulijn, J. A. & Mul, G. Enabling electrocatalytic Fischer–Tropsch synthesis from carbon dioxide over copper-based electrodes. *Catal. Lett.* **123,** 186–192 (2008).
33. Zhang, Y.-J. & Peterson, A. A. Oxygen-induced changes to selectivity-determining steps in electrocatalytic CO_2 reduction. *Phys. Chem. Chem. Phys.* **17,** 4505–4515 (2015).
34. Nie, X., Griffin, G. L., Janik, M. J. & Asthagiri, A. Surface phases of $Cu_2O(111)$ under CO_2 electrochemical reduction conditions. *Catal. Commun.* **52,** 88–91 (2014).
35. Le, M. *et al.* Electrochemical reduction of CO_2 to CH_3OH at copper oxide surfaces. *J. Electrochem. Soc.* **158,** E45–E49 (2011).
36. Roldan Cuenya, B. Metal nanoparticle catalysts beginning to shape-up. *Acc. Chem. Res.* **46,** 1682–1691 (2012).
37. Menard, L. D., Xu, F., Nuzzo, R. G. & Yang, J. C. Preparation of TiO_2-supported Au nanoparticle catalysts from a Au_{13} cluster precursor: ligand removal using ozone exposure versus a rapid thermal treatment. *J. Catal.* **243,** 64–73 (2006).
38. Zou, J.-J., Liu, C.-J. & Zhang, Y.-P. Control of the metal-support interface of NiO-loaded photocatalysts via cold plasma treatment. *Langmuir* **22,** 2334–2339 (2006).
39. Ono, L. K. & Roldan Cuenya, B. Formation and thermal stability of Au_2O_3 on gold nanoparticles: size and support effects. *J. Phys. Chem. C* **112,** 4676–4686 (2008).
40. Zhou, Y. *et al.* Dopant-induced electronic structure modification of HOPG surfaces: implications for high activity fuel cell catalysts. *J. Phys. Chem. C* **114,** 506–515 (2009).
41. Kortlever, R., Shen, J., Schouten, K. J. P., Calle-Vallejo, F. & Koper, M. T. Catalysts and reaction pathways for the electrochemical reduction of carbon dioxide. *J. Phys. Chem. Lett.* **6,** 4073–4082 (2015).
42. Nie, X., Esopi, M. R., Janik, M. J. & Asthagiri, A. Selectivity of CO_2 reduction on copper electrodes: the role of the kinetics of elementary steps. *Angew. Chem. Int. Ed. Engl.* **52,** 2459–2462 (2013).
43. Calle-Vallejo, F. & Koper, M. Theoretical considerations on the electroreduction of CO to C2 species on Cu (100) electrodes. *Angew. Chem. Int. Ed. Engl.* **125,** 7423–7426 (2013).
44. Montoya, J. H., Shi, C., Chan, K. & Norskov, J. K. Theoretical insights into a CO dimerization mechanism in CO_2 electroreduction. *J. Phys. Chem. Lett.* **6,** 2032–2037 (2015).
45. Dutta, A., Kuzume, A., Rahaman, M., Vesztergom, S. & Broekmann, P. Monitoring the chemical state of catalysts for CO_2 electroreduction: an in operando study. *ACS Catal.* **5,** 7498–7502 (2015).
46. Baruch, M. F., Pander, III J. E., White, J. L. & Bocarsly, A. B. Mechanistic insights into the reduction of CO_2 on tin electrodes using *in situ* ATR-IR spectroscopy. *ACS Catal.* **5,** 3148–3156 (2015).
47. Chen, Y. & Kanan, M. W. Tin oxide dependence of the CO_2 reduction efficiency on tin electrodes and enhanced activity for tin/tin oxide thin-film catalysts. *J. Am. Chem. Soc.* **134,** 1986–1989 (2012).
48. Lee, S., Kim, D. & Lee, J. Electrocatalytic production of C3-C4 compounds by conversion of CO_2 on a chloride-induced bi-phasic Cu_2O-Cu catalyst. *Angew. Chem. Int. Ed. Engl.* **127,** 14914–14918 (2015).

Acknowledgements

We acknowledge Anatoly Frenkel for helpful discussions regarding the XAFS analysis. This work was funded by the German Federal Ministry of Education and Research (Bundesministerium für Bildung und Forschung, BMBF) under grant #03SF0523 - 'CO2EKAT' and the Cluster of Excellence RESOLV at RUB (EXC 1069) funded by the Deutsche Forschungsgemeinschaft. In addition, financial support from the US National Science Foundation (NSF-Chemistry 1213182 an NSF-DMR 1207065) and the Office of Basic Energy Sciences of the US Department of Energy (DE-FG02-08ER15995) is greatly appreciated. Work at beamline BL2-2 at SSRL was supported in part by the U.S. DOE Grant No. DE-SC0012653 to the Synchrotron Catalysis Consortium. E.A.S. acknowledges support to the Center for Functional Nanomaterials, which is a U.S. DOE Office of Science Facility, at Brookhaven National Laboratory under Contract No. DE-SC0012704. C.S.B. and J.C.Y. acknowledges support by the Office of Basic Energy Sciences of the US Department of Energy (DE-FG02-03ER15476) and the US National Science Foundation (NSF CHE 1534630) in collaboration with UT Austin. Transmission electron microscopy work was performed at the National Center for Electron Microscopy at the Molecular Foundry at Lawrence Berkeley National Laboratory which is supported by the Office of Science, Office of Basic Energy Sciences, of the U.S. Department of Energy under Contract No. DE-AC02-05CH11231.

Author contributions

H.M. synthesized the samples, and A.S.V. performed all electrochemical measurements. I.Z. performed the SEM measurements. K.K. and E.A.S. prepared cross-sections of the samples for TEM. C.S.B. and J.C.Y. performed TEM and EDS measurements, and performed SAED and porosity analysis. H.M., I.S., I.Z., Y.C., and B.R.C. performed the XAFS measurements, H.M. and B.R.C. analysed and interpreted the XAFS data. A.S.V. and P.S. analysed the electrochemical data. H.M., A.S.V., P.S. and B.R.C. wrote the manuscript. B.R.C. designed the experiment.

Additional information

Competing financial interests: The authors declare no competing financial interests.

23

Impeding $^{99}Tc(IV)$ mobility in novel waste forms

Mal-Soon Lee[1], Wooyong Um[2,3], Guohui Wang[2], Albert A. Kruger[4], Wayne W. Lukens[5], Roger Rousseau[1] & Vassiliki-Alexandra Glezakou[1]

Technetium (^{99}Tc) is an abundant, long-lived radioactive fission product whose mobility in the subsurface is largely governed by its oxidation state. Tc immobilization is crucial for radioactive waste management and environmental remediation. Tc(IV) incorporation in spinels has been proposed as a novel method to increase Tc retention in glass waste forms during vitrification. However, experiments under high-temperature and oxic conditions show reoxidation of Tc(IV) to volatile pertechnetate, Tc(VII). Here we examine this problem with *ab initio* molecular dynamics simulations and propose that, at elevated temperatures, doping with first row transition metal can significantly enhance Tc retention in magnetite in the order Co > Zn > Ni. Experiments with doped spinels at 700 °C provide quantitative confirmation of the theoretical predictions in the same order. This work highlights the power of modern, state-of-the-art simulations to provide essential insights and generate theory-inspired design criteria of complex materials at elevated temperatures.

[1] Fundamental and Computational Sciences Directorate, Pacific Northwest National Laboratory, Richland, Washington 99352, USA. [2] Energy and Environment Directorate, Pacific Northwest National Laboratory, Richland, Washington 99352, USA. [3] Pohang University of Science and Technology, Pohang 37673, South Korea. [4] United States Department of Energy, Office of River Protection, Richland, Washington 99352, USA. [5] Lawrence Berkeley National Laboratory, Berkeley, California 94720, USA. Correspondence and requests for materials should be addressed to W.U. (email: Wooyong.Um@pnnl.gov) or to V.-A.G. (email: Vanda.Glezakou@pnnl.gov).

Technetium (^{99}Tc) is an abundant long-lived radioactive fission product present in used nuclear fuel and waste generated from nuclear fuel reprocessing. Owing to its long half-life (2.1×10^5 years) and relatively high fission yield ($\sim 6\%$), ^{99}Tc can generate the greatest radiation dose in the vicinity of a waste repository, and for a much longer time compared with other fission products, such as ^{90}Sr and ^{137}Cs (with half-life ~ 30 years)[1,2]. In addition, TcO_4^- is highly soluble and weakly adsorbed in the near-field, while Tc(IV) is highly adsorbable to geological materials and clays[3]. Thus, migration of Tc from a waste repository may be prevented by immobilizing Tc(IV) in durable waste forms, such as glass or ceramic materials[4–6]. Although $Tc(VII)O_4^-$ is the most stable Tc species under aerobic conditions, it is highly volatile at glass vitrification temperatures ($\sim$1,200 °C), leading to poor Tc retention in the final waste glass[4–7]. Retention of Tc in the glass is generally improved by reducing conditions since Tc(IV) is less volatile[6,8,9]. Tc(VII) may be effectively reduced to Tc(IV) by Fe(II) in oxide and sulfide minerals or by Fe(II) adsorbed to mineral surfaces such as iron oxides or aluminium oxides[9–17]. However, retention of Tc is still limited because of re-oxidation of Tc(IV) back to Tc(VII) (refs 18–21). Consequently, simply reducing Tc(VII) to Tc(IV) before vitrification is unlikely to stabilize Tc and prevent its volatilization as Tc(VII). An alternative approach would be to trap Tc(IV) in the lattice of a metal oxide by co-precipitation. Spinels are attractive targets for Tc stabilization during vitrification because of their physical and chemical stability under the high temperatures used in preparing borosilicate glasses[4,6]. In this respect, efficient incorporation and high retention of Tc by glass-incorporated spinels is very important for radioactive waste management and offers substantial economic benefit because of reduction in the amount of glass needed to immobilize ^{99}Tc.

Magnetite (Fe_3O_4) has a cubic inverse spinel structure, where the oxygen anions form a slightly distorted face-centred-cubic sublattice and the iron cations occupy tetrahedral and octahedral interstitial sites. In the [001] direction, two types of layer stacking occur: *A* layers with tetrahedral Fe(III) and *B* layers with O and octahedral Fe(II)/Fe(III) (see Fig. 1). Marshall *et al.*[8] showed that Tc(VII) can be reduced to Tc(IV) and incorporated into the magnetite structure under high pH conditions (pH 10.5 – 13.1). They also observed that Tc(IV) incorporation occurred at the octahedral sites and remobilization of Tc(IV) was limited during subsequent air oxidation. Kobayashi *et al.* observed Tc(IV) incorporation into the magnetite structure at pH 6 and pH 7.5 (ref. 22). However, magnetite oxidizes to maghemite ($\gamma - Fe_2O_3$) in oxic conditions or under high temperature through maghemitization, where all the Fe(II) atoms oxidize to Fe(III), while the oxygen sublattice remains unchanged[23]. When maghematization takes place, iron atoms diffuse towards the surface, leaving octahedral cation vacancies[23,24]. As a result, maghematization could lead to re-oxidation of Tc(IV) because of the increase in Fe(III), a highly efficient oxidizing agent[8]. Sidhu *et al.*[23] suggested that incorporation of trace elements into magnetite stabilizes Fe(II) and suppresses maghematization by decreasing electron mobility. The majority of experimental studies on Tc retention are conducted at low temperatures, while theoretical studies employ static structural models that neglect temperature effects. Under these conditions, these studies cannot address Tc volatilization during vitrification that leads to poor Tc retention in the glass waste form. Thus, elucidation of high-temperature effects is important for understanding Tc retention by magnetite at elevated temperatures. *Ab initio* molecular dynamics (AIMD) simulations can describe the temperature effects on the change in structure, bonding and associated change in the oxidation states of Tc and Fe, which ultimately affects Tc retention. Here our simulations indeed show that leaching of Tc is accompanied with re-oxidation of Tc(IV) to Tc(VII) at high temperatures, but it can be suppressed by doping. We propose that inclusion of first transition metal dopants (Co, Zn and Ni) significantly improves Tc retention in magnetite at high temperature. Quantitative confirmation is further provided by X-ray absorption near edge structure (XANES) measurements and gravimetric analysis.

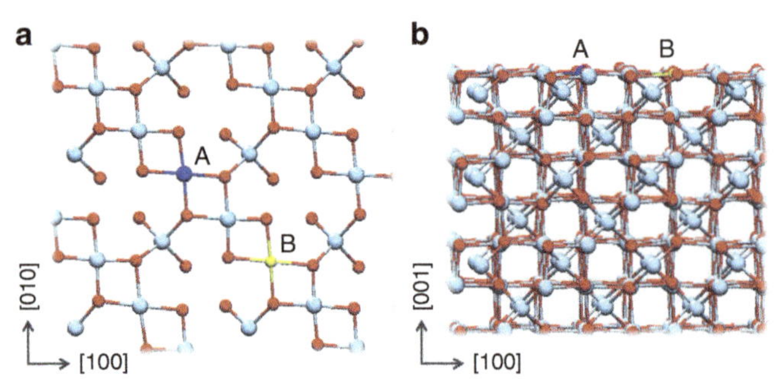

Figure 1 | The B-truncated (octahedral Fe) $Fe_3O_4(001)$ structure. (**a**) Top view and (**b**) side view of surface structure. Red and cyan circles represent oxygen and iron, respectively. A (blue circle) can be either Fe or Tc and B (yellow circle) can be either Fe or an impurity atom (Ni/Zn/Co).

Results

Temperature effects on Tc(IV)-incorporated spinel. The $Fe_3O_4(001)$ surface has been studied extensively[25–30]. Pentcheva *et al.*[28] compiled a phase diagram for the $Fe_3O_4(001)$ surface in an *ab initio* thermodynamics study showing that the most stable surface structure is a *B*-terminated surface with octahedral iron and oxygen atoms forming a wave-like structure along the (001) direction. On the basis of these results, we generated a $Fe_3O_4(001)$ model of a *B*-terminated surface (Fig. 1) that we let fully relax. Below the Verwey transition (125 K), the surface possesses a permanent dipole that has been shown to drive the formation of surface defects in these types of material[25]. However, above this temperature, magnetite is metallic, and the surface dipole is quenched[31] and is not likely to affect surface charge defects. Details on the computational models and methods can be found in the Methods section.

To assess the temperature effects on Tc retention in magnetite, we replaced one octahedral Fe on the surface with Tc and performed AIMD simulations at two different temperatures, 25 and 600 °C, representing the ambient and the lower-end temperature range of the vitrification process, respectively. Figure 2a shows the calculated atomic density profiles of the different species with Tc scaled by 5 for clarity. The dotted grey line denotes the edge of the magnetite surface defined by the average position of the topmost oxygen atoms. At 25 °C, Tc stays within the top surface layer for the duration of the simulation ($\sim$20 ps). Computation of pair distribution functions, *g*(R), reveals that, on average, surface-incorporated Tc has five nearest O neighbours with an average Tc–O distance 1.98 Å (Fig. 2b upper panel, Supplementary Fig. 1 and Supplementary Table 1). Additional exploratory simulations with Tc in an inner lattice position show a *g*(R) maximum at 2.01 Å for the Tc–O distance, compatible with the reduced Tc(IV) in magnetite (Supplementary Table 2). According to X-ray absorption fine structure (XAFS) analysis, the Tc(IV)–O distance is $\sim$2.0 Å (see Supplementary Table 2 and (refs 22, 32) for addi-

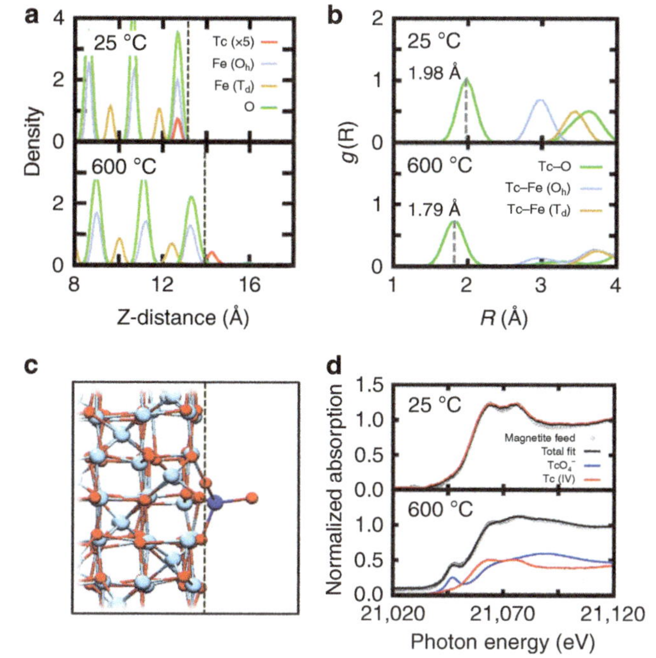

Figure 2 | Structural properties and XANES spectra in the presence of Tc. (**a**) Atomic-density profile showing atomic arrangement along the *z*-direction at 25 and 600 °C obtained from AIMD simulations, where dotted vertical line denotes the magnetite surface. (**b**) Pair distribution function g(R) obtained from AIMD simulation trajectories at 25 and 600 °C. (**c**) Snapshot of the structure at 600 °C from AIMD trajectories where a blue circle represents Tc, red for O and cyan for Fe. The dotted vertical line denotes the magnetite surface. (**d**) Normalized XANES spectra at 25 and 600 °C.

tional structural parameters). These observations imply that (i) within the surface layer the oxidation state of Tc is essentially Tc(IV) and (ii) at 25 °C (the glass-feed stage) reduced Tc(IV) is the prevalent oxidation state. However, the completely opposite picture emerges at high temperatures, 600 °C or higher. Tc moves above the surface, dragging coordinating surface oxygens along with it (see Supplementary Movie 1). The local Tc geometry is consistent with a tetrahedral $Tc(VII)O_4^-$ species, with two or three of the coordinating oxygens dynamically connected to Fe atoms on the surface (Fig. 2c). Analysis of *g*(R) for Tc–O pairs shows a peak at 1.79 Å (Fig. 2b, lower panel), an almost 10% reduction compared with the Tc(IV)–O distance at 25 °C. This change is compatible with the shorter Tc(VII) – O distances of ~1.75 Å as determined by XAFS (Supplementary Table 2 and refs 22, 33). From this observation, we infer, that beginning at 600 °C, Tc oxidation is in process, commensurate with the tetrahedrally coordinated Tc transitioning to TcO_4^-. In addition to the system with Tc at the surface, we also examined a system with Tc at an inner lattice site at 600 °C. The calculated *g*(R) gives the distance between Tc and O as 2.01 Å, consistent with the reduced Tc(IV) in magnetite. As shown in Supplementary Fig. 2, Tc remains in the same layer throughout the simulation timescale. When comparing the energetics of the configurations with Tc below or at the surface layer, the energy with Tc below the surface is 2.5 eV higher than when Tc is at the surface. This implies that there is a thermodynamic driving force that will eventually move Tc out to the surface.

Experimentally, Tc-magnetite samples, heated from room temperature to 600 °C and then cooled back to room temperature, were analysed to determine the Tc oxidation state using

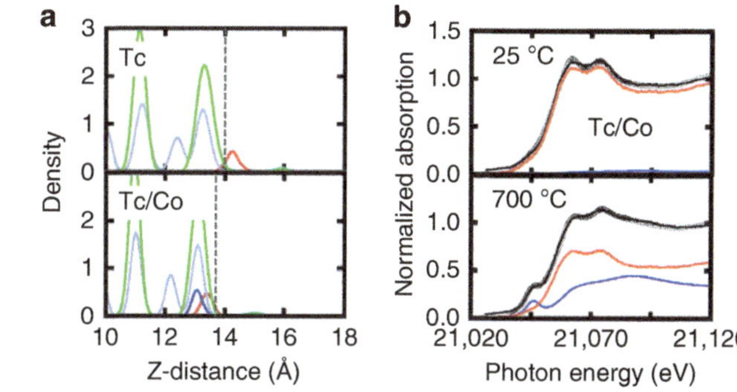

Figure 3 | Atomic density and normalized XANES spectra with Co dopant. (**a**) Atomic density profiles with and without Co dopant at 600 °C. Red lines represent Tc, blue line for doped atom, grey lines for Fe and green lines for O. (**b**) Normalized XANES spectra for the Co-doped magnetite at 25 °C as made and treated at 700 °C. Colour codes used are the same as those shown in Fig. 2.

XANES as shown in Fig. 2d. In the figure, the grey diamonds and black line indicate the measured data and a linear combination fit, respectively, for Tc-magnetite samples, while the red and blue lines represent the contribution from Tc(IV) and Tc(VII), respectively. At 25 °C, the spectrum of the sample shows only Tc(IV) (feed, red) but no Tc(VII) (blue), indicating that all Tc in the sample is in its reduced form. In the sample heated to 600 °C, however, the spectrum shows a mixture of both Tc(IV) and Tc(VII). All these observations are compatible with the simulations.

Effects of dopants on Tc retention. To simulate the effect of dopants on the Tc redox chemistry and immobilization, we modified the magnetite by substituting one surface Fe atom with Ni, Zn or Co (~1% wt each) at a site close to Tc. This choice was motivated by earlier experiments by Sidhu *et al.*[23], who observed stabilization of Fe(II) and suppression of maghematization when first row transition metal dopants were present in magnetite even at concentrations ~1 wt %. The atomic density profiles along the surface normal from AIMD at 600 °C in the presence of the doping elements are shown in Fig. 3 (Co) and Supplementary Fig. 3 (Ni and Zn), exhibiting an increase in Tc retention in the order Co > Zn > Ni.

In the case of Ni, the Tc population is bi-modal where Tc remains mostly on top of the surface with only a small population within the top surface layer. In the case of Zn, the bi-modal Tc distribution is shifted towards a larger Tc population within the surface. Analysis of trajectories also shows that the distance between Tc and the coordinating O fluctuates between 1.71 and 1.92 Å, compatible with an equilibrium between Tc(VII) and Tc(IV) oxidation states (see Supplementary Fig. 4). This behaviour implies that Ni and Zn only partially, and to a similar degree, hinder Tc oxidation. Finally, in the presence of Co, Tc remains almost in its entirety within the surface at all times indicative of a Tc(IV) state. We examined Tc(IV) stabilization in the presence of Co by conducting a simulation starting with TcO_4^- on top of the surface. As shown in Supplementary Movie 2, Tc(VII) rapidly migrates into the surface becoming Tc(IV), within 1.5 ps of simulation time.

To validate these findings, we prepared three different magnetite samples doped with ~10% wt of Ni, Zn and Co. Details on the preparation of samples can be found in the Methods section. The samples were heated at 700 °C in a furnace for 1 h, and the remaining Tc was measured (see Supplementary Table 3). Gravimetric measurement showed that doping with Co resulted in the highest Tc retention (29% wt) compared with less than half that amount for Zn (12% wt) and ~1/8 of that for Ni (4% wt). No detectable amount of Tc was found in the Tc-magnetite sample prepared without dopant and treated at 700 °C. We also performed XANES measurements for the samples prepared at 25 and 700 °C (see Supplementary Methods for details) and confirmed our theoretical prediction of the highest Tc retention with Co dopant at high temperature, as shown in Supplementary Table 4 and Fig. 3b and Supplementary Fig. 5.

Equilibrium constants and free energy estimates. To best connect with the experimental observations, we determined the ratio of the equilibrium populations between the two different oxidation states of Tc(IV) ($[Tc_{in}]$) and Tc(VII) ($[Tc_{out}]$). This can be achieved by integrating the area under the atomic density profiles for Tc in Fig. 3. An equilibrium constant between the two populations, determined as the ratio $K_{eq} = [Tc_{in}]/[Tc_{out}]$, was used to calculate the Gibbs free energy for this equilibrium from the relation $\Delta G = -RT \ln K_{eq}$, where R is the gas constant and T is the absolute temperature. Negative values indicate that the equilibrium favours a higher population of Tc(IV). Table 1 summarizes the computed values of K_{eq} and ΔG, as well as the measured Tc retention for the different doping agents Ni, Zn and Co. These results show a remarkable agreement between the theoretical prediction and experimental validations, not only in terms of relative order but also in magnitude. The underlying reason is based on the increase in the reducing capacity of the Tc-containing spinels upon doping. This can be quantified by the difference in energy between the Fermi level, E_F, and the highest occupied molecular orbital (HOMO) of Tc d states, ΔE_{gap}, see last column in Table 1. The calculated total and projected density of states (DOS) of the d-band for Tc and dopant are shown in Supplementary Fig. 6. Whereas in the case of Ni only a marginal stabilization of the Tc d-states occurs (small ΔE_{gap}), in the case of Zn and Co, a much higher stabilization takes place that ultimately hinders Tc re-oxidation.

Discussion

In conclusion, we propose that standard reduction potentials of transition metal ions relative to those of parent spinel, combined with their available oxidation states, can be a useful diagnostic tool for identifying appropriate additives. The reduction potential for magnetite ranges from +0.22 to +0.66 V (ref. 34), while those for Co^{2+}, Ni^{2+} and Zn^{2+} are −0.28, −0.26 and −0.76 V, respectively[35], and in principle Co^{2+} and Ni^{2+} should have similar and limited effect upon Tc retention, while Zn^{2+} should have a more pronounced influence. However, Co^{2+} with a wide range of redox values towards Co^{3+} (refs. 35, 36), it greatly increases the overall reducing capacity of the spinel material. This is reflected in the increased stabilization of the Tc d-states, see Table 1. Both simulations and experiment show that cobalt is by far the most effective additive for Tc retention compared with the undoped magnetite. We postulate that Tc retention, during the glass vitrification, can be controlled by balancing the redox capacity of oxide materials and doping agents. The current study underscores the impact of complex models incorporating both electronic structure and temperature effects that reveal the critical variables needed for predictive materials' design.

Methods

Density functional theory (DFT) parameters. Spin-polarized DFT simulations were performed with periodic boundary conditions (3D PBC) as implemented in the CP2K package[37]. The Perdew, Burke and Ernzerhof (PBE) generalized gradient approximation was used for the exchange-correlation functional[38]. The core electrons were described by the norm-conserving pseudopotentials[39], while the valence wave functions were expanded in terms of double-zeta quality basis sets optimized for condensed systems to minimize linear dependencies and superposition errors[40]. An additional auxiliary plane wave basis set with a 500-Ry cutoff was used to calculate the electrostatic terms. The GGA + U scheme was used to provide more accurate electronic structure for the localized d-orbitals. The Hubbard parameter (U–J) of 3.5 eV was taken for the Fe 3d states, which results in a work function of 5.32 eV, in good agreement with that obtained by Pentcheva *et al.*[28] Owing to large supercell simulations, the Γ-point approximation was used for the Brillouin zone integration.

Computational models. To study Tc incorporation in magnetite with and without dopants, we used a $2 \times 2 \times 2$ supercell in all simulations to minimize periodic images. Optimization of the bulk structure of magnetite had a cell parameter of 8.391 Å, which agrees well with experimental data (8.390 Å (ref. 41)). Using this optimized cell parameter, we constructed a magnetite(001) surface model terminated at an octahedral Fe sublattice, since it is known to be the most stable surface structure in magnetite. A more recent surface model was also considered[42], but was found not to have significant impact on the present problem, see SI. Our model system consisted of a symmetric slab with seven octahedral and six tetrahedral Fe sublattices (384 atoms) with a vacuum region of 12.5 Å between slabs. To study Tc incorporation, one surface octahedral Fe was replaced with Tc, followed by structural optimization. We also optimized a structure with one octahedral Fe in the third layer replaced by Tc. For the doping studies, we substituted a surface Fe atom with Co, Ni or Zn (~1 wt%) at a lattice position close to Tc. In all simulations, we fixed the atomic positions of the four bottom atomic layers.

AIMD simulations. AIMD simulations were performed with and without Tc at 25 °C and with the dopants Co/Ni/Zn at 600 °C, with the Nosé–Hoover thermostat for NVT ensemble and a time step of 1.0 fs. Each simulation was equilibrated for at least 20–28 ps, and the last 10–12 ps of the trajectories was used for the analysis. Owing to the big computational cost of high-temperature simulations, we chose lower range of vitrification temperatures (600 °C), while experiments were performed at somewhat higher temperatures (~700 °C).

Spinel synthesis and XAFS analysis. Ni-, Zn- or Co-doped Tc-incorporated magnetite was synthesized at high pH (> 13). Three solutions of 0.05 M Ni, Zn and Co in distilled deionized water (DDI) were prepared using analytical-grade $NiCl_2$, $ZnCl_2$ and $CoCl_2$. Technetium solution (0.001 M) was prepared by spiking 10,000 p.p.m. $NaTcO_4$ stock solution into 1 M NaOH solution. Synthesized $Fe(OH)_2$ dry powder (0.09 g) was mixed with 5 ml of $NiCl_2$, $ZnCl_2$ or $CoCl_2$ solution in 20-ml poly vials and shaken on an orbital shaker (120 r.p.m.) for 24 h at room temperature (RT). After 24 h shaking, 15 ml of the Tc-spiked 1-M NaOH was added to each vial and heated in an oven at 75 °C for 72 h. After cooling to RT, the precipitates were separated using 0.45-μm filters and washed using ~120 ml DDI

Table 1 | Equilibrium constants and free energy estimates considering doping effects.

Doping	K_{eq}	ΔG (kJ mol^{-1})	Exp. retention (%wt)	ΔE_{gap}(eV)
Tc	0.15	14.2	-	0.15
Tc/Ni	0.56	4.3	4	0.35
Tc/Zn	2.79	−7.7	12	1.03
Tc/Co	21.80	−23.1	29	1.12

Tc implies system without dopant. Relative populations determined as a ratio from the computed K_{eq} values. Experimental values correspond to the amount of Tc remaining in the doped magnetite after exposure to 700 °C. ΔE_{gap} represents the energy difference between the Fermi level E_F and the Tc HOMO energy from the projected DOS.

water immediately after filtering. The collected solid precipitates were air-dried at RT overnight and stored in glass vials. Strong microwave-assisted digestion with a solution consisting of 16 M HNO_3 (17%), 12 M HCl (7%), 32 M HF (3.3%), 0.5 g H_3BO_3 (1.5%) and DDI water (71.2%) on a volume basis was used to determine the total Tc concentration in the final solid samples. For the 600 °C XANES sample, ~5 g of Tc-magnetite was mixed with other basic glass feeds in a Pt crucible and heated in a furnace to 1,000 °C at 5 °C increase per minute. After air quenching, the final glass was pulverized and used for XANES analysis. Additional Tc XAFS samples were also prepared for Ni-, Zn- or Co-doped Tc-incorporated magnetite at room temperature without basic glass feeds and treated at 700 °C inside an oven. The XAFS spectra were collected at room temperature on Beamline 4-1 at the Stanford Synchrotron Radiation Laboratory. A Si(220) double-flat crystal monochromator was used, and the energy was calibrated by using the first inflection point of the Tc K edge spectrum of the Tc(VII) standard ($KTcO_4$) defined as 21.044 keV. The XAFS spectra of Tc standards and Tc-magnetite samples were collected in transmission and fluorescence mode, respectively, at RT using a 13-element germanium detector. Data reduction and analysis were performed using the software IFEFFIT and Athena/Artemis[43] after detector dead-time correction. The XANES spectra for Tc samples were fit using a linear combination of the XANES spectra of $KTcO_4$ as the Tc(VII) standard spectrum and $TcO_2 \bullet 2H_2O$ as the Tc(IV) standard spectrum, see Supplementary Fig. 7 and Supplementary Methods for more details.

References

1. Kotegov, K. V., Pavlov, O. N. & Shvedov, V. P. in *Advances in Inorganic Chemistry and Radiochemistry* (eds Emeléus, H. J. & Sharpe, A. G.) Vol. 11, 1–90 (Academic Press, 1968).
2. Luykx, F. in *Technetium in the Environment* (eds Desmet, G. & Myttenaere, C.) 21–27 (Springer, 1986).
3. Chen, F., Burns, P. C. & Ewing, R. C. Near-field behavior of ^{99}Tc during the oxidative alteration of spent nuclear fuel. *J. Nucl. Mater.* **278,** 225–232 (2000).
4. Childs, B., Poineau, F., Czerwinski, K. & Sattelberger, A. The nature of the volatile technetium species formed during vitrification of borosilicate glass. *J. Radioanal. Nucl.* Ch. **306,** 417–421 (2015).
5. Darab, J. G. & Smith, P. A. Chemistry of technetium and rhenium species during low-level radioactive waste vitrification. *Chem. Mater.* **8,** 1004–1021 (1996).
6. Muller, I. S., McKeown, D. A. & Pegg, I. L. Structural behavior of Tc and I ions in nuclear waste glass. *Proc. Mater. Sci.* **7,** 53–59 (2014).
7. Icenhower, J. P., Qafoku, N. P., Martin, W. & Zachara, J. M. *The Geochemistry of Technetium: a Summary of the Behavior of an Artificial Element in the Natural Environment.* PNNL-18139. (Pacific Northwest National Laboratory, Richland, Washington, USA, 2008).
8. Marshall, T. A. *et al.* Incorporation and retention of 99-Tc(IV) in magnetite under high pH conditions. *Environ. Sci. Technol.* **48,** 11853–11862 (2014).
9. Zachara, J. M. *et al.* Reduction of pertechnetate [Tc(VII)] by aqueous Fe(II) and the nature of solid phase redox products. *Geochim. Cosmochim. Acta* **71,** 2137–2157 (2007).
10. Cui, D. & Eriksen, T. E. Reduction of pertechnetate in solution by heterogeneous electron transfer from Fe(II)-containing geological material. *Environ. Sci. Technol.* **30,** 2263–2269 (1996).
11. Cui, D. & Eriksen, T. E. Reduction of pertechnetate by ferrous iron in solution: influence of sorbed and precipitated Fe(II). *Environ. Sci. Technol.* **30,** 2259–2262 (1996).
12. Fan, D. *et al.* Reductive sequestration of pertechnetate ($^{99}TcO_4^-$) by nano zerovalent iron (nZVI) transformed by abiotic sulfide. *Environ. Sci. Technol.* **47,** 5302–5310 (2013).
13. Farrell, J., Bostick, W. D., Jarabek, R. J. & Fiedor, J. N. Electrosorption and reduction of pertechnetate by anodically polarized magnetite. *Environ. Sci. Technol.* **33,** 1244–1249 (1999).
14. Fredrickson, J. K. *et al.* Reduction of TcO_4^- by sediment-associated biogenic Fe(II). *Geochim. Cosmochim. Acta* **68,** 3171–3187 (2004).
15. Fredrickson, J. K. *et al.* Oxidative dissolution potential of biogenic and abiogenic TcO_2 in subsurface sediments. *Geochim. Cosmochim. Acta* **73,** 2299–2313 (2009).
16. Peretyazhko, T. *et al.* Heterogeneous reduction of Tc(VII) by Fe(II) at the solid–water interface. *Geochim. Cosmochim. Acta* **72,** 1521–1539 (2009).
17. Peretyazhko, T. *et al.* Reduction of Tc(VII) by Fe(II) sorbed on Al (hydr)oxides. *Environ. Sci. Technol.* **42,** 5499–5506 (2008).
18. Burke, I. T. *et al.* Reoxidation behavior of technetium, iron, and sulfur in estuarine sediments. *Environ. Sci. Technol.* **40,** 3529–3535 (2006).
19. Fan, D. *et al.* Oxidative remobilization of technetium sequestered by sulfide-transformed nano zerovalent iron. *Environ. Sci. Technol.* **48,** 7409–7417 (2014).
20. Szecsody, J. E., Jansik, D. P., McKinley, J. P. & Hess, N. J. Influence of alkaline co-contaminants on technetium mobility in vadose zone sediments. *J. Environ. Radioact.* **135,** 147–160 (2014).
21. Um, W. *et al.* Immobilization of 99-technetium (VII) by Fe(II)-goethite and limited reoxidation. *Environ. Sci. Technol.* **45,** 4904–4913 (2011).
22. Kobayashi, T., Scheinost, A. C., Gaona, X. & Altmaier, M. Redox behavior of Tc(VII)/Tc(IV) under various reducing conditions in 0.1 M NaCl solutions. *Radiochim. Acta* **101,** 323–332 (2013).
23. Sidhu, P. S., Gilkes, R. J. & Posner, A. M. Mechanism of the low temperature oxidation of synthetic magnetites. *J. Inorg. Nucl. Chem.* **39,** 1953–1958 (1977).
24. Jolivet, J. P. & Tronc, E. Interfacial electron transfer in colloidal spinel iron oxide. conversion of Fe_3O_4-γFe_2O_3 in aqueous medium. *J. Colloid Interf. Sci.* **125,** 701 (1988).
25. Bliem, R. *et al.* Subsurface cation vacancy stabilization of the magnetite (001). *Surf. Sci.* **346,** 1215–1218 (2014).
26. Chambers, S. A., Thevuthasan, S. & Joyce, S. A. Surface structure of MBE-grown Fe_3O_4(001) by X-ray photoelectron diffraction and scanning tunneling microscopy. *Surf. Sci.* **450,** L273–L279 (2000).
27. Gaines, J. M. *et al.* An STM study of Fe_3O_4(100) grown by molecular beam epitaxy. *Surf. Sci.* **373,** 85–94 (1997).
28. Pentcheva, R. *et al.* Jahn-Teller stabilization of a "polar" metal oxide surface: Fe_3O_4(001). *Phys. Rev. Lett.* **94,** 126101 (2005).
29. Stanka, B., Hebenstreit, W., Diebold, U. & Chambers, S. A. Surface reconstruction of Fe_3O_4(001). *Surf. Sci.* **448,** 49–63 (2000).
30. Tarrach, G., Bürgler, D., Schaub, T., Wiesendanger, R. & Güntherodt, H. J. Atomic surface structure of Fe_3O_4(001) in different preparation stages studied by scanning tunneling microscopy. *Surf. Sci.* **285,** 1–14 (1993).
31. Verwey, E. J. W. Electronic conduction of magnetite (Fe_3O_4) and its transition point at low temperatures. *Nature* **144,** 327–328 (1939).
32. Lukens, W. W., Bucher, J. J., Edelstein, N. M. & Shuh, D. K. Products of pertechnetate radiolysis in highly alkaline solution: structure of $TcO_2 \cdot xH_2O$. *Environ. Sci. Technol.* **36,** 1124–1129 (2002).
33. Allen, P. G. *et al.* Technetium speciation in cement waste forms determined by X-ray absorption fine structure spectroscopy. *Radiochim. Acta* **76,** 77–86 (1997).
34. Pang, S. C., Chin, S. F. & Anderson, M. A. Redox equilibria of iron oxides in aqueous-based magnetite dispersions: Effect of pH and redox potential. *J. Colloid. Interf. Sci.* **311,** 94–101 (2007).
35. CRC. *Handbook of Chemistry and Physics* 75th edn (CRC Press, 1994-1995).
36. Hamdani, M., Singh, R. N. & Chartier, P. Co_3O_4 and Co-based spinel oxides bifunctional oxygen electrodes. *Int. J. Electrochem. Sci.* **5,** 556–577 (2010).
37. VandeVondele, J. *et al.* Quickstep: fast and accurate density functional calculations using a mixed gaussian and plane waves approach. *Comput. Phys. Commun.* **167,** 103–128 (2005).
38. Perdew, J. P., Burke, K. & Ernzerhof, M. Generalized gradient approximation made simple. *Phys. Rev. Lett.* **77,** 3865–3868 (1996).
39. Goedecker, S., Teter, M. & Hutter, J. Separable dual-space Gaussian pseudopotentials. *Phys. Rev. B* **54,** 1703–1710 (1996).
40. VandeVondele, J. & Hutter, J. Gaussian basis sets for accurate calculations on molecular systems in gas and condensed phases. *J. Chem. Phys.* **127,** 114105 (2007).
41. Wright, J. P., Attfield, J. P. & Radaelli, P. G. Charge ordered structure of magnetite Fe_3O_4 below the Verwey transition. *Phys. Rev. B* **66,** 214422 (2002).
42. Bliem, R. *et al.* Adsorption and incorporation of transition metals at the magnetite Fe_3O_4 (001) surface. *Phys. Rev. B* **92,** 075440 (2015).
43. Ravel, B. & Newville, M. ATHENA, ARTEMIS, HEPHAESTUS: data analysis for X-ray absorption spectroscopy using IFEFFIT. *J. Synchrotron Radiat.* **12,** 537–541 (2005).

Acknowledgements

This work was supported by the US Department of Energy, Office of River Protection, Waste Treatment and Immobilization Plant Federal Project and the Office of Basic Energy Science, Division of Chemical Sciences, Geosciences and Biosciences (R.R. and V.-A.G.). PNNL is a multiprogramme national laboratory operated for DOE by Battelle. Computational resources were provided by PNNL's Platform for Institutional Computing (PIC), the W. R. Wiley Environmental Molecular Science Laboratory (EMSL), a national scientific user facility sponsored by the Department of Energy's Office of Biological and Environmental Research located at PNNL and the National Energy Research Scientific Computing Center (NERSC) at Lawrence Berkeley National Laboratory. Part of this work was supported by the US Department of Energy, Office of Science, Basic Energy Sciences, Chemical Sciences, Biosciences, and Geosciences Division, Heavy Element Chemistry Program (W.W.L.) and was performed at Lawrence Berkeley National Laboratory under contract No. DE-AC02-05CH11231. Tc K-edge XAFS spectra were obtained at the Stanford Synchrotron Radiation Lightsource, SLAC National Accelerator Laboratory, which is supported by the U.S. Department of Energy, Office of Science, Office of Basic Energy Sciences under Contract No. DE-AC02-76SF00515. We thank Dong-Sang Kim and Michael J. Schweiger of PNNL for their constructive feedback and in-depth discussions associated with the development of this plan. The high-temperature experiments conducted by Steven A. Luksic and Chuck Z. Soderquist in PNNL are greatly appreciated. Nathan Johnson (PNNL) created the cover and feature images in collaboration with V.-A.G.

Author contributions

All authors provided input for the manuscript. M.-S.L. contributed to the planning and executed the simulations and analysed the data, W.U., G.W. and W.W.L. performed the experiments and related data analysis. R.R. provided pseudopotentials for the calculations and contributed to the analysis of the data. V.-A.G. planned and supervised the research. M.-S.L., R.R. and V.-A.G. jointly wrote the manuscript with input from all authors.

Additional information

Competing financial interests: The authors declare no competing financial interests.

24

Activated desorption at heterogeneous interfaces and long-time kinetics of hydrocarbon recovery from nanoporous media

Thomas Lee[1,2], Lydéric Bocquet[1,2,3] & Benoit Coasne[1,2,4]

Hydrocarbon recovery from unconventional reservoirs (shale gas) is debated due to its environmental impact and uncertainties on its predictability. But a lack of scientific knowledge impedes the proposal of reliable alternatives. The requirement of hydrofracking, fast recovery decay and ultra-low permeability—inherent to their nanoporosity—are specificities of these reservoirs, which challenge existing frameworks. Here we use molecular simulation and statistical models to show that recovery is hampered by interfacial effects at the wet kerogen surface. Recovery is shown to be thermally activated with an energy barrier modelled from the interface wetting properties. We build a statistical model of the recovery kinetics with a two-regime decline that is consistent with published data: a short time decay, consistent with Darcy description, followed by a fast algebraic decay resulting from increasingly unreachable energy barriers. Replacing water by CO_2 or propane eliminates the barriers, therefore raising hopes for clean/efficient recovery.

[1] MultiScale Materials Science for Energy and Environment, Joint CNRS-MIT Laboratory, UMI CNRS 3466, Massachusetts Institute of Technology, Cambridge, Massachusetts 02139, USA. [2] Department of Civil and Environmental Engineering, Massachusetts Institute of Technology, Cambridge, Massachusetts 02139, USA. [3] Laboratoire de Physique Statistique, UMR CNRS 8550, Ecole Normale Supérieure, 75005 Paris, France. [4] Laboratoire Interdisciplinaire de Physique, CNRS and Université Grenoble Alpes, UMR CNRS 5588, 38000 Grenoble, France. Correspondence and requests for materials should be addressed to L.B. (email: lyderic.bocquet@ens.fr) or to B.C. (email: benoit.coasne@ujf-grenoble).

Despite its increasing role in today's energy market, hydrocarbon extraction from gas shale remains poorly understood with many specificities left unexplained. Owing to their ultra-low permeability, typically six orders of magnitude below that of conventional reservoirs, gas and oil recovery from these unconventional reservoirs requires severe stimulation techniques such as hydrofracking. Moreover, different wells display a broad, unexpected variability in hydrocarbon production, which rapidly declines over several months, typically algebraically in time[1–6]. Typically, shale gas reservoirs consist of a collection of kerogen pockets, the host nanoporous organic material containing the hydrocarbons, distributed throughout the mineral shale rock (sketched in Fig. 1)[7]. On hydraulic fracturing, these nanoporous kerogen reservoirs connect to the macroscopic fracture network and release their hydrocarbon content as the pressure in the fracking fluid is decreased. This picture strongly differs from standard oil recovery from conventional reservoirs, which is usually described within the framework of fluid dynamics in porous media, involving a combination of Darcy's law and percolation models accounting for the disordered nature of the fluid pathways through the rocks[8]. These approaches fail to account for the nanoscale porosity of the kerogen pockets, which leads to strong adsorption effects and an unavoidable breakdown of continuum hydrodynamics as the atom granularity of the fluid becomes non-negligible[9,10]. Some corrections have been proposed to account for this breakdown by modifying Darcy's law for slippage through the Klinkenberg effect. While such a formulation accounts for experimental data on gas flow in low-permeability shales[11], the molecular origin of slippage corrections in this context is not evident owing to the strong attractive molecular interactions between methane and kerogen. Beyond such pitfalls, the dispersed texture of kerogen within the mineral matrix raises the question of the unexplored role of interfacial and wettability effects at their boundaries on hydrocarbon desorption and long-time recovery. One may anticipate that this question is also relevant to a much broader range of situations involving interface-dominated multiphase flow across nanoporous materials, as is ubiquitous in catalysis, adsorption, membrane technology and electrochemistry, for example, supercapacitors[12–15].

In this article, we tackle this question by coming back to the microscopic mechanism at stake and climb up the scales from the nanoporous kerogen to the production level. We accordingly address the problem of desorption at wet heterogeneous surfaces and long-time hydrocarbon kinetics at two levels. First, we explore hydrocarbon desorption from a nanoporous membrane mimicking kerogen. Using advanced molecular simulation techniques, we show that, in the presence of the pressure-transmitting (fracking) water, methane desorption is an activated process dominated by interfacial effects, with a wettability-dependent free-energy barrier. In a second step, we demonstrate that such an activated desorption from the nanoporous reservoirs deeply affects the long-time recovery of the hydrocarbons. As a practical implication of the present results, we show that such a multiscale approach involving retarded interfacial transport allows us to explain the unexpectedly fast decline and variable production rates observed in shale gas wells.

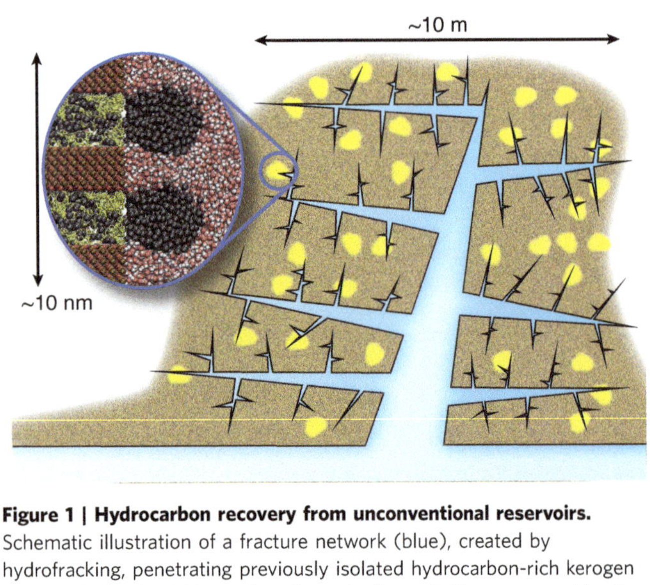

Figure 1 | Hydrocarbon recovery from unconventional reservoirs. Schematic illustration of a fracture network (blue), created by hydrofracking, penetrating previously isolated hydrocarbon-rich kerogen pockets (yellow) within a mineral matrix (brown). Here we consider the post-fracking situation in which water within the hydrofracking network is in contact with the kerogen surface. Extraction of the hydrocarbon requires formation of a nucleus with a high interfacial energy. The zoomed image illustrates such a scenario, in which a methane nucleus (dark grey) forms at a kerogen surface (yellow) adjacent to hydrophilic mineral surfaces present in shales (here quartz, with Si and O atoms as red and golden spheres). Considering other inorganic phases such as clays will lead to the same consistent picture of interfacial activated transport as they have similar wetting properties towards methane and water. However, local variations in surface chemistry and geometry will determine the magnitudes of the energy barriers preventing extraction, which will have a broad range of values due to the heterogeneous, multiscale texture of the shale.

Results

Activated interfacial transport. We have considered several models of kerogen, accounting for its main features, that is, a porous carbon material with nanometric pores[16]: a disordered hydrophobic nanoporous kerogen, an ordered carbon material, as well as a composite system capturing the hydrophobic/hydrophilic interface associated with shale (Fig. 2a and Supplementary Figs 1 and 2). In the following we will focus on the ordered system, consisting of a hydrophobic nanomembrane represented here as an array of carbon nanotubes (CNTs) of radius *r*. The CNTs are arranged in a triangular lattice, with the void between tubes capped at both ends by a graphene sheet. Despite its simplicity, this robust model captures the main physical ingredients at play in hydrocarbon desorption from nanoporous kerogen through its wet external interface towards the fracture network, while allowing for a systematic variation of the geometrical parameters of the porosity. This is key to gaining fundamental understanding of the mechanism at play. Kerogens are hydrophobic materials with oxygen-to-carbon ratio from a few % up to $\sim$10%, therefore making our approximation of a pure carbonaceous phase relevant (molecular simulations have confirmed the hydrophobicity of such carbon-rich phases, including the specific case of the disordered matrix considered in this work[17]). As a result, while the exact chemistry will slightly affect adsorption energies, it will not modify the activated mechanism observed in the present work. As for the nanopore size considered in our work, it is consistent with available experimental data that provide evidence for kerogen's significant nanoporosity. Indeed, while several adsorption-based techniques are available to finely characterize the porosity in kerogens, they all lead to pore-size distributions with significant nanoporosity[16,18,19] (such nanoporosity has been also evidenced from small-angle neutron scattering[16]). We emphasize furthermore that we confirmed that all different models, ordered or disordered, lead to similar conclusions.

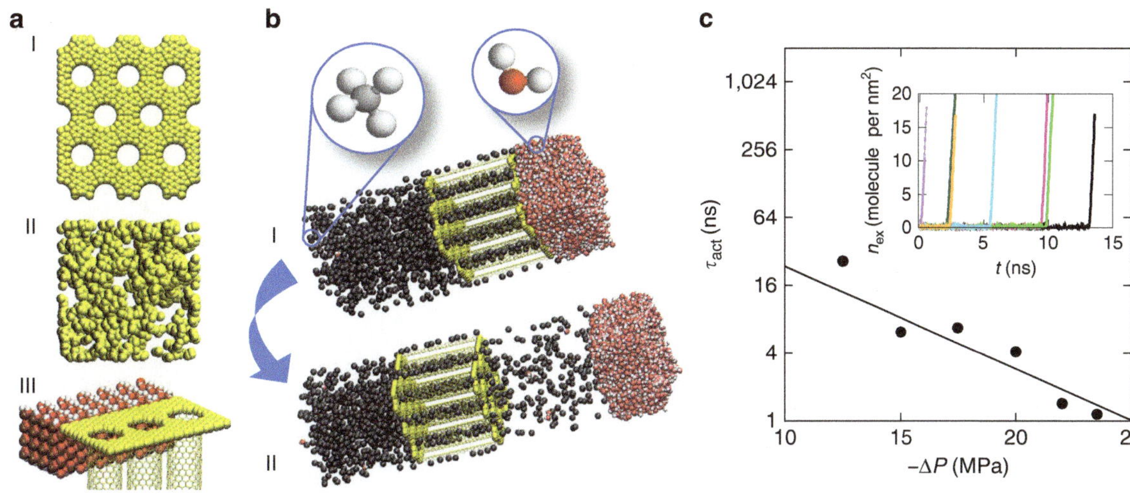

Figure 2 | Interfacial transport in nanoporous media under applied pressure difference. (**a**) Three model kerogens were investigated: (I) an ordered CNT array; (II) a disordered nanoporous carbon material; and (III) a composite membrane containing hydrophilic (quartz) and hydrophobic (CNTs) regions. (**b**) Methane (dark grey) is initially confined within a CNT membrane (yellow) arranged in a triangular lattice. (I) The left side of the membrane is in contact with a reservoir of methane held at constant pressure $P_\uparrow = 25$ MPa through the use of a piston (not drawn). The right side of the membrane is covered by a thick film of liquid water (red and white) with a pressure maintained constant at $P_\downarrow$ through the use of a second solid piston (not drawn). At a time $t = 0$, a pressure difference $\Delta P < 0$ is applied to extract methane by decreasing $P_\downarrow$ (II). Different nanotube radii r and pore spacings D were considered (here $r = 0.59$ nm and $D = 1.70$ nm). (**c**) Average time τ_{act} until an escape event as a function of the applied pressure difference ΔP, with the solid line indicating an exponential fit to the points ($\tau_{act} \sim \exp(a\Delta P)$ with $a = 0.21 \pm 0.04$ MPa^{-1}). The insert shows different equivalent simulations used to estimate τ_{act}; under exactly identical temperature and pressure conditions (here $\Delta P = -15$ MPa and $T = 423$ K) but different (here 7) initial configurations, the amount, n_{ex}, of methane extracted from the membrane per unit of surface area is monitored as a function of time t.

The left side of the membrane is in contact with a reservoir of methane held at constant pressure ($P_\uparrow = 25$ MPa) through the use of a piston. The external surface on the right side of the nanomembrane is covered by a thick film of liquid water, which is left after fracking. The pressure of this fracking water is maintained constant at a pressure $P_\downarrow$ through the use of a second piston, initially set to 25 MPa before decreasing its value to trigger desorption. While further work is needed to fully characterize the distribution of kerogen in gas shales along with its connections with cracks and fractures, we believe that our model provides a simple yet representative picture of kerogen's nanoporosity and its interface with the external surface. In particular, the use of a wet kerogen interface in our model can be justified as follows. First, considering that kerogen is embedded within hydrophilic minerals such as clay, quartz, pyrite and so on, the most stable configuration corresponds to water adsorbed at this interface while the gas/oil remains trapped in kerogen (this is established in the present paper by means of free-energy calculations for such composite systems, which lead to even larger activation energies). Even for pure kerogen interfaces, the free-energy calculations below show that the stable configuration corresponds to water adsorbed at this interface while methane remains trapped in kerogen's nanoporosity through strong adsorption/confinement effects. Second, even if many kerogen pockets are not in contact with water and therefore empty rapidly on pressure drop, the long-time recovery behaviour will be driven by activated interfacial transport of gas at wet kerogen pockets in contact with water located in the fracture network. As discussed at the end of this paper, the fact that activated interfacial transport potentially describes large-scale observations further supports a model of wet kerogen external surfaces.

We first investigated methane desorption in this molecular model using molecular dynamics simulations, as well as free-energy calculations performed using the umbrella-sampling formalism. Details regarding the models and simulations can be found in the Methods section. Methane desorption from the nanoporous membrane depicted in Fig. 2b was investigated under temperature and pressure relevant to shale reservoir conditions ($T = 423$ K and $P \sim 25$ MPa). The inset to Fig. 2c shows the amount n_{ex} of methane extracted from the pores as a function of time t for different, yet equivalent, starting configurations; t is the time after inducing a pressure drop by decreasing the pressure $P_\downarrow$ on the right-hand side of the membrane. Despite the pressure difference $\Delta P = -15$ MPa imposed across the nanoporous medium, methane remains trapped for long times until it gets extracted while water desorbs from the external surface, with considerable variation in the time before the onset of extraction. This is a typical signature of an activated process. As shown in Fig. 2c, the average timescale τ_{act} required to observe methane desorption in the presence of the liquid film at the external surface decreases exponentially with the pressure difference $\Delta P < 0$:

$$\tau_{act} \sim \exp(\upsilon^* \Delta P / k_B T) \quad (1)$$

with υ^* a molecular volume; under the conditions of Fig. 2, $\upsilon^* = 1.2$ nm^3. Such a scaling indicates that fluid desorption through an external surface covered by another (immiscible) fluid is an activated process, possibly inducing important retardation effects in recovery. Counterintuitively, despite such an activated desorption mechanism, fluid extraction occurs at pressure differences ΔP, which are still much lower than the Laplace pressure needed to form an oil (methane) hemispherical bubble at the pore mouth (radius r) into the external water film: $P_L = \gamma_{OW}/r$. For the conditions considered in Fig. 2c, $P_L \sim 100$ MPa; this is well above the observed extraction pressures, in the range of 10–20 MPa. This indicates that extraction is actually promoted by thermal fluctuations, which are relevant here due to the nanoscale dimensions of the porous matrix.

To probe the origin of the energy barrier observed in fluid recovery through a wet external surface, we combined molecular dynamics simulations with free-energy calculations in the framework of the umbrella-sampling technique described in

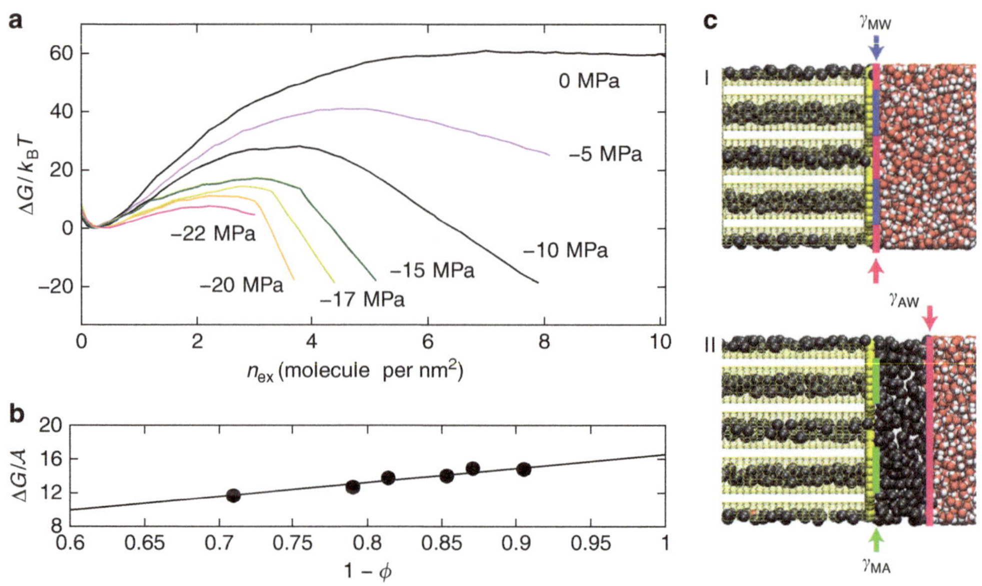

Figure 3 | Activated desorption across wet external surface. (**a**) Free energy $\Delta G/k_BT$ calculated using umbrella sampling as a function of extracted methane per unit area, n_{ex}, for several pressure differences ΔP (indicated in the graph). The free energy is given relative to the local minimum at low n_{ex}. The pore radius is $r=0.59$ nm and pore spacing $D=1.7$ nm. (**b**) Change in free energy per unit area $\Delta G/\mathcal{A}$ (in mJ m^{-2}) between the confined (I) and extracted (II) states when $\Delta P=0$ as a function of the solid fraction of the surface, $1-\phi$. The pore radius r and spacing D of each point are listed in Supplementary Table 3 (systems a and d-h in the table). The straight line indicates a fit to $\Delta G/\mathcal{A}=-\mathcal{S}(1-\phi)$ with $\mathcal{S}=-16.6\pm0.4$ mJ m^{-2}. (**c**) Typical molecular configurations corresponding to the confined (I, low n_{ex}) and extracted (II, high n_{ex}) states. The dark grey spheres are for methane molecules while the red and white spheres correspond to water molecules (the nanoporous membrane is shown in yellow). For each system, we also show the different interfaces: MW; AW; and MA.

detail in the Methods section. Figure 3a shows the free energy $\Delta G/k_BT$ as a function of the amount n_{ex} of extracted methane (per unit surface) for different pressure differences ΔP. When $\Delta P=0$, the stable state corresponds to methane remaining trapped in the porous membrane, with a minute amount of methane solubilized in the adsorbed water film ($n_{ex}^0 \sim 0.2$ molecule per nm^2). In contrast, large n_{ex}, which correspond to situations where methane desorbs from the porous membrane, are not favourable, and the corresponding free energy increases beyond n_{ex}^0 and then plateaus at $n_{ex}=6$ molecules per nm^2. As expected from the data in Fig. 2c, the free energy for $\Delta P<0$ exhibits a maximum, although the extracted state is thermodynamically favourable. At large n_{ex} the free energy decreases linearly with n_{ex}, approximately according to $dG/dn_{ex}=-k_BT\ln(f_\uparrow/f_\downarrow)$, where $f_\downarrow$ and $f_\uparrow$ are the fugacities of methane on the downstream and upstream sides of the membrane, respectively. The activated behaviour observed in Fig. 2c is robust as it is also observed using a more realistic disordered nanoporous membrane, which captures the morphological and topological pore disorder in kerogen (sketched in Fig. 2a (II), results in Supplementary Fig. 1). Such a behaviour was also found for a composite hydrophobic/hydrophilic (carbon/silica) membrane, which corresponds to a simple yet physical description of chemical heterogeneities in gas shales (sketched in Fig. 2a (III), results in Supplementary Fig. 2). In particular, while we found that the free-energy barrier increases when more hydrophilic surfaces are considered, it is drastically decreased on applying a pressure drop ΔP. This implies that activated transport of hydrocarbon across wet external surfaces remains relevant even when more complex models of gas shales are considered.

Free-energy calculations for different nanotube radii r and spacings D demonstrate that the free-energy barrier ΔG^* scales with the fraction of the surface occupied by the external surface area, $1-\phi$, where ϕ is the membrane porosity (Fig. 3b and Supplementary Fig. 3). This result suggests that the free-energy difference corresponds to the interfacial free-energy cost of replacing the membrane–water (MW) interface (state I in Fig. 3c) by membrane–methane (MA) and methane–water (AW) interfaces (state II in Fig. 3c). The corresponding surface contribution to the free-energy barrier is

$$[\Delta G]^*_{surf} = \mathcal{A}\times[-\gamma_{MW}(1-\phi)+\gamma_{MA}(1-\phi)+\gamma_{AW}(1-\phi)] \equiv -\mathcal{S}\mathcal{A}(1-\phi) \quad (2)$$

where $\mathcal{A}$ is the cross-sectional area of the membrane and γ_{ij} is the surface tension of the interface between i and j ($i,j=$ methane (A); membrane (M); or water (W)); and $\mathcal{S}=\gamma_{MW}-\gamma_{MA}-\gamma_{AW}$ is the spreading parameter of a methane bubble formed in water at the wet external membrane surface. The trapped state should be favoured when $\mathcal{S}<0$ (ref. 20). For non-vanishing pressure drops, one expects a supplementary term $v^*\Delta P$ to add to $[\Delta G]^*_{surf}$—with $v^*=l^*\times\mathcal{A}$ a molecular volume corresponding to a wetting molecular film, in line with the previous findings from equation (1). The prediction in equation (2) is found to be in good qualitative and quantitative agreement with the molecular dynamics results in Fig. 3b. Indeed, we performed independent molecular simulations to estimate the surface tensions of the three interfaces using molecular dynamics simulations described in the Methods section. These calculations lead to $\gamma_{MA}\sim16$ mJ m^{-2}, $\gamma_{MW}\sim82$ mJ m^{-2} and $\gamma_{AW}\sim116$ mJ m^{-2}, and therefore a spreading parameter $\mathcal{S}\approx-18$ mJ m^{-2}, which is in good agreement with the value estimated from the linear fit in Fig. 3b, $\mathcal{S}=-16.6$ mJ m^{-2}. The fact that $\mathcal{S}<0$ indicates that the confined state, that is, when methane is trapped in kerogen with a water film adsorbed at kerogen's external surface, is thermodynamically stable.

The linear dependence of ΔG with the lateral area of the membrane surface $\mathcal{A}$ points to the fact that the critical nucleus corresponding to the transition state, as shown in Fig. 3a, extends laterally beyond the maximum lateral size of the simulation box.

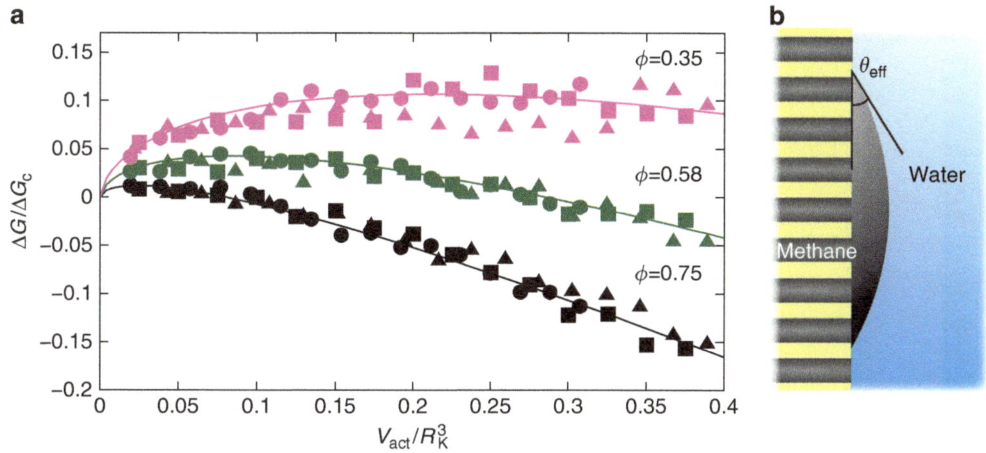

Figure 4 | Surface energy minimization and critical nucleus. (**a**) Free energy ΔG of the methane nucleus as a function of its volume, V_{act}, as calculated by surface energy minimization for different pressure differences (10, 12.5 and 15 MPa for circles, squares and triangles, respectively). Different pore surface fractions ϕ (indicated in the graph) are considered. The free energy is normalized by $\Delta G_c = \gamma_{AW} R_K^2$ and the volume by R_K^3, with the Kelvin radius $R_K = \gamma_{AW}/|\Delta P|$. In this plot, for a given surface geometry—as characterized by the corresponding porosity ϕ—data for various pressures drops collapse onto a single curve. Solid lines show the predicted free energy assuming a spherical cap geometry with an effective contact angle θ_{eff} determined by a fit to the Surface Evolver results. For each geometry, the fitted θ_{eff} is within 2° of the value predicted using the Cassie-Baxter equation for wetting on heterogeneous surfaces (Supplementary Table 1). (**b**) Schematic representation of the geometry of the contact angle formed by a methane spherical cap (grey phase) at the interface between a nanoporous membrane (yellow) and water (blue). θ_{eff} is the contact angle as described in the Cassie-Baxter equation.

Therefore, to apprehend the detailed activation process, we extended our investigation using a mesoscale thermodynamic description on the basis of the ingredients identified in the previous molecular approach. Our model considers the free-energy cost to create a methane bubble on the wet heterogeneous membrane. As in the classical nucleation theory, the nucleus shape is obtained by minimizing the surface energy at fixed volume. To account for the full complexity of the heterogeneous nanoporous surface, we performed calculations using the Surface Evolver programme[21] (details in the Methods section), which we compare with analytical estimates. In these calculations, we used the various surface tensions determined previously using molecular dynamics simulations. Figure 4a shows the free energy ΔG of the methane nucleus as a function of its volume V_{act} under various conditions, in terms of pressure differences ΔP and pore geometry. Here we normalized the free energy and the volume by characteristic quantities $\Delta G_K = \gamma_{AW} \times R_K^2$ and $V_K = R_K^3$, with the Kelvin radius $R_K = \gamma_{AW}|\Delta P|$.

For each volume V_{act} and pressure difference ΔP, the solution of the free-energy minimization corresponds to a nearly spherical methane cap having a contact angle θ_{eff} (Fig. 4b). Interestingly, θ_{eff} is very close to the solution of the Cassie–Baxter equation, which describes the effective contact angle θ_{eff} on the porous surface as a linear combination of the contact angles on the solid ($\theta_{solid} = 32°$) and on the porous domains ($\theta_{pore} = 0°$):

$$\cos\theta_{eff} = \phi + (1-\phi)\cos\theta_{solid}. \quad (3)$$

It is interesting to note that the notion of Cassie–Baxter composite wetting extends here to the description of free-energy barriers and transition states on heterogeneous surfaces.

Using the spherical cap approximation, a straightforward calculation shows that the corresponding free energy of the spherical cap is given by

$$\Delta G[V_{act}] = \psi(\theta_{eff}) \times \gamma_{AW} V_{act}^{2/3} + V_{act}\Delta P \quad (4)$$

where $\psi(\theta_{eff})$ is a geometrical term that depends only on the effective contact angle θ_{eff}, defined by the expression given in the Methods section. In the limit of small contact angle θ_{eff}, one has $\psi(\theta_{eff}) \sim \theta_{eff}^{4/3}$. We show in Fig. 4a that the free energy predicted using the spherical cap approximation, with the values of the effective contact angle θ_{eff} obtained by a fit to the Surface Evolver results using equation (4), are in very good agreement with the complete numerical calculations of the nucleus shape. The fitted contact angles are within 2° of those predicted by the Cassie–Baxter equation (values in Supplementary Table 1), the slight difference resulting from spatial distortions induced by the line tension contribution exerted at the contact line between the three coexisting phases (kerogen, methane and water). The free-energy barrier for methane desorption is then obtained by maximizing $\Delta G[V_{act}]$, leading to $\Delta G^* = \kappa(\theta_{eff}) \times \gamma_{AW} R_K^2$ with $R_K = \gamma_{AW}/|\Delta P|$; the geometrical term, given in the Methods section, takes the form $\kappa(\theta_{eff}) \sim \theta_{eff}^4$ in the limit of small θ_{eff}.

Long-time kinetics of methane recovery. Altogether, the microscopic and mesoscale approaches above point to activated desorption at heterogeneous interfaces, and allow quantitative estimates for the free-energy barrier for hydrocarbon extraction from nanoporous media. These physical ingredients are expected to deeply impact the dynamics at large scales, but they have not been included up to now in the description of hydrocarbon recovery. Several key features emerge from the above description that allow the identification of crucial limiting steps in hydrocarbon extraction. First, the possible range of energy barriers, which depend on the porosity and pressure difference, is found to be of about a few tens of $k_B T$ for standard recovery conditions $\Delta P \sim -15$ MPa. In the framework of the nucleation theory, the activation time is given by an Arrhenius law $\tau_{act} = \tau_0 \exp(\Delta G^*/k_B T)$ with $\tau_0 \sim 10^{-13}$ to 10^{-12} s a typical microscopic attempt time. This leads to timescales τ_{act} of the order of a month to years, which are relevant to the typical production declines observed in shale gas recovery[2]. A second important insight from the approach above is the strong dependence of the energy barrier on the effective contact angle, which scales as θ_{eff}^4 for small θ_{eff}. Returning to the picture of a collection of kerogen pockets dispersed in a mineral matrix (Fig. 1), one expects, therefore, a broad distribution of effective contact angles for the various individual reservoirs, due to wetting and geometrical variability. In turn, this induces an even broader distribution of free-energy barriers ΔG^* because of the scaling relation $\Delta G^* \propto \theta_{\mathrm{eff}}^4 \gamma_{\mathrm{AW}}^3$. Such a variability in the free-energy barriers, which are expected to be drastically affected by the local physical chemistry of the kerogen and the presence of surfactants, leads to activation times τ_{act} that are also widely distributed. Although in a different context, this picture shares ingredients with the long-time kinetics for capillary condensation in granular materials, leading to logarithmic ageing[22]. Typically, for a given time t, only the reservoirs with an activation time τ_{act} smaller than t have desorbed. The recovered amount is calculated in terms of the number of active reservoirs. The overall gas volume $\mathcal{V}(t)$ extracted at a time t is accordingly

$$\mathcal{V}(t) = \mathcal{N}\left\langle \int_0^t \mathrm{d}t_{\mathrm{i}} \upsilon(t - t_{\mathrm{i}}) p_{\mathrm{act}}(t_{\mathrm{i}}) \right\rangle_{\mathrm{pockets}} \tag{5}$$

where $p_{\mathrm{act}}(t_{\mathrm{i}}) = 1 - \exp[-t_{\mathrm{i}}/\tau_{\mathrm{act}}[\Delta G^*]]$ is the probability of a pocket with a particular energy barrier being overcome at a specific time; $\mathcal{N}$ is the total number of gas pockets; and $\upsilon(t)$ is the (time-dependent) volume of extracted gas once a barrier has been overcome. The latter increases from zero to a maximum of, say, V_0 over a time τ_{h}, which is the typical time to empty a single reservoir.

To estimate $\mathcal{V}(t)$, a specific distribution of the effective contact angle and surface tension should be used to estimate the distribution of energy barriers ΔG^*. The crucial point is, however, the broad variability of these parameters among reservoirs (for a given ΔP). To simplify the analysis and obtain analytical predictions, we assume that they follow a simple exponential distribution but the precise form is not critical as discussed below. Using $\Delta G^* = \kappa(\theta_{\mathrm{eff}})\gamma_{\mathrm{AW}}^3/|\Delta P|^2$, the energy barrier distribution is accordingly in the form

$$p_{\Delta G^*}[\Delta G^*] = \frac{\alpha}{k_{\mathrm{B}}T}\exp[-\alpha\Delta G^*/k_{\mathrm{B}}T] \tag{6}$$

with $\alpha = \frac{k_{\mathrm{B}}T}{\kappa(\theta_0)\gamma_0^3}|\Delta P|^2$; θ_0 and γ_0 fix the typical range spanned by these parameters over the reservoirs. While such an exponential distribution represents merely one possible form to estimate $\mathcal{V}(t)$, the specific distribution considered for ΔG^* is not crucial; the key ingredient in our model to predict the short-time and long-time algebraic decays is the existence of an energy barrier, taking values over a broad interval. Such an energy barrier introduces a typical activation time τ_{act}, which defines a short-time $t < \tau_{\mathrm{act}}$ and a long-time $t > \tau_{\mathrm{act}}$ regimes.

We solve this problem for two regimes that depend on the time τ_{h} to empty a single reservoir (details of the steps involved are provided in the Methods section). For $t < \tau_{\mathrm{h}}$, we must consider the dynamical process during this emptying, and one expects $\upsilon(t) \propto \sqrt{t}$ as predicted from a classical boundary-limited flow applied to a single pocket using Darcy transport leading to diffusion-like equation (5). We predict that the rate of recovery will then scale as $Q(t) \propto 1/t^{\frac{1}{2}+\alpha}$. In the long-time regime $t > \tau_{\mathrm{h}}$, the finite emptying time can be neglected and the need to overcome ever-larger energy barriers then limits the rate of recovery. Accordingly, we predict

$$Q(t) \propto \frac{1}{t^{1+\alpha}} \tag{7}$$

where $\alpha = \frac{k_{\mathrm{B}}T}{\kappa(\theta_0)\gamma_0^3}|\Delta P|^2$ is a non-universal exponent, which strongly depends on thermodynamic conditions (T, ΔP) but also on the specific interactions with the gas shale components through upper limits θ_0 and γ_0 on the local effective contact angle and surface tension. Beyond the typical time τ_{h} needed to empty a single reservoir, the number of active reservoirs decays rapidly as the energy barrier that must be overcome to activate them becomes unreachable. The statistical model above shows that the activated kinetics of hydrocarbon recovery departs from the overall classical boundary-limited flow, which predicts $Q(t) \sim t^{-1/2}$ for a single reservoir. As shown by equation (7), the recovery is predicted to exhibit a faster decline for long times, with an exponent of the algebraic decline of the order of unity, although dependent on the pressure protocol used to trigger recovery and on the local characteristics of the well under investigation.

Discussion

The statistical model developed in our paper predicts a two-regime scenario for the production decline with strong dependence on the fracking fluid through its wetting properties and miscibility in hydrocarbon. Rigorous validation of this critical prediction against field-scale data requires more work, including experimental investigation on simple systems before moving to gas shale. Moreover, shale production data display a wide variety of length and timescales associated with complex phenomena (geomechanics, transport and so on), which makes direct comparison with our model premature unless intermediate validation steps are added. Nevertheless, at this stage, it is important to check that our prediction, that is, activated transport across wet kerogen interfaces, is compatible with real data. The Supplementary Discussion section contains a discussion of a large collection of field scale data on gas production over time for some typical examples of unconventional wells from different shale plays. The statistical model presented in our paper is consistent with the general experimental behaviour; the two-step algebraic decline predicted by our statistical model, with a more rapid decline at long times than at short times, is compatible with short- and long-time extraction rates previously identified in field scale data. Moreover, by including ingredients such as interfacial and physical chemistry effects, such a multiscale model also accounts qualitatively for the effect of changing the fracking fluids (non-water fracking fluids tend to have smaller decline exponents).

The field-scale data gathered in Supplementary Fig. 4 also indicate that in general the presence of two regimes is clearer in hydrofracked wells compared with those stimulated with other fracking fluids containing little or no water. Such a dependence on the fracking fluid is consistent with the activated recovery found in the present work; indeed, while hydrofracking requires the system to overcome energy barriers to initiate hydrocarbon extraction, such a behaviour is not expected for fluids that are miscible with the hydrocarbon fluid such as liquid petroleum gas and CO_2 in specific temperature and pressure ranges. To validate this conjecture, we carried out a series of additional simulations in which water was replaced with CO_2. After an equilibration stage, during which an additional force field prevents methane from leaving the pores and CO_2 from entering, a pressure gradient is applied and the system is monitored over time. Multiple repeats were performed using systems with equivalent initial states to test for the presence of an energy barrier. These results, plotted in Fig. 5, show that unlike the water simulations (Fig. 2c), no retardation in the transport of methane out of the membrane was observed, therefore suggesting that no energy barrier exists to inhibit extraction in this case. Furthermore, CO_2 reliably replaces methane within the pores, as shown by the blue markers in Fig. 5. This presents a win-win strategy in which CO_2 as a fracking fluid reduces the environmental impact of the process while allowing efficient CO_2 capture within the shale reservoir at the end of the process. While CO_2 is already used for conventional reservoirs in the framework of enhanced oil recovery, limitations of CO_2 as a fracking fluid have been identified such as its low viscosity, high

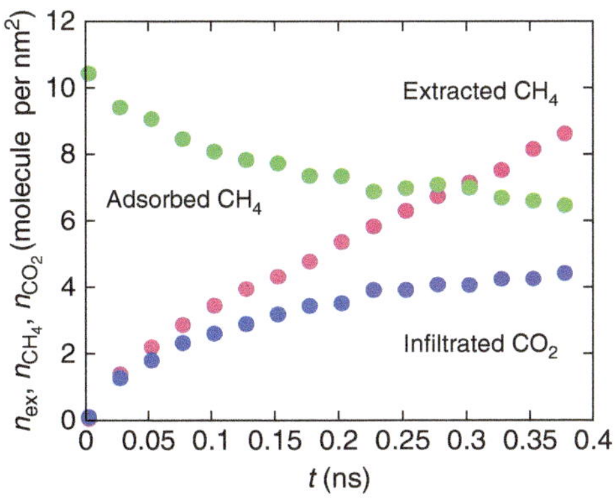

Figure 5 | CO_2 as an alternative fracking phase. The number of methane molecules extracted (magenta), adsorbed in the pores (green), and the number of fracking phase CO_2 molecules injected into the nanoporous membrane (blue) over 0.2 ns of unbiased molecular dynamics simulation. The pressure difference is $\Delta P = -20$ MPa. The symbols correspond to the average values over seven equivalent simulations, that is, 'repeats', with different initial conditions. When CO_2 is used as the fracking fluid, methane immediately desorbs from the pores with no evidence of an energy barrier limiting its recovery, in contrast to Fig. 2c.

compressibility and poor proppant carrier properties. However, the role of CO_2 proposed here, as an alternative to hydrofracking, is fundamentally different[23]; CO_2-fracking eliminates activated interfacial transport at the external surface of kerogen. By replacing water with propane, we also observed that this alternative fracking fluid leads to non-activated interfacial transport owing to its favourable interactions with the confined hydrocarbon (results not shown). However, while CO_2 replaces methane in kerogen, propane was found to be recovered together with the hydrocarbon phase on extraction. This suggests that different recovery strategies, that is, allowing CO_2 capture or efficient energy extraction without hydrocarbon loss, can be envisaged by playing with the different surface interactions at play through the choice of the fracking fluid. Despite the benefits of using fracking fluids that eliminate activated transport on shale gas extraction, further investigation is required to include possible swelling effects as CO_2-fracking, for instance, is known to swell kerogen and reduce shale permeability.

Our novel framework emphasizes that new paradigms must be envisioned to understand hydrocarbon extraction from unconventional reservoirs. These new insights into transport at the nanoscale suggest new leads for the industry and pave the way for the rational adjustment or re-design of existing processes to minimize the retardation due to these interfacial effects. In shale gas extraction, control can be obtained over the surface tensions, and therefore the energy barriers, by altering the composition of the pressure-transmitting fracking fluids.

Beyond shale gas, we expect that such activated desorption phenomena will be of paramount importance for any field involving nanoporous media in which nanoscale fluid interfaces are present. While structural defects at the external surface of nanoporous materials have been identified as limiting steps in transport and reactivity in confined geometries[15,24–26], retardation effects arising from the extraction of a liquid phase into an immiscible liquid wetting the external surface are unprecedented. Such effects are related to, but distinct from, fluid–fluid and fluid–solid interfaces, which are known to resist or drive transport in nanopores via the Laplace pressure[27,28]. Manipulation of interfacial parameters in such systems allow envisioning rational control over transport inhibition in a variety of contexts such as membranes, catalysis and chromatography.

Methods

Simulation model. We used the open-source LAMMPS software to carry out molecular dynamics simulations, employing a velocity-Verlet algorithm with a timestep of 2 fs. Figure 2b illustrates the geometry of the simulations containing a porous membrane. We define the coordinate system such that the axis of the pores was aligned with the z axis, with the methane reservoir at the more negative end. The membrane pores are composed of CNTs with a zig-zag configuration, arranged into a triangular lattice with the rows of the lattice aligned in the x direction. Graphene sheets at the ends of the pores block the voids between the CNTs, with the centre of each nanotube aligned with the centre of an graphene ring. The nanotubes ends are cleared by removing atoms in the graphene sheet located within $r + 1.42$ Å of the CNT centre, the extra 1.42 Å equal to the carbon–carbon bond length, ensuring a reasonable spacing between carbon atoms. We apply periodic boundary conditions in the x and y directions, and 'shrink-wrapped' boundary conditions in the non-periodic z direction, allowing the system to expand and contract as necessary. The initial state is created by running the simulation at 25 MPa for 1 ns while preventing the movement of methane out or water into the pores using an additional repulsive force field at the pore opening.

We use a Lennard–Jones potential to model the non-electrostatic forces between all particles, with the form

$$U_{ij}(r_{ij}) = 4\varepsilon_{ij}\left(\left(\frac{\sigma_{ij}}{r_{ij}}\right)^{12} - \left(\frac{\sigma_{ij}}{r_{ij}}\right)^{6}\right); r_{ij} < r_c \qquad (8)$$

where r_{ij} is the separation between two particles and $r_c = 13.5$ Å is the force cutoff distance for interactions not involving piston atoms (which we describe later). The values of the σ_{ij} and ε_{ij} parameters for different types of interacting particles are summarized for particles of the same type in Supplementary Table 2. For interactions between unlike particles, we use Lorentz–Berthelot combining rules, such that for particles of type α and β, $\varepsilon_{ij} = \sqrt{\varepsilon_\alpha \varepsilon_\beta}$ and $\sigma_{ij} = \frac{1}{2}(\sigma_\alpha + \sigma_\beta)$. The exception to these combining rules is the methane–piston interaction, which we describe below.

We use a united atom description of methane with the TraPPE force field[29]. Each methane molecule is represented by a single Lennard–Jones particle, with interaction parameters as listed in Supplementary Table 2. This model accurately describes the liquid–vapour coexistence curve and critical temperature of methane. We include at least 40 methane molecules per square nanometre of membrane cross-sectional area.

We employ the simple point charge (SPC) water model[30], with the charges and Lennard–Jones parameters given in Supplementary Table 2. The bond angle is fixed at 109.47° and the hydrogen–oxygen bond length at 1 Å using the SHAKE algorithm[31]. For the electrostatic interactions, we truncate the force at 9 Å. Owing to the lack of periodicity in the z dimension, Ewald sum methods cannot be used for the long-range electrostatic forces. While the use of a cutoff affects the value of the surface tensions, this simplified water molecule still served the purpose of providing a polar fluid, which is immiscible with methane. We included at least 100 water molecules per square nanometre of membrane lateral area.

For simulations involving CO_2, we use the 'elementary physical mode' described by Harris and Yung[32], with interatomic interaction parameters as described in Supplementary Table 2. The carbon–oxygen bond length is fixed at 1.149 Å using the SHAKE algorithm, while constraining the bond angle θ by a harmonic potential $\frac{1}{2}k_\theta(\theta - \pi)$ with $k_\theta = 1{,}236\,\mathrm{kJ\,mol^{-1}\,rad^{-2}}$.

A Langevin thermostat acting on the methane and oxygen atoms with a friction coefficient of 0.01 $\mathrm{fs^{-1}}$ holds the system temperature at 423 K. The thermostat acts only in the x–y plane so as not to interfere with the transport of the fluid along the axis of the pores.

Opposing pistons with a graphene-like structure apply a prescribed pressure to each side of the membrane. Piston atoms are constrained such that they only move along the z axis. At each time step, the force on each piston atom is set to the average force of all atoms in the piston—causing them to move in unison—plus an additional component corresponding to the external pressure on that piston. We define the interaction between piston and fluid atoms by a Lennard–Jones potential truncated at the minimum energy ($r_c = 2^{1/6}\sigma_{ij}$), and therefore purely repulsive. To prevent methane accumulation at the downstream piston (in contact with the water phase), we use a large σ_{ij} value of 6.8 Å for methane–piston interactions.

Activation time determination. In the simulations used to generate the data shown in Fig. 2c we used a membrane with a pore radius $r = 0.59$ nm and pore spacing $D = 1.70$ nm, and simulation dimensions in the periodic dimensions x and y of 1.704 and 2.951 nm, respectively, such that the system contained two nanotubes. After the initialization process described above, we equilibrated the system with no pressure difference for 2 ns, then linearly decreased the pressure on the left (positive z) side of the membrane containing the water over 10 ps. The simulation was run until escape was observed (and for a short time after). We report results for six pressure differences, for each of which we simulated seven trials with different but equivalent initial conditions. The activation times for the individual trials are shown in Supplementary Fig. 5.

Umbrella-sampling molecular dynamics simulations. We used umbrella sampling to measure the free energy as a function of the amount of extracted methane and gain insight into the origin of the energy barrier. Umbrella sampling is a method commonly used to study the thermodynamics of rare events inhibited by energy barriers[33]. An order parameter must first be identified describing the process of interest. We used the number of extracted methane molecules per unit area as an order parameter, n_{ex}, which we have approximated using a sigmoidal function to make the variable continuous and differentiable:

$$n_{ex} = \sum_{i=1}^{N_{CH_4}} \left(1 + e^{-(z_i - z_p)/\lambda}\right)^{-1} \quad (9)$$

where z_i is the z coordinate of the ith methane particle, $z_p = 1.8$ Å relative to the centres of the carbon atoms in the external graphene surface and $\lambda = 0.25$ Å. We associate a biasing potential $\omega(n_{ex})$ with the order parameter to force the system into an ordinarily unlikely state. By measuring the probability distribution of n_{ex} in the biased simulation, $P^B(n_{ex})$, we deduce the unbiased free energy from the relationship

$$\Delta G(n_{ex}) = k_B T \ln P^B(n_{ex}) - \omega(n_{ex}) \quad (10)$$

In practice, rather than attempt to find a bias potential, which allows the entire domain of n_{ex} to be sampled, it is more practical to run many simulation 'windows'. In each window we applied a simple harmonic bias potential so as to sample a particular range of n_{ex}, with the form

$$\omega_i(n_{ex}) = \frac{1}{2} K (n_{ex} - n_i)^2 \quad (11)$$

where K and n_i determine the strength and centre of the bias for the ith window. To implement such a bias in a molecular dynamics simulation, the force on each methane particle resulting from the bias must be described as a function of the particle position along the z axis:

$$F_{bias}(z) = -\frac{d\omega(z)}{dz} \quad (12)$$

$$= -\frac{dn_{ex}}{dz}\frac{d\omega(z)}{dn_{ex}} \quad (13)$$

$$= -K\frac{dn_{ex}}{dz}(n_{ex} - n_i) \quad (14)$$

The need for n_{ex} to be differentiable with respect to the particle position z motivates the use of the logistic form in equation (9), for which the derivative is well known and straightforward to implement:

$$\frac{dn_{ex}}{dz} = \frac{e^{(z - z_p)/\lambda}}{\lambda\left(1 + e^{(z - z_p)/\lambda}\right)^2} \quad (15)$$

To calculate the full unbiased probability distribution P^U, we used a weighted average of the unbiased probabilities in each window, P_i^U, according to the weighted histogram analysis method[33]. The weightings were calculated so as to minimize the statistical error in P^U:

$$P^U(n_{ex}) = \sum_{i}^{\text{windows}} N_i e^{-(\omega_i(n_{ex}) - F_i)/k_B T} P_i^U(n_{ex}) \quad (16)$$

where P_i^U and N_i are the measured unbiased probability and number of samples, respectively, in the ith window. The F_i terms were calculated according to

$$e^{-F_i/k_B T} = \int dn_{ex} P^U(n_{ex}) e^{-\omega_i(n_{ex})/k_B T} \quad (17)$$

We found a self-consistent solution by starting from equation (16) with $F_i = 0$ for all windows, then iterating between equation (17) and (16) until convergence[33]. We considered the solution to have converged when the maximum value of $(1 - F_{old}/F_i)^2$ for any window was $< 10^{-15}$, where F_{old} is the value of F_i in the previous iteration.

After unblocking the pores and applying the bias, we equilibrated the initial simulation window at $n_i = 0$ for at least 1 ns. To measure $P^B(n_{ex})$, we sampled the system every 0.2 ps for at least 1 ns. We generated additional windows by restarting a simulation equilibrated with a similar n_{ex}, equilibrating for at least 0.2 ns. We typically simulated windows separated in n_i by 0.4 molecule per nm^2, though when close to a free-energy maximum we sometimes required additional windows separated by 0.1 molecule per nm^2, and with larger spring constant. We list the membrane parameters for the simulations used to produce the results in Fig. 3a in Supplementary Table 3 (system a).

Disordered membrane. To test the generality of the effects observed for the model CNT membrane, we carried out umbrella-sampling simulations in a system with a 5 × 5 × 5 nm disordered porous carbon membrane obtained using an atom-scale reconstruction technique. The pore size, chemical composition (including sp^2/sp^3 hybridization), and morphological disorder of this structure are comparable with those of kerogen[34–37].

Heterogeneous membrane. In complex porous materials such as gas shale, a wide variety of geometries and surface chemistries may be present. We have explored the impact of such heterogeneities using a model composite membrane consisting of nanoporous hydrophobic component of width 3.12 nm, similar to the ordered hydrophobic membrane already described, and a hydrophilic component represented by a quartz surface of width 2.16 nm. This membrane is pictured in Fig. 2a(III) and the upper-left inset in Supplementary Fig. 2. The length of the simulation domain in the direction parallel to the stripes was 5.106 nm. The quartz surface was prepared by cutting a bulk quartz lattice[38] along the (100) plane, then attaching protons to the resulting dangling oxygen bonds and allowing them to relax over a short NVE simulation. The quartz atoms are frozen during the simulations. The charge and Lennard–Jones parameters for the silicon, oxygen and hydrogen from which the quartz is composed are listed in Supplementary Table 4. The water was completely wetting on the quartz, while showing a contact angle of 133° on the graphene component (see Supplementary Fig. 6 and below in the Methods section), higher than typically observed for graphene because only a single layer was used. The free energy as a function of extracted methane for $\Delta P = 0$ and two additional pressure differences as measured using umbrella sampling is shown in Supplementary Fig. 2.

Free-energy barrier dependence on porosity. To test the linear scaling with $1 - \phi$, we used umbrella sampling to determine the free energy as a function of n_{ex} for different combinations of pore diameter and spacing listed in Supplementary Table 3 (systems a, d–h). To calculate the porosity, a correction term c was added to the pore radius to correct for the finite size of the carbon atoms, such that $\phi = \frac{2\pi}{\sqrt{3}}\frac{(r-c)^2}{D^2}$. The results confirm the linear relationship in equation (2), as seen in Fig. 3, with the best fit found for $S = 16.55 \pm 0.41$ mJ m^{-2} and $c = 0.10 \pm 0.03$ nm.

Surface tension calculation. To verify the proportionality with the spreading parameter in equation (2), we measured the surface tensions of the three interfaces of interest in a separate set of molecular dynamics simulations. We prepared three systems: one in which a water phase and methane phase are held in contact under pressure from two graphene pistons, much like the simulations already described, but without any membrane separating the phases; another with only a water phase between the pistons; and another with only methane. The systems are 5-nm wide in the x and y directions, and contain 7,200 water molecules and/or 910 methane molecules. The potentials used to model interactions between the piston atoms and other species are identical to those used for the membrane atoms in umbrella sampling simulations, listed in Supplementary Table 2, with the force truncated at 13.5 Å. We equilibrated the systems for 2 ns before beginning the surface tension measurement. In the methane–water mixed system we initially equilibrated for an additional 1 ns with a planar force separating the phases. We simulated the methane–water systems for 15 ns, water-only for 18.4 ns and methane-only for 10 ns, taking samples of the configuration every 1 ps.

We can express the surface tension between phases A and B using the Kirkwood and Buff approach[39], as an integral across the interface of the difference between the pressure components normal and tangential to that interface, $p_N(z)$ and $p_T(z)$:

$$\gamma_{AB} = \int_{z_A}^{z_B} (p_N(z) - p_T(z))dz \quad (18)$$

where the z axis is normal to the plane containing the interfaces, and z_A and z_B correspond to points within the bulk of the A and B phases. The normal and tangential pressures can be written in terms of the pressure tensor elements, $p_N = p_{zz}$ and $p_T = \frac{1}{2}(p_{xx} + p_{yy})$.

The pressure tensor components themselves contain a kinetic component, equivalent to that of an ideal gas, plus a second component taking into account the interactions between molecules. The diagonal elements of the pressure tensor at a position z, $p_{\alpha\alpha}(z)$, where $\alpha = x$, y or z, can be calculated using the expression

$$p_{\alpha\alpha}(z) = k_B T \rho(z) - \frac{1}{A}\left\langle \sum_{i=1}^{N-1}\sum_{j=i+1}^{N}\sum_{a=1}^{n_i}\sum_{b=1}^{n_j} \frac{(\alpha_i - \alpha_j)(\alpha_{ia} - \alpha_{jb})}{r_{iajb}}\frac{dU(r_{iajb})}{dr_{iajb}}\xi(z, z_i, z_j)\right\rangle \quad (19)$$

where the sums are over each interatomic interaction between each pair of molecules, each molecule being composed of n_i atoms. α_i and α_{ia} denote the α coordinate of the centre of mass of the ith molecule, and the ath atom of the ith molecule, respectively. r_{iajb} is the separation between atoms ia and jb, $\rho(z)$ is the fluid density and A is the simulation cross-sectional area in the x–y plane. We count the frozen graphene atoms as individual molecules when calculating p_{zz}, but we neglect them in the calculation of p_{xx} and p_{yy} due to the symmetry in these dimensions.

The function $\xi(z, z_i, z_j)$ depends on the choice of the contour drawn between particles i and j along which the contribution of that interaction is distributed. There is no unique definition for this contour[40]. We use the Irving and Kirkwood convention[41], which has been shown to agree with other methods[42]. The contour in this model is a straight line connecting the two interacting molecules, such that $\xi(z, z_i, z_j)$ takes the form:

$$\xi(z, z_i, z_j) = \frac{1}{|z_j - z_i|} H\left(\frac{z - z_i}{z_j - z_i}\right) H\left(\frac{z_j - z}{z_j - z_i}\right) \quad (20)$$

where $H(x)$ is the Heavyside step function. In practice, we divide the system along the z axis into slabs of width 1 Å, and split the contribution to the pressure tensor of each interaction equally between any slabs between or containing the two interacting particles.

At equilibrium, the normal component of the pressure tensor must in principle be constant and equal to the pressure applied by the pistons. In practice the length of time for which the system must be sampled before this limit is reached for the water phase is very long. We have made the assumption that the normal pressure would eventually converge in our calculation of the surface tension.

The tangential and normal pressure profiles calculated using this method are shown in Supplementary Fig. 7. The measurements of γ_{MA} and γ_{MW} in the mixed system are consistent with the more precise measurements in the single-component systems.

Surface energy minimization with Surface Evolver. To overcome the limitation imposed by the small molecular dynamics simulation box size and construct a model for the critical nucleus, we have employed a mesoscale thermodynamic approach. We use the energy minimization algorithm implemented in the open-source Surface Evolver programme[21] to determine the nucleus geometry at a range of volume.

Surface Evolver represents surfaces by mesh of triangular facets. In a standard minimization step (default 'g' command), the force on each vertex is calculated based on the gradient of the free energy, and then the vertices move according to the strength and direction of that force. Alternatively, the 'Hessian seek' command allows the Hessian matrix of second derivatives to be used to find the minimum energy configuration in the direction of motion as determined by the forces. These two techniques can be alternated while minimizing the energy. Further details of the minimization algorithm can be found in the Surface Evolver manual.

In the absence of a pressure difference, the total energy of the system can be described by

$$\Delta G = \mathcal{A}_{AW}\gamma_{AW} + \int d\mathcal{A}_{base}\gamma_{base}(x, y) \tag{21}$$

where $\gamma_{base}(x,y)$ is a periodic function of circular domains representing the pores, arranged in a triangular lattice. The function is equal to $-\gamma_{AW}$ within these circular domains, and $\gamma_{MA} - \gamma_{MW}$ everywhere else.

The surface tensions calculated using the molecular dynamics simulations described above were used for the minimization calculations.

During the evolution of a surface in Surface Evolver it is usual to start with a crude approximation to the final geometry, then alternating between moving towards the minimum energy, and refining the surface by splitting existing facet in half. We restricted the refinement of the base of the nucleus such that new vertices are only created on the contact line, since additional vertices within the base perimeter are redundant, experiencing no net force.

The initial geometry consists of 50 vertices arranged on the substrate in a circle to form the contact line. Each of these contact line vertices are connected by an edge to their two neighbours, to a vertex in the centre of the circle, and to a vertex positioned above the centre of the circle at the height required for the target nucleus volume. Vertices located on the contact line are constrained such that they only move in the plane of the surface. Four initial contact radii were tested for each volume, and all but the lowest energy result discarded.

Surface Evolver minimizations of drop or bubble geometries on patterned surface can be challenging due to a tendency for the system to become stuck in local minima. If care is not taken a situation can arise in which even the liquid–vapour interface obviously fails to adopt a physically reasonable shape. We have found an effective strategy to avoid such problems is to alternately evolve the surface with the contact line vertices fixed in place or allowed to move freely. We used a combination of the regular linear gradient decent method (the default 'g' command) and the Hessian seek method ('Hessian_seek' command), described above. Motion of the vertices in both cases is multiplied by a 'scale factor' (<1), automatically calculated by the software, which dampens the motion as an energy minimum is approached. If the scale factor approaches zero, the surface stops evolving, while not necessarily having found the minimum energy configuration.

The algorithm described in Supplementary Note 1 was used within Surface Evolver to minimize the interfacial free energy.

Cassie-Baxter model of the nucleus. Here we describe the derivation of the thermodynamic model to predict the size and energy of the critical nucleus, which we compare with the Surface Evolver results in Fig. 4.

Consider a methane nucleus with volume V_{act} on a porous surface. We make the simplifying assumption that the nucleus adopts an idealized spherical cap geometry with base radius R and an effective contact angle θ_{eff}. The base radius R, contact angle, volume and curved surface area $\mathcal{A}_{AW}$ (methane–water interface) are related by

$$V_{act} = \pi\beta_v(\theta_{eff})R^3 \tag{22}$$

$$\mathcal{A}_{AW} = \pi\beta_a(\theta_{eff})R^2 \tag{23}$$

where $\beta_v(\theta_{eff}) = (2 - 3cos\,\theta_{eff} + cos^3\theta_{eff})/(3sin^3\theta_{eff})$ and $\beta_a(\theta_{eff}) = 2(1 - cos\,\theta_{eff})/sin^2\theta_{eff}$. In the limit of low θ_{eff}, $\beta_v(\theta_{eff}) = \theta_{eff}/4$ and $\beta_a = 1$.

Assuming the length scale of the pattern is significantly smaller than the radius of the nucleus, the total free energy can be written

$$\begin{aligned}\Delta G = \; & \gamma_{AW}\mathcal{A}_{AW} \\ & - \phi\gamma_{AW}\mathcal{A}_{base}(V_{act}, \theta_{eff}) \\ & + (1-\phi)(\gamma_{MA} - \gamma_{MW})\mathcal{A}_{base}(V_{act}, \theta_{eff}) \\ & + V_{act}\Delta P\end{aligned} \tag{24}$$

In the case of the simple geometry used in the molecular dynamics and surface evolver calculations, $\phi = \frac{2\pi r^2}{\sqrt{3}D^2}$. The free energy can be rewritten in terms of the contact angle of the methane on the solid phase:

$$\begin{aligned}\Delta G = \; & \gamma_{AW}\mathcal{A}_{AW} \\ & - \gamma_{AW}\mathcal{A}_{base}(\phi + (1-\phi)cos\,\theta_{solid}) \\ & + V_{act}\Delta P\end{aligned} \tag{25}$$

into which we may substitute the expressions for the area, volume and the Cassie–Baxter effective contact angle, to produce an expression for the free energy in terms of γ_{AW}, V_{act}, θ_{eff} and ΔP:

$$\Delta G(V_{act}) = \psi(\theta_{eff})\gamma_{AW}V_{act}^{2/3} + V_{act}\Delta P \tag{26}$$

where $\psi(\theta_{eff}) = \pi^{1/3}\beta_v^{-2/3}(\beta_a - \cos\theta_{eff})$. In the limit of low θ_{eff}, $\psi(\theta_{eff}) = (2\pi)^{1/3}\theta_{eff}^{4/3}$. It is straightforward then to determine the maximum in the free energy:

$$\Delta G^* = \kappa(\theta_{eff})\gamma_{AW}R_K^2 \tag{27}$$

where we introduce the Kelvin radius $R_K = \gamma_{AW}/|\Delta P|$ and a geometric factor $\kappa(\theta_{eff}) = \frac{4}{27}\psi(\theta_{eff})^3$. In the small angle limit $\kappa(\theta_{eff}) = \frac{8}{27}\pi\theta_{eff}^4$.

Measuring hydrophobicity of the membrane. To confirm that the model membrane is hydrophobic, we carried out a molecular dynamics simulation of a water drop on a graphene substrate. We simulated 700 water molecules represented by the SPC model described above section in an initially cubic lattice. During the initial 0.5 ns, a velocity rescaling thermostat ramped the temperature linearly from 400 to 300 K—a process we find allows the drop to relax more rapidly. The drop momentum in the xy plane was reset to zero every 20 ps during this first 0.5 ns. After the initial relaxation, the temperature was held at 300 K using Langevin thermostat. After equilibrating for a further 3.5 ns the system was sampled every 1 fs over a 2-ns period so as to measure the contact angle.

The radial density profile of the drop, measured relative to the centre of mass, is shown in Supplementary Fig. 6. To obtain the position of the surface as a function of z and $r_s(z)$, the system is divided into slabs parallel to the substrate and 0.5 Å thick in z. In each slab we fit the radially averaged density to the function $\rho(r) = \frac{\rho_L}{2}\tanh(2(r - r_s(z))/w)$ via r_s, the position of the interface; and ρ_L, the bulk density, fixing $w = 0.1$ nm. To measure the contact angle, a circle is fitted to $r_s(z)$ in the central region of the drop, avoiding deviations due to contact line tension near the base and low density regions near the top.

Statistical model of long-time recovery. Here we describe with additional detail the derivation of the statistical model for long-time recovery kinetics. We begin by considering the shale as containing a large number of trapped gas pockets, $\mathcal{N}$, each containing a volume of gas V_0, and each of which must overcome some energy barrier ΔG^* before the gas within may be recovered. This scenario is illustrated schematically in Fig. 1. These pockets have a wide distribution of energy barriers as a result of variations in the local geometry and surface chemistry of the shale. The variation in energy barriers leads in turn to variations in activation time τ_{act}, which could potentially span many orders of magnitude.

The probability of a pocket with a particular energy barrier being overcome at a particular time is given by $p_{act}(t_i) = e^{-t_i/\tau_{act}}/\tau_{act}$. Once the energy barrier associated with a pocket has been overcome, let the volume recovered from that pocket vary with time according to $v(t) = V_0\Phi(t)$, which increases from zero to a maximum of, V_0 over a time τ_h.

We can write the total recovered volume at a given time in terms of an ensemble average over all the gas pockets of the probable amount of gas extracted from a given pocket at that time, given by equation (5).

Given a distribution of activation times, p_τ, we can write equation (5) as an integral:

$$\mathcal{V}(t) = \mathcal{N}V_0\int d\tau_{act}p_\tau(\tau_{act})\int_0^t dt_i\Phi(t - t_i)\frac{e^{\frac{-t_i}{\tau_{act}}}}{\tau_{act}} \tag{28}$$

To explore the scaling of gas recovery with time according to equation (28), we assume an exponential distribution of energy barriers:

$$p_{\Delta G^*}[\Delta G^*] = \frac{\alpha}{k_B T}e^{-\alpha\frac{\Delta G^*}{k_B T}} \tag{29}$$

This simply represents one possible distribution, and the precise form is not critical to our analysis. The corresponding distribution of activation times is

$$p_\tau(\tau_{act}) = \frac{\alpha\tau_0^\alpha}{\tau_{act}^{1+\alpha}} \tag{30}$$

where we define $\alpha = \frac{k_B T}{\kappa(\theta_0)\gamma_0}$; θ_0 and γ_0 fix the typical range spanned by these parameters over the reservoirs.

Taking this distribution of barriers and substituting into equation (28), we find the expression

$$\begin{aligned} \mathcal{V}(t) &= \mathcal{N} V_0 \alpha \int_0^t dt_i \Phi(t - t_i) \int_{\tau_0}^{\infty} d\tau_{act} \frac{e^{-\frac{t_i}{\tau_{act}}}}{\tau_{act}^{2+\alpha}} \\ &= \mathcal{N} V_0 \alpha \int_0^t dt_i \Phi(t - t_i) \frac{\Gamma(1+\alpha) - \Gamma(1+\alpha, t_i)}{t_i^{\alpha+1}} \end{aligned} \tag{31}$$

where $\Gamma(\alpha,t)$ is the incomplete gamma function. The solution to this expression is obtained for the two regimes $t < \tau_h$ and $t > \tau_h$ by using the Laplace transform for the convolution to get the asymptotics, with the results $Q(t) \propto 1/t^{\frac{1}{2}+\alpha}$ and $Q(t) \propto 1/t^{1+\alpha}$, respectively, as discussed above.

References

1. Cueto-Felgueroso, L. & Juanes, R. Forecasting long-term gas production from shale. *Proc. Natl Acad. Sci. USA* **110,** 19660–19661 (2013).
2. Baihly, J. D., Altman, R. M., Malpani, R. & Luo, F. Shale gas production decline trend comparison over time and basins. In *SPE Annual Technical Conference and Exhibition*. SPE-135555-MS (Society of Petroleum Engineers, 2010).
3. Kerr, R. A. Natural gas from shale bursts onto the scene. *Science* **328,** 1624–1626 (2010).
4. Bažant, Z. P., Salviato, M., Chau, V. T., Visnawathan, H. & Zubelewicz, A. Why fracking works. *J. Appl. Mech.* **81,** 101010 (2014).
5. Monteiro, P. J. M., Rycroft, C. H. & Barenblatt, G. I. A mathematical model of fluid and gas flow in nanoporous media. *Proc. Natl Acad. Sci. USA* **109,** 20309–20313 (2012).
6. Patzek, T. W., Male, F. & Marder, M. Gas production in the Barnett shale obeys a simple scaling theory. *Proc. Natl Acad. Sci. USA* **110,** 19731–19736 (2013).
7. Loucks, R. G., Reed, R. M., Ruppel, S. C. & Jarvie, D. M. Morphology, genesis, and distribution of nanometer-scale pores in siliceous mudstones of the Mississippian Barnett shale. *J. Sedimen. Res.* **79,** 848–861 (2009).
8. Sahimi, M. Flow phenomena in rocks: from continuum models to fractals, percolation, cellular automata, and simulated annealing. *Rev. Mod. Phys.* **65,** 1393–1534 (1993).
9. Bocquet, L. & Charlaix, E. Nanofluidics, from bulk to interfaces. *Chem. Soc. Rev.* **39,** 1073 (2010).
10. Mehmani, A., Prodanovic, M. & Javadpour, F. Multiscale, multiphysics network modeling of shale matrix gas flows. *Transp. Porous Med.* **99,** 377–390 (2013).
11. Heller, R., Vermylen, J. & Zoback, M. Experimental investigation of matrix permeability of gas shales. *AAPG Bull.* **98,** 975–995 (2014).
12. Coasne, B., Galarneau, A., Gerardin, C., Fajula, F. & Villemot, F. Molecular simulation of adsorption and transport in hierarchical porous materials. *Langmuir* **29,** 7864–7875 (2013).
13. Valiullin, R. *et al.* Exploration of molecular dynamics during transient sorption of fluids in mesoporous materials. *Nature* **443,** 965–968 (2006).
14. Bonilla, M. R., Valiullin, R., Kärger, J. & Bhatia, S. K. Understanding adsorption and transport of light gases in hierarchical materials using molecular simulation and effective medium theory. *J. Phys. Chem. C* **118,** 14355–14370 (2014).
15. Kärger, J. *et al.* Microimaging of transient guest profiles to monitor mass transfer in nanoporous materials. *Nat. Mater.* **13,** 333–343 (2014).
16. Clarkson, C. R. *et al.* Pore structure characterization of North American shale gas reservoirs using USANS/SANS, gas adsorption, and mercury intrusion. *Fuels* **103,** 606–616 (2013).
17. Billemont, P., Coasne, B. & De Weireld, G. Adsorption of carbon dioxide, methane, and their mixtures in porous carbons: effect of surface chemistry, water content, and pore disorder. *Langmuir* **29,** 3328–3338 (2013).
18. Rexer, C. R. *et al.* High-pressure methane adsorption and characterization of pores in Posidonia shales and isolated kerogens. *Energy Fuels* **28,** 2886–2901 (2014).
19. Bustin, R. M. & Bustin, A. M. M. Impact of shale properties on pore structure and storage characteristics. In *Proceedings of the SPE Shale Gas Production Conference* SPE-119892-MS (Society of Petroleum Engineers, 2008).
20. de Gennes, P. Wetting: statics and dynamics. *Rev. Mod. Phys.* **57,** 827–863 (1985).
21. Brakke, K. A. The surface evolver. *Exp. Math.* **1,** 141–165 (1992).
22. Bocquet, L., Charlaix, E., Ciliberto, S. & Crassous, J. Moisture-induced ageing in granular media and the kinetics of capillary condensation. *Nature* **396,** 735–737 (1998).
23. Sheng, J. J. Enhanced oil recovery in shale reservoirs by gas injection. *J. Nat. Gas Sci. Eng.* **22,** 252–259 (2015).
24. Karwacki, L. *et al.* Morphology-dependent zeolite intergrowth structures leading to distinct internal and outer-surface molecular diffusion barriers. *Nat. Mater.* **8,** 959–965 (2009).
25. Heinke, L., Gu, Z. & Wöll, C. The surface barrier phenomenon at the loading of metal-organic frameworks. *Nat. Commun.* **5,** 4562 (2014).
26. Hibbe, F. *et al.* The nature of surface barriers on nanoporous solids explored by microimaging of transient guest distributions. *J. Am. Chem. Soc.* **133,** 2804–2807 (2011).
27. Huber, P. Soft matter in hard confinement: phase transition thermodynamics, structure, texture, diffusion and flow in nanoporous media. *J. Phys. Condens. Matter* **27,** 103102 (2015).
28. Xue, Y., Markmann, J., Duan, H., Weissmüller, J. & Huber, P. Switchable imbibition in nanoporous gold. *Nat. Commun.* **5,** 4237 (2014).
29. Martin, M. G. & Siepmann, J. I. Transferable potentials for phase equilibria. 1. United-atom description of *n*-alkanes. *J. Phys. Chem. B* **102,** 2569–2577 (1998).
30. Berendsen, H. J. C., Grigera, J. R. & Straatsma, T. P. The missing term in effective pair potentials. *J. Phys. Chem.* **91,** 6269–6271 (1987).
31. Ryckaert, J.-P., Ciccotti, G. & Berendsen, H. J. Numerical integration of the cartesian equations of motion of a system with constraints: molecular dynamics of n-alkanes. *J. Comp. Phys.* **23,** 327–341 (1977).
32. Harris, J. G. & Yung, K. H. Carbon dioxides liquid-vapor coexistence curve and critical properties as predicted by a simple molecular model. *J. Phys. Chem.* **99,** 12021–12024 (1995).
33. Kästner, J. Umbrella sampling. *WIREs Comput. Mol. Sci.* **1,** 932–942 (2011).
34. Firouzi, M., Rupp, E. C., Liu, C. W. & Wilcox, J. Molecular simulation and experimental characterization of the nanoporous structures of coal and gas shale. *Int. J. Coal Geol.* **121,** 123–128 (2014).
35. Orendt, A. M. *et al.* Three-dimensional structure of the siskin green river oil shale kerogen model: a comparison between calculated and observed properties. *Energy Fuels* **27,** 702–710 (2013).
36. Kelemen, S. R. *et al.* Direct characterization of kerogen by x-ray and solid-state 13 C nuclear magnetic resonance methods. *Energy Fuels* **21,** 1548–1561 (2007).
37. Pikunic, J., Llewellyn, P., Pellenq, R. & Gubbins, K. E. Argon and nitrogen adsorption in disordered nanoporous carbons: a simulation and experiment. *Langmuir* **21,** 4431–4440 (2005).
38. Kihara, K. An X-ray study of the temperature dependence of the quartz structure. *Eur. J. Mineral.* **2,** 63–78 (1990).
39. Kirkwood, J. G. & Buff, F. P. The statistical mechanical theory of surface tension. *J. Chem. Phys.* **17,** 338 (1949).
40. Schofield, P. & Henderson, J. R. Statistical mechanics of inhomogeneous fluids. *Proc. R. Soc. Lond. A* **379,** 231–246 (1982).
41. Irving, J. H. & Kirkwood, J. G. The statistical mechanical theory of transport processes. IV. The equations of hydrodynamics. *J. Chem. Phys.* **18,** 817 (1950).
42. Blokhuis, E. M. & Bedeaux, D. Pressure tensor of a spherical interface. *J. Chem. Phys.* **97,** 3576 (1992).

Acknowledgements

This work was supported by the X-Shale project enabled through MIT's Energy Initiative in collaboration with Shell and Schlumberger. Additional support was provided by the ICoME2 Labex (ANR-11-LABX-0053) and the A*MIDEX projects (ANR-11-IDEX-0001-02) co-funded by the French programme 'Investissements d'Avenir' managed by ANR, the French National Research Agency. L. B. also acknowledges partial support from ANR-14-CE05-0037. We thank Roland Pellenq, Franz Ulm, L. Joly and B. Rotenberg for fruitful discussions. We also thank J. Baihly, R. L. Kleinberg and D. Pomerantz from Schlumberger for very helpful comments about fundamental and practical aspects of shale gas recovery.

Author contributions

All authors contributed to the design of the molecular simulation method. T.L. carried out the numerical simulations and analysis, with input from L.B. and B.C. All authors contributed to the development of the thermodynamic model for nucleation. All authors contributed to the writing of the manuscript.

Additional information

Competing financial interests: The authors declare no competing financial interests.

Light-melt adhesive based on dynamic carbon frameworks in a columnar liquid-crystal phase

Shohei Saito[1,2], Shunpei Nobusue[3,†], Eri Tsuzaka[3], Chunxue Yuan[3,†], Chigusa Mori[3], Mitsuo Hara[4], Takahiro Seki[4], Cristopher Camacho[3,5], Stephan Irle[3,6] & Shigehiro Yamaguchi[3,6]

Liquid crystal (LC) provides a suitable platform to exploit structural motions of molecules in a condensed phase. Amplification of the structural changes enables a variety of technologies not only in LC displays but also in other applications. Until very recently, however, a practical use of LCs for removable adhesives has not been explored, although a spontaneous disorganization of LC materials can be easily triggered by light-induced isomerization of photoactive components. The difficulty of such application derives from the requirements for simultaneous implementation of sufficient bonding strength and its rapid disappearance by photoirradiation. Here we report a dynamic molecular LC material that meets these requirements. Columnar-stacked V-shaped carbon frameworks display sufficient bonding strength even during heating conditions, while its bonding ability is immediately lost by a light-induced self-melting function. The light-melt adhesive is reusable and its fluorescence colour reversibly changes during the cycle, visualizing the bonding/nonbonding phases of the adhesive.

[1] Department of Chemistry, Graduate School of Science, Kyoto University, Kyoto 606-8502, Japan. [2] PRESTO, Japan Science and Technology Agency, Kyoto 606-8502, Japan. [3] Department of Chemistry, Graduate School of Science, Nagoya University, Nagoya 464-8602, Japan. [4] Department of Molecular Design and Engineering, Graduate School of Engineering, Nagoya University, Nagoya 464-8603, Japan. [5] School of Chemistry, University of Costa Rica, San Pedro Montes de Oca, San José 11501-2060, Costa Rica. [6] Institute of Transformative Bio-Molecules, Nagoya University, Nagoya 464-8602, Japan. † Present addresses: Department of Materials Engineering Science, Graduate School of Engineering Science, Osaka University, Osaka 560-8531, Japan (S.N.); College of Materials Science and Engineering, Tongji University, Shanghai 201804, China (C.Y.). Correspondence and requests for materials should be addressed to S.S. (email: s_saito@kuchem.kyoto-u.ac.jp).

Liquid crystals (LCs) are versatile condensed materials. Their applications are not limited to LC displays, but covering lasers, photovoltaics, light-emitting diodes (LEDs), field-effect transistors, nonlinear optics, biosensors, switchable windows and nanophotonics[1–3]. Photochemical control of LC materials, as well as polymers and crystals, has long been an attractive subject due to the promising applications[4–8], for example, optical switching and image storage[9–13], photoalignment technique[14–20] and mechanical force induction with photoresponsive actuators[21–30]. In these studies, the unique characteristics of LC materials, the spontaneous amplification of disorganized/ordered phase has been utilized. In some designed systems, ordered packing structures of rod-shaped LC molecules are destroyed by doping a guest molecule with a bent shape[4]. When the guest component is introduced by *in situ* photoisomerization, as often observed in LC azobenzene derivatives, the bulk LC material shows an instant isothermal photoinduced phase transformation, depending on the concentration of the guest dopant. In this context, photoactive LC is a most suitable platform for realizing a quick photomelting (that is, photochemical reaction-triggered isothermal phase transformation into a fluid mixture), while some crystals show a slow photomelting behaviour[31–35]. In spite of this advantage, the application of photomelting LCs to light-removable adhesives has not been explored until very recent reports on azobenzene smectic LCs[36,37], although some polymer materials are developed as photo-deactivatable resins based on various operation principles such as photoinduced crosslinking[38–40], photoacid-catalysed chain modification[41], photochemical cleavage of polymer chains[42] and thermal cleavage of supramolecular chains by light irradition[43]. The light-melting function offers a new manufacturing technique not realized with conventional hot-melt adhesives[44] as long as the following essential requisites are fulfilled: first, adequate strength for a temporary bond (>1 MPa) even under heating conditions; second, significant reduction of the bonding strengths by light irradiation; and third, quick photoresponse for the separation of bonded materials. It is a formidable challenge to overcome all these essential requisites simultaneously.

Here we report a LC material that satisfies all of the above-mentioned requisites for the light-melt adhesives, namely, a shear strength over 1 MPa up to 110 °C for bonding glass plates, an 85% reduction of the strength by ultraviolet irradiation, and an instant photomelting of the LC film in a few seconds (Supplementary Movie 1). Moreover, this material is reusable as an adhesive, and the transformation between the LC and melted phases is associated with an informative colour change in fluorescence.

Results

Molecular design of photoresponsive LC. Our molecular design of the LC material **1** is based on a unique photoresponsive carbon framework[45,46] with typical dendritic moieties (Fig. 1)[47,48]. The photoresponsive framework is composed of rigid anthracene wings and a flexible joint of cyclooctatetraene[49–51]. This hybrid design of rigidity and flexibility confers two important characteristics on the material. First, the V-shaped molecule with rigid aromatic wings has strong stacking ability to form a columnar array in the condensed phases, which now results in high cohesive force of the LC material for realizing high-temperature resistant bonding. Second, the flexible framework changes its conformation into a flat shape on photoexcitation in the LC phase, which allows the photodimerization of the anthracene moiety[52,53]. This photoresponse leads to the light-induced melting of the LC phase, accompanied by a separation of bonded glass plates.

Thermal and structural analysis. The dendritic peripheral chains of **1** caused a LC phase between 65 and 140 °C (Fig. 2a). The LC phase showed only little signs of fluidity, although a shear-induced alignment was confirmed by the polarized optical microscopy (POM; Supplementary Fig. 1). By means of the X-ray diffraction (XRD) analysis, the LC phase was assigned to be a rectangular columnar structure, in which the V-shaped molecules align on top of each other and the stacked arrays are located side by side (Supplementary Figs 2–4). The formation of the columnar structure was supported by the single-crystal X-ray structure analysis of the corresponding derivative **2**, which has no peripheral chains (Fig. 2b; Supplementary Fig. 5). In spite of its non-planarity, the V-shaped framework constructs the twofold π-stacked array with both anthracene wings tightly stacked. The interfacial distance $d(\pi-\pi)$ between the stacked anthracene moieties was observed to be 3.50 Å. The strong intermolecular interaction was also supported by the very large enthalpy change[54] over 30 kJ mol^{-1} in the phase transition of **1** ~140 °C between the columnar LC and isotropic liquid phases (Fig. 2a), which is responsible for the sufficient cohesion strength of this LC material.

Photoresponse of the LC material. By the ultraviolet exposure using a 365-nm LED, a LC thin film of **1** between glass plates turned into liquid. The temperature was kept at 100 °C on a hot-stage microscopy. In the POM analysis under crossed Nicols, the bright image of the 5-μm-thick LC film immediately disappeared with 160 mW cm^{-2} irradiance and, instead, a dark field was observed only in the irradiated area, indicative of a photoinduced phase transformation into an isotropic liquid (Fig. 2c). Since this transformation did not take place in the solid phase and significant photochemical reaction was not observed after the photoirradiation at 50 °C (Supplementary Fig. 6), the LC nature is important to induce this photoresponse efficiently. Observation with a thermography camera confirmed a negligible local temperature increase caused by ultraviolet irradiation (Supplementary Fig. 7). The analyses of the resulting liquid revealed that the dimer and trace of oligomers of **1** were produced by light, while unreacted monomer was still present in the melted mixture (Fig. 2d; Supplementary Fig. 8). The photodimer was isolated as a main product and its less symmetric structure was determined based on the ^{1}H and ^{13}C nuclear magnetic resonance analyses (Supplementary Figs 9 and 10). The isolated photodimer showed thermal back reaction, gradually affording the monomer at 130 °C in a solution (Supplementary Fig. 11). Accordingly, on heating the melted mixture at 160 °C for 15 min, the bright POM image of the monomer LC phase was recovered at 100 °C (Supplementary Fig. 12). The recovery of the initial LC structure

1 (R = $C_{12}H_{25}$)
2 (R = CH_3)

Figure 1 | Molecular structures of the photoresponsive liquid crystal 1 and its derivative 2. Rigid anthracene units (blue wings) are fused with a flexible cyclooctatetraene ring (eight-membered ring). Dendritic moieties (orange fans) are attached to the flapping core framework.

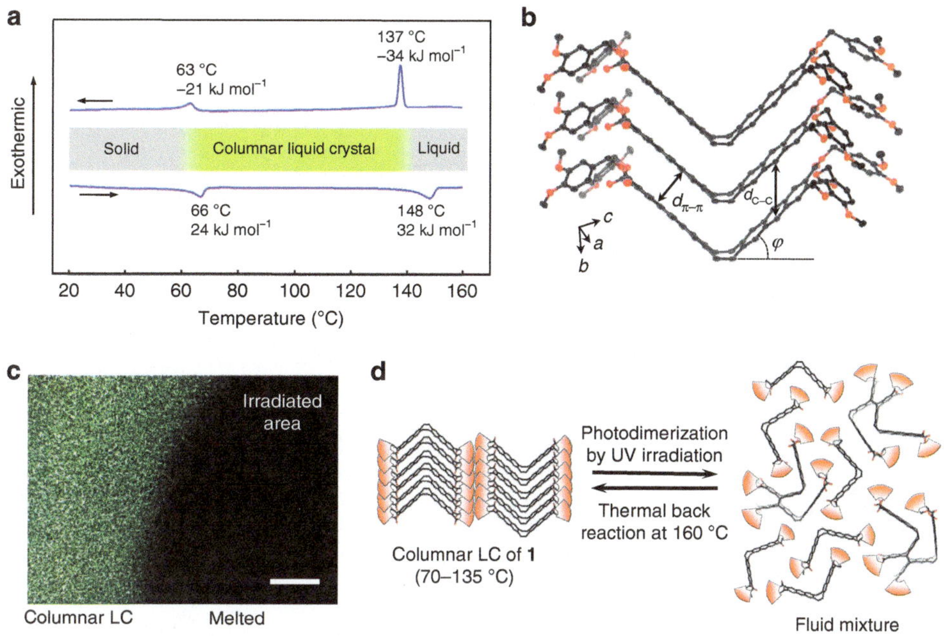

Figure 2 | Photoinduced melting in a columnar LC phase. (**a**) Differential scanning calorimetry (DSC) traces of **1** at 2 °C min^{-1} rate of cooling (top) and heating (bottom). (**b**) Crystal packing structure of a derivative **2** with no dendritic peripheral chain. Interfacial distance of the π-stacked anthracene moieties, $d(\pi\text{-}\pi) = 3.50$ Å. Intermolecular distance between the photoreactive carbon sites, $d(\text{C-C}) = 4.84$ Å. Bent angle of the V-shaped molecule, $\varphi = 43.8°$. (**c**) POM image of the LC film of **1** (bright) and its photoirradiated area (dark) under the crossed Nicols. Scale bar, 500 μm. (**d**) Isothermal photoinduced melting of **1** in the range of 70–135 °C (left to right), in which the columnar LC phase of **1** is photochemically transformed into a fluid mixture mainly composed of unreacted **1** and its photodimer product. Heating the melted mixture at 160 °C induces a thermal back reaction of the photodimer into the monomer **1**, which recovers the columnar LC phase when the temperature is set again in the range of 70–135 °C (right to left).

was confirmed in the grazing incidence XRD analysis of a spin-coated film (Supplementary Fig. 15). Apparently, the photo-induced *in situ* generation of the dimer molecules destabilizes the columnar LC phase of **1** because of their unsuitable shape for the columnar packing[4]. Complete consumption of all monomer units was not necessary for the isothermal photomelting function (Supplementary Fig. 8), presumably because the destruction of the columnar packing is spontaneously induced based on the amplification of the disorganized phase.

Perfomance of the light-melt adhesive. Practical performance of **1** as a light-melt adhesive was next demonstrated (Fig. 3). Two non-treated glass plates were bonded by the film of **1**. The glass plates with an attached weight were still bound together over 70 °C by heating with a blow-dryer, at which temperature the phase transition of **1** into the LC phase completed. Once exposed to the ultraviolet LED during the heating, the glass plates were separated in a few seconds and the attached weight dropped (Supplementary Movie 1). However, the ultraviolet exposure at 25 °C did not result in the glass separation even after the 60-s irradiation, which consists with the no significant photoreactive properties in the solid phase of **1**. This inert feature at room temperature is favourable for the avoidance of an unexpected failure under ambient light. Only 2–3 mg of the adhesive was sufficient for bonding two glass plates with 4-cm^2 area, from which 8-kg water bottles were suspended (Fig. 3b). Ultimate shear strengths were measured for uniformly prepared 130-μm-thick films (Supplementary Fig. 16). When non-treated glass plates were used, the shear strengths of the film were 1.6 MPa at 25 °C in the solid state, 1.2 MPa at 100 °C, 1.1 MPa at 110 °C and 0.9 MPa at 120 °C in the LC state (Fig. 3c). Notably, the bonding strength still exceeds 1 MPa up to 110 °C, indicating the significant thermal resistance. This strength is sufficient for the temporary bonding of removable adhesives[44]. The shear strength was independent of the hydrophilicity of the glass surface (Fig. 3d). Therefore, the cohesive force, closely related to the intermolecular interaction of **1**, plays a key role in determining the bonding strength of this system rather than the adhesion force, the interaction between the glass surface and the adhesive material (Supplementary Fig. 17)[44].

Photoinduced melting near the interface. Despite the strong bonding properties, only after 320 mJ cm^{-2} light exposure at 100 °C the shear strength markedly decreased to 0.2 MPa (Fig. 3c). When the photoirradiation was carried out using a hand-held ultraviolet lamp (365 nm, 3.2 mW cm^{-2}), it took ~100 s to induce the photoseparation. On the other hand, quick detachment was achieved in few seconds using ultraviolet LED (365 nm, 160 mW cm^{-2}). Since the condensed molecules of **1** efficiently absorb the ultraviolet light, the photoinduced detachment took place near the interface between the LC film and the irradiated glass plate. The transmittance exponentially decreased dependent on the thickness of spin-coated films of **1** (Fig. 4a; Supplementary Figs 18 and 19). More than 95% of the 365-nm light was absorbed within 3-μm depth of the film. Consistent with this insight, total irradiation dose required for the glass separation using a 0.5-kg weight was almost constant (320 mJ cm^{-2}) regardless of the film thickness over 5 μm (Fig. 4b). From a manufacturing point of view, it is important that the amount of the adhesive residue on the irradiated glass plate is always small even when the thickness of the film is larger (Supplementary Fig. 20).

Reusability and visualization technology. The reusability of the adhesive **1** was also demonstrated. After the photoirradiation, the adhesive was heated at 160 °C for 15 min. Then, the shear

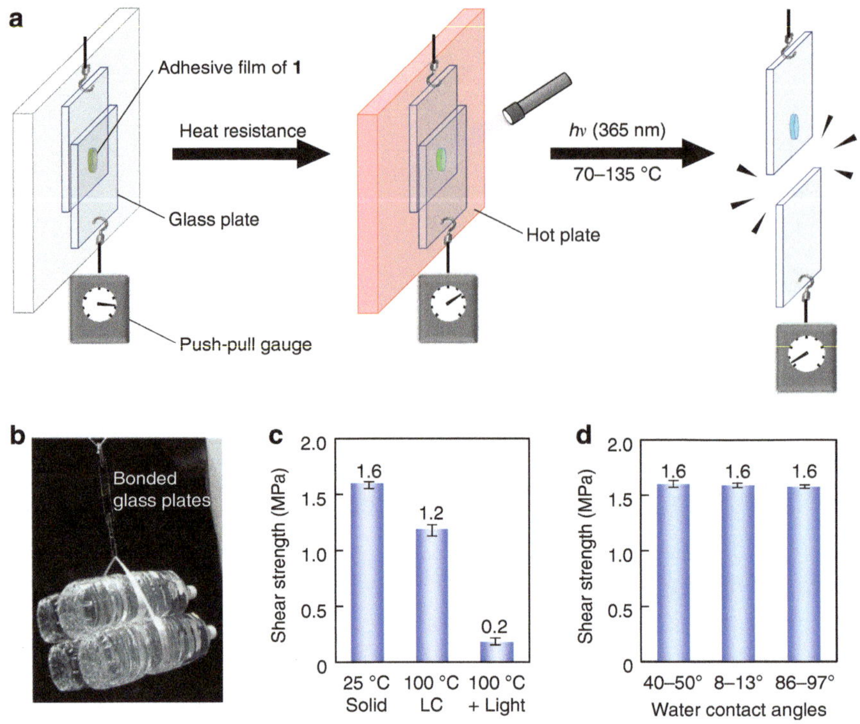

Figure 3 | Light-melt adhesive properties. (**a**) High-temperature resistant bonding and photoinduced separation of two glass plates stuck with the adhesive film of **1**. (**b**) Demonstration of the strong adhesive function of **1**. (**c**,**d**) Ultimate shear strengths of the 130-μm-thick film of **1** depending on the phase of **1** (**c**) and on the hydrophilicity of the glass surface (**d**). See the Methods section for the preparation of the uniform films as well as the hydrophilic and hydrophobic glass surfaces.

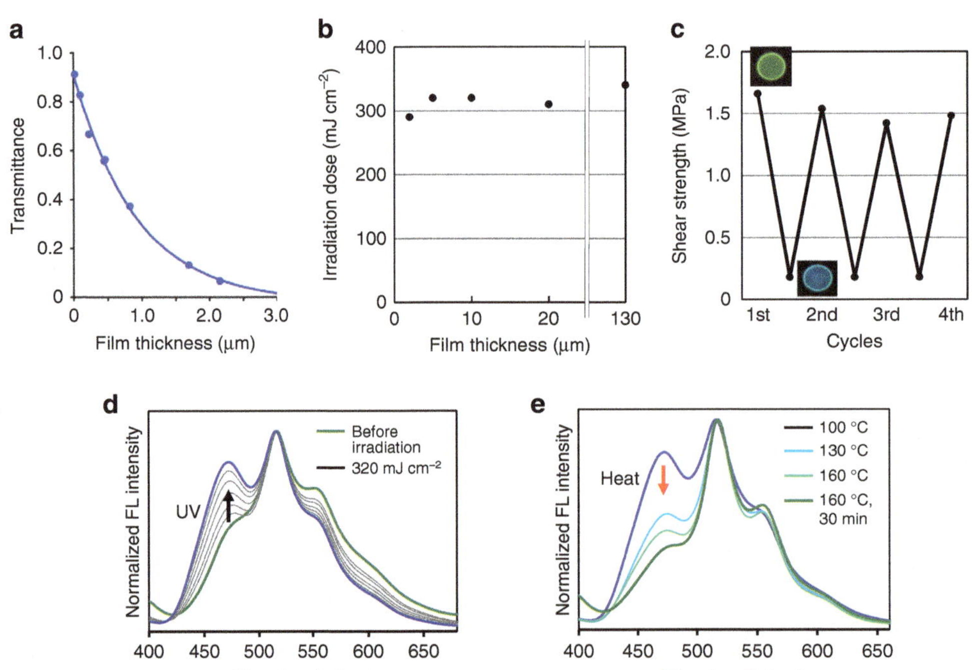

Figure 4 | Photoresponse and thermal recovery of the film. (**a**) Transmittance of 365-nm light dependent on the film thickness of **1**. (**b**) Required irradiation dose for glass separation at 100 °C using the films of **1** with different thickness. (**c**) Reusability of the adhesive **1**. The room-temperature shear strengths before and after 320 mJ cm^{-2} ultraviolet exposure at 100 °C in the recycling processes. The inset photographs show the fluorescent film in the corresponding stage. (**d**,**e**) Fluorescence spectral change during the ultraviolet irradiation on the 5-μm-thick film of **1** (**d**) and its thermal restoring steps (**e**). The film fluorescence at 100 °C before (green line) and after (blue line) light irradiation at 3.2 mW cm^{-2} for 100 s (**d**) and the film fluorescence before (blue line) and after (green line) heating at 160 °C for 30 min (**e**).

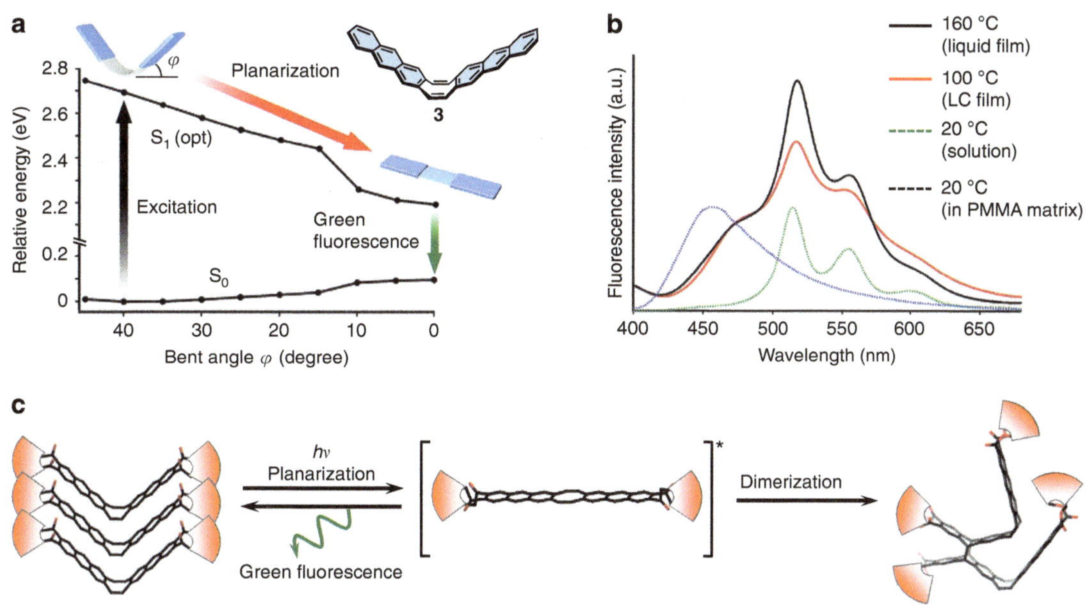

Figure 5 | Interpretation of the photoresponse mechanism. (**a**) Calculated potential energy diagram for the ground state (S_0) and lowest excited state (S_1) of the carbon framework **3** with fixed bent angle φ. The relaxed potential energy surface scan was performed for the S_1 state of **3** at the TD-PBE0/def-SV(P) level. (**b**) Fluorescence spectra of a thin film of **1** in the liquid phase (160 °C, black line) and in the LC phase (100 °C, red line). Excitation at 365 nm. Fluorescence spectra of **1** in CH_2Cl_2 solution (green dotted line) and in polymethyl methacrylate (PMMA) matrix (blue dotted line) are shown for comparison. The fluorescence in PMMA matrix was measured at 20 °C, which is lower than the glass transition temperature of the PMMA (T_g = 105 °C). (**c**) The reaction process of the photodimerization of **1** in the LC phase.

strength ~1.5 MPa at 25 °C and the quick photoresponse function at 100 °C were recovered at least four times through this recycling process (Fig. 4c). In addition, the process was accompanied by the fluorescence colour change (Fig. 4d,e), which is useful for the contactless investigation of the adhesive bonding ability. Namely, green fluorescence is a sign of the adhesive recovery in the initializing step, while blue emission of the melted phase informs that the glass plates can be separated. The fluorescence colour change of the LC film observed during the ultraviolet irradiation and the following thermal restoring steps were well explained by the formation of the photodimer of **1** and the recovery of the monomer **1**, respectively (Supplementary Figs 13 and 14).

Discussion

The mechanism of the photoreaction was studied taking into account the behaviour of **1** in the singlet excited state (S_1; Fig. 5). In general, photodimerization of anthracene units proceeds in S_1 (ref. 55) only when the reactive sites (the central carbon atoms that form new C–C bonds) approach each other within a distance of 4.2 Å in the condensed phase[53]. By means of the XRD analysis of **1** at 100 °C, the corresponding distance *d*(C–C) in the columnar packing of the V-shaped molecules was estimated to be ca. 4.7 Å, resulting from the slipped alignment of the anthracene moieties (Supplementary Figs 2 and 3), which does not allow the photodimerization. As the significant photoreaction of **1** was not detected at 50 °C (Supplementary Fig. 6), the columnar packing structure is supposed to be fixed in the solid phase (Supplementary Fig. 4). In contrast, the observed photoreactivity in the LC phase indicated that the packing structure was perturbed by the photoexcitation and the dimerization occurred in S_1. This interpretation was supported both by theoretical calculations and experimental results. Structural optimization in S_1 was performed on the main nanocarbon framework **3**, suggesting that the bent molecule spontaneously changes its conformation in S_1 into a planar structure, because the S_1 energy is significantly lowered as the molecular shape becomes flat (Fig. 5a; Supplementary Figs 21 and 22). Fluorescence behaviour of the newly synthesized compound **3** was consistent with the theoretical result. The observed Stokes shift in solution (5,320 cm^{-1}) was far larger than those in frozen glass or in polymethyl methacrylate matrix (Supplementary Fig. 23). To elucidate whether the dynamic conformational change of **1** occurs not only in solution but also in the LC phase, the fluorescence spectrum of the LC film was measured. As a result, both the LC film of **1** at 100 °C and the neat liquid of pure compound **1** at 160 °C showed the same green emission band at 515 nm, which was also observed in its solution phase with the large Stokes shift (Fig. 5b; Supplementary Fig. 24). This correspondence indicates that the photoinduced planarization of the molecular framework still took place under these conditions, whereas the dynamic motion was suppressed in polymethyl methacrylate matrix and the V-shaped form of **1** emitted a blue fluorescence band at 460 nm. The blue emission at 475 nm observed in the LC and neat liquid film of **1** can be interpreted as the fluorescence of the shallow V-shaped structure with small bent angle φ, which was provided because the conformational planarization in S_1 was partly suppressed due to the increased viscosity of the LC and neat liquid phases when compared with the low-viscous solution. Overall, the full reaction mechanism of the photomelting process is interpreted as follows (Fig. 5c): near the interface between the LC film of **1** and the glass substrate, the photoexcitation of the stacked V-shaped molecule induces its conformational change into the planar form. As a consequence, the intermolecular distance between the reactive sites of the anthracene moieties is significantly perturbed. Some molecular pairs dimerize photochemically in the S_1 excited state, while other excited molecules relax back to the singlet ground state (S_0) accompanied by the green fluorescence emitted from the planar form. The photoproducts induce disorganization of the columnar LC phase, leading to the melting behaviour with the loss of bonding strength.

In conclusion, a strong light-melt adhesive based on a photoresponsive columnar LC material has been developed. Tight π-stacking of the V-shaped nanocarbon frameworks resulted in the sufficient bonding properties of the adhesive film even at high temperature. Remarkable photoinduced decrease in the shear strength was realized by the *in situ* dimerization of the LC molecule and the following disorganization of the columnar structure. The fast melting response achieved a quick separation of bonded glass plates leaving the small adhesive residue. We envision that composite materials with the light-melt function will further improve the performance in manufacturing processes, which will accelerate the on-demand photoseparation technology complementary to the other switchable adhesion approaches[56].

Methods

Synthesis. Compounds **1** and **2** were synthesized from a naphthalene dimer[45] having the cyclooctatetraene ring at its centre. Double acene elongation reaction[57] was employed with the corresponding benzyl fumarates. Compound **3** was synthesized through the [2+2+2+2] cycloaddition[58] of a terminal diyne precursor, 2,3-bis(2-propynyl)naphthalene, and the following 2,3-dichloro-5,6-dicyano-1,4-benzoquinone (DDQ) oxidation (Supplementary Methods). The ^{1}H and ^{13}C nuclear magnetic resonance spectra have been displayed for all new compounds (Supplementary Figs 25–30).

Structural analysis of the LC phase. XRD measurement of **1** in a Lindemann glass capillary was performed at 100 °C using Cu Kα radiation (Supplementary Fig. 2). The LC phase was assigned to be a columnar rectangular phase of *C2/m* with lattice parameters $a = 68$ Å and $b = 59$ Å (Supplementary Fig. 3). A small peak at $2\theta = 18.86°$, reproducibly observed with the broad signal of the alkyl chain diffraction, was assigned as (001). This signal corresponds to the average interval of the stacking molecules, d(C–C) = 4.7 Å in the LC phase of **1**, which is comparable to that in the crystal packing of **2** (Fig. 2b).

Single-crystal X-ray structure analysis. Yellow prism crystals of **2** were obtained by slow diffusion of ethanol into a solution of **2** in anisole. The measurement was performed at 103 K (Supplementary Fig. 5). The structures were solved by a direct method and refined by least-squares calculations on F^2 for all independent reflections (SHELXL-2013)[59]. Total 27,088 reflections were collected, among which 5,795 reflections were independent ($R_{int} = 0.0312$). Monoclinic crystal system, space group *C2/c* (#15), $a = 54.39(5)$, $b = 4.843(4)$, $c = 23.157$ (18) Å, $\beta = 111.690$ (9) °, $V = 5{,}667$ (8) Å^3, $Z = 4$, $T = 103$ K, $R_1 = 0.0367$ ($I > 2\sigma(I)$), $wR_2 = 0.0979$ (all the data), Goodness of fit (GOF) = 1.063.

Evaluation of the light irradiance. Ultraviolet irradiation was carried out using a UV-400 series UV-LED (Keyence, UV-50H type, 365 nm) equipped with a UV-L3 lens unit or using a hand-held ultraviolet lamp LUV-16 (AS ONE, 365 nm). Irradiance from these light sources on the sample was measured in advance to be 160 and 3.2 mW cm^{-2}, respectively, using an ultraviolet irradiance metre UIT-150 (USHIO).

Measurement of the ultimate shear strength. The 130-μm-thick films of **1** were uniformly prepared between glass plates (Supplementary Fig. 16). After the film preparation, one side of the glass plates was connected to a fixed laboratory stand with a strap, while the other side was connected to a digital force gauge. Each measurement was conducted twice and the average values were shown with the error bar. In Fig. 3c, the strength in the LC phase was measured after annealing the sample on the hot plate at 100 °C for 15 min. Ultraviolet light irradiation was performed at 100 °C for 100 s using a hand-held ultraviolet lamp (365 nm, 3.2 mW cm^{-2}), and then the shear strength was measured at the same temperature. In Fig. 2d, non-treated borosilicate glass plates showed a water contact angle of 40–50°. The glass plates with hydrophilic surface (8–13°) were prepared by sonication in a saturated solution of KOH in EtOH for 30 min followed by three times sonication in distilled water for 30 min, while the glass plates with hydrophobic surface (86–97°) were prepared by the exposure of this hydrophilic surface to a HMDS (1,1,1,3,3,3-hexamethyldisilazane) vapour in a sealed tube for >24 h.

Measurement of the light transmittance. To a quartz glass plate (15 × 10 mm) placed on a spin coater, a solution of **1** in $CHCl_3$ with different concentrations (1, 2, 3, 5 and 10 wt%) was dropped so that the solution can cover the entire surface of the glass plate. The sample was rotated on a spin coater at 1,000 r.p.m. for 30 s. The film thickness was determined by the atomic force microscopy analysis to be 14, 96, 220, 460 and 820 nm, respectively (Supplementary Fig. 18). The films with 1,690- and 2,150-nm thicknesses were prepared by overlapping the two and three films coated with the 10 wt% $CHCl_3$ solution of **1**. In consideration of a transmittance loss, the absorbance of the samples at the 365-nm wavelength showed a linear fitting against the film thickness with the coefficient of determination $R^2 = 0.997$ (Supplementary Fig. 19).

Theoretical calculations. Ground- and excited-state density functional theory calculations were performed using the TURBOMOLE program (Fig. 5a; Supplementary Figs 21 and 22): *TURBOMOLE*, ver. 6.3, 2012; a development of University of Karlsruhe and Forschungszentrum Karlsruhe GmbH, 1989 – 2007, TURBOMOLE GmbH, since 2007, available at http://www.turbomole.com.

References

1. Goodby, J. W. *et al.* (eds). *Handbook of Liquid Crystals* (Wiley-VCH, 2014).
2. Li, Q. (ed). *Liquid Crystals Beyond Displays* (John Wiley & Sons, 2012).
3. Fleischmann, E. K. & Zentel, R. Liquid-crystalline ordering as a concept in materials science: from semiconductors to stimuli-responsive devices. *Angew. Chem. Int. Ed.* **52,** 8810–8827 (2013).
4. Ikeda, T. Photomodulation of liquid crystal orientations for photonic applications. *J. Mater. Chem.* **13,** 2037–2057 (2003).
5. Browne, W. R. & Feringa, B. L. Making molecular machines work. *Nature Nanotech.* **1,** 25–35 (2006).
6. Zhao, Y. & Ikeda, T. (eds). *Smart Light-Responsive Materials* (John Wiley & Sons, 2009).
7. Irie, M., Fukaminato, T., Matsuda, K. & Kobatake, S. Photochromism of diarylethene molecules and crystals: memories, switches, and actuators. *Chem. Rev.* **114,** 12174–12277 (2014).
8. White, T. J. & Broer, D. J. Programmable and adaptive mechanics with liquid crystal polymer networks and elastomers. *Nat. Mater.* **14,** 1087–1098 (2015).
9. Tazuke, S., Kurihara, S. & Ikeda, T. Amplified image recording in liquid crystal media by means of photochemically triggered phase transition. *Chem. Lett.* **16,** 911–914 (1987).
10. Ikeda, T. & Tsutsumi, O. Optical switching and image storage by means of azobenzene liquid-crystal films. *Science* **268,** 1873–1875 (1995).
11. Huck, N. P. M., Jager, W. F., de Lange, B. & Feringa, B. L. Dynamic control and amplification of molecular chirality by circular polarized light. *Science* **273,** 1686–1688 (1996).
12. Hrozhyk, U. A., Serak, S. V., Tabiryan, N. V. & Bunning, T. J. Optical tuning of the reflection of cholesterics doped with azobenzene liquid crystals. *Adv. Funct. Mater.* **17,** 1735–1742 (2007).
13. White, T. J., McConney, M. E. & Bunning, T. J. Dynamic color in stimuli-responsive cholesteric liquid crystals. *J. Mater. Chem.* **20,** 9832–9847 (2010).
14. Ichimura, K., Suzuki, Y., Seki, T., Hosoki, A. & Aoki, K. Reversible change in alignment mode of nematic liquid crystals regulated photochemically by command surfaces modified with an azobenzene monolayer. *Langmuir* **4,** 1214–1216 (1988).
15. Gibbons, W. M., Shannon, P. J., Sun, S.-T. & Swetlin, B. J. Surface-mediated alignment of nematic liquid crystals with polarized laser light. *Nature* **351,** 49–50 (1991).
16. Schadt, M., Schmitt, K., Nozinkov, V. & Chigrinov, V. Surface-induced parallel alignment of liquid crystals by linearly polymerized photopolymers. *Jpn. J. Appl. Phys.* **31,** 2155–2164 (1992).
17. Ichimura, K. Photoalignment of liquid-crystal systems. *Chem. Rev.* **100,** 1847–1873 (2000).
18. Yaroshchuk, O. & Reznikov, Y. Photoalignment of liquid crystals: basics and current trends. *J. Mater. Chem.* **22,** 286–300 (2012).
19. Seki, T., Nagano, S. & Hara, M. Versatility of photoalignment techniques: from nematics to a wide range of functional materials. *Polymer* **54,** 6053–6072 (2013).
20. Fukuhara, K., Nagano, S., Hara, M. & Seki, T. Free-surface molecular command systems for photoalignment of liquid crystalline materials. *Nat. Commun.* **5,** 3320 (2014).
21. Finkelmann, H., Nishikawa, E., Pereira, G. G. & Warner, M. A new opto-mechanical effect in solids. *Phys. Rev. Lett.* **87,** 015501 (2001).
22. Yu, Y., Nakano, M. & Ikeda, T. Directed bending of a polymer film by light. Miniaturizing a simple photomechanical system could expand its range of applications. *Nature* **425,** 145 (2003).
23. Li, M.-H., Keller, P., Li, B., Wang, X. & Brunet, M. Light-driven side-on nematic elastomer actuators. *Adv. Mater.* **15,** 569–572 (2003).
24. Camacho-Lopez, M., Finkelmann, H., Palffy-Muhoray, P. & Shelley, M. Fast liquid-crystal elastomer swims into the dark. *Nat. Mater.* **3,** 307–310 (2004).
25. Eelkema, R. *et al.* Nanomotor rotates microscale objects. A molecular motor in a liquid-crystal film uses light to turn items thousands of times larger than itself. *Nature* **440,** 163 (2006).

26. Kobatake, S., Takami, S., Muto, H., Ishikawa, T. & Irie, M. Rapid and reversible shape changes of molecular crystals on photoirradiation. *Nature* **446,** 778–781 (2007).
27. Yamada, M. *et al.* Photomobile polymer materials: towards light-driven plastic motors. *Angew. Chem. Int. Ed.* **47,** 4986–4988 (2008).
28. van Oosten, C. L., Bastiaansen, C. W. M. & Broer, D. J. Printed artificial cilia from liquid-crystal network actuators modularly driven by light. *Nat. Mater.* **8,** 677–682 (2009).
29. Iamsaard, S. *et al.* Conversion of light into macroscopic helical motion. *Nat. Chem.* **6,** 229–235 (2014).
30. Priimagi, A., Barrett, C. J. & Shishido, A. Recent twists in photoactuation and photoalignment control. *J. Mater. Chem. C* **2,** 7155–7162 (2014).
31. Uchida, K. *et al.* Photoinduced reversible formation of microfibrils on a photochromic diarylethene microcrystalline surface. *Angew. Chem. Int. Ed.* **45,** 6470–6473 (2006).
32. Norikane, Y., Hirai, Y. & Yoshida, M. Photoinduced isothermal phase transitions of liquid-crystalline macrocyclic azobenzenes. *Chem. Commun.* **47,** 1770–1772 (2011).
33. Hoshino, M. *et al.* Crystal melting by light: X-ray crystal structure analysis of an azo crystal showing photoinduced crystal-melt transition. *J. Am. Chem. Soc.* **136,** 9158–9164 (2014).
34. Uchida, E., Azumi, R. & Norikane, Y. Light-induced crawling of crystals on a glass surface. *Nat. Commun.* **6,** 7310 (2015).
35. Baroncini, M. *et al.* Photoinduced reversible switching of porosity in molecular crystals based on star-shaped azobenzene tetramers. *Nat. Chem.* **7,** 634–640 (2015).
36. Akiyama, H. & Yoshida, M. Photochemically reversible liquefaction and solidification of single compounds based on a sugar alcohol scaffold with multi azo-arms. *Adv. Mater.* **24,** 2353–2356 (2012).
37. Akiyama, H. *et al.* Photochemically reversible liquefaction and solidification of multiazobenzene sugar-alcohol derivatives and application to reworkable adhesives. *ACS Appl. Mater. Interfaces* **6,** 7933–7941 (2014).
38. Boyne, J. M., Millan, E. J. & Webster, I. Peeling performance of a novel light switchable pressure-sensitive adhesive. *Int. J. Adhes. Adhes.* **21,** 49–53 (2001).
39. Ebe, K., Seno, H. & Horigome, K. UV curable pressure-sensitive adhesives for fabricating semiconductors. I. development of easily peelable dicing tapes. *J. Appl. Polym. Sci.* **90,** 436–441 (2003).
40. Trenor, S. R., Long, T. E. & Love, B. J. Development of a light-deactivatable PSA via photodimerization. *J. Adhes.* **81,** 213–229 (2005).
41. Inui, T., Sato, E. & Matsumoto, A. Pressure-sensitive adhesion system using acrylate block copolymers in response to photoirradiation and postbaking as the dual external stimuli for on-demand dismantling. *ACS Appl. Mater. Interfaces* **4,** 2124–2132 (2012).
42. Asadirad, A. M., Boutault, S., Erno, Z. & Branda, N. R. Controlling a polymer adhesive using light and molecular switch. *J. Am. Chem. Soc.* **136,** 3024–3027 (2014).
43. Heinzmann, C., Coulibaly, S., Roulin, A., Fiore, G. L. & Weder, C. Light-induced bonding and debonding with supramolecular adhesives. *ACS Appl. Mater. Interfaces* **6,** 4713–4719 (2014).
44. da Silva, L. F. M., Öchsner, A. & Adams, R. D. (eds). *Handbook of Adhesion Technology* (Springer-Verlag, 2011).
45. Yuan, C. *et al.* A π-conjugated system with flexibility and rigidity that shows environment-dependent RGB luminescence. *J. Am. Chem. Soc.* **135,** 8842–8845 (2013).
46. Yuan, C. *et al.* Hybridization of a flexible cyclooctatetraene core and rigid aceneimide wings for multiluminescent flapping π systems. *Chem. Eur. J* **20,** 2193–2200 (2014).
47. Kato, T., Mizoshita, N. & Kishimoto, K. Functional liquid-crystalline assemblies: self-organized soft materials. *Angew. Chem. Int. Ed.* **45,** 38–68 (2006).
48. Rosen, B. M. *et al.* Dendron-mediated self-assembly, disassembly, and self-organization of complex systems. *Chem. Rev.* **109,** 6275–6540 (2009).
49. Paquette, L. A. The current view of dynamic change within cyclooctatetraenes. *Acc. Chem. Res.* **26,** 57–62 (1993).
50. Nishinaga, T., Ohmae, T. & Iyoda, M. Recent studies on the aromaticity and antiaromaticity of planar cyclooctatetraene. *Symmetry* **2,** 76–97 (2010).
51. Nishiuchi, T. & Iyoda, M. Bent π-conjugated systems composed of three-dimensional benzoannulenes. *Chem. Rec.* **15,** 329–346 (2015).
52. Breton, G. W. & Vang, X. Photodimerization of anthracene: a $[4\pi_s + 4\pi_s]$ photochemical cycloaddition. *J. Chem. Educ.* **75,** 81–82 (1998).
53. Bhola, R. *et al.* A two-dimensional polymer from the anthracene dimer and triptycene motifs. *J. Am. Chem. Soc.* **135,** 14134–14141 (2013).
54. Acree, Jr W. E. & Chickos, J. S. Phase change enthalpies and entropies of liquid crystals. *J. Phys. Chem. Ref. Data.* **35,** 1051–1330 (2006).
55. Okuda, M. & Katayama, K. Photochemical reaction dynamics measured using the near-field heterodyne transient grating method. *J. Phys. Chem. A* **112,** 4545–4549 (2008).
56. Kamperman, M. & Synytska, A. Switchable adhesion by chemical functionality and topography. *J. Mater. Chem.* **22,** 19390–19401 (2012).
57. Lin, C.-H., Lin, K.-H., Pal, B. & Tsou, L.-D. Iterative synthesis of acenes via homo-elongation. *Chem. Commun.* 803–805 (2009).
58. Wender, P. A. & Christy, J. P. Nickel(0)-catalyzed [2+2+2+2] cycloadditions of terminal diynes for the synthesis of substituted cyclooctatetraenes. *J. Am. Chem. Soc.* **129,** 13402–13403 (2007).
59. Sheldrick, G. M. Crystal structure refinement with SHELXL. *Acta Crystallogr. C Struct. Chem.* **C71,** 3–8 (2015).

Acknowledgements

This work was supported by PRESTO 'Molecular technology and creation of new functions' from Japan Science and Technology Agency (JST) to S.S., by Grant-in-Aid for Scientific Research on Innovative Areas 'Photosynergetics' (MEXT KAKENHI Grant Number JP15H01083) to S.S. and by Grant-in-Aid for Young Scientists (A) (JSPS KAKENHI Grant Number JP15H05482) to S.S. We thank T. Kato and T. Nishimura (The University of Tokyo) for the fluorescence measurement of the LC thin film. We thank T. Hikage (Nagoya University) for the XRD measurement of the LCs. We thank M. Kotani, S. Suzuki and J. Watanabe (Denka Company Limited) for the valuable advice in the evaluation of shear strength. This work was partly supported by Nanotechnology Platform Program (Molecule and Material Synthesis) of the Ministry of Education, Culture, Sports, Science and Technology (MEXT), Japan.

Author contributions

S.S. conceived the research and performed the single-crystal X-ray structure analysis: S.N., E.T., and C.M. synthesized the compounds: S.N. evaluated the properties of the adhesive: M.H. and T.S. performed the grazing incidence XRD and atomic force microscopy measurements: C.Y. contributed to the establishment of the synthetic schemes: C.C. and S.I. conducted theoretical calculations and interpreted the results: S.S., S.N. and M.H. co-wrote the paper, and T.S., S.I. and S.Y. proofread it: S.S. designed and directed all the project.

Additional information

Competing financial interests: The authors declare no competing financial interests.

26

Identification of a new electron-transfer relaxation pathway in photoexcited pyrrole dimers

Simon P. Neville[1,†], Oliver M. Kirkby[2], Nikolas Kaltsoyannis[2,†], Graham A. Worth[1] & Helen H. Fielding[2]

Photoinduced electron transfer is central to many biological processes and technological applications, such as the harvesting of solar energy and molecular electronics. The electron donor and acceptor units involved in electron transfer are often held in place by covalent bonds, π-π interactions or hydrogen bonds. Here, using time-resolved photoelectron spectroscopy and *ab initio* calculations, we reveal the existence of a new, low-energy, photoinduced electron-transfer mechanism in molecules held together by an NH$\cdots\pi$ bond. Specifically, we capture the electron-transfer process in a pyrrole dimer, from the excited π-system of the donor pyrrole to a Rydberg orbital localized on the N-atom of the acceptor pyrrole, mediated by an N-H stretch on the acceptor molecule. The resulting charge-transfer state is surprisingly long lived and leads to efficient electronic relaxation. We propose that this relaxation pathway plays an important role in biological and technological systems containing the pyrrole building block.

[1] School of Chemistry, University of Birmingham, Edgbaston, Birmingham B15 2TT, UK. [2] Department of Chemistry, University College London, 20 Gordon Street, London WC1H 0AJ, UK. † Present addresses: Department of Chemistry, University of Ottawa, 10 Marie Curie, Ottawa, Ontario, Canada K1N 6N5 (S.P.N.); School of Chemistry, The University of Manchester, Oxford Road, Manchester M13 9PL, UK (N.K.). Correspondence and requests for materials should be addressed to G.A.W. (email: g.a.worth@bham.ac.uk) or to H.H.F. (email: h.h.fielding@ucl.ac.uk).

Electron-transfer (ET) reactions in which an electron moves from a donor atom or molecule to an acceptor atom or molecule are ubiquitous in nature and technology[1,2]; important examples include the ET chains that drive photosystems I and II (ref. 3) and nanoscale electronic devices[4]. ET processes are often mediated by covalently bound molecular bridges that hold the donor and acceptor in place; for example, in bridged zinc porphyrin-quinone complexes, which are structurally related to the photosynthetic reaction centre, excitation of the porphyrin results in rapid ET to the quinone accompanied by large-scale vibrations of the bridges[5,6]. Photoinduced ET can also take place between non-covalently bonded donors and acceptors; for example, between π-stacked nucleobases[7]. Recently, there has been a great deal of interest in ET reactions occurring between hydrogen-bonded donors and acceptors accompanied by proton transfer. This class of ET reaction has been shown to provide an efficient mechanism for excited state deactivation following absorption of UV photons. Such proton-coupled ET reactions are important in the photostability of proteins, DNA and other essential bio-macromolecules[8] and in the photorepair cycles of UV-damaged DNA[9]. Proton-coupled ET has also been found to play an important role in the photostability of hydrogen-bonded heterocycles, such as the pyrrole dimer[10,11].

The pyrrole molecule is a common motif in biology; for example, in tryptophan and porphyrins. It is the key to the photoactivation of the phytochrome enzyme[12]. Pyrrole is also a basic building block in many biologically and technologically important systems, such as polyamide DNA-binding agents in chemical biology[13,14], dye-sensitized solar cells[15,16], the polypyrrole conducting polymer[17] and self-assembled architectures containing donor–acceptor units[18]. Moreover, it is a prototypical heteroaromatic molecule that undergoes highly efficient non-radiative relaxation on a $\pi\sigma^*$ potential surface[19–21]. Dimerization of pyrrole has been found to open up additional relaxation pathways, such as a high-energy channel involving proton-coupled ET[10,11].

In this article, we report the observation of a new, low-energy relaxation channel in pyrrole dimers in the gas-phase, involving electron transfer between the two pyrrole molecules. Using time-resolved photoelectron spectroscopy, we find that at high gas pressures, when dimers are present, a spectroscopic feature appears on a sub-picosecond timescale that is not observed at low pressures when only monomers are present. High level *ab initio* calculations show that the lowest excited state of the dimer corresponds to a charge-transfer (CT) state and that the ionization energy of this state is consistent with the spectral feature associated with the dimers. Analysis of the orbitals and molecular structures shows that ET takes place from a π-orbital on the donor pyrrole molecule to a Rydberg orbital localized on the N-atom of the acceptor pyrrole molecule. Time-resolved photoelectron spectra recorded following excitation at different wavelengths, together with calculations of potential energy surfaces, suggest that the N–H bond of the acceptor pyrrole must stretch for ET to take place; in the framework of Marcus theory, the N–H bond of the acceptor pyrrole stretches to bring the dimer to a geometry in which there is a strong vibronic coupling between neutral and CT states. As with the high-energy, proton-coupled ET process identified by Slavíček and co-workers[10,11], this new channel competes with dynamics on the $\pi\sigma^*$ state. A surprising feature of this CT complex is its lifetime. One might assume that, since the dimer is composed of identical units, the back transfer is fast and the CT complex has a negligible lifetime; however, we do not find any evidence for back transfer, which suggests that the CT complex is sufficiently long lived to potentially form a pyrrole ion pair. The elucidation of this new ET mechanism in the pyrrole dimer is significant and may provide a basis for practical applications, such as the rational design of new photoactive materials.

Results

Photoelectron spectra of pyrrole and pyrrole clusters. He carrier gas was passed through a reservoir of pyrrole held at 25 °C and the mixture was expanded through a 50 µm diameter, 1-mm-long nozzle. The distribution of pyrrole cluster ions formed by 1 + 1 ionization of the clusters was 100% monomer when using 0.4 bar He and ~89% monomer, 10% dimer and 1% trimer when using 1.8 bar He.

The spectral signatures of the pyrrole dimer were identifed by comparing photoelectron spectra obtained with different backing pressures of He gas. Figure 1 shows the 240 nm (5.17 eV) 1 + 1 multiphoton photoelectron spectra, recorded with helium carrier gas pressures of 0.4 and 1.8 bar. At 0.4 bar, the spectrum is identical to the one-photon, vacuum UV photoelectron spectrum of pyrrole[22], indicating that it is predominantly a 1 + 1 multiphoton photoelectron spectrum of the monomer. The maximum electron kinetic energy (eKE) possible following 1 + 1 photoionization of pyrrole, calculated using the central wavelength of the laser pulse and the experimental adiabatic ionization energy (8.207 eV) (refs 23,24), is 2.13 eV and is marked on Fig. 1. The photoelectron spectrum at 1.8 bar has an additional feature ~2.45 eV, which lies above the maximum eKE possible following 1 + 1 photoionization of the monomer. The intensity of this feature is independent of laser power but increases with

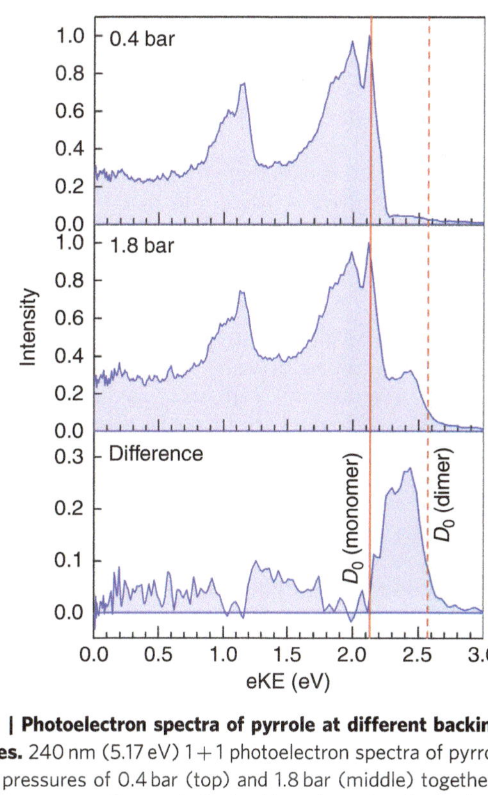

Figure 1 | Photoelectron spectra of pyrrole at different backing pressures. 240 nm (5.17 eV) 1 + 1 photoelectron spectra of pyrrole with He backing pressures of 0.4 bar (top) and 1.8 bar (middle) together with the difference spectrum (bottom) obtained by subtracting the 0.4 bar spectrum from the 1.8 bar spectrum. The 0.4 and 1.8 bar spectra have been normalized to the maximum intensity peak at 2.13 eV. The solid and dashed red lines mark the maximum eKEs possible from 1 + 1 ionization of the pyrrole monomer and dimer, respectively, calculated using the central wavelength of the laser pulses; the experimental adiabatic ionization potential of the monomer[23] and the EOM-IP-CCSD/aug-cc-pVDZ calculated vertical ionization energy of the dimer (Table 2).

increasing He backing pressure. At higher pressures, pyrrole clusters are formed and this feature can be attributed to photoionization of these clusters. Since our molecular beam at 1.8 bar is dominated by monomers and the concentration of clusters decreases exponentially with size, it seems most likely that this feature arises from photoionization of pyrrole dimers. To test this suggestion, we have calculated the vertical ionization energies (VIEs) of the monomer, dimer and trimer at the EOM-IP-CCSD/aug-cc-pVDZ level of theory. For the monomer, the lowest VIE is 8.15 eV (ref. 21), in good agreement with the experimental adiabatic ionization energy of 8.207 eV (refs 23,24). For the dimer, the lowest VIE from the T-shaped minimum energy geometry is 7.77 eV. The pyrrole trimer has three stable conformers and the minimum energy form is predicted by theory and experiment to have a cyclic triangular shape with C_{3h} symmetry[25,26]. From this conformer, the lowest VIE of the trimer is 8.10 eV, very close to that calculated for the monomer. The next most stable trimer structure is 0.16 eV (15 kJ mol^{-1}) higher in energy and will not contribute significantly to our experiments. The same trend in VIEs has been calculated at the MP2 level[27]. The optimized geometries for the dimer and trimer at the MP2 level are given in Supplementary Tables 2 and 3.

Using these calculated VIEs, the maximum eKEs for electrons generated by 240 nm 1 + 1 photoionization of the dimer and trimer are 2.57 and 2.24 eV, respectively, confirming our suggestion that the feature observed around 2.45 eV in the spectrum recorded with 1.8 bar He backing pressure is attributed to 1 + 1 photoionization of the pyrrole dimer. The 240 nm 1 + 1 photoelectron spectrum of the dimer can be obtained by subtracting the 0.4 bar spectrum (mostly monomer) from the 1.8 bar spectrum (monomer and clusters) and this is presented in the lower panel of Fig. 1. It should be noted that this is only approximate since the normalization of the two individual spectra is rather arbitrary and there will be some contributions from larger clusters. Nonetheless, it is clear that the photoelectron spectrum of the dimer is dominated by a relatively sharp and symmetric feature centred ~2.4 eV and that the broad feature that lies under the monomer photoelectron spectrum is significantly weaker than the monomer photoelectron spectrum.

Electronic structure of the pyrrole dimer. The pyrrole dimer is known to have a T-shaped structure with C_s symmetry[25,26], in which the two pyrrole molecules are held together by an NH$\cdots\pi$ bond (Fig. 2a), a motif found in protein structures[28,29]. This non-typical hydrogen bond is also responsible for the self-assembly of pyrrole dimers in apolar solvents[30]. The monomer whose N–H bond points into vacuum is labelled A and the monomer whose N–H bond is directed towards the π electron system of A is labelled B. The centres of mass of the monomers are calculated to be separated by $R = 4.06$ Å, the planes of the monomers are at angle of $\phi = 51.5°$, and the centre of mass vector has an angle of $\theta = 13.6°$ from the norm of the plane of monomer A, in agreement with the results of gas-phase experiments[31,32] and other *ab initio* calculations[11,33], including recent dispersion-corrected density functional theory (DFT) calculations[34]. The latter study also employed molecular dynamics simulations to show that the NH$\cdots\pi$ bond persists in liquid pyrrole.

The results of electronic structure calculations for the pyrrole dimer are summarized in Tables 1 and 2 and the energy level diagram in Fig. 2a. At all levels of theory, both the S_1 and S_2 states of the neutral dimer are found to be of A″ symmetry and have 3*s* character at the Franck–Condon point. The $S_1(\pi_B 3s_A)$ state has CT character, with the dominant configuration corresponding to excitation from a π-orbital localized on monomer B to the 3*s* orbital localized on the N-atom of monomer A (Fig. 2b). In contrast, the $S_2(\pi_A 3s_A)$ state is dominated by a configuration corresponding to excitation from a π-orbital localized on monomer A to the 3*s* orbital localized on the same monomer (Fig. 2b), reminiscent of the $A_2(\pi 3s)$ state of the pyrrole monomer[35].

The $D_1(\pi_A)$ and $D_2(\pi_A)$ states of the pyrrole dimer cation have A′ and A″ symmetries, respectively, and are dominated by configurations corresponding to removal of an electron from π-orbitals localized on monomer A. From a consideration of the dominant configurations and the symmetries of these states, the dimer cation states $D_1(\pi_A)$ and $D_2(\pi_A)$ correlate with the $D_1(\pi)$ and $D_0(\pi)$ states of the monomer cation, which have symmetries B_1 and A_2 respectively, and correspond to removing an electron from the HOMO-1 π-orbitals. The $D_0(\pi_B)$ dimer cation state is dominated by a configuration corresponding to removal of an electron from a π-orbital localized on monomer B and lies below the lowest two monomer-like states with a calculated VIE of 7.77 eV.

The photoionization cross-sections corresponding to ionization to the $D_0(\pi_B)$ and $D_2(\pi_A)$ cation states from the $S_1(\pi_B 3s_A)$ and $S_2(\pi_A 3s_A)$ dimer neutral states are plotted in Fig. 2c. The $S_1(\pi_B 3s_A)$ state is predicted to ionize preferentially to the $D_0(\pi_B)$ state in the region of the probe photon energy (4.13 eV), while the $S_2(\pi_A 3s_A)$ state is predicted to ionize preferentially to the higher lying $D_2(\pi_A)$ state. The cross-section for ionization to the $D_1(\pi_A)$ state of the cation is close to zero for ionization from either neutral excited state, similar to the negligible photoionization cross-section found for the $S_1(\pi 3s)$ state of the pyrrole monomer to the monomer cation $D_1(\pi)$ state[21].

Time-resolved photoelectron spectra. Figure 3a–d shows the time-resolved 1 + 1′ photoelectron spectra recorded using 1.8 bar He carrier gas (monomers and clusters) with pump wavelengths of 249.5 nm (4.97 eV), 245 nm (5.06 eV), 240 nm (5.17 eV) and 200 nm (6.20 eV) and a probe wavelength of 300 nm (4.13 eV). It is worth noting that the eKE axes are not directly comparable with those in Fig. 1 because different wavelengths are employed in the 1 + 1 and 1 + 1′ photoionization schemes.

For each photoexcitation wavelength, the integrated areas of the photoelectron spectra have been scaled to the total integrated photoelectron signals at the corresponding pump–probe delays and plotted as a contour map. No obvious variation in anisotropy parameter with pump–probe delay was observed. The lifetimes determined from the time-resolved photoelectron spectra are presented in Table 3. To gain insight into the flow of excited-state population, the time-resolved photoelectron spectra were fit to sums of the exponential decays listed in Table 3 and convoluted with the pump–probe cross-correlation $g(t)$,

$$S(\text{eKE}, t) = \sum_i C_i(\text{eKE}) e^{-t/\tau_i} \otimes g(t). \qquad (1)$$

The coefficients C_i(eKE) represent the contribution of the *i*th decay at a given eKE. The spectra of the fit coefficients C_i(eKE) are plotted in Fig. 3e–h. Positive values of C_i(eKE) represent exponential decay on a timescale τ_i and negative values represent exponential growth on a timescale τ_i.

Only a single lifetime, $\tau_1 = 47$ fs, was required to fit the 249.5 nm data. The spectrum associated with this lifetime (Fig. 3e) has positive amplitudes everywhere, indicating a rapid decay of population out of the photoionization window. Following excitation at 245 and 240 nm, the spectra associated with the τ_1 timescale have positive amplitudes in the range 0–1.3 eV but negative amplitudes ~1.45 eV (Fig. 3f,g). This suggests an evolution along the excited potential energy surface from a region with a photoelectron spectrum with lower eKE to a

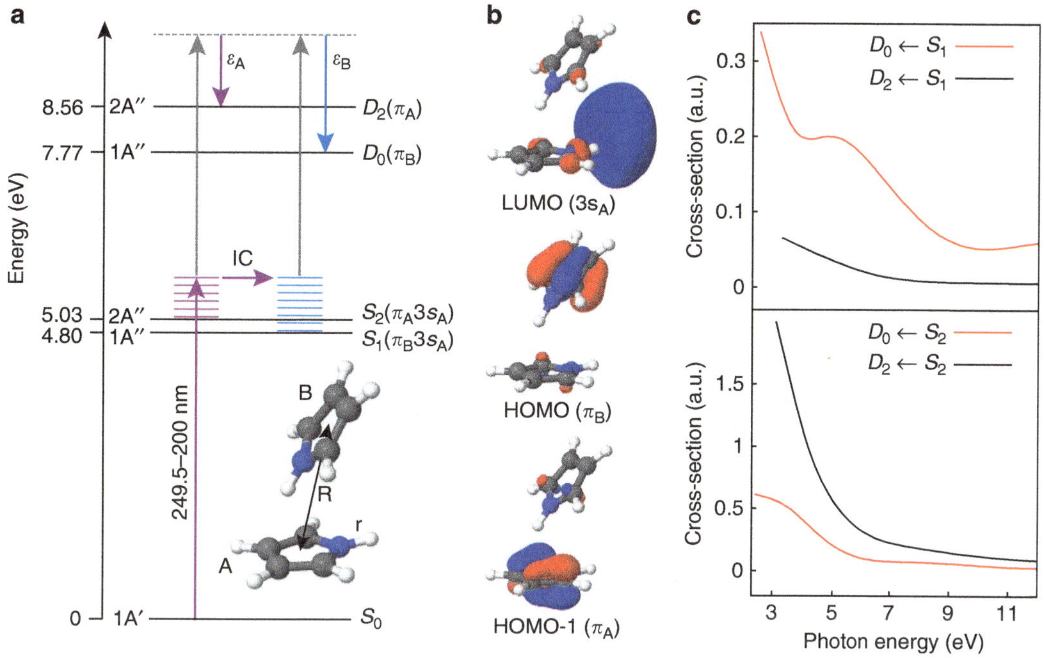

Figure 2 | Electronic structure of the pyrrole dimer. (**a**) Energy level diagram showing the ordering of the lowest electronic states of the pyrrole dimer calculated at the CASPT2(8,8) level and the cation calculated at the EOM-IP-CCSD/aug-cc-pVDZ level. Upward arrows represent pump (purple) and probe (grey) laser pulses and the horizontal lines represent vibrational energy in the $S_1(\pi_B 3s_A)$ and $S_2(\pi_A 3s_A)$ states. The horizontal arrow represents IC from the $S_2(\pi_A 3s_A)$ state to the $S_1(\pi_B 3s_A)$ state. Downward arrows represent the eKE of the emitted photoelectrons following $S_1(\pi_B 3s_A) \rightarrow D_0(\pi_B)$ and $S_2(\pi_A 3s_A) \rightarrow D_2(\pi_A)$ ionization. Inset: ground-state minimum energy geometry of the pyrrole dimer calculated at the SCS-MP2/aug-cc-pVDZ level. The monomer whose N-H bond points into vacuum is labelled A and the other monomer is labelled B. (**b**) Molecular orbitals involved in transitions from S_0 to the $S_1(\pi_B 3s_A)$ and $S_2(\pi_A 3s_A)$ states of the neutral dimer. (**c**) Photoionization cross-sections, as a function of photon energy, for ionization from the $S_1(\pi_B 3s_A)$ and $S_2(\pi_A 3s_A)$ states to the $D_0(\pi_B)$ and $D_2(\pi_A)$ cation states, calculated at the EOM-IP-CCSD/aug-cc-pVDZ level. IC, internal conversion.

Table 1 | VEEs of the pyrrole dimer.

State	Configuration	ΔE(CASSCF)	ΔE(CASPT2)	ΔE(EOM-CCSD)	ΔE(DFT/MRCI)
S_1, 1A″	$\pi_B \rightarrow 3s_A$	4.84	4.80	5.19	4.79
S_2, 2A″	$\pi_A \rightarrow 3s_A$	5.18	5.03	5.56	5.24

DFT, density functional theory; MRCI, multi-reference configuration interaction; VEE, vertical excitation energy.
VEEs and dominant configurations of the first two excited states of the pyrrole dimer calculated at the CASSCF(8,8), CASPT2(8,8), EOM-CCSD and DFT/MRCI levels using the aug-cc-pVDZ basis. VEEs are given in units of eV.

Table 2 | VIEs of the pyrrole dimer.

State	Configuration	VIE
D_0, 1A″	$\pi_B \rightarrow \psi_k$	7.77
D_1, 1A″	$\pi_A \rightarrow \psi_k$	8.52
D_2, 2A″	$\pi_A \rightarrow \psi_k$	8.56

VIE, vertical ionization energy.
VIEs and dominant configurations of the first three cationic states of the pyrrole dimer, calculated at the EOM-IP-CCSD level using the aug-cc-pVDZ basis set. VIEs are given in units of eV.

region with a photoelectron spectrum with higher eKE. At 245 and 240 nm, spectra associated with a second ultrafast timescale, $\tau_2 \approx 190$–270 fs, are centred around 1.45 eV with positive amplitude, suggesting that once populated, the region of the excited potential energy surface with a photoelectron spectrum with higher eKE decays out of the photoionization window with a slightly longer timescale τ_2. This sequential population and decay is clearly visible as a delayed rise and fall of photoelectron signal ~1.45 eV in Fig. 3b,c as well as in plots showing the integrated photoelectron signals corresponding to τ_1 and τ_2 (Fig. 3j,k). At 240 nm, a third lifetime is required to fit the data, $\tau_3 \approx 2.5$ ps. The spectrum associated with this timescale has very low amplitude, which is either the signature of larger clusters or may be attributed to a small fraction of the population remains trapped on the excited potential energy surface of the dimer within the photoionization window and decaying on a much slower timescale. At 200 nm, the situation is similar to that for 240 nm. The spectrum associated with the τ_1 timescale has positive amplitudes across the range 0–2.5 eV, but there is a dip in the amplitude ~1.45 eV that coincides with the spectrum associated with the τ_2 timescale that has positive amplitude. This is consistent with a flow of population from a region of the excited potential energy surface that has a broad photoelectron spectrum with maximum eKE ~2.5 eV to a region of the photoelectron spectrum that has a maximum eKE ~1.45 eV. A third lifetime, $\tau_3 \approx 1.1$ ps, is also required to fit the 200 nm photoelectron spectra. Again, the spectrum associated with this timescale has low amplitude; however, it is larger than the equivalent spectrum for 240 nm and has a small dip ~1.5 eV. Again, this is either a signature of larger clusters or it may be attributed to population that remains trapped on the excited potential energy surface of

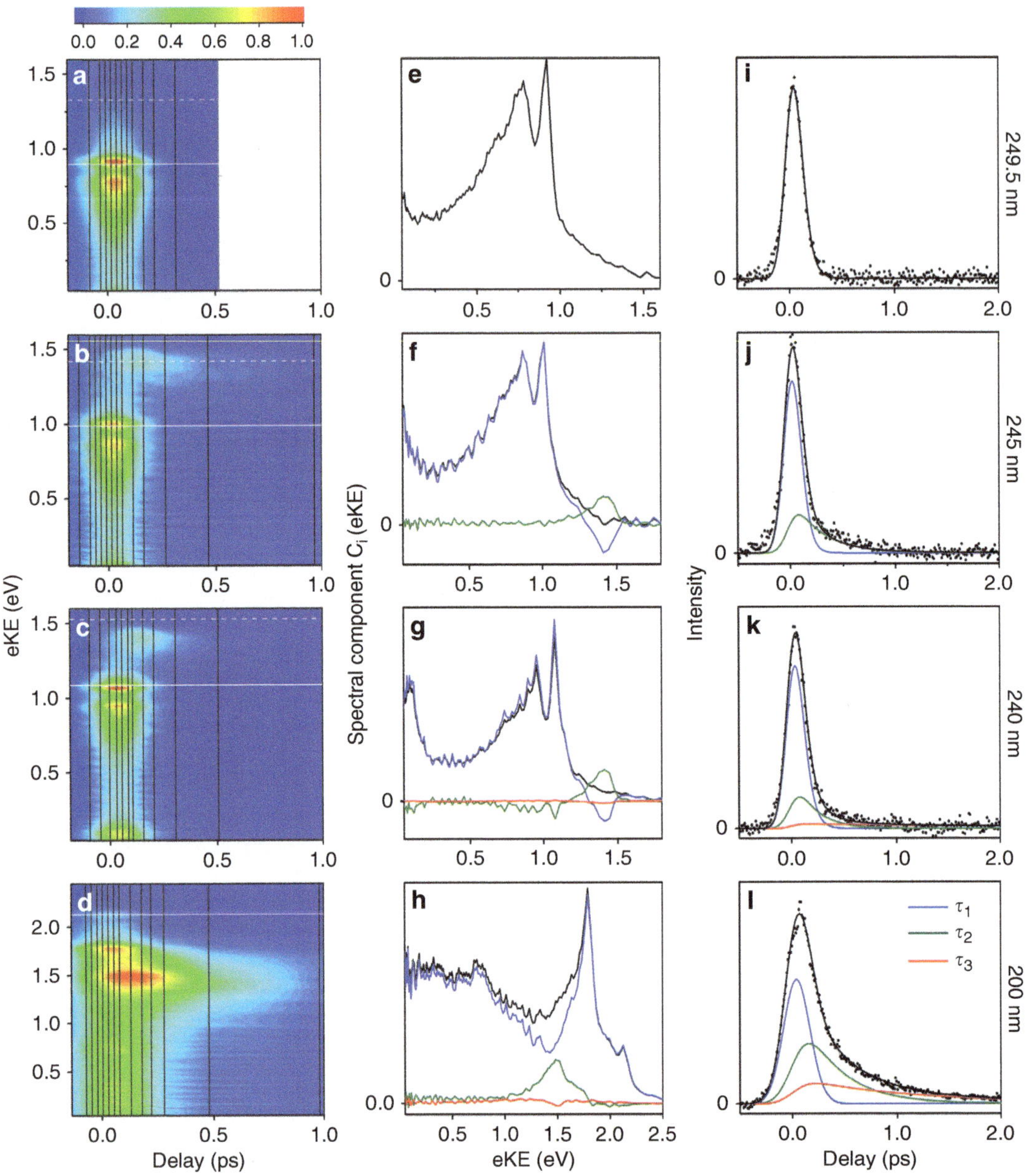

Figure 3 | Excited state dynamics of pyrrole at 1.8 bar. (a-d) Contour plots showing experimental time-resolved photoelectron spectra following excitation at 249.5 nm (4.97 eV), 245 nm (5.06 eV), 240 nm (5.17 eV) and 200 nm (6.20 eV). Individual plots were normalized to their maximum photoelectron signals. The vertical black lines mark the pump-probe delays at which photoelectron spectra were recorded and the shading was smoothed using linear interpolation. The horizontal white lines mark the maximum eKEs possible from 1 + 1′ ionization, calculated using the central wavelengths of the pump and probe laser pulses and the experimental adiabatic ionization potentials of the pyrrole monomer, 8.207 eV (solid lines)[23], and EOM-IP-CCSD/aug-cc-pVDZ calculated vertical ionization energies of the pyrrole dimer (dashed lines), 7.77 eV (this work). **(e-h)** Spectral components of $C_i(eKE)$ extracted from the time-resolved photoelectron spectra using the decay times listed in Table 3 and equation (1). The sum of $C_i(eKE)$ (black lines) represents the combined photoelectron spectrum of the initially excited states of the pyrrole monomer and dimer. **(i-l)** Integrated photoelectron spectra of the initially populated momomer and $S_2(\pi_A 3s_A)$ state of the dimer (blue) that decay on a timescale <60 fs, the $S_1(\pi_B 3s_A)$ state of the dimer (green) whose population rises on a timescale <60 fs and subsequently decays on a timescale 190-360 fs, the longer-lived excited state population (red) and the total integrated photoelectron spectrum (black), together with experimental data (scattered points).

the dimer and flows into the region of the photoelectron spectrum with maximum eKE ~1.45 eV.

Discussion

The time-resolved photoelectron spectra are dominated by the excited-state dynamics of the pyrrole monomer. Recently, Wu *et al.* used time-resolved photoelectron spectroscopy to probe the relaxation dynamics of the pyrrole monomer following excitation in the range 242–217 nm (5.1–5.7 eV) (ref. 21). They found that photoexcitation in this range resulted in ultrafast decay of the system from the ionization window on a timescale <30 fs, in agreement with other femtosecond studies[36–39]. Our time-resolved photoelectron spectra of the pyrrole monomer using 0.4 bar He carrier gas and the lifetimes (Supplementary Table 1) are in agreement with those of Wu *et al.* The main difference between the time-resolved photoelectron spectra recorded at 0.4 and 1.8 bar is the feature around 1.45 eV, which has already been assigned to the dimer and is observed to

Table 3 | Summary of pyrrole lifetimes and pump-probe cross-correlation measurements.

λ_{pump}/nm	τ_1/fs	τ_2/fs	τ_3/ps	$g(t)$/fs
249.5	47 ± 1			180 ± 1
245	46 ± 8	270 ± 40		179 ± 8
240	57 ± 8	190 ± 30	2.5 ± 1.5	179 ± 9
200	31 ± 1	360 ± 120	1.1 ± 0.2	289 ± 3

The 1/e lifetimes and pump-probe cross-correlation, $g(t)$, extracted from time-resolved photoelectron spectra of pyrrole using 1.8 bar He carrier gas with different pump wavelengths. The errors quoted represent 2 s.d.

have a delayed rise on an ultrafast timescale (<60 fs) and subsequent decay on a slightly longer timescale (190–360 fs).

To a first approximation, the photoelectron spectrum of the dimer can be understood using conservation of energy. Assuming the excess vibrational energy for a given electronic state, $E_{vib} = h\nu_{pump} - E(S_n)$, is conserved during photoionization, $eKE \approx h\nu_{probe} - [E(D_n) - E(S_n)]$. Thus, from the vertical excitation energies (VEEs) of the neutral dimer calculated at the CASPT2 level (Table 1) and the VIEs (Table 2), ionization from the $S_2(\pi_A 3s_A)$ state of the dimer to the $D_2(\pi_A)$ state is predicted to produce photoelectrons with eKE ~ 0.6 eV, coinciding with the photoelectron spectrum of the monomer. The $S_1(\pi_B 3s_A)$ state is predicted to ionize preferentially to the $D_0(\pi_B)$ cation and is predicted to produce photoelectrons with eKE ~ 1.2 eV. Thus, our photoelectron spectra show that the $S_2(\pi_A 3s_A)$ state of the dimer, which has a similar photoelectron spectrum to the $S_1(\pi 3s)$ state of the monomer, is populated directly and has a lifetime that is indistinguishable from that of the $S_1(\pi 3s)$ state of the monomer. The $S_1(\pi_B 3s_A)$ CT state of the dimer is populated indirectly, from the $S_2(\pi_A 3s_A)$ state on a timescale of <60 fs and subsequently decays on a slightly longer timescale (190–360 fs).

For the dimer, the calculated VEEs (Table 1), confirm that excitation at 249.5, 245 and 240 nm will result predominantly in population of the $S_2(\pi_A 3s_A)$ state, which has an oscillator strength an order of magnitude higher than the $S_1(\pi_B 3s_A)$ state (Supplementary Table 4). Excitation at 200 nm is likely to result in transitions to higher lying optically bright $\pi \rightarrow \pi^*$ or $\pi \rightarrow 3p$ states, predominantly localized on monomer A (Supplementary Table 4). The potential energy surfaces of the S_0, $S_1(\pi_B 3s_A)$ and $S_2(\pi_A 3s_A)$ states along the N–H dissociation coordinate $\boldsymbol{r}$ of monomer A and the vector $\boldsymbol{R}$ connecting the centres of mass of the two monomers are presented in Fig. 4. In the N–H dissociation limit, the adiabatic S_1 state correlates with the diabatic $S_2(\pi_A 3s_A)$ state, while the adiabatic S_2 state correlates with the diabatic $S_1(\pi_B 3s_A)$state. This change in character is represented on the plot by a change in colour on the adiabatic curves along the N–H stretching coordinate $\boldsymbol{r}$. In a diabatic picture, the $S_1(\pi_B 3s_A)$ CT state is essentially bound with respect to N–H dissociation and the $S_2(\pi_A 3s_A)$ state has only a shallow barrier to dissociation. However, in the adiabatic picture, the N–H bond becomes bound in the $S_2(\pi_A 3s_A)$ state due to the avoided crossing with the lower $S_1(\pi_B 3s_A)$ CT state. The $S_1(\pi_B 3s_A)$ state is found to be weakly bound with respect to the monomerization coordinate $\boldsymbol{R}$, with a barrier to monomerization of around 0.1 eV, whereas the $S_2(\pi_A 3s_A)$ state potential is found to be repulsive along this $\boldsymbol{R}$ coordinate. Motions along the $\boldsymbol{r}$ and $\boldsymbol{R}$ coordinates will be the key initial motions because the N–H bond in pyrrole is photolabile and the cluster is only weakly bound; however, motions along other degrees of freedom may also be involved in the electronic relaxation pathway. For example, motion along the N out-of-plane coordinate and some ring stretches are known be involved in the relaxation dynamics of pyrrole[35] so may also play a role in the relaxation of the pyrrole dimer. Motion along a tilting coordinate that brings the monomers into co-planarity may play a role at longer timescales since it is known that in the benzene dimer the monomers are co-planar[40] and such a geometry might stabilize the CT state in the pyrrole dimer. Nonetheless, these motions will not change the major character of the initial dynamics which will be dominated by motions along the N–H stretch and monomerization coordinates.

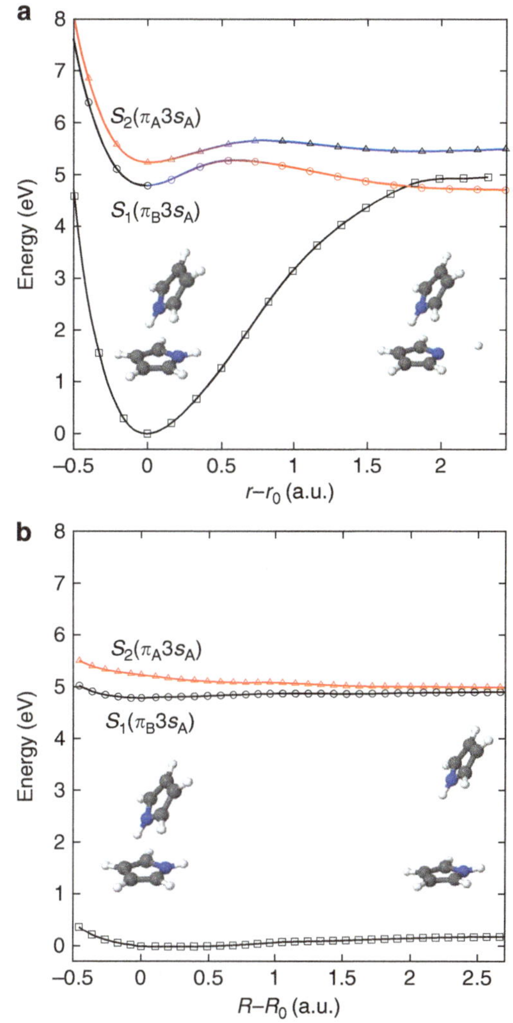

Figure 4 | Electron-transfer process between non-covalently bonded pyrrole molecules. Adiabatic potential energy curves for the S_0 (black), $S_1(\pi_B 3s_A)$ (blue) and $S_2(\pi_A 3s_A)$ (red) states calculated at the DFT/MRCI/aug-cc-pVDZ level, (**a**) along the N-H dissociation coordinate of monomer A and (**b**) along the vector $\boldsymbol{R}$ connecting the centres of masses of the two monomers.

At 249.5 nm, a single ultrafast decay out of the photoionization window of the predominantly populated $S_2(\pi_A 3s_A)$ state is observed. Since there is only one timescale, which is indistinguishable from that of the monomer, it seems that monomer separation together with ultrafast N–H dissociation of monomer A is likely to be the predominant decay channel of the dimer at this excitation energy. At this wavelength, excitation is to the bottom of the $S_2(\pi_A 3s_A)$ potential well plotted as a function of the N–H bond length; however, as soon as the dimers start to separate along the dissociative R monomerization coordinate, the barrier to N–H dissociation will approach that of the monomer and thus the dimer will display only monomer-like behaviour.

At 245 and 240 nm, ultrafast decay out of the photoionization window is still observed, but there is a delayed rise and subsequent decay of the higher-energy dimer band around 1.45 eV (Fig. 3b,c), which is consistent with a competing process populating the $S_1(\pi_B 3s_A)$ state. At 200 nm, a similar ultrafast population and subsequent decay of the $S_1(\pi_B 3s_A)$ state is observed, also ~1.45 eV because of the propensity for conservation of excess vibrational energy during photoionization. These observations suggest that a crossing point between the $S_1(\pi_B 3s_A)$ and $S_2(\pi_A 3s_A)$ states becomes accessible between 249.5–245 nm (4.97–5.06 eV) and crossing occurs to form the CT complex on a timescale that is faster than that of monomer separation in the $S_2(\pi_A 3s_A)$ state. This population is likely to occur at the avoided crossing between the $S_1(\pi_B 3s_A)$ and $S_2(\pi_A 3s_A)$ states along the N–H bond stretch (Fig. 4). However, as discussed above, motions along other coordinates are also expected to play a role in the dynamics, possibly turning this avoided crossing into a conical intersection (although as the states have the same symmetry this would be an accidental intersection) and moving towards co-planarity. Nonetheless, these motions will not change the basic picture of initial N–H stretching followed by crossing to form the CT state that must occur.

It can also be inferred that the decay time $\tau_2 \sim 190$–360 fs (Table 3) is the lifetime of the dimer CT complex. However, how the CT complex relaxes cannot be ascertained from our experiments and calculations and various pathways are possible. Significantly, there is no experimental evidence for re-crossing from the CT complex back to the neutral dimer excited state. This is surprising given the equivalent electronegativities of the monomers. Another possibility might be N–H dissociation in the $S_1(\pi_B 3s_A)$ state, but it should be noted that the barrier to H-atom loss in the $S_1(\pi_B 3s_A)$ state (0.5 eV, calculated at the DFT/multi-reference configuration interaction (MRCI) level), is significantly larger than that of the monomer (0.25 eV (ref. 35)). The most plausible pathways are: direct crossing to the ground-state, as suggested by Poterya *et al.*[10]; slow separation to form a monomer pair as the barrier to monomerization calculated to exist in the $S_1(\pi_B 3s_A)$ state at the DFT/MRCI level is very small (Fig. 4); or formation of a co-planar excimer ion pair as seen in benzene[40] if the tilt motion becomes important during separation.

In conclusion, the dynamics of the excited states of the pyrrole dimer have been studied using a combination of time-resolved photoelectron spectroscopy and *ab initio* calculations. It has been revealed that the formation of pyrrole dimers opens up a new relaxation pathway involving ET from the excited π-system of one pyrrole molecule to a Rydberg orbital localized on the N-atom of the other pyrrole molecule and that the ET is mediated by stretching of the N–H bond on the acceptor molecule leading to efficient curve crossing. Furthermore, our results reveal that the resulting CT complex has a lifetime of a few 100 fs and that it may form a pyrrole ion pair. We propose that this relaxation pathway will play a role in the photochemistry and photophysics of biologically and technologically important systems containing the pyrrole building block or indeed other NH–π bonded molecular units.

Methods

Cluster source. Helium carrier gas (0.4–1.8 bar) was passed through a reservoir of liquid pyrrole held at room temperature (25 °C) outside the vacuum chamber and carried into the vacuum chamber through 1/16″ tubing to a 1-mm-long nozzle of 50 μm diameter. The vapour pressure of pyrrole at 25 °C is 8.35 mm Hg and the concentration of pyrrole in 1.8 bar He is estimated to be 0.6%. The molecular beam passed through a 1 mm skimmer, located ~40 mm from the nozzle, into the interaction region of a photoelectron velocity map imaging spectrometer[41–43]. To confirm the formation of clusters, we recorded ion images following two-photon ionization at 249.5 nm. The time-of-flight from the interaction region to the detector depends on the mass of the ion, $t_{\mathrm{ToF}} \propto \sqrt{m}$, and we observe images of dimer ions displaced by $x_{\mathrm{py}}\sqrt{2}$ and trimer ions displaced by $x_{\mathrm{py}}\sqrt{3}$, where x_{py} is the displacement along the molecular beam axis of the pyrrole monomer ions, which is perpendicular to the time-of-flight axis. With a carrier gas pressure of 0.4 bar, 100% of the ion signal corresponded to $m/z \equiv C_4H_5N^+$. At 1.8 bar, the cluster ion distribution at the detector is exponential with around 89% monomer, 10% dimer and 1% trimer (Supplementary Fig. 1 and Supplementary Note 1) and the average cluster ion size at the detector is $\bar{N} \approx 1$. We note that these measurements are lower limits for the fractions of clusters in the interaction region because they do not account for fragmentation of cluster ions on the way to the detector. Nonetheless, our observations are consistent with those of Poterya *et al.*[10] and Profant *et al.*[27] who measured average cluster sizes $\bar{N} = 3$ following supersonic expansion of pyrrole through a 2-mm-long conical nozzle of 60 μm diameter, with a He backing pressure of 1.5 bar and pyrrole reservoir and nozzle temperatures 8 and 9 °C, respectively, substantially lower than those in our experiments.

Time-resolved photoelectron spectra. The time-resolved photoelectron spectra were recorded with helium gas pressures of 1.8 bar. After collimation by a 1 mm skimmer, the molecular beam was intersected by femtosecond pump (249.5–200 nm) and probe (300 nm) laser pulses. Pump–probe cross-correlation full-width half-maximum measurements were in the range 180–290 fs. The pump and probe pulses were focussed to a spot size of diameter ~50 μm after being attenuated to <1 μJ per pulse ($<10^{11}$ W cm^2) to minimize multiphoton processes and space-charge effects. For each excitation wavelength, a set of ~15 photoelectron images with pump–probe delays in the range −0.25 to 1 ps were recorded, together with the total integrated photoelectron signal. Photoelectron velocity distributions were recovered from the raw photoelectron images using the pBasex image inversion algorithm[44] and the energy scale was calibrated by recording the 2+1 resonance-enhanced multiphoton ionization spectrum of Xe at 249.6 nm (ref. 45). The resolution of this instrument is ~3.5%. To extract decay times from a set of time-resolved photoelectron spectra, the total integrated areas of the photoelectron spectra recorded at each pump–probe delay were scaled to the total integrated photoelectron signal intensity. A least-squares fitting procedure using the Levenberg–Marquardt optimization algorithm was used to fit integrated portions of the set of scaled spectra to sums of exponentially decaying profiles convoluted with a Gaussian cross-correlation function, representing the cross-correlation of the pump and probe laser pulses $g(t)$ (refs 43,46),

$$S(t) = \sum_i c_i e^{-t/\tau_i} \otimes g(t). \tag{2}$$

c_i is the intensity of the ith decay with $1/e$ decay time τ_i. The residuals from the fits are shown in Supplementary Fig. 2.

Computational methodology. Ground-state minimum energy geometries of the pyrrole monomer, dimer and trimer were optimized at the spin-component-scaled MP2 (SCS-MP2)[47] level using the aug-cc-pVDZ basis. Ionization potentials for the lowest cationic states were calculated at the EOM-IP-CCSD/aug-cc-pVDZ level. For the dimer, vertical excitation energies of the low-lying excited states at the Franck–Condon point geometry were calculated using the EOM-CCSD, CASSCF, CASPT2 and DFT/MRCI methods and the aug-cc-pVDZ basis set. The DFT/MRCI calculations[48] were performed using the BH-LYP functional. The CASSCF and CASPT2 calculations were carried out using eight electrons in eight orbitals and the aug-cc-pVDZ basis set, that is, CAS(8,8)/aug-cc-pVDZ and CASPT2(8,8)/aug-cc-pVDZ. The active space was the four occupied orbitals $21a'(\pi_A)$, $22a'(\pi_A)$, $13a''(\pi_B)$ and $14a''(\pi_A)$, and four virtual orbitals $23a'(3s_A)$, $24a'(3p_{Z,A})$, $25a'(\pi_A^*)$ and $15a''(\pi_A^*)$. The subscripts A and B denote the monomer on which a given orbital is localized. The CAS calculations were performed using MOLPRO[49], the EOM-CCSD calculations using QChem[50] and the MP2 and DFT/MRCI calculations using Turbomole[51].

Photoionization cross-sections, $P_I^{\alpha k}(E)$, corresponding to ionization from the low-lying excited neutral states were calculated under the assumption of vertical excitation and the applicability of Fermi's golden rule,

$$\begin{aligned} P_I^{\alpha k}(E) &\propto \left|\left\langle \Phi_I^{(N)} | \hat{\mu} | \Phi_\alpha^{(N-1)} \psi_k \right\rangle\right|^2 \delta(E - E_I^\alpha - E_k) \\ &\approx \left|\left\langle \phi_I^\alpha | \hat{\mu} | \psi_k \right\rangle\right|^2 \delta(E - E_I^\alpha - E_k) \end{aligned} \tag{3}$$

Here, the $\Phi_I^{(N)}$ and Φ_α^{N-1} correspond to the N-electron neutral and (N-1)-electron cationic states, respectively, while ψ_k denotes the photoelectron orbital (taken as a Coulomb partial wave with kinetic energy $E_k = k^2/2$). The energy of the applied laser pulse is denoted by E, whilst E_I^α denotes the vertical energy difference between the netural and cationic states, and $\hat{\mu}$ is the molecular dipole operator. The one-electron quantity ϕ_I^α is a so-called Dyson orbital, obtained from the overlap of the neutral and cationic states[52].

Dyson orbitals corresponding to ionization from both the S_1 and S_2 states to each of the D_0, D_1 and D_2 states were calculated at the EOM-CCSD/EOM-IP-CCSD level using the aug-cc-pVDZ basis set. Evaluation of the photoionization matrix elements $\langle \phi_I^\alpha | \hat{\mu} | \psi_k \rangle$ was performed numerically on a grid and with isotropic averaging over molecular orientations using the ezDyson program[53].

References

1. Delor, M., Sazanovich, I. V., Towrie, M. & Weinstein, J. A. Probing and exploiting the interplay between nuclear and electronic motion in charge transfer processes. *Acc. Chem. Res.* **48,** 1131–1139 (2015).
2. Zhang, W. & Gaffney, K. J. Mechanistic studies of photoinduced spin crossover and electron transfer in inorganic complexes. *Acc. Chem. Res.* **48,** 1140–1148 (2015).
3. Romero, E. *et al.* Quantum coherence in photosynthesis for efficient solar-energy conversion. *Nat. Phys.* **10,** 676–682 (2014).
4. Sukegawa, J. *et al.* Electron transfer through rigid organic molecular wires enhanced by electronic and electronvibration coupling. *Nat. Chem.* **6,** 899–905 (2014).
5. Dreuw, A., Worth, G. A., Cederbaum, L. S. & Head-Gordon, M. Ultrafast photoinitiated long-range electron transfer in cyclophane-bridged zincporphyrin—quinone complexes via conical intersections. *J. Phys. Chem. B* **108,** 19049–19055 (2004).
6. Delor, M. *et al.* Toward control of electron transfer in donor-acceptor molecules by bond-specific infrared excitation. *Science* **346,** 1492–1495 (2014).
7. Zhang, Y. *et al.* Photoinduced electron transfer in DNA: charge shift dynamics between 8-oxo-guanine anion and adenine. *J. Phys. Chem. B* **119,** 7491–7502 (2015).
8. Domcke, W. & Sobolewski, A. L. Spectroscopy meets theory. *Nat. Chem.* **5,** 257 (2013).
9. Zhong, D. Electron transfer mechanisms of DNA repair by photolyase. *Annu. Rev. Phys. Chem.* **66,** 691–715 (2015).
10. Poterya, V., Profant, V., Fárnk, M., Slavíček, P. & Buck, U. Experimental and theoretical study of the pyrrole cluster photochemistry: closing the $\pi\sigma^*$ dissociation pathway by complexation. *J. Chem. Phys.* **127,** 064307 (2007).
11. Slavíček, P. & Fárnk, M. Photochemistry of hydrogen bonded heterocycles probed by photodissociation experiments and *ab initio* methods. *Phys. Chem. Chem. Phys.* **13,** 12123–12137 (2011).
12. Ulijasz, A. T. *et al.* Structural basis for the photoconversion of a phytochrome to the activated Pfr form. *Nature* **463,** 250–254 (2010).
13. White, S., Szewczyk, J. W., Turner, J. M., Baird, E. E. & Dervan, P. B. Recognition of the four Watson-Crick base pairs in the DNA minor groove by synthetic ligands. *Nature* **391,** 468–471 (1998).
14. Pazos, E., Mosquera, J., Vázquez, M. E. & Mascareñas, J. L. DNA recognition by synthetic constructs. *ChemBioChem.* **12,** 1958–1973 (2011).
15. Rasmussen, S. C. & Evenson, S. J. Dithieno[3,2-b:2,3-d]pyrrole-based materials: synthesis and application to organic electronics. *Prog. Polym. Sci.* **38,** 1773–1804 (2013).
16. Wang, Z. *et al.* Organic dyes containing dithieno[2,3-d:2,3-d]thieno[3,2-b:3, 2-b]dipyrrole core for efficient dye-sensitized solar cells. *J. Mater. Chem. A* **3,** 4865–4874 (2015).
17. MacDiarmid, a. G. Synthetic metals: a novel role for organic polymers (Nobel Lecture). *Angew Chem. Int. Ed. Engl.* **40,** 2581–2590 (2001).
18. Souto, M. *et al.* Self-assembled architectures with segregated donor and acceptor units of a dyad based on a monopyrrolo-annulated TTF-PTM radical. *Chemistry* **21,** 8816–8825 (2015).
19. Sobolewski, A. L., Domcke, W., Dedonder-Lardeux, C. & Jouvet, C. Excited-state hydrogen detachment and hydrogen transfer driven by repulsive $\pi\sigma^*$ states: a new paradigm for nonradiative decay in aromatic biomolecules. *Phys. Chem. Chem. Phys.* **4,** 1093–1110 (2002).
20. Ashfold, M. N. R., Cronin, B., Devine, A. L., Dixon, R. N. & Nix, M. G. D. The role of $\pi\sigma^*$ excited states in the photodissociation of heteroaromatic molecules. *Science* **312,** 1637–1640 (2006).
21. Wu, G. *et al.* Excited state non-adiabatic dynamics of pyrrole: a time-resolved photoelectron spectroscopy and quantum dynamics study. *J. Chem. Phys.* **142,** 074302 (2015).
22. Klasinc, L., Sabljić, A., Kluge, G., Rieger, J. & Scholz, M. Chemistry of excited States. Part 13. Assignment of lowest π-ionizations in photoelectron spectra of thiophen, furan and pyrrole. *J. Chem. Soc. Perkin Trans. 2* **2,** 539–542 (1982).
23. Derrick, P., Åsbrink, L., Edqvist, O. & Lindholm, E. Photoelectron-spectroscopical study of the vibrations of furan, thiophene, pyrrole and cyclopentadiene. *Spectrochim. Acta* **27,** 2525–2537 (1971).
24. van den Brom, A. J. *et al.* Photodissociation and photoionization of pyrrole following the multiphoton excitation at 243 and 364.7 nm. *Phys. Chem. Chem. Phys.* **7,** 892–899 (2005).
25. Gómez-Zavaglia, A. & Fausto, R. Self-aggregation in pyrrole: matrix isolation, solid state infrared spectroscopy, and DFT study. *J. Phys. Chem. A* **108,** 6953–6967 (2004).
26. Dauster, I., Rice, C. A., Zielke, P. & Suhm, M. A. N-H $\cdots\pi$ interactions in pyrroles: systematic trends from vibrational spectroscopy in clusters. *Phys. Chem. Chem. Phys.* **10,** 2827–2835 (2008).
27. Profant, V., Poterya, V., Fárnk, M., Slavíček, P. & Buck, U. Fragmentation dynamics of size-selected pyrrole clusters prepared by electron impact ionization: forming a solvated dimer ion core. *J. Phys. Chem. A* **111,** 12477–12486 (2007).
28. Worth, G. & Wade, R. The aromatic—(i + 2) amine interaction in peptides. *J. Phys. Chem.* **99,** 17473–17482 (1995).
29. Chakrabarti, P. & Bhattacharyya, R. Geometry of nonbonded interactions involving planar groups in proteins. *Prog. Biophys. Mol. Biol.* **95,** 83137 (2007).
30. Stefov, V., Pejov, L. & Soptrajanov, B. Experimental and quantum chemical study of pyrrole self-association through N H p hydrogen bonding. *J. Mol. Struct.* **649,** 231–243 (2003).
31. Columberg, G. & Bauder, A. Pure rotational spectrum, quadrupole coupling constants and structure of the dimer of pyrrole. *J. Chem. Phys.* **106,** 504–510 (1997).
32. Lukeš, V., Breza, M. & Biskupic, S. Regular article interaction energy anisotropy of the pyrrole dimer : *ab initio* theoretical study. *Theor. Chem. Acc.* **101,** 319–324 (1999).
33. Kabeláč, M., Hobza, P. & Špirko, V. The structure and vibrational dynamics of the pyrrole dimer. *Phys. Chem. Chem. Phys.* **11,** 3885–3891 (2009).
34. Gao, W., Feng, H., Xuan, X. & Chen, L. A theoretical study of N—H $\cdots\pi$ h-bond interaction of pyrrole: from clusters to the liquid. *Mol. Phys.* **110,** 2151–2161 (2012).
35. Neville, S. P. & Worth, G. A. A reinterpretation of the electronic spectrum of pyrrole: a quantum dynamics study. *J. Chem. Phys.* **140,** 034317 (2014).
36. Roberts, G. M. *et al.* Probing ultrafast dynamics in photoexcited pyrrole: timescales for 1 ps* mediated H-atom elimination. *Faraday Discuss* **163,** 95–116 (2013).
37. Montero, R. *et al.* Femtosecond evolution of the pyrrole molecule excited in the near part of its UV spectrum femtosecond evolution of the pyrrole molecule excited in the near part of its UV spectrum. *J. Chem. Phys.* **2012,** 064317 (2012).
38. Montero, R., Ovejas, V., Fernández-Fernández, M., Peralta Conde, A. & Longarte, A. Revisiting the relaxation dynamics of isolated pyrrole. *J. Chem. Phys.* **141,** 014303 (2014).
39. Lippert, H., Ritze, H.-H., Hertel, I. V. & Radloff, W. Femtosecond time-resolved hydrogen-atom elimination from photoexcited pyrrole molecules. *ChemPhysChem.* **5,** 1423–1427 (2004).
40. Diri, K. & Krylov, A. I. Electronic states of the benzene dimer: a simple case of complexity. *J. Phys. Chem. A* **116,** 653–662 (2012).
41. Minns, R. S., Parker, D. S. N., Penfold, T. J., Worth, G. A. & Fielding, H. H. Competing ultrafast intersystem crossing and internal conversion in the channel 3 region of benzene. *Phys. Chem. Chem. Phys.* **12,** 15607–15615 (2010).
42. Spesyvtsev, R., Kirkby, O. M., Vacher, M. & Fielding, H. H. Shedding new light on the role of the Rydberg state in the photochemistry of aniline. *Phys. Chem. Chem. Phys.* **14,** 9942–9947 (2012).
43. Kirkby, O. M. *et al.* Comparing the electronic relaxation dynamics of aniline and d_7-aniline following excitation at 272–238 nm. *Phys. Chem. Chem. Phys.* **17,** 16270–16276 (2015).
44. Garcia, G. A., Nahon, L. & Powis, I. Two-dimensional charged particle image inversion using a polar basis function expansion. *Rev. Sci. Instrum.* **75,** 4989–4996 (2004).
45. Hansen, J. E. & Persson, W. Revised analysis of singly ionized xenon, Xe II. *Phys. Scripta* **36,** 602–643 (1987).
46. Spesyvtsev, R., Kirkby, O. M. & Fielding, H. H. Ultrafast dynamics of aniline following 269–238 nm excitation and the role of the $S_2(\pi 3s/\pi\sigma^*)$ state. *Faraday Discuss* **157,** 165–179 (2012).
47. Grimme, S. Improved second-order MøllerPlesset perturbation theory by separate scaling of parallel- and antiparallel-spin pair correlation energies. *J. Chem. Phys.* **118,** 9095–9102 (2003).
48. Grimme, S. & Waletzke, M. A combination of Kohn Sham density functional theory and multi-reference configuration interaction methods. *J. Chem. Phys.* **111,** 5645–5655 (1999).
49. Werner, H.-J. *et al.* MOLPRO, v.2012.1. https://www.molpro.net (2012).
50. Shao, Y. *et al.* Advances in molecular quantum chemistry contained in the Q-Chem 4 program package. *Mol. Phys.* **113,** 184–215 (2015).
51. TURBOMOLE v.6.2. TURBOMOLE GmbH. http://www.turbomole.com (2010).
52. Oana, C. M. & Krylov, A. I. Dyson orbitals for ionization from the ground and electronically excited states within equation-of-motion coupled-cluster formalism: theory, implementation, and examples. *J. Chem. Phys.* **127,** 234106 (2007).
53. Gozem, S. & Krylov, A. I. ezDyson v.3. http://iopenshell.usc.edu/downloads/ezdyson (2005).

Acknowledgements

We are grateful to Michael Parkes for his help in the preparation of Fig. 3. This work was enabled by support from EPSRC grants EP/G055270/1 and GR/T20304/01.

Author contributions

H.H.F. conceived and designed the experiments; O.M.K. performed the experiments; N.K. performed preliminary calculations; G.A.W. and S.P.N. designed the calculations presented in the paper; and S.P.N. performed the calculations. All authors discussed the results of the experiments and calculations. H.H.F., G.A.W. and S.P.N. wrote the manuscript.

Additional information

Competing financial interests: The authors declare no competing financial interests.

Array of nanosheets render ultrafast and high-capacity Na-ion storage by tunable pseudocapacitance

Dongliang Chao[1], Changrong Zhu[1], Peihua Yang[1], Xinhui Xia[2], Jilei Liu[1], Jin Wang[3], Xiaofeng Fan[4], Serguei V. Savilov[5], Jianyi Lin[3], Hong Jin Fan[1] & Ze Xiang Shen[1,3]

Sodium-ion batteries are a potentially low-cost and safe alternative to the prevailing lithium-ion battery technology. However, it is a great challenge to achieve fast charging and high power density for most sodium-ion electrodes because of the sluggish sodiation kinetics. Here we demonstrate a high-capacity and high-rate sodium-ion anode based on ultrathin layered tin(II) sulfide nanostructures, in which a maximized extrinsic pseudocapacitance contribution is identified and verified by kinetics analysis. The graphene foam supported tin(II) sulfide nanoarray anode delivers a high reversible capacity of $\sim$1,100 mAh g^{-1} at 30 mA g^{-1} and $\sim$420 mAh g^{-1} at 30 A g^{-1}, which even outperforms its lithium-ion storage performance. The surface-dominated redox reaction rendered by our tailored ultrathin tin(II) sulfide nanostructures may also work in other layered materials for high-performance sodium-ion storage.

[1]School of Physical and Mathematical Sciences, Nanyang Technological University, Singapore 637371, Singapore. [2]State Key Laboratory of Silicon Materials, Department of Materials Science and Engineering, Zhejiang University, Hangzhou 310027, China. [3]Energy Research Institute @ NTU, Nanyang Technological University, Singapore 639798, Singapore. [4]College of Materials Science and Engineering, Jilin University, Changchun 130012, China. [5]Department of Chemistry, Moscow State University, Moscow 119992, Russia. Correspondence and requests for materials should be addressed to Z.X.S. (email: zexiang@ntu.edu.sg).

While open framework materials have proven effective in constructing kinetically favourable sodium (Na)-ion channels as cathodes for sodium-ion batteries (SIBs)[1–7], the sluggish Na-ion transport and severe volume expansion currently limit the rate performance and stability in most anode materials. The typical alloying materials (for example, Sn, Ge, Pb, Sb) possess high capacities ($Na_{15}Sn_4$: 847 mAh g^{-1}, Na_3Ge: 1,108 mAh g^{-1}, $Na_{15}Pb_4$: 484 mAh g^{-1}, Na_3Sb: 660 mAh g^{-1}) for Na storage but have severe volume expansion/contraction during the Na alloying/dealloying ($\sim$360 – 420%) (refs 8–10). To address this pulverization issue, one effective approach is to design integrated electrodes in which nanosized active materials are grafted to a secondary matrix[8,10]. Compared with tin (Sn) metal anodes, tin-based oxides and chalcogenides can store Na^+ through a combined electrochemical conversion and alloying mechanisms, giving rise to higher theoretical capacities (SnO_2: 1,378 mAh g^{-1}, SnS_2: 1,136 mAh g^{-1}, SnS: 1,022 mAh g^{-1}) (ref. 11). Sulfides are typically more reversible than oxides due to relatively weaker M – S ionic bonds compared with M–O bonds, resulting in kinetically more favourable and higher first-cycle efficiency of tin chalcogenides[11]. Moreover, with merits of high electrical conductivity (0.193 – 0.0083 S cm^{-1}), higher capacity and earth abundancy, tin(II) sulfide (SnS) is considered to be a very promising anode material for SIBs. SnS has a smaller lattice expansion (242%) in the sodiation/desodiation process than SnS_2 (324%) (ref. 11). This correlates to a two-structure phase reaction in SnS (from orthorhombic-SnS to cubic-Sn to orthorhombic-$Na_{15}Sn_4$) compared with the three-structure transformation in SnS_2 (from hexagonal-SnS_2 to tetragonal-Sn to orthorhombic-$Na_{15}Sn_4$ (ref. 11). Despite the high capacities in Sn-based alloying materials, high rate capability and fast-charging have not yet been reported, to the best of our knowledge.

Compared with the diffusion-controlled process (insertion, conversion and alloying) in conventional Li/Na-ion storage battery materials, capacitive charge storage has the advantage of rendering high charging rate and therefore high power. In particular, pseudocapacitance refers to underpotential deposition, faradaic charge-transfer reactions including surface or near-surface redox reactions and bulk fast ion intercalation[12–16]. Pseudocapacitance can be intrinsic or extrinsic to a material[12]; intrinsic ones (such as RuO_2, MnO_2 and Nb_2O_5) display the capacitive characteristics for a wide range of particle sizes and morphologies, whereas extrinsic ones (such as $LiCoO_2$, MoO_2 and V_2O_5) emerge only when they are made into nanoscale dimensions to maximize reaction sites on the surface[14–17]. So far, enhanced pseudocapacitive contributions have been realized in some insertion and conversion Li-ion battery (LIB) and SIB materials with high-rate performance (see summary in Supplementary Table 1)[12,18–20]. However, it has not yet been implemented in alloying-type materials where the challenge is to realize high-capacity materials that can accommodate fast kinetics.

Here we demonstrate and prove by quantitative kinetics analysis, the pseudocapacitive contribution to the high capacity of Na-ion storage in few-layered SnS nanosheet arrays directly grown on a graphene foam (GF) backbone (Fig. 1). To the best of our knowledge, our purposely engineered SnS nanohoneycomb structure exhibits the highest reversible capacity, rate capability compared with the reported carbon allotrope, metal/alloy and metal oxides/sulfides as SIB anodes. Excitingly, due to the maximized pseudocapacitive contribution, its rate performance in Na-ion storage even exceeds that for Li ion. Our result may bring a paradigm shift in SIB anode materials to layered metal sulfides, and also afford deeper understanding as well as other nanoscale engineering strategies to boost the performance of SIBs.

Results

Structure and growth mechanism. Figure 1a–c illustrates the fabrication procedure of the flexible GF-supported SnS electrodes by a rapid one-step *in situ* hot bath route (details are described in the Methods section). To demonstrate the morphology-dependent property, we obtained three types of samples with gradient morphologies that are denoted as nanowall (NW), nanoflakes (NF) and nanohoneycomb (NH). Herein, it is important to control the precursor concentration and nucleation rates to achieve desirable SnS nanostructure. Our synthesis leads to a full and uniform coverage of the substrate by SnS nanostructures, for which the size distribution can also be tailored (Supplementary Fig. 1). The SnS presents average lateral sizes of 400–500, 400–500 and 50–70 nm and thickness of $\sim$150, 10 and 5 nm for NW, NF and NH, respectively. The growth of the SnS nanostructures is also proved substrate friendly (Ni foam, carbon cloth and ITO, see Supplementary Fig. 2). A growth mechanism is proposed based on the crucial roles of the substrate and ethanol, which involves (i) the hydrolysis of thioacetamide and (ii) the *in situ* metathesis reactions, self-assembly and oriented crystallization processes (details are provided in Supplementary Fig. 3 and Supplementary Note 1)[5]. The structure evolution and purity are verified by X-ray diffraction , energy-dispersive spectroscopy and Raman measurements (Supplementary Figs 4 and 5). The surface chemical-bonding state of GF-SnS electrode is also detected by X-ray photoelectron spectroscopy (XPS) and presented in Supplementary Fig. 6. The existence of C–S bonds is confirmed by spectra of both S 2p (163.7 eV) and C 1s (285.7 eV). XPS results suggest that the SnS might be chemically bonded with the GF matrix besides physical deposition[19] (details are provided in Supplementary Note 2).

Transmission electron microscopy (TEM) and high-resolution transmission electron microscopy (HRTEM) images in Fig. 2 further confirm the crystallographic orientation and unique nanosheet-on-microstructure three-dimensional (3D) porous nanowall, nanoflake and nanohoneycomb architectures. For SnS NW (Fig. 2a,b), the lattice-resolved HRTEM image shows interplanar spacing of 2.9 Å for the (101) planes of SnS. Further, the inset fast Fourier transform spots also reveal the existence of (101), (002) and (100) facets in the [010] zone axis, demonstrating that the layers of SnS are stacked along the [010] direction (see illustration in Fig. 1e). For SnS NF (Fig. 2c,d), the interconnected nanoflakes are regular with a periodic stacking of fringes ($\sim$15 layers) enforced by van der Waals interactions along the [010] direction. The layer distance is measured as $\sim$6.2 Å, which is slightly larger than the layer-to-layer spacing in reported work[21]. The lateral view of the nanoflake in Fig. 2d also illustrates an interplanar spacing of 2.9 Å for the (101) planes of SnS with a crystal grain size $\sim$20 nm. The results also confirm the formation of nanosheets by stacking of (010) facets. Similarly, SnS NH also presents intrinsic corrugations and lamellar structure in the nanosheets with interlayer spacing $\sim$6.2 Å, but with thinner thickness ($\sim$6 layers). Interestingly, numerous tiny nanoclusters ($\sim$5 nm) and nanocavities (3–5 nm) are also observed from the cross-section of the wrinkles in Fig. 2f. Meanwhile, HRTEM and fast Fourier transform pattern reveals clear lattices with spacings of 3.4 and 2.8 Å for (120) and (040) planes, respectively, under [001] zone axis. A schematic illustration of the SnS laminar structure seen from [001] zone axis can be seen from Fig. 1d.

Na-ion storage performance. Our designed battery electrodes allow electron and sodium-ion transfer through the GF–SnS network without the necessity of extra binders, conductive additives or metal Cu current collectors, which are essential for

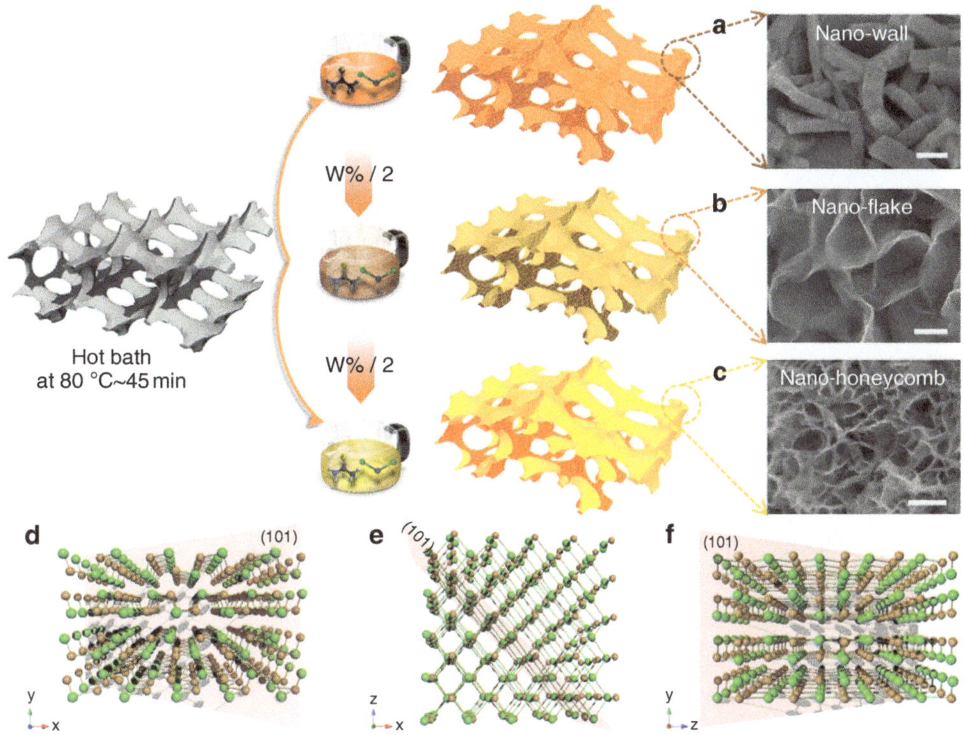

Figure 1 | Synthesis and structure of SnS nanostructures. (**a-c**) SnS nanostructures synthesized in different solution concentrations for (**a**) nano-wall; (**b**) nano-flake; (**c**) nano-honeycomb. Scale bar, 200 nm. (**d-f**) Schematic illustrations of the SnS laminar structure viewed along the [001], [010] and [100] zone axis with inserted (101) planes.

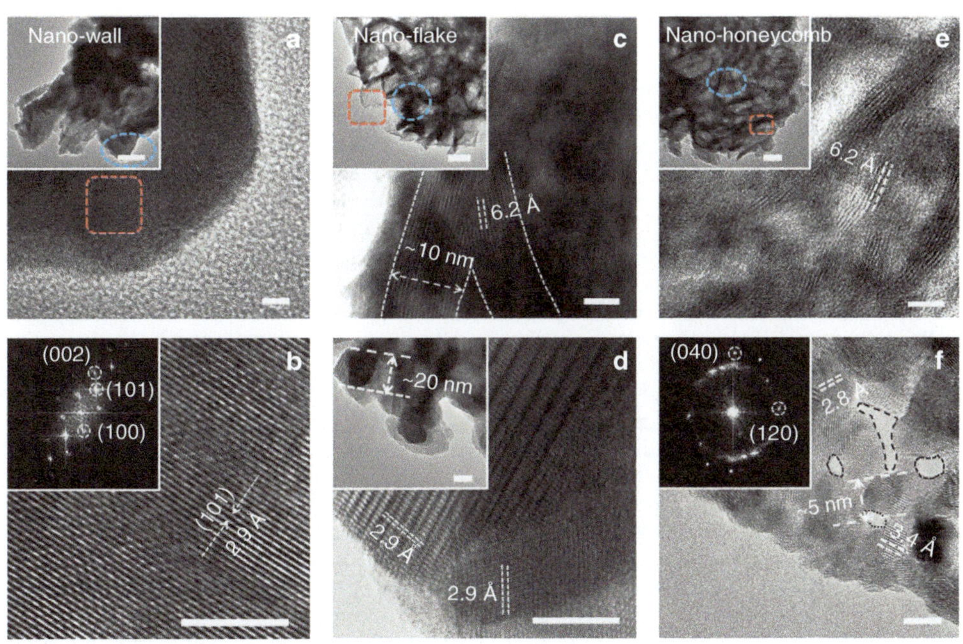

Figure 2 | TEM images of SnS nanostructures. (**a,b**) HRTEM images for the nanowall SnS structure. Scale bar, 5 nm. Inset a: low magnification TEM image (scale bar, 100 nm) and (**b**) fast Fourier transform (FFT) pattern along [010] zone axis. (**c,d**) HRTEM images for nanoflake SnS structure. Scale bar, 5 nm. Insets **c,d**: low magnification TEM images. Scale bars of 400 and 10 nm, respectively. (**e,f**) HRTEM images for nanohoneycomb SnS structure. Scale bar, 5 nm. Inset **e**: low magnification TEM image. Scale bar, 30 nm. Inset **f**: FFT pattern in the [001] zone axis. The dashed loops denote the mesopores.

exploring intrinsic sodium-storage properties of active material and increasing the energy/power densities of the full cell[5,22]. Herein, GF serves as both a lightweight 3D porous current collector (for electron transfer) and compressible/flexible backbone. The porous nanoarray feature prevents the aggregation and expansion of SnS during charge/discharge cycles. During sodiation, electrolyte can enter the interval between nanoarrays on both outside and inner surface of GF, so that the Na ion and electrons can react with the SnS nanoarrays effectively.

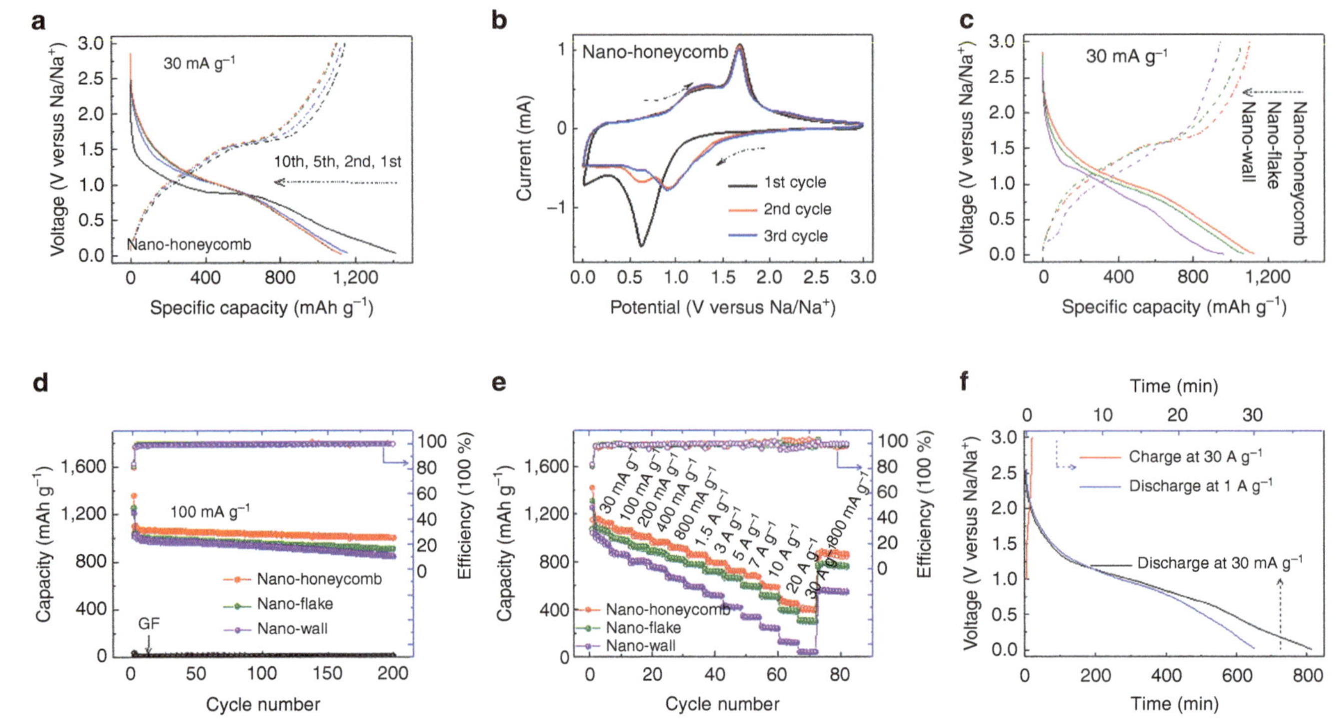

Figure 3 | Electrochemical Na^+ storage performance of SnS electrodes. (**a**) Galvanostatic charge/discharge profiles during the first 10 cycles of the nanohoneycomb electrode. (**b**) CV curves of the first three cycles of the SnS nanohoneycomb electrode at a scan rate of $0.2\,mV\,s^{-1}$. (**c**) Galvanostatic charge/discharge profiles of SnS electrodes after five cycles activation. (**d**) Long-term cycling performances of SnS electrodes and the pure graphene foam electrode at a current density of $100\,mA\,g^{-1}$. (**e**) Rate performances of SnS electrodes at various current densities from 30 to 30,000 $mA\,g^{-1}$. (**f**) Fast charging (charge at $30\,A\,g^{-1}$ in 1 min, discharge at $30\,mA\,g^{-1}$ and $1\,A\,g^{-1}$ with ~13 h and 30 min, respectively) properties of the nanohoneycomb electrode.

The reaction of SnS with sodium undergoes two steps: conversion followed by alloying according to equations (1) and (2) (ref. 11):

$$SnS + 2Na^+ + 2e^- \leftrightarrow Sn + Na_2S \quad (1)$$

$$4Sn + 15Na^+ + 15e^- \leftrightarrow Na_{15}Sn_4. \quad (2)$$

From the galvanostatic discharge-charge results (Fig. 3a and Supplementary Fig. 7), the first discharge presents a steep decrease and remains relatively flat in voltage. This may be due to the following two possible reasons: (1) deactivated surface and related high electrochemical polarization during the first sodiation; (2) transition from crystalline SnS to metal Sn nanograins and Na_2S (equation (1)) (refs 11,23). However, the subsequent curves after the first three cycles almost overlap, indicating a stable surface state, structure and electrochemical reversibility after the initial activation process. The detailed reaction processes could be disclosed by the cyclic voltammetric (CV) curves (Fig. 3b). In the first cathodic scan, two prominent peaks were observed at around $0.6-0.7$ and $0.01-0.1$ V, respectively. The former is associated with both the conversion reaction (equation (1)) and the alloying reaction (equation (2), Na_xSn, $x \sim 0.75$) because it is difficult to distinguish the conversion and alloying peaks especially in the first discharge process[11,24]. Obviously this peak also includes the contribution from solid electrolyte interphase (SEI) formation since its intensity significantly decays in the subsequent cycles. The peak at $0.01-0.1$ V is regarded as the reaction between Na and Na_xSn ($x \sim 3.75$) alloy due to the multi-step Na–Sn alloying feature[11]. In the anodic scan, a serial of small peaks below 1.4 V (~0.3, 0.7, $1.2-1.4$ V) correspond to the multi-step dealloying reaction of Na_xSn (ref. 24); whereas the distinct peak at ~1.7 V could be attributed to the reversible conversion reaction from Sn to SnS, which was also observed in SnS@G[11], 3D porous interconnected SnS (ref. 25), and proved by *ex situ* Raman results of the electrodes after charging and discharging processes (Supplementary Fig. 8).

The first sodiation and desodiation capacities of GF–SnS NH are 1,416 and 1,147 $mAh\,g^{-1}$, respectively, calculated based on active materials. A rather high first-cycle coulombic efficiency of >80% is observed. A reversible capacity more than 1,100 $mAh\,g^{-1}$ could be maintained in the following cycles. We note this value is higher than all reported ones from sodium alloys, oxides, sulfides and carbonaceous anodes so far (see comprehensive comparison in Supplementary Table 2), to the best of our knowledge. Note that the pure GF contributed negligibly to the capacity: ~30 $mAh\,g^{-1}$ for the first discharge process and <10 $mAh\,g^{-1}$ in the following cycles. Compared with NH electrode, NF presents a similar behaviour due to their similar microtopography, whereas the NW one exhibits a suppressed capacity and higher polarization. More importantly, the voltage profile in NW electrode shows more obvious plateaus, which is a typical feature of diffusion-controlled charge storage of battery materials[16]. The improvements in the coulombic efficiency and high reversible capacity of GF–SnS NH electrode should be correlated to the reversible formation/decomposition of the polymeric film on the surface of SnS NH (refs 11,26), its distinct disorder two-dimensional structure, ultrathin-layered mesoporous SnS nanocrystals, and as a result, its unique electrochemical mechanism (capacitive contribution), which is to be discussed below. The GF–SnS electrodes deliver excellent capacity retention from the third cycle onwards. After 200 cycles, the capacity retains at 1,010 $mAh\,g^{-1}$ for the NH electrode with well-preserved microstructure (Supplementary Fig. 9).

The rate capability is a crucial indicator for large scale application of batteries, such as regenerative braking and fast recharging of electric vehicles and cellphones. The drawback of low power becomes particularly evident in high capacity (that is, high energy density) materials[14]. It is found that the NH electrode has the best rate capability, in addition to consistently highest capacity, among the three GF–SnS electrodes (see Fig. 3e). For a 1,000-fold increase in current density (from 30 to 30 A g^{-1}), a discharge capacity of more than 400 mAh g^{-1} (in 1 min) could still be retained. If this electrode were used to power a cellphone system, it is estimated that the battery could be charged in 1 min and discharged in ~13 h at 30 mA g^{-1} (Fig. 3f). On the basis of a comprehensive summary (Supplementary Table 2), this is the best rate capability among all reported anode materials for SIBs, to the best of our knowledge. The preliminary result of full-cell fabrication (see Supplementary Fig. 10, $Na_3(VO)_2(PO_4)_2F$ cathode\\SnS anode) demonstrates the potential commercial application of our SnS electrodes to be considered as an anode material for SIBs, although further optimization is urgent to improve the skill in full-cell fabrication and cycling stability.

Kinetics and quantitative analysis. To explain the high-rate performance, we analysed the redox pseudocapacitance-like contribution in the GF–SnS electrodes by investigating the kinetics of the SnS electrodes (Fig. 4) to separate the diffusion-controlled capacity and capacitive capacity[15,27]. Resulting from the stepwise sodiation mechanism, CV curves with similar shapes at various scan rates from 0.2 to 0.8 mV s^{-1} (Supplementary Fig. 11) display two broad cathodic peaks as the scan rate increases. As cation intercalation reaction can be ruled out from our SnS electrode, we mainly consider the below three charge-storage mechanisms: the diffusion-controlled faradaic contribution from conversion and alloying reaction, the faradaic contribution from charge transfer with surface/subsurface atoms (that is, extrinsic pseudocapacitance effect), and the non-faradaic contribution from electrical double-layer effect[17,19,20].

The ratios of Na-ion capacitive contribution can be further quantitatively quantified by separating the current response i at a fixed potential V into capacitive effects (proportional to the scan rate v) and diffusion-controlled reactions ($k_2v^{1/2}$), according to Dunn[15,28]:

$$i(V) = k_1 v + k_2 v^{1/2} \quad (3)$$

By determining both k_1 and k_2 constants, we can distinguish the fraction of the current from surface capacitance and Na^+ semi-infinite linear diffusion. Fig. 4a shows the typical voltage profile for the capacitive current (red region) in comparison with the total current. A dominating capacitive contribution (~84%) is obtained for the NH electrode. As the scan rate increases, the role of capacitive contribution further enlarges (Fig. 4b) with a maximum value of ~95% at 5 mV s^{-1}. By similar analysis, the pseudocapacitive contribution is found more than 80% for the NF electrode, but only around 60% for the NW one at 0.8 mV s^{-1} (Supplementary Figs 12 and 13). This is unsurprising since the pseudocapacitive contribution should play a critical role for smaller particle size with high surface area (~154 m^2 g^{-1}) and/or high porosity (mesoporous from 7 to 37 nm, see Supplementary Fig. 14)[15,18]. Finally, the thin-film electric conductivity (Supplementary Fig. 15) and electrochemical impedance spectra (Supplementary Fig. 16) suggest that the NH electrode have a favourable charge transfer kinetics compared with NW electrode.

Comparison with Li-ion storage capability. It is widely believed that Na^+ transport and storage are more sluggish with more severe lattice expansion than the Li^+ one because of the larger radius of Na ions[1,29–32]. So far, the performance in SIB is generally worse than that in LIBs when the same electrode material is used, including capacity, high rate capability and polarization. Herein, we carefully compare the performance of SnS NH anode for both Na^+ and Li^+ tests.

To avoid the influence on Li uptakes in GF backbone, in our comparison experiment we used the SnS NH grown on Ni foam as the same electrode. Figure 5 shows the results. Strikingly, one can see the rate capacity for sodiation/desodiation is superior to Li^+ uptakes, particularly in the high-rate regions. At current of 30 A g^{-1}, the electrode delivers a Na^+ discharge capacity of ~410 mAh g^{-1} compared with ~105 mAh g^{-1} for the Li^+ electrode. In the galvanostatic charge/discharge processes (Fig. 5b), the Li^+ electrode shows two distinct plateaus at 1.2 – 1.4 and 0.01 – 0.4 V versus Li/Li^+ corresponding to their CV curves (Supplementary Fig. 17a), suggesting a lower fraction of capacitive contribution. The first discharge plateau during lithiation is attributed to the conversion reaction from SnS→Sn and the second one is the alloying reaction-forming $Li_{15}Sn_4$ phase[33,34]. The polarization from 30 mA g^{-1} to 7 A g^{-1} during lithiation (~340 mV) is twice to that in sodiation (~170 mV). In addition, the capacitive fraction for Li^+ storage is 75% (inset in Fig. 5b), whereas 85% for the Na^+ storage of the same electrode (Supplementary Fig. 18). Similar higher Na^+ capacitive contribution than that of Li^+ had also been observed in $Li_4Ti_5O_{12}$ spinel thin film electrode but with much lower discharge capacity[20]. Finally, the sodiation discharge curves have more moderate and continuous operation voltage than the lithiation ones (Fig. 5b), which is favourable to achieving high energy density of full cells and avoiding dendrite growth[19].

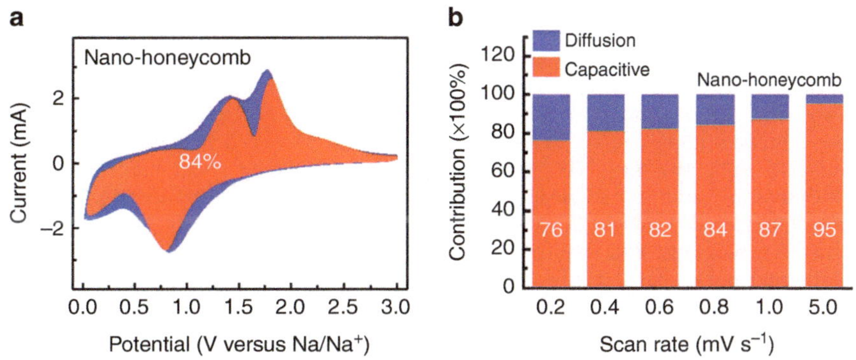

Figure 4 | Kinetics and quantitative analysis of the Na^+ storage mechanism. (**a**) Capacitive (red) and diffusion-controlled (blue) contribution to charge storage of nanohoneycomb at 0.8 mV s^{-1}. (**b**) Normalized contribution ratio of capacitive (red) and diffusion-controlled (blue) capacities at different scan rate.

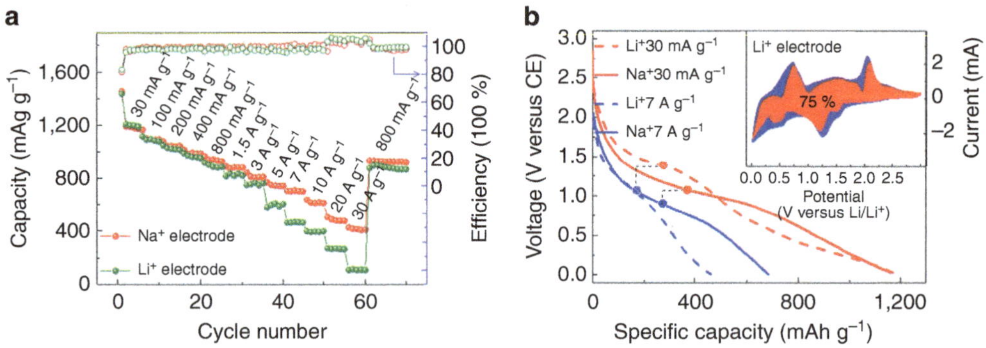

Figure 5 | Comparison between Na-ion and Li-ion storage. (**a**) Rate performance comparison of the nanohoneycomb electrode at various current densities from 30 to 30,000 mA g^{-1}. (**b**) Galvanostatic profiles of Na, Li electrodes at rates of 30 mA g^{-1} and 7 A g^{-1} after activation. Inset: capacitive (red) and diffusion-controlled (blue) contribution to charge storage at 0.8 mV s^{-1} during Li uptake. All of the batteries were tested in the same voltage range of 0.01–3 V versus Na/Na^+ and Li/Li^+ for Na-ion and Li-ion batteries, respectively.

Discussion

The results presented above demonstrate sodium-ion storage with both high capacity and high rate capability rendered by tunable extrinsic pseudocapacitance in our GF-supported SnS nanosheets. The electrode architecture provides, to the best of our knowledge, the highest reported reversible capacity of 1,100 mAh g^{-1} at 30 mA g^{-1}. Even at a high current density of 30 A g^{-1} (1,000-fold increase), the capacity is retained at 400 mAh g^{-1}, which is higher than that of Li^+ electrode ($\sim$105 mAh g^{-1} at 30 A g^{-1}). As the diffusion time of ions (t) is proportional to the square of the diffusion length (L), $t \approx L^2/D$, a short lithium diffusion time of $\sim$0.01 s is obtained on the basis of the ultrathin SnS architectures. As a consequence, similar to supercapacitors, the limiting factor for high rate charge/discharge is the transfer of ions and electrons to the surface of nanosheets rather than the conventional solid-state diffusion. As schematically shown in Fig. 6, the strongly solvating groups (carbonyl groups) in the organic electrolyte will arrange in an appropriate manner as a preferred solvation shell of the cations (Li^+ or Na^+) (refs 35–37), and it has been proven that sodium ion presents a weaker solvation shell with smaller de-solvation energy and lower activation barrier for sodiation transport compared with the lithium ion[35,38]. In additionally, higher mobility and conductivity of Na^+ solutions also contribute to the ion transfer in the electrolyte[39,40]. As a result, faster sodiation/desodiation kinetics are possible for SIBs as compared with LIBs[35,39–41].

The surface-dominated extrinsic pseudocapacitance is identified as a major energy-storage mechanism in favour of high capacity and fast Na^+ uptakes (Fig. 6). First, the chemically bonded GF–SnS hybrid demonstrates excellent structure stability and electronic/ionic conductivity through the network, which have also been shown to be a prerequisite for the extrinsic pseudocapacitance in nanosized MoO_2 (ref. 16). Moreover, the nanoscale dimension, especially thickness of the electrode materials, has been emphasized to be an important factor on the rate properties and corresponding redox capacitive contribution[20,27,42,43]. Compared with the SnS NW, the few layered architecture and mesoporous iso-oriented nanocrystals nature of the NH enables an interior Na^+ or electrolyte access into the van der Waals gaps of nanosheets, and thus results in both the exterior and interior parts participating in the electrochemical reaction[15]. This feature also facilitates ion access and shortens the ion diffusion.

Our kinetics analysis verifies the surface-dominated redox reaction mechanism in the Na-ion storage process, whereas the battery-type diffusion contribution is suppressed. This is closely correlated to the engineered thin-sheet structure of SnS NH, and may account for the superior Na^+ storage performance (high rate capacity) of the SnS NH to its Li^+ one. This strategy renders increased power density with maintained high energy density. This encouraging result may accelerate further development of SIBs by smart nanoengineering of the electrode materials.

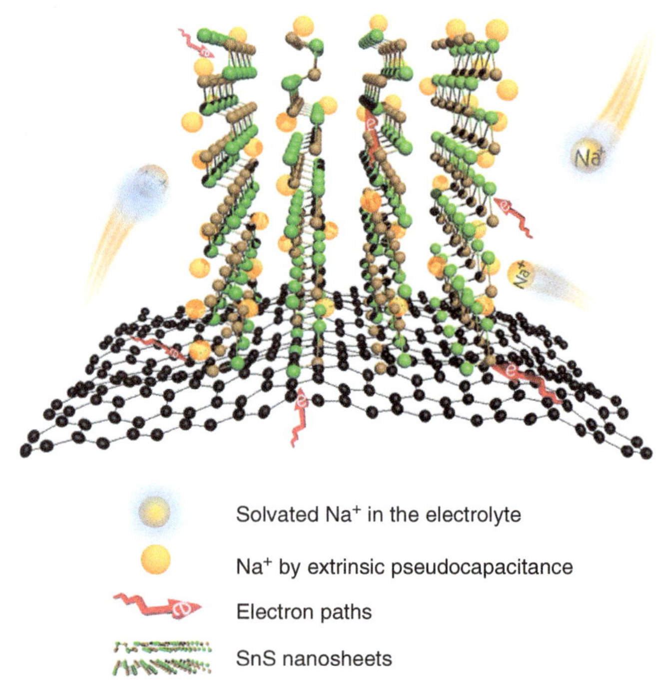

Figure 6 | Schematic illustration of high-rate charge storage of SnS architecture. During transfer of ions and electrons, the solvated Na^+/electron can easily enter into the open spaces between neighbouring ultrathin nanosheets on both the outside and inner surface of the graphene foam. After de-solvation of Na^+ on the surface of the layered SnS architecture, rapid sodiation takes place by the surface-dominated extrinsic pseudocapacitance.

Methods

Synthesis and characterization. SnS nanostructures were fabricated by a facile hot bath method. First, precursors with three sets of different concentrations, namely, high, medium and low concentrations of tin(II) chloride dehydrate: thioacetamide, respectively for nanowall (100:300 mM), nanoflake (50:150 mM) and nanohoneycomb (25:75 mM), were dissolved in 50 ml ethanol at 80 °C. Then 3D GFs(2×5 cm^2, $\sim$0.8 mg cm^{-2}, prepared by chemical vapour deposition method according to our previous result[44]) or other type of substrates such as Ni foam, carbon cloth and ITO glass were immersed into the above reaction solutions and kept for 45 min. Finally, the samples were collected and rinsed with distilled water and ethanol in turn three times, and dried at 150 °C in vacuum to obtain 3D GF-supported SnS free-standing electrodes. The SnS loading was ~1.0 mg cm^{-2} for nanoflake and nanohoneycomb electrode, and 1.2 mg cm^{-2} for nanowall electrode.

The crystal structures of the samples were identified using X-ray diffraction (RigakuD/Max-2550 with Cu-Kα radiation). Raman spectra were obtained with a WITec-CRM200 Raman system (WITec, Germany) with a laser wavelength of 532 nm (2.33 eV). The morphologies of the samples were characterized by field emission scanning electron microscopy. The structures of the samples were investigated by HRTEM(JEOL JEM-2010F at 200 kV). The XPS measurements were performed by a VG ESCALAB 220i-XL system using a monochromatic Al Kα1 source (1,486.6 eV). The thin-film electric conductivity was measured by four-point probe sheet resistance. The surface area of the SnS electrode was determined by N_2 adsorption/desorption isotherms.

Electrochemical measurements. Standard CR2032-type coin cells were assembled in an argon-filled glove box (Mbraun, Germany) with the as-fabricated GF-supported SnS nanoarrays as the working electrode (without any binder or additives). For SIB fabrication, the metallic sodium foil as the counter-electrode, 1 M $NaPF_6$ in ethylene carbonate (EC)–diethyl carbonate (DEC)–fluoroethylene carbonate (FEC) (1:1:0.03 in volume) as the electrolyte, and glass fibre as the separator. For the LIB case, except for the metallic lithium foil as the counter-electrode and 1 M $LiPF_6$ as the solute, the other parameters are the same with the SIB fabrication. For full-cell testing, the cathode-active material was $Na_3(VO)_2(PO_4)_2F$ nanoparticle. The one-time discharge/charge cycled SnS NH served as the anode. The weight ratio between anode and cathode active material was ~0.15:1. The specific capacity was calculated based on the mass of the cathode-active material. The CV measurements were carried out using CHI660 electrochemical workstation. Electrochemical impedance spectroscopy was recorded on Solartron 1470E, the amplitude of the sine perturbation signal was 5 mV, and the frequency was scanned from the highest (10 kHz) to the lowest (5 mHz). Galvanostatic charge discharge cycles were tested by Neware battery tester at different current densities at room temperature.

References

1. Larcher, D. & Tarascon, J. M. Towards greener and more sustainable batteries for electrical energy storage. *Nat. Chem.* **7**, 19–29 (2015).
2. Li, C. *et al.* An FeF3 · 0.5H2O polytype: a microporous framework compound with intersecting tunnels for Li and Na batteries. *J. Am. Chem. Soc.* **135**, 11425–11428 (2013).
3. Yabuuchi, N. *et al.* P2-type Na(x)[Fe(1/2)Mn(1/2)]O2 made from earth-abundant elements for rechargeable Na batteries. *Nat. Mater.* **11**, 512–517 (2012).
4. Fang, Y. *et al.* Mesoporous amorphous $FePO_4$ nanospheres as high-performance cathode material for sodium-ion batteries. *Nano Lett.* **14**, 3539–3543 (2014).
5. Chao, D. *et al.* Graphene quantum dots coated VO_2 arrays for highly durable electrodes for Li and Na ion batteries. *Nano Lett.* **15**, 565–573 (2015).
6. Lim, S. Y. *et al.* Role of intermediate phase for stable cycling of $Na_7V_4(P_2O_7)_4PO_4$ in sodium ion battery. *Proc. Natl Acad. Sci. USA* **111**, 599–604 (2013).
7. Liu, H., Xu, J., Ma, C. & Meng, Y. S. A new O_3-type layered oxide cathode with high energy/power density for rechargeable Na batteries. *Chem. Commun.* **51**, 4693–4696 (2015).
8. Wu, L. *et al.* Electrochemical properties and morphological evolution of pitaya-like Sb@C microspheres as high-performance anode for sodium ion batteries. *J. Mater. Chem. A* **3**, 5708–5713 (2015).
9. Kohandehghan, A. *et al.* Activation with Li enables facile sodium storage in germanium. *Nano Lett.* **14**, 5873–5882 (2014).
10. Xie, X. *et al.* Sn@CNT nanopillars grown perpendicularly on carbon paper: a novel free-standing anode for sodium ion batteries. *Nano Energy* **13**, 208–217 (2015).
11. Zhou, T. *et al.* Enhanced sodium-ion battery performance by structural phase transition from two-dimensional hexagonal-SnS2 to orthorhombic-SnS. *ACS Nano* **8**, 8323–8333 (2014).
12. Augustyn, V., Simon, P. & Dunn, B. Pseudocapacitive oxide materials for high-rate electrochemical energy storage. *Energy Environ. Sci.* **7**, 1597–1614 (2014).
13. Augustyn, V. *et al.* High-rate electrochemical energy storage through Li+ intercalation pseudocapacitance. *Nat. Mater.* **12**, 518–522 (2013).
14. Simon, P., Gogotsi, Y. & Dunn, B. Where do batteries end and supercapacitors begin? *Science* **343**, 1210–1211 (2014).
15. Brezesinski, T., Wang, J., Tolbert, S. H. & Dunn, B. Ordered mesoporous alpha-MoO3 with iso-oriented nanocrystalline walls for thin-film pseudocapacitors. *Nat. Mater.* **9**, 146–151 (2010).
16. Kim, H. S., Cook, J. B., Tolbert, S. H. & Dunn, B. The development of pseudocapacitive properties in nanosized-MoO_2. *J. Electrochem. Soc.* **162**, A5083–A5090 (2015).
17. Li, S. *et al.* Surface capacitive contributions: towards high rate anode materials for sodium ion batteries. *Nano Energy* **12**, 224–230 (2015).
18. Chen, Z. *et al.* High-performance sodium-ion pseudocapacitors based on hierarchically porous nanowire composites. *ACS Nano* **6**, 4319–4327 (2012).
19. Chen, C. *et al.* Na(+) intercalation pseudocapacitance in graphene-coupled titanium oxide enabling ultra-fast sodium storage and long-term cycling. *Nat. Commun.* **6**, 6929 (2015).
20. Yu, P., Li, C. & Guo, X. Sodium storage and pseudocapacitive charge in textured $Li_4Ti_5O_{12}$ thin films. *J. Phys. Chem. C* **118**, 10616–10624 (2014).
21. Gou, X.-L., Chen, J. & Shen, P.-W. Synthesis, characterization and application of SnSx (x = 1, 2) nanoparticles. *Mater. Chem. Phys.* **93**, 557–566 (2005).
22. Sun, Y. *et al.* Direct atomic-scale confirmation of three-phase storage mechanism in $Li_4Ti_5O_{12}$ anodes for room-temperature sodium-ion batteries. *Nat. Commun.* **4**, 1870 (2013).
23. He, K. *et al.* Sodiation kinetics of metal oxide conversion electrodes: a comparative study with lithiation. *Nano Lett.* **15**, 5755–5763 (2015).
24. Ma, C. Z. *et al.* Investigating the energy storage mechanism of SnS_2-rGO composite anode for advanced Na-ion batteries. *Chem. Mater.* **27**, 5633–5640 (2015).
25. Zhu, C. B. *et al.* A general strategy to fabricate carbon-coated 3D porous interconnected metal sulfides: case study of SnS/C nanocomposite for high-performance lithium and sodium ion batteries. *Adv. Sci.* **2**, 1500200 (2015).
26. Luo, B. *et al.* Two dimensional graphene–SnS_2 hybrids with superior rate capability for lithium ion storage. *Energy Environ. Sci.* **5**, 5226–5230 (2012).
27. Muller, G. A., Cook, J. B., Kim, H. S., Tolbert, S. H. & Dunn, B. High performance pseudocapacitor based on 2D layered metal chalcogenide nanocrystals. *Nano Lett.* **15**, 1911–1917 (2015).
28. Bard, A. J. & Faulkner, L. R. *Electrochemical Method: Fundamentals and Applications* (John Wiley & Sons, 1980).
29. Kim, H. *et al.* Aqueous rechargeable Li and Na ion batteries. *Chem. Rev.* **114**, 11788–11827 (2014).
30. Kundu, D., Talaie, E., Duffort, V. & Nazar, L. F. The emerging chemistry of sodium ion batteries for electrochemical energy storage. *Angew. Chem. Int. Ed. Engl.* **54**, 3431–3448 (2015).
31. Xu, X. *et al.* *In situ* investigation of Li and Na ion transport with single nanowire electrochemical devices. *Nano Lett.* **15**, 3879–3884 (2015).
32. Wang, Y. *et al.* A zero-strain layered metal oxide as the negative electrode for long-life sodium-ion batteries. *Nat. Commun.* **4**, 2365 (2013).
33. Lu, J., Nan, C., Li, L., Peng, Q. & Li, Y. Flexible SnS nanobelts: facile synthesis, formation mechanism and application in Li-ion batteries. *Nano Res.* **6**, 55–64 (2012).
34. Tripathi, A. M. & Mitra, S. The influence of electrode structure on the performance of an SnS anode in Li-ion batteries: effect of the electrode particle, conductive support shape and additive. *RSC Adv.* **5**, 23671–23682 (2015).
35. Ponrouch, A. *et al.* Non-aqueous electrolytes for sodium-ion batteries. *J. Mater. Chem. A* **3**, 22–42 (2015).
36. Ponrouch, A., Marchante, E., Courty, M., Tarascon, J. M. & Palacin, M. R. In search of an optimized electrolyte for Na-ion batteries. *Energy Environ. Sci.* **5**, 8572–8583 (2012).
37. Abe, T., Fukuda, H., Iriyama, Y. & Ogumi, Z. Solvated Li-ion transfer at interface between graphite and electrolyte. *J. Electrochem. Soc.* **151**, A1120–A1123 (2004).
38. Jonsson, E. & Johansson, P. Modern battery electrolytes: ion-ion interactions in Li^+/Na^+ conductors from DFT calculations. *Phys. Chem. Chem. Phys.* **14**, 10774–10779 (2012).
39. Yabuuchi, N., Kubota, K., Dahbi, M. & Komaba, S. Research development on sodium-ion batteries. *Chem. Rev.* **114**, 11636–11682 (2014).
40. Zhang, Y., Narayanan, A., Mugele, F., Cohen Stuart, M. A. & Duits, M. H. G. Charge inversion and colloidal stability of carbon black in battery electrolyte solutions. *Colloids Surf. A* **489**, 461–468 (2016).
41. Ong, S. P. *et al.* Voltage, stability and diffusion barrier differences between sodium-ion and lithium-ion intercalation materials. *Energy Environ. Sci.* **4**, 3680–3688 (2011).
42. Come, J. *et al.* Electrochemical kinetics of nanostructured Nb_2O_5 electrodes. *J. Electrochem. Soc.* **161**, A718–A725 (2014).
43. McDowell, M. T. *et al.* *In situ* observation of divergent phase transformations in individual sulfide nanocrystals. *Nano Lett.* **15**, 1264–1271 (2015).
44. Chao, D. *et al.* A V_2O_5/conductive-polymer core/shell nanobelt array on three-dimensional graphite foam: a high-rate, ultrastable, and freestanding cathode for lithium-ion batteries. *Adv. Mater.* **26**, 5794–5800 (2014).

Acknowledgements

Z.X.S. acknowledge the financial supported by Ministry of Education, Tier 1 (Grant number: M4011424.110), Tier 2 (Grant number: M4020284.110); H.J.F. acknowledges the financial supported by MOE AcRF Tier 1 (RG104/14, RG98/15). We also acknowledge support from the Energy Research Institute @NTU (ERI@N). We thank Professor Bruce Dunn, University of California, Los Angeles for useful discussions.

Author contributions

D.L.C. and C.R.Z. conceived the experiment. D.L.C., C.R.Z., P.H.Y., X.H.X., J.L.L. and J.W. conducted the material synthesis, characterization and electrochemical measurement. X.F.F., S.V.S., J.Y.L., H.J.F. and Z.X.S. were involved in discussion on the kinetics and quantitative analysis. D.L.C., C.R.Z., H.J.F. and Z.X.S. wrote the manuscript.

Additional information

Competing financial interests: The authors declare no competing financial interests.

Permissions

All chapters in this book were first published in NC, by Nature Publishing Group; hereby published with permission under the Creative Commons Attribution License or equivalent. Every chapter published in this book has been scrutinized by our experts. Their significance has been extensively debated. The topics covered herein carry significant findings which will fuel the growth of the discipline. They may even be implemented as practical applications or may be referred to as a beginning point for another development.

The contributors of this book come from diverse backgrounds, making this book a truly international effort. This book will bring forth new frontiers with its revolutionizing research information and detailed analysis of the nascent developments around the world.

We would like to thank all the contributing authors for lending their expertise to make the book truly unique. They have played a crucial role in the development of this book. Without their invaluable contributions this book wouldn't have been possible. They have made vital efforts to compile up to date information on the varied aspects of this subject to make this book a valuable addition to the collection of many professionals and students.

This book was conceptualized with the vision of imparting up-to-date information and advanced data in this field. To ensure the same, a matchless editorial board was set up. Every individual on the board went through rigorous rounds of assessment to prove their worth. After which they invested a large part of their time researching and compiling the most relevant data for our readers.

The editorial board has been involved in producing this book since its inception. They have spent rigorous hours researching and exploring the diverse topics which have resulted in the successful publishing of this book. They have passed on their knowledge of decades through this book. To expedite this challenging task, the publisher supported the team at every step. A small team of assistant editors was also appointed to further simplify the editing procedure and attain best results for the readers.

Apart from the editorial board, the designing team has also invested a significant amount of their time in understanding the subject and creating the most relevant covers. They scrutinized every image to scout for the most suitable representation of the subject and create an appropriate cover for the book.

The publishing team has been an ardent support to the editorial, designing and production team. Their endless efforts to recruit the best for this project, has resulted in the accomplishment of this book. They are a veteran in the field of academics and their pool of knowledge is as vast as their experience in printing. Their expertise and guidance has proved useful at every step. Their uncompromising quality standards have made this book an exceptional effort. Their encouragement from time to time has been an inspiration for everyone.

The publisher and the editorial board hope that this book will prove to be a valuable piece of knowledge for researchers, students, practitioners and scholars across the globe.

List of Contributors

Hongning Chen and Qingli Zou
Department of Mechanical and Automation Engineering, The Chinese University of Hong Kong, Shatin, N.T. 999077, Hong Kong SAR, China

Zhuojian Liang and Yi-Chun Lu
Department of Mechanical and Automation Engineering, The Chinese University of Hong Kong, Shatin, N.T. 999077, Hong Kong SAR, China
Shun Hing Institute of Advanced Engineering, The Chinese University of Hong Kong, Shatin, N.T. 999077, Hong Kong SAR, China

Hao Liu and Quan Li
Department of Physics, The Chinese University of Hong Kong, Shatin, N.T. 999077, Hong Kong SAR, China

Xinjian Shi
Department of Chemical and Biomolecular Engineering, Yonsei University, 50 Yonsei-ro, Seodaemun-gu, Seoul 120-749, Republic of Korea
Department of Mechanical Engineering, Stanford University, Stanford, California 94305, USA

Hokyeong Jeong, Seung Jae Oh, Il Yong Choi and Jong Kyu Kim
Department of Materials Science and Engineering, Pohang University of Science and Technology, 77 Cheongam-ro, Nam-gu, Pohang 790-784, Korea

Ming Ma and Jeong Kwon
Department of Chemical Engineering, School of Chemical Engineering and SKKU Advanced Institute of Nano Technology (SAINT), Sungkyunkwan University, Chunchun-dong, Suwon 440-746, Korea

Kan Zhang and Jong Hyeok Park
Department of Chemical and Biomolecular Engineering, Yonsei University, 50 Yonsei-ro, Seodaemun-gu, Seoul 120-749, Republic of Korea

In Taek Choi and Hwan Kyu Kim
Global GET-Future Lab, Department of Advanced Materials Chemistry, Korea University, 2511 Sejong-ro, Sejong 339-700, Korea

Parisa Pourhossein and Ryan C. Chiechi
Stratingh Institute for Chemistry, and Zernike Institute for Advanced Materials, University of Groningen, Nijenborgh 4, Groningen 9747 AG, The Netherlands

Ratheesh K. Vijayaraghavan and Stefan C. J. Meskers
Molecular Materials and Nanosystems and Institute for Complex Molecular Systems, Eindhoven University of Technology, PO Box 513, Eindhoven 5600 MB, The Netherlands. Department of Chemical Sciences, Indian Institute of Science Education and Research Kolkata, Mohanpur 741246, India

Lingzheng Bu, Xiaoqing Huang and Jianlin Yao
College of Chemistry, Chemical Engineering and Materials Science, Soochow University, Jiangsu 215123, China

Shaojun Guo
Department of Materials Science and Engineering, College of Engineering, Peking University, Beijing 100871, China
Department of Energy and Resources Engineering, College of Engineering, Peking University, Beijing 100871, China

Xu Zhang and Gang Lu
Department of Physics and Astronomy, California State University, Northridge 91330, California, USA

Xuan Shen and Dong Su
Center for Functional Nanomaterials, Brookhaven National Laboratory, Upton, New York 11973, USA

Xing Zhu and Jun Guo
Testing and Analysis Center, Soochow University, Jiangsu 215123, China

Johannes Ihli, Fiona C. Meldrum, Yi-Yeoun Kim, Anna S. Schenk and Alexander N. Kulak
School of Chemistry, University of Leeds,Woodhouse Lane, Leeds LS2 9JT, UK

Jesse N. Clark
Stanford PULSE Institute, SLAC National Accelerator Laboratory, 2575 Sand Hill Road, Menlo Park, California 94025, USA
Center for Free-Electron Laser Science (CFEL), Deutsches Elektronensynchrotron (DESY) Notkestrasse 85, 22607 Hamburg, Germany

Alexander S. Côté and Dorothy M. Duffy
Department of Physics and Astronomy, University College London, Gower Street, London WC1E 6BT, UK

Timothy P. Comyn
Institute for Materials Research, University of Leeds, Leeds LS2 9JT, UK

Oliver Chammas
School of Physics and Astronomy, University of Leeds, Leeds LS2 9JT, UK

Ross J. Harder
Advanced Photon Source, Argonne, Illinois 60439, USA

Ian K. Robinson
London Centre for Nanotechnology, University College London, 17–19 Gordon Street, LondonWC1H 0AH, UK

Andrew S. Crampton and Marian D. Rötzer
Chair of Physical Chemistry, Catalysis Research Center, Chemistry Department, Technische Universität München, Lichtenbergstrae 4, Garching 85748 Germany

Claron J. Ridge
Chair of Physical Chemistry, Catalysis Research Center, Chemistry Department, Technische Universität München, Lichtenbergstrae 4, Garching 85748, Germany
Air Force Research Laboratory, Energetic Materials Branch, 2306 Perimeter Road, Eglin AFB, Florida 32542, USA

Florian F. Schweinberger and Ueli Heiz
Chair of Physical Chemistry, Catalysis Research Center, Chemistry Department, Technische Universität München, Lichtenbergstrae 4, Garching 85748, Germany

Bokwon Yoon and Uzi Landman
School of Physics, Georgia Institute of Technology, Atlanta, Georgia 30332-0430, USA

Zhaojun Zhang and Bina Fu
State Key Laboratory of Molecular Reaction Dynamics and Center for Theoretical and Computational Chemistry, Dalian Institute of Chemical Physics, Chinese Academy of Sciences, Dalian 116023, China

Tianhui Liu
State Key Laboratory of Molecular Reaction Dynamics and Center for Theoretical and Computational Chemistry, Dalian Institute of Chemical Physics, Chinese Academy of Sciences, Dalian 116023, China
Department of Chemical Physics, University of Science and Technology of China, Hefei 230026, China

Xueming Yang
State Key Laboratory of Molecular Reaction Dynamics and Center for Theoretical and Computational Chemistry, Dalian Institute of Chemical Physics, Chinese Academy of Sciences, Dalian 116023, China
Department of Chemical Physics, University of Science and Technology of China, Hefei 230026, China
Center for Advanced Chemical Physics and 2011 Frontier Center for Quantum Science and Technology, University of Science and Technology of China, Hefei 230026, China

Dong H. Zhang
State Key Laboratory of Molecular Reaction Dynamics and Center for Theoretical and Computational Chemistry, Dalian Institute of Chemical Physics, Chinese Academy of Sciences, Dalian 116023, China
Center for Advanced Chemical Physics and 2011 Frontier Center for Quantum Science and Technology, University of Science and Technology of China, Hefei 230026, China

Kirsten M. Ø. Jensen
Department of Applied Physics and Applied Mathematics, Columbia University, New York, New York 10027, USA

Pavol Juhas
Condensed Matter Physics and Materials Science Department, Brookhaven National Laboratory, Upton, New York 11973, USA

Marcus A. Tofanelli, Christine L. Heinecke and Christopher J. Ackerson
Department of Chemistry, Colorado State University, Fort Collins, Colorado 80523, USA.

Gavin Vaughan
European Synchrotron Radiation Facility, 38043 Grenoble, France.

Simon J. L. Billinge
Department of Applied Physics and Applied Mathematics, Columbia University, New York, New York 10027, USA
Condensed Matter Physics and Materials Science Department, Brookhaven National Laboratory, Upton, New York 11973, USA

Aline Nonat and Loïc J. Charbonnière
Laboratoire d'Ingénierie Moléculaire Appliquée áí Analyse, IPHC, UMR 7178 CNRS, Université de Strasbourg, ECPM, Bât R1N0, 25 rue Becquerel, Strasbourg, 67087, France

Chi Fai Chan, Wai-Kwok Wong and Ka-Leung Wong
Department of Chemistry, Hong Kong Baptist University, Hong Kong SAR, Hong Kong

Tao Liu
Laboratoire d'Ingénierie Moléculaire Appliquée áí Analyse, IPHC, UMR 7178 CNRS, Université de Strasbourg, ECPM, Bât R1N0, 25 rue Becquerel, Strasbourg, 67087, France
Department of Chemistry, Hong Kong Baptist University, Hong Kong SAR, Hong Kong

Carlos Platas-Iglesias
Centro de Investigaciones Cientı ´ficas Avanzadas, Departamento de Química Fundamental, Universidade da Coruña, Campus da Zapateira, Rúa da Fraga 10, A Coruña 15008, Spain

Zhenyu Liu
State Key Laboratory of Chirosciences, Department of Applied Biology and Chemical Technology, Hong Kong Polytechnic University, Kowloon, Hong Kong

Wing-Tak Wong
State Key Laboratory of Chirosciences, Department of Applied Biology and Chemical Technology, Hong Kong Polytechnic University, Kowloon, Hong Kong
Department of Applied Biology and Chemical Technology, Hong Kong Polytechnic University, Hong Kong SAR, Hong Kong

Jonathan Martens, Josipa Grzetic and Giel Berden
Radboud University, Institute for Molecules and Materials, FELIX Laboratory, Toernooiveld 7c, 6525ED Nijmegen, The Netherlands

Jos Oomens
Radboud University, Institute for Molecules and Materials, FELIX Laboratory, Toernooiveld 7c, 6525ED Nijmegen, The Netherlands
2Vańt Hoff Institute for Molecular Sciences, University of Amsterdam, Science Park 908, 1098XH Amsterdam, The Netherlands

Sebastián A. Dıáz and Thomas M. Jovin
Laboratory of Cellular Dynamics, Max Planck Institute for Biophysical Chemistry, am Fassberg 11, 37077 Göttingen, Germany

Florencia Gillanders
Laboratory of Cellular Dynamics, Max Planck Institute for Biophysical Chemistry, am Fassberg 11, 37077 Göttingen, Germany
Departamento de Química Orgánica, Facultad de Ciencias Exactas y Naturales, Universidad de Buenos Aires, CIHIDECAR, CONICET, 1428 Buenos Aires, Argentina

Elizabeth A. Jares-Erijman
Departamento de Química Orgánica, Facultad de Ciencias Exactas y Naturales, Universidad de Buenos Aires, CIHIDECAR, CONICET, 1428 Buenos Aires, Argentina

Jianhui Wang and Ching Hua Chiang
Department of Chemical System Engineering, University of Tokyo, 7-3-1, Hongo, Tokyo 113-8656, Japan

Yuki Yamada and Atsuo Yamada
Department of Chemical System Engineering, University of Tokyo, 7-3-1, Hongo, Tokyo 113-8656, Japan
Elements Strategy Initiative for Catalysts and Batteries, Kyoto University, 1-30, Goryo-Ohara, Kyoto 615-8245, Japan

Keitaro Sodeyama
Elements Strategy Initiative for Catalysts and Batteries, Kyoto University, 1-30, Goryo-Ohara, Kyoto 615-8245, Japan
JST PRESTO, 4-1-8 Honcho Kawaguchi, Saitama 332-0012, Japan
International Center for Materials Nanoarchitectonics, National Institute for Materials Science, 1-1, Namiki, Tsukuba 305-0044, Japan

Yoshitaka Tateyama
Elements Strategy Initiative for Catalysts and Batteries, Kyoto University, 1-30, Goryo-Ohara, Kyoto 615-8245, Japan
International Center for Materials Nanoarchitectonics, National Institute for Materials Science, 1-1, Namiki, Tsukuba 305-0044, Japan

Landon Oakes
Department of Mechanical Engineering, Vanderbilt University, Nashville, Tennessee 37235, USA
Interdisciplinary Materials Science Program, Vanderbilt University, Nashville, Tennessee 37235, USA
Rachel Carter, Trevor Hanken and Adam P. Cohn
Department of Mechanical Engineering, Vanderbilt University, Nashville, Tennessee 37235, USA

Keith Share
Department of Mechanical Engineering, Vanderbilt University, Nashville, Tennessee 37235, USA

Benjamin Schmidt
Vanderbilt Institute of Nanoscale Science and Engineering, Vanderbilt University, Nashville, Tennessee 37235, USA

Cary L. Pint
Department of Mechanical Engineering, Vanderbilt University, Nashville, Tennessee 37235, USA
Interdisciplinary Materials Science Program, Vanderbilt University, Nashville, Tennessee 37235, USA
3Vanderbilt Institute of Nanoscale Science and Engineering, Vanderbilt University, Nashville, Tennessee 37235, USA

Guoxing Li, Jinhua Sun, Wenpeng Hou, Shidong Jiang, Yong Huang and Jianxin Geng
Technical Institute of Physics and Chemistry, Chinese Academy of Sciences, 29 hongguancun East Road, Haidian District, Beijing 100190, China

Feifei Shi
Department of Mechanical Engineering, University of California, Berkeley, California 94720, USA
Materials Sciences Division, Lawrence Berkeley National Laboratory, Berkeley, California 94720, USA

Zhichao Song and Kyriakos Komvopoulos
Department of Mechanical Engineering, University of California, Berkeley, California 94720, USA

Philip N. Ross
Materials Sciences Division, Lawrence Berkeley National Laboratory, Berkeley, California 94720, USA

Gabor A. Somorjai
Materials Sciences Division, Lawrence Berkeley National Laboratory, Berkeley, California 94720, USA
Department of Chemistry, University of California, Berkeley, California 94720, USA

Robert O. Ritchie
Department of Mechanical Engineering, University of California, Berkeley, California 94720, USA
Materials Sciences Division, Lawrence Berkeley National Laboratory, Berkeley, California 94720, USA
Department of Materials Science and Engineering, University of California, Berkeley, California 94720, USA

Kai He, Jing Li, Xiqian Yu, Qingping Meng, Enyuan Hu, Ke Sun, Xiao-Qing Yang, Yimei Zhu, Hong Gan and Eric A. Stach
Brookhaven National Laboratory, Upton, New York 11973, USA

Sen Zhang, Hongseok Yun and Christopher B. Murray
Department of Chemistry, University of Pennsylvania, Philadelphia, Pennsylvania 19104,USA

Yizhou Zhu and Yifei Mo
Department of Materials Science and Engineering, University of Maryland, College Park, Maryland 20742, USA

Bao Qiu, Yonggao Xia and Zhaoping Liu
Ningbo Institute of Materials Technology and Engineering (NIMTE), Chinese Academy of Sciences, Zhejiang 315201, China

Minghao Zhang, Danna Qian, Haodong Liu, Sunny Hy and Ying Shirley Meng
Department of Nano Engineering, University of California San Diego (UCSD), La Jolla, California 92093, USA

Lijun Wu and Yimei Zhu
Department of Condensed Matter Physics and Materials Science, Brookhaven National Laboratory, Upton, New York 11973, USA

Jun Wang
MEET Battery Research Center/Institute of Physical Chemistry, University of Müenster, Corrensstrasse 46, Müenster 48149, Germany

Yan Chen and Ke An
Chemical and Engineering Materials Division, Oak Ridge National Laboratory, Oak Ridge, Tennessee, 37830 USA

Xue Lu Wang, Hua Gui Yang and Wen Qi Fang
Key Laboratory for Ultrafine Materials of Ministry of Education, School of Materials Science and Engineering, East China University of Science and Technology, Shanghai 200237, China

Wenqing Liu, Yanhong Song and Daxiu Wei
Department of Physics, Shanghai Key Laboratory of Magnetic Resonance, East China Normal University, Shanghai 200062, China

Yan-Yan Yu and Xue-Qing Gong
Key Laboratory for Advanced Materials, Centre for Computational Chemistry, Research Institute of Industrial Catalysis, East China
University of Science and Technology, Shanghai 200237, China

Ye-Feng Yao
Department of Physics, Shanghai Key Laboratory of Magnetic Resonance, East China Normal University, Shanghai 200062, China
NYU-ECNU Institute of Physics at NYU Shanghai, 3663 Zhongshan Road North,
Shanghai 200062, China

Adam Gali
Institute for Solid State Physics and Optics, Wigner Research Centre for Physics, Hungarian Academy of Sciences, PO Box 49, H-1525 Budapest, Hungary
Department of Atomic Physics, Budapest University of Technology and Economics, Budafoki út 8, H-1111 Budapest, Hungary

Tamás Demján
Institute for Solid State Physics and Optics, Wigner Research Centre for Physics, Hungarian Academy of Sciences, PO Box 49, H-1525 Budapest, Hungary
Institute of Physics, Loránd Eötvös University, Pázmány Péter sétány 1/A, H-1117 Budapest, Hungary

Márton Vörös4,5,
Materials Science Division, Argonne National Laboratory, Lemont, Illinois 60439, USA
Institute for Molecular Engineering, University of Chicago, Chicago, Illinois 60637, USA

Gerg+o Thiering
Institute for Solid State Physics and Optics, Wigner Research Centre for Physics, Hungarian Academy of Sciences, PO Box 49, H-1525 Budapest, Hungary
Department of Atomic Physics, Budapest University of Technology and Economics, Budafoki út 8, H-1111 Budapest, Hungary

Elena Cannuccia
Aix-Marseille Université, CNRS, PIIM UMR 7345, 13397 Marseille, France

Andrea Marini
Institute for Material Science (ISM) of the National Research Council (CNR), Via Salaria Km 29.3, CP 10, 00016 Monterotondo Stazione, Italy
Mario González-Jiménez, Gopakumar Ramakrishnan, Klaas Wynne and Adrian J. Lapthorn
School of Chemistry,WestCHEM, University of Glasgow, Glasgow G12 8QQ, UK

Thomas Harwood and Elizabeth M. Ellis
Institute of Pharmacy and Biomedical Sciences, University of Strathclyde, Glasgow G4 0RE, UK

Sharon M. Kelly
Institute of Molecular Cell and Systems Biology, University of Glasgow, Glasgow G12 8QQ, UK

Shougo Higashi
Department of Materials Science and Engineering, Stanford University, 476 Lomita Mall, McCullough Building 343, Stanford, California 94305, USA
Smart Design of Materials and Process Research-Domain, Toyota Central R&D Laboratories, Inc., 41-1 Nagakute, Aichi 480-1192, Japan

Seok Woo Lee
Department of Materials Science and Engineering, Stanford University, 476 Lomita Mall, McCullough Building 343, Stanford, California 94305, USA
School of Electrical and Electronic Engineering, Nanyang Technological University, 50 Nanyang Avenue, Singapore 639798, Singapore

Jang Soo Lee
Department of Materials Science and Engineering, Stanford University, 476 Lomita Mall, McCullough Building 343, Stanford, California 94305, USA

Kensuke Takechi
Smart Design of Materials and Process Research-Domain, Toyota Central R&D Laboratories, Inc., 41-1 Nagakute, Aichi 480-1192, Japan

Yi Cui
Department of Materials Science and Engineering, Stanford University, 476 Lomita Mall, McCullough Building 343, Stanford, California 94305, USA
SLAC National Accelerator Laboratory, Stanford Institute for Materials and Energy Sciences, 2575 Sand Hill Road, Menlo Park, California 94025, USA

Hemma Mistry
Department of Physics, University of Central Florida, Orlando, Florida 32816, USA
Department of Physics, Ruhr-University Bochum, 44780 Bochum, Germany

Ana Sofia Varela and Peter Strasser
Department of Chemistry, Chemical Engineering Division, Technical University Berlin, 10623 Berlin, Germany

Cecile S. Bonifacio and Judith C. Yang
Chemical and Petroleum Engineering and Physics, University of Pittsburgh, Pittsburgh, Pennsylvania 15261, USA

Ioannis Zegkinoglou, Ilya Sinev, Yong-Wook Choi and Beatriz Roldan Cuenya
Department of Physics, Ruhr-University Bochum, 44780 Bochum, Germany

Kim Kisslinger and Eric A. Stach
Center for Functional Nanomaterials, Brookhaven National Laboratory, Upton, New York 11973, USA

Mal-Soon Lee, Roger Rousseau and Vassiliki-Alexandra Glezakou
Fundamental and Computational Sciences Directorate, Pacific Northwest National Laboratory, Richland,Washington 99352, USA

Wooyong Um
Energy and Environment Directorate, Pacific Northwest National Laboratory, Richland, Washington 99352, USA
Pohang University of Science and Technology, Pohang 37673, South Korea

Guohui Wang
Energy and Environment Directorate, Pacific Northwest National Laboratory, Richland, Washington 99352, USA

Albert A. Kruger
United States Department of Energy, Office of River Protection, Richland, Washington 99352, USA

Wayne W. Lukens
Lawrence Berkeley National Laboratory, Berkeley, California 94720, USA

Thomas Lee
MultiScale Materials Science for Energy and Environment, Joint CNRS-MIT Laboratory, UMI CNRS 3466, Massachusetts Institute of Technology, Cambridge, Massachusetts 02139, USA
Department of Civil and Environmental Engineering, Massachusetts Institute of Technology, Cambridge, Massachusetts 02139, USA

Lydéric Bocquet
MultiScale Materials Science for Energy and Environment, Joint CNRS-MIT Laboratory, UMI CNRS 3466, Massachusetts Institute of Technology, Cambridge, Massachusetts 02139, USA
Department of Civil and Environmental Engineering, Massachusetts Institute of Technology, Cambridge, Massachusetts 02139, USA
Laboratoire de Physique Statistique, UMR CNRS 8550, Ecole Normale Supe´rieure, 75005 Paris, France

Benoit Coasne
MultiScale Materials Science for Energy and Environment, Joint CNRS-MIT Laboratory, UMI CNRS 3466, Massachusetts Institute of Technology, Cambridge, Massachusetts 02139, USA
Department of Civil and Environmental Engineering, Massachusetts Institute of Technology, Cambridge, Massachusetts 02139, USA
Laboratoire Interdisciplinaire de Physique, CNRS and Universite´ Grenoble Alpes, UMR CNRS 5588, 38000 Grenoble, France

Shohei Saito
Department of Chemistry, Graduate School of Science, Kyoto University, Kyoto 606-8502, Japan
PRESTO, Japan Science and Technology Agency, Kyoto 606-8502, Japan

Shunpei Nobusue, Eri Tsuzaka, Chunxue Yuan and Chigusa Mori
Department of Chemistry, Graduate School of Science, Nagoya University, Nagoya 464-8602, Japan

Mitsuo Hara and Takahiro Seki
Department of Molecular Design and Engineering, Graduate School of Engineering, Nagoya University, Nagoya 464-8603, Japan

Cristopher Camacho
Department of Chemistry, Graduate School of Science, Nagoya University, Nagoya 464-8602, Japan
School of Chemistry, University of Costa Rica, San Pedro Montes de Oca, San Jose´ 11501-2060, Costa Rica

Stephan Irle and Shigehiro Yamaguchi
Department of Chemistry, Graduate School of Science, Nagoya University, Nagoya 464-8602, Japan
Institute of Transformative Bio-Molecules, Nagoya University, Nagoya 464-8602, Japan

Simon P. Neville
School of Chemistry, University of Birmingham, Edgbaston, Birmingham B15 2TT, UK
Department of Chemistry, University of Ottawa, 10 Marie Curie, Ottawa, Ontario, Canada K1N 6N5 (S.P.N.); School of Chemistry, The University of Manchester, Oxford Road, Manchester M13 9PL, UK (N.K.)

Graham A. Worth
School of Chemistry, University of Birmingham, Edgbaston, Birmingham B15 2TT, UK.
Department of Chemistry, University of Ottawa, 10 Marie Curie, Ottawa, Ontario, Canada K1N 6N5 (S.P.N.); School of Chemistry, The University of Manchester, Oxford Road, Manchester M13 9PL, UK (N.K.)

Oliver M. Kirkby and Helen H. Fielding
Department of Chemistry, University College London, 20 Gordon Street, London WC1H 0AJ, UK

Nikolas Kaltsoyannis
Department of Chemistry, University of Ottawa, 10 Marie Curie, Ottawa, Ontario, Canada K1N 6N5 (S.P.N.); School of Chemistry, The University of Manchester, Oxford Road, Manchester M13 9PL, UK (N.K.)

Department of Chemistry, University College London, 20 Gordon Street, London WC1H 0AJ, UK
Department of Chemistry, University of Ottawa, 10 Marie Curie, Ottawa, Ontario, Canada K1N 6N5 (S.P.N.); School of Chemistry, The University of Manchester, Oxford Road, Manchester M13 9PL, UK (N.K.)

Dongliang Chao, Changrong Zhu, Peihua Yang, Hong Jin Fan and Jilei Liu
School of Physical and Mathematical Sciences, Nanyang Technological University, Singapore 637371, Singapore

Xinhui Xia
State Key Laboratory of Silicon Materials, Department of Materials Science and Engineering, Zhejiang University, Hangzhou 310027, China

Jin Wang and Jianyi Lin
Energy Research Institute @ NTU, Nanyang Technological University, Singapore 639798, Singapore

Xiaofeng Fan
College of Materials Science and Engineering, Jilin University, Changchun 130012, China

Serguei V. Savilov
Department of Chemistry, Moscow State University, Moscow 119992, Russia

Ze Xiang Shen
School of Physical and Mathematical Sciences, Nanyang Technological University, Singapore 637371, Singapore
Energy Research Institute @ NTU, Nanyang Technological University, Singapore 639798, Singapore

Index

N

O

P

Q

R

S

T

U

V

X